AF772769

Environmental Protection: Policies and Management

Environmental Protection: Policies and Management

Edited by Chuck Lancaster

SYRAWOOD
PUBLISHING HOUSE
New York

Published by Syrawood Publishing House,
750 Third Avenue, 9th Floor,
New York, NY 10017, USA
www.syrawoodpublishinghouse.com

Environmental Protection: Policies and Management
Edited by Chuck Lancaster

© 2017 Syrawood Publishing House

International Standard Book Number: 978-1-68286-473-9 (Hardback)

This book contains information obtained from authentic and highly regarded sources. Copyright for all individual chapters remain with the respective authors as indicated. All chapters are published with permission under the Creative Commons Attribution License or equivalent. A wide variety of references are listed. Permission and sources are indicated; for detailed attributions, please refer to the permissions page and list of contributors. Reasonable efforts have been made to publish reliable data and information, but the authors, editors and publisher cannot assume any responsibility for the validity of all materials or the consequences of their use.

The publisher's policy is to use permanent paper from mills that operate a sustainable forestry policy. Furthermore, the publisher ensures that the text paper and cover boards used have met acceptable environmental accreditation standards.

Trademark Notice: Registered trademark of products or corporate names are used only for explanation and identification without intent to infringe.

Cataloging-in-Publication Data

Environmental protection : policies and management / edited by Chuck Lancaster.
p. cm.
Includes bibliographical references and index.
ISBN 978-1-68286-473-9
1. Environmental protection. 2. Environmental policy. 3. Environmental management.
4. Pollution prevention. 5. Sustainable development. 6. Environmental monitoring. I. Lancaster, Chuck.
TD170 .E58 2017
363.7--dc23

Printed in China.

TABLE OF CONTENTS

PREFACE

Environmental protection is an extensive field of study. The rapidly increasing pollution and waste have resulted in severe degradation of the environment. Thus, effective strategies and plans need to be developed and implemented in order to conserve the environment from further damage. This book attempts to understand the multiple branches that fall under the discipline of environmental protection and how such concepts have practical applications. It includes some of the vital pieces of work being conducted across the world, on various topics related to this area of study. For all readers who are interested in environmental protection, the case studies included in this book will serve as excellent guide to develop a comprehensive understanding.

Various studies have approached the subject by analyzing it with a single perspective, but the present book provides diverse methodologies and techniques to address this field. This book contains theories and applications needed for understanding the subject from different perspectives. The aim is to keep the readers informed about the progress in the field; therefore, the contributions were carefully examined to compile novel researches by specialists from across the globe.

Indeed, the job of the editor is the most crucial and challenging in compiling all chapters into a single book. In the end, I would extend my sincere thanks to the chapter authors for their profound work. I am also thankful for the support provided by my family and colleagues during the compilation of this book.

Editor

1

Pollution indicators and pathogenic microorganisms in wastewater treatment: Implication on receiving water bodies

Tomilola Debby Olaolu[1], Oghenerobor Benjamin Akpor[2, *], Charity Omeche Akor[2]

[1]Biochemistry Unit, Department of Biological Sciences, Landmark University, Omu-Aran, Kwara State, Nigeria
[2]Microbiology Unit, Department of Biological Sciences, Landmark University, Omu-Aran, Kwara State, Nigeria

Email address:
Olaolu.tomilola@lmu.edu.ng (T. D. Olaolu), akpor.oghenerobor@lmu.edu.ng (O. B. Akpor), omeche.charity@lmu.edu.ng (C. O. Akor)

Abstract: There is the indication that human deaths as a result of water-borne diseases exceed five million people per year, with over half of the diseases being microbial intestinal infections, especially cholera and diarrhea. A number of pathogenic microorganisms, regarded as water pollution indicators have been observed as the agents of such. These water pollution indicators are present in feces, sewage and can survive as long as pathogenic organisms. These pathogenic microorganisms cause several waterborne infections and diseases like bacterial (cholera, salmonellosis, shigellosis and several diseases associated with pathogenic strains of *E. coli*), viral (ranging from a mild febrile illness to myocarditis, meningoencephalitis, poliomyelitis, herpangina, hand-foot-and mouth disease and neonatal multi-organ failure), protozoan (cryptosporidiosis, diarrhea encaphilitis, giardiasis, amoebiasis) and fungal (candidiasis, blastomycosis, cryptococcusis, aspergilosis). These biological contaminants that cause several water-borne diseases can however be removed from water through physical (ultraviolet radiation, solar radiation and boiling) and chemical disinfection methods (chlorination, chloramination and ozonation). The aim of this paper was to review the microbial indicators and pathogenic microorganisms in water and wastewater. The paper also discussed the treatment strategies for microbial-contaminated water and wastewaters.

Keywords: Pollution, Wastewater, Pathogens, Microbial Indicators

1. Introduction

Untreated and improperly treated wastewater is known to contain pathogens which can cause disease outbreak, hence the need for adequate treatment before discharge into receiving water bodies. The treatment of wastewater is carried out to minimize and eliminate potential health risks. The major groups of pathogens that are of importance to wastewater are either bacteria, viruses, fungi or protozoa (FAO, 1993).

Bacteria, which are the most common pathogens in water, gain entrance into water mostly through fecal contamination (Sharma, 2013). Pathogenic or potentially pathogenic bacteria are normally absent from a healthy intestine unless infection occurs. In the case of an infection, large numbers of pathogenic bacteria will be passed in the faeces. Bacterial pathogens cause typhoid, paratyphoid and other *Salmonella* type diseases. Some of the bacteria pathogens that are found in wastewater include *Salmonella*, *Shigella*, pathogenic strain of *Escherichia coli*, *Yersinia, Campylobacter, Vibrio* (Bitton, 2005; Cabral, 2010). Viruses are reported to be the most hazardous pathogens found in wastewater. They are more resistant to treatment, more infectious and difficult to detect in wastewater. Their entrance into water is also through fecal contamination. The viral pathogens that are of importance in wastewater are enteroviruses (including polioviruses), hepatovirus, reoviruses and diarrhoea-causing viruses, especially rotavirus (Bosch, 1998). In the case of protozoa, their detection in wastewater is often as cysts and oocysts. Many species of protozoa can infect humans through ingestion, causing diarrhea and dysentery. The two major protozoan pathogens in water are *Cryptosporidium* and *Giardia* (FAO, 1993; Toze, 1997; Yates, 2013).

Since there is a wide range of pathogens that are present in wastewater and owing to the inherent difficulties of analysis, thus making the direct determination of the numbers impractical on a routine basis, the microbiological analysis of water depends majorly on detecting the presence of indicator organisms. Indicator organisms are present in large numbers in wastewater, although they may not necessarily be pathogenic. Their detection suggests that human contamination of the water has occurred and that more dangerous organisms could be present. An indicator organism should be applicable in all types of water, always be present or absent when pathogens are present or absent, respectively and should lend itself to routine quantitative testing procedures without interference from or confusion of results because of extraneous organisms. Most indicator organisms are are present in feces and sewage and can survive as long as pathogenic organisms. The major factors that are considered in classifying an organism as an indicator organism includes its association with fecal contamination, its relationship with the pathogen concerned and efficient and simple testing procedures. Indicator organisms are therefore referred to as basic tools for the measurement of water quality, they provide evidences for the presence or absence of pathogenic organisms in water (Plutzer and Torokne, 2012; Department of Health, 2013).

The aim of this paper was to review the major pathogenic and indicator organisms that are of importance in wastewater effluents. The paper also reviewed the various treatment methods for the removal of microbial pathogens from wastewater effluents with particular emphasis on the merits and demerits of each treatment method.

2. Microbial Pathogens in Wastewater

The major microbial pathogens in water are bacteria, viruses, fungi and protozoan parasites. Bacteria pathogens are mostly present in feces and a wide variety can be present in wastewater due to fecal contamination. The discharge of untreated or inadequately treated wastewater into the environment can have negative impact on human health due to the release of pathogenic microorganisms into water which could lead to serious health diseases (Rosario *et al.*, 2009). Water that is contaminated with microbial pathogens is a medium for several waterborne diseases, such as cholera, typhoid fever, shigellosis, salmonellosis, campylobacteriosis, giardiasis, cryptosporidiosis and Hepatitis A (WHO, 2004). Several pathogenic organisms in contaminated water are the basic causes of gastrointestinal illnesses in human. Some of the pathogens are known to cause several outbreaks of diseases by releasing toxins in the human body (Krauss and Griebler, 2011). The major groups of water borne pathogens and the diseases they cause are shown in Table 1.

Bacterial pathogens

The major water related diseases caused by bacteria are cholera, salmonellosis and shigellosis. The majority of bacterial pathogens in wastewater are found in the gastrointestinal tract of humans and animals. They mostly belong to the following genera: Vibrio, Salmonella, Shigella, Leptospira and pathogenic strains of *Escherichia coli*, Yersinia and Campylobacter (Cabral, 2010; Yates, 2013). These pathogens cause diseases, such as cholera, salmonellosis, shigellosis/bacillary dysentery and diseases caused by pathogenic strains of E.coli.

Table 1. *Major Water Borne Pathogens and Diseases they Cause*

Microorganisms	Pathogenic organism	Diseases caused
Bacteria	*Salmonella*	Typhoid, bacillary dysentery
	E.coli (enterotoxigenic)	Gastroenteritis
	Yersinia	Gastroenteritis
	Campylobacter	Gastroenteritis
	Vibrio	Cholera
	Leptospira	Leptospirosis
Viruses	Polio virus	Paralysis
	Rota virus	Infantile gastroenteritis
	Hepatitis A virus	Infectious hepatitis
	Norwalk virus	Gastroenteritis
	Adenovirus	Conjunctivitis
	Reovirus	Respiratory disease
	Echovirus	Aseptic Meningitis
Protozoa, helminths and other parasites	*Giardia lamblia*	Diarrhea, malabsorption
	Entamoeba coli	Diarrhea, ulceration
	Entamoeba histolytica	Amoebic dysentery
	Cryposporidium parvum	Diarrhea
	Ascaris lumbricoides	Ascariasis
	Ancylostoma	Anemia
	Nectar	Anemia
	Trichuris	Diarrhea, anemia
	Tenia solium	Teniasis
	Trichuris trichuria	Trichuriasis

Source: (Yates, 2013)

Cholera, an infection of the small intestine caused by *Vibrio cholera* is primarily transmitted through the consumption of contaminated water (Cabral, 2010; Michelle, 2013). The incubation period of the disease ranges from 1–3 days. The disease is characterized by an acute and very intense diarrhea that can exceed one liter per hour. The bacteria secrets exotoxins that attach to the intestinal cells, causing loss of water, potassium, chloride and carbonate ions from the cells of the mucous membrane leading to diarrhea in patient (Farmer and Brenner, 2003; Cabral, 2010).

Salmonellosis is a disease caused by Salmonella species. The two types of salmonellosis caused by these pathogens are typhoid and paratyphoid fever, which are associated with ingestion of fecal-contaminated water and gastroenteritis (Le Minor, 2003; Michelle, 2013). Salmonellosis of newborns and infants presents diverse clinical symptoms, from grave typhoid-like illness with septicemia to mild or asymptomatic infection (Cabral, 2010).

The causative agent of shigellosis in human is Shigella species. The disease transmission through water is due to their high surviving nature. In the epithelial cell of the intestinal tract of humans, Shigella produces a high level of cytotoxic shiga toxin. The incubation period of the disease ranges from 1–4 days and usually begins with fever, anorexia, fatigue and malaise (Strockbine and Maurelli, 2005; Emch and Yunus, 2008; Cabral, 2010). .

The groups of pathogenic E. coli which cause diseases in human health include the enterotoxigenic E. coli (ETEC), enterohemorrhagic E. coli (EHEC) and enteroinvasive E. coli (EIEC). The ETEC strains are the common cause of infantile gastroenteritis, commonly associated with children below 5 years of age in developing countries and accounts for several million cases of diarrhea disease and several ten of thousand deaths each year (Bettelheim, 2003; Scheutz and Strockbine, 2005; WHO, 2006; Cabral, 2010). The EHEC strain (serotype O157:H7) has an incubation period of 3-4 days and symptoms occur for 7-10 days and 2 -7 % leads to renal failure. The EIEC strains are like the Shigella. They invade and replicate in the intestinal cells of humans. The disease symptom is characterized by diarrhea, abdominal cramps, vomiting, fever, chills, malaise and also appearance of blood and mucous in stools of patient. They are frequently associated with diseases such as gastroenteritis and enterocolitis dysentery (Bettelheim, 2003; Scheutz and Strockbine, 2005; WHO, 2010).

Viral pathogens

Viral pathogens, especially the human enteric viruses, which comprise of the highest number of water borne pathogens, have been found in water as a result of fecal contamination. The enteric viruses include the families Picornaviridae (poliovirus, enterovirus, coxsackievirus, hepatitis A virus and echovirus), Caliciviridae, Reoviridae, Adenoviridae, and Coronaviridae (Krauss and Griebler, 2011). Viruses are among the most important and potentially most hazardous pollutants in wastewater. Generally, viruses are indicated to be more resistant to treatment, more infectious, more difficult to detect and require smaller doses to cause infections. The presence of viruses in polluted waters could lead to a broad range of asymptomatic to severe gastrointestinal, respiratory, skin, eye, nose and ear infection. The enteroviruses and adenoviruses are the most common cause of symptoms of illness in healthy individuals and are often present in polluted water (Environmental News Network, 2007). The spectrum of diseases caused by enteroviruses is broad and ranges from a mild febrile illness to myocarditis, meningoencephalitis, poliomyelitis, herpangina, hand-foot-and mouth disease and neonatal multi-organ failure. Most infections, particularly in children, are asymptomatic, but still lead to the excretion of large numbers of the viruses, which may cause clinical disease in other individuals (WHO, 2006). The major disease caused by the enteric virus is acute gastroenteritis in both adult and adolescence. The hepatitis A virus is a very small enteric virus and is transmitted through fecally contaminated water by infected persons. The symptoms of disease include inflamed liver, followed by lassitude anorexia, weakness, fever, jaundice and nausea (Yates, 2013).

Protozoan pathogens

Protozoa are mostly associated with gastrointestinal disease, dysentery and ulceration of the liver and intestine. Protozoans can cause life threatening diseases in unborn children, immunocompressed individuals and even the elderly. They are also related to diseases, such as plasmodia, schistosomes, giardiasis, amoebiasis, cryptosporidiosis. The presence of free-living amoeba in water can cause severe diseases in man, for example the Naegleria fowleri, which is the primary cause of amoebic meningoencaphalitis (US National Library of Medicine National Institute of Health, 2003).

The most common protozoan pathogens that are reportedly present in wastewater treatment systems are classified into three groups, the flagellates (Peranema), the sarcodines (Arcella, Euglypha) and the ciliates (Trachelophyllum, Litonotus, Aspidisca, Euplotes, Carchesium, Epistylis, Opercularia). Others include the zoonotic Cryptosporidium pavum, Giardia lamblia, Toxoplasma gondiiis, Entamoeba histolystica, and also free-living amoeba species e.g. Acanthamoeba specie and Naegleria fowleri (Amaral et al, 2004; Krauss and Griebler, 2011).

The two most common parasitic protozoa that are found in wastewater are Giardia and Cryptosporidium. When Giardia cysts gain access to the stomach, they adhere to intestinal wall, leading to giardiasis, a disease, though not fatal but is characterized by diarrhea, abdominal cramps, weight loss, nausea and general gastrointestinal distress (Toze, 1997). Similarly, *Cryptosporidium parvum* is a parasitic protozoan which causes cryptosporidiosis in individuals who have ingested the oocyst in contaminated water (WHO, 2006). When cryptosporidiosis oocysts attach to the gastrointestinal tract of humans, they produce symptoms that include headache, abdominal cramps, nausea, vomiting, diarrhea and fever. Also, *Entamoeba hystolistica* can cause diarrhea encaphilitis and even dysentery to infected individuals (ExotoxNet, 1997).

Fungal pathogens

The most common pathogenic fungi that have been isolated from wastewater include species of Aspergillus, Candida, Rhizopus, Penicillium, Drechslera, and Rhodotorula. Several infections that are known to be caused by fungi are classified according to the site of initial infection, which are superficial mycoses, subcutaneous mycoses and systemic mycoses (Faryal and Hameed, 2005; Department of Environmental and Rural Affairs, 2011).

The majority of fungal diseases are mostly related to water used for recreational activities, such as bathing, swimming and washing. Some pathogenic fungal species could cause fatal nervous system infection through constant exposure to fungal spores can also cause respiratory allergies. They could also produce toxins which can be fatal to health when ingested. The diseases caused by fungi are known as mycosis. Most waterborne fungi remain in spore form and immunocompromised individuals are mostly at risk of infection. The major water-related fungal diseases are candidiasis, causeds by Candida albicans, blastomycosis, caused by Blastomyces dermatitidis, cryptococcusis, caused by Crptococcus neoforman, aspergilosis, caused by Aspergillus fumigatus, coccidiomycosis, caused by Coccidioides immitis and paracoccidioidomycoses, caused by Paracocciodoides brasiliensis (Warrington, 2011).

3. Microbial Indicators in Wastewater

Indicator organisms in wastewater are organisms whose presence suggests the presence of a pathogen in wastewater. The density of an indicator organism is always associated with health hazards and several sources of pollution. It is indicated that for an organism to qualify as an indicator organism of a particular pathogen, it must be continuously and totally related to the source of the pathogen and be abundant enough to provide appropriate and exact mass concentration of the level of pathogen in relation to high risk of illness. Also, an indicator organism should have resistant ability to disinfectants, environmental stress and toxic materials that may be present at the source of the pathogen (Berg, 1978; Galveston Bay Centre, 2002).

Bacteria indicators are the most important indicators of faecal contamination. These indicators include members of the Enterobacteriacea family, which the total and the fecal coliforms and *Escherichia coli*. Other bacterial indicators are the fecal streptococci (*Streptococcus* and *Enterocoocus*) and *Clostridium* (Krauss and Griebler, 2011). Generally, the coliforms are gram negative, non-spore forming, oxidase negative, rod shaped, and facultative anaerobic bacteria. They are lactose fermenters with gas production at 35 – 37 °C in a medium with bile salt (WHO, 2001). They can thrive at high temperatures, high pH (up to 9.6), and high salt concentration (up to 6.5% sodium chloride) and are considered as the most valuable indicator of fecal contamination. Although it is indicated that several species of the four *Enterobacteriaceae* genera *Escherichia* sp., *Klebsiella* sp., *Enterobacter* sp. and *Citrobacter* sp. have the ability of giving positive results to tests on environmental waters, the fecal coliform groups are the most widely used. The wide usage of the fecal coliforms is due to their continuous association in fecal waste of man and animals (Cabral, 2010).

In wastewaters, coliforms are seen in abundance, ranging from one to ten million colony forming units in every 100 mL of primary treated effluent (Galveston Bay Information Centre, 2002). Of the total coliform group, *E. coli* is the most numerous in mammalian feces, hence is considered the most specific indicator of fecal pollution (National Health and Medical Research Council, 2003). *Escherichia coli*, a member of the coliform bacteria population which could serve as an indication of fecal contamination, is a natural and essential part of the bacterial flora in the gut of humans and animals. Although some *E. coli* strains are pathogenic and play roles in intestinal and urinary tract infections, the majorities of *E. coli* strains are nonpathogenic and reside harmlessly in the colon (Scheutz and Strockbine, 2005).

Another group of indicators of bacterial pollution in water and wastewater are the fecal streptococci. The fecal streptococci are gram-positive, catalase-negative cocci. They consist of two main genera, *Enterococcus* and *Streptococcus*, with the *Enterococcus* indicated to be the preferred indicator of fecal pollution. These organisms are normally found in the gastro-intestinal tract of warm-blooded animals; hence their presence in water indicates fecal contamination. The fecal streptococci and fecal coliform are used to differentiate human fecal contamination from that of warm blooded animals (WHO, 2001). Some species of fecal streptococci include *Enterococcus faecalis*, *Enterococcus faecium*, *Enterococcus durams*, *Enterococcus hirae*, *Streptococcus bovis* and *Streptococcus equines* (Galveston Bay Information Centre, 2002).

Other indicators of bacterial pollution are *Clostridium perfringes* and *Bifidobacteria. Clostridium perfringens* is an anaerobic gram-positive, endospore-forming, rod-shaped, sulfite-reducing bacterium. It is found in the colon and represents approximately 0.5 % of the fecal microflora that is commonly found in human and animal feces (Bitton, 2005). The *Bifidobacteria* are gram positive rods, obligate anaerobes, non-spore formers and non-motile. They are all catalase negative, lactose fermenters and are numerous in feaces of warm-blooded animals. In addition, they are resistant to high temperature environment and can survive longer in an environment than enteric pathogens. Although they are indicated to be highly numbered in human feces, because of their sensitivity to oxygen, their role as useful indicators of fecal pollution in water is limited (Galveston Bay information Centre, 2002).

The major indicators of viral pollution in water and wastewaters are the phages, which are viruses that affect bacteria. The phages are used as models for human enteric viruses in water quality testing (Ashbolt *et al.*, 2001). Generally, phages have been considered microbial indicators of viral pollution as they have similar characteristics with the human enteric virus. The coliphages are viruses that infect coliforms. They are also known as phages of coliforms and have been used to indicate human enteric viruses. The most used coliphages are the T-type (DNA that contains tailed phages) and F-RNA coliphages (RNA phages which can cause infection through the F-Pilli also known as sex factor). Because of their similar morphological structure and behaviour, they have been regarded as the best indicators for fecal pollution than fecal indicator.

Three groups of bacteriophages have been considered as indicators: somatic coliphages, male-specific RNA coliphages (FRNA phages/F+ coliphages) and phages infecting *Bacteroides fragilis* (Leclerc et al., 2000; Berger and Oshiro, 2002). The somatic coliphages infect mostly *E. coli*, although some can infect other enterobacteriaceae. They have been reportedly used as water quality indicators in estuaries, seawater, freshwater, potable water, wastewater and biosolids (Mocé-Llivina et al., 2003). The F+ coliphages include the families Inoviridae (FDNA) and leviviridae (FRNA). These coliphages are either single-stranded DNA or RNA and infect *E. coli* cells that contain the F plasmid, which is the plasmid that codes for the F or sex pilus to which the phage attach. Their consideration as indicators of wastewater pollution is due to their high numbers in wastewaters and their relatively high resistance to chlorination (Nasser et al., 1993; Yahya and Yanko, 1992).

Furthermore, diatoms are used to indicate the general quality of wastewater, with respect to nutrient enrichment. This is because they provide several valuable interpretations with respect to changes in water quality, such as turbidity, conductivity, chemical oxygen demand, biological oxygen demand, dissolved oxygen and chloride (Rey et al., 2002).

4. Treatment Strategies for Microbial-Contaminated Water and Wastewater

The two fundamental reasons for the treatment of water and wastewater are to safeguard public health and prevent the pollution of receiving water. The strategies for the treatment of microbial-contaminated water and wastewater are grouped into two: physical (solar radiation and ultraviolet radiation) and chemical disinfection (chlorination, chloramination, ozonation) processes. Each of these processes involves the destruction or inactivation of microorganisms. Although, no disinfection process is known to provide 100 % effectiveness, any good process should be characterized by the ability to penetrate and destroy pathogenic organisms during water treatment process and should be harmless to humans and the environment. Also, it must be cheap to acquire and maintain, easy and safe to handle, store and transport. In addition, during the disinfection process, toxic residuals, mutagens and carcinogens should not be produced (LeChevallier et al., 2004).

Physical disinfection processes

The two main physical disinfection processes are solar and ultraviolet (UV) radiations. Ultraviolet disinfection involves the exposure of contaminated water to UV light that has been generated by an electronic discharge through mercury vapour. The radiations from the UV light penetrates cells of pathogenic organisms in water and destroys their cell genetic material, thereby preventing reproduction (EPA, 1999; Tech Brief, 1999). Ultraviolet radiation is an effective method for the removal of microorganisms, such as *Cryptosporidium* and *Giardia*. Because UV radiation is not associated with the production of harmful by-products and known to improve the taste, colour and odour of wastewater, it is considered a safe method of primary disinfection (Trojan Technologies, 2013).

The advantage of a UV disinfection process is that it is an effective means of inactivating virus, spores and cysts. It does not produce toxic by-product and odour and does not also pose any danger of overdosing (Andrew, 2005). Despite the advantages, UV radiation is disadvantageous because it does not leave any residual behind after disinfection; hence there is the possibility of microbial regrowth. Another disadvantage is that because the method lacks any mechanism for determination of its effectiveness, it may not be effective in water that is highly turbid (Environmental Health Directorate, Department of Health, 2010). Generally, UV disinfection method is a recommended treatment method for several pathogenic organisms, such as *Burkholderia* sp., *Clostridium perfringens, Escherichia coli* 0157:H7, *Salmonella typhi, Shigella dysenteriaeI,* Calciviruses, Hepatitis A virus, Rotavirus, *Cryptosporidium* sp., *Giardia* sp.and *Toxoplasma gondii* (EPA, 2012).

A solar disinfection method is a thermal inactivation and photo-oxidation used to inactivate pathogenic organisms in contaminated water of low turbidity (CDC, 2012a). This method involves the use of the synergetic combination of optical and thermal processes for the treatment of contaminated water that is already contained in a transparent container for the inactivation of microorganisms such as bacteria, virus and protozoa. It is one of the simplest, inexpensive treatments available. It entails filling plastic containers with contaminated water and exposing to the sun for a whole day, thereby exposing microbial cells to ultraviolet radiation from sunlight, which leads to the destruction of their DNA (CDC, 2008; Dawney and Pearce, 2012).

The advantages of this method are its effectiveness against bacteria, protozoa and viruses, its proven reduction of diarrheal disease incidence in users, acceptability to users due to its simplicity and low cost. Also, the method causes minimal change in taste of the water and there is a reduced likelihood of recontamination since the water is served directly from the small, narrow-necked bottles with caps in which it was disinfected (CDC, 2008). The drawbacks of the method are that it requires pretreatment (filtration or flocculation) of waters of higher turbidity and the acceptability concerns because of the limited volume of water that can be treated at once. Also, the length of time required to treat water is long (CDC, 2012b).

Chemical disinfection processes

There are three commonly used chemical disinfection processes: chloramination, chlorination and ozonation. Chloramination, refers to the use of chlorine and ammonia to disinfect water and is considered one of the best technologies for water disinfection. Chloramines are formed from the combination of chlorine and ammonia. The disinfection method is by the frequent addition of ammonia to water that already contains free chlorine at pH 8.4. The method is effective and kills microbial cells by penetrating their cell walls, thereby obstructing their metabolism. It is documented that chloramination has served as an effective means for removing microbes such as coliform bacteria, heterotrophic bacteria as well as Legionella bacteria (San Fransisco Public Utilities Commission, 2013). The advantages of chloramination are its ability to improve the odour, taste, smell and flavour of water and it remains active for a very long period of time. The drawback of the use of chloramine for disinfection is it's less effectiveness in killing some pathogenic microorganisms like E. coli, rotavirus and polio (Brockovich, 2010).

In the case of chlorination, the method involves the use of chlorine to disinfect water. The chlorine could either be in solid (calcium hypochlorite) or liquid (sodium hypochlorite) and gas. When added to water, chlorine helps in the reduction of pathogens by affecting the reproduction and metabolism of microorganisms present in the water (Michigan State University, 2010). This method involves introduction of chlorine, (about 0.2 – 0.4 mg/L) into contaminated water at pH of between 5.5–7.5 (Lenntech, 2013). Chlorination disinfection method is a well-established technology and leaves a residual behind, which can prolong disinfection after the treatment process, thereby preventing microbial regrowth. The disadvantages of chlorination are the high corrosiveness of chlorine and the production of harmful by-products (trihalomethanes) during disinfection process. Also, apart from the fact that some microorganisms are resistant to chlorine when in low dosage, the method is indicated to increase the chloride content of wastewater (EPA, 1999).

For ozonation, it involves the use of ozone for disinfection. Ozone is considered a very effective oxidizing agent for the treatment of water. The method is indicated to have good bactericidal and virucidal activities. It is considered the most effective disinfectant that can kill microorganisms including bacteria, viruses and protozoans (OzoneFac, 2013). Ozone is considered as the most powerful oxidant which can inactivate microorganisms in contaminated waters and it is more effective than chlorine (Promolife, 2011). The advantage of ozone is that, when compared to chlorination and chloramination, it is considered more effective in killing microorganisms and fast in activity. Also, it is not associated with the production of harmful by-products and helps in the removal of taste and odour from water. The disadvantage is that it that it is not effective in killing cysts and some other larger organisms. Besides, the cost of the equipment used for ozone generation is high and there is always the possibility of microbial re-growth. In addition, apart from the fact that ozone itself can be harmful to human health, the process requires electricity byproducts like bromate which can also cause harm to humans (Guides Network, 2013).

5. Conclusion

This paper which was aimed at reviewing the microbial pathogens, indicators of microbial pollutants and treatment strategies for microbial-contaminated water and wastewater has revealed the following:

- The presence of pathogenic organisms in water and wastewater has led to several water-related diseases.
- Because of the enormous time and cost involved in the detection of some of these microbial pollutants, indicator organisms are used.
- The advantages of these indicator organisms are that they are not pathogenic, and directly detect the presence of fecal contamination from warm-blooded animals.
- The various disinfection process enumerated above is beneficial not only to wastewater treatment but also drinking water.

This review was able to provide an insight to the major microbial pathogens and indicators of microbial pollution in water and wastewater. It has also shown the effectiveness of the various treatment processes for the elimination from and inactivation of microbial contaminants in water and wastewater.

References

[1] Amaral AL, da Motta M, Pons, MN, Vivier H., Roche N, Mota M and Ferreira EC. 2004. Survey of protozoa and metazoan populations in wastewater treatment plants by image analysis and discriminant analysis. Environmetrics, 15: 381 – 390

[2] Ashbolt N, Grabow W and Snozzi M. 2001. Water quality: guidance, standards and health: indicators of microbial water quality. IWA Publishing: London

[3] Berg G. 1978. The indicator system. In: Indicators of viruses in water and food (ed. G. Berg). Ann Arbor Science Publishers: Ann Arbor

[4] Berger SB, and Oshiro RK. 2002. Source water protection: Microbiology of source water, pp. 2967-2978, In: Encyclopedia of Environmental Microbiology, (Bitton G ed.). Wiley-Interscience: New York

[5] Bettelheim KA. 2003. The genus Escherichia. In: The Prokaryotes: An Evolving Electronic Resource for the Microbiological Community, 3th ed.; Dworkin, M., Falkow, S.,

[6] Bitton G. 2005. Microbial indicators of fecal contamination: application to microbial source tracking. Department of Environmental Engineering Sciences, University of Florida. Gainesville. Report submitted to the Florida Stormwater Association 719 East Park Avenue, Tallahassee, 32301.

[7] Bosch A. 1998. Human enteric viruses in the water environment: a minireview. International Journal of Microbiology, 1:191-196

[8] Brockovich E. 2010. Chloramination of drinking water. Available at: http://www.ErinBrockovichChloramination.html.

[9] Cabral JPS. 2010. Water Microbiology. Bacterial pathogens and water. International Journal Research and Public Health, 7:3657-3703

[10] CDC. 2008. Household water treatment options in developing countries: solar disinfection. Centre for Disease Control and Prevention (CDC) Documents. Available at www.cdc.gov/../options-sodis.pdf

[11] CDC. 2012a. A guide to drinking water treatment and sanitation for Backcountry and travel use. Center for Disease Control and Prevention (CDC) Publication. Available at: http://www.cdc.gov/healthywater/drinking/public/water_disinfection.html

[12] CDC. 2012b. Disinfection with chlorine and chloramines.Solar disinfection. Center for Disease Control and Prevention, (CDC) Publication. Available at: http://www.cdc.gov/healthywater/drinking/travel/backcountry_water_treatment.html.

[13] Dawney B and Pearce J. 2012. Optimizing the solar water disinfection (SODIS) method for decreasing turbidity with NaCl. Journal of Water, Sanitation and Hygiene for Development 2(2): 87-94

[14] Department of Health, (2013). Indicator organism. Available at: http://www.baltimorecountymd.gov/Agencies/health/environmentalhealth/watersampling/terms/term6.html

[15] Department of Environment Food and Rural Affairs. 2011. A review of fungi in drinking water and the implications for human health. Available at: http://dwi.defra.gov.uk/research/completed-research/reports/DWI70-2-255.pdf

[16] Emch, M.; Ali, M.; Yunus, M. 2008. Risk Areas and Neighborhood-Level Risk Factors for Shigella dysenteriae 1 and Shigella flexneri. Health Place, 14: 96–105

[17] Environmental News Network. 2007. Viruses in water: The imaginative in pursuit of the fugitive. Society for General Microbiology. Available at: http://www.sgm.ac.uk

[18] EPA. 1999. Ultraviolet disinfection. Water technology fact sheet. EPA water Resources Centre, Washington DC.

[19] EPA. 2012. Containment and disposal of large amounts of contaminated water: A support guide for water utilities. Office of water (4608T). Available at http://www.epa.gov/watersecurity

[20] Exotoxnet. 1997. Microorganisms, Bacteria and Viruses. Available at: http://extoxnet.orst.edu/faqs/extoxnet.htm

[21] FAO. 1993. Quality control of waste water for irrigated crop production:Health risks associated with wastewater. Natural Resources Management and Environmental Development, 2: 10-17.

[22] Farmer JJ and Hickam-Brenner FW. 2003. The Genus Vibrio and Phtotobacterium. In The Prokaryotes: An Evolving Electronic Resource for the Microbiological Community, electronic release 3.14, 3th ed.; Dworkin, M., Falkow, S., Rosenberg, E., Eds.; Springer-Verlag: New-York

[23] Faryal R and Abdul H. 2005. Isolation and characterization of various fungal strains from textile effluent for their use in bioremediation. Pakistan Journal of Botany, 37(4): 1003-1008

[24] Galveston Bay Centre, (2002). Indicator organisms and water quality criteria. Available at: http://www.Indicator organisms and water quality-2.pdf

[25] Guides Network. 2013. Ozone wastewater treatment. Available at: http://www.water-pollution.org.uk/ozonewastewatertreatment.html

[26] Krauss S, and Griebler C. 2011. Pathogenic Microorganisms and Viruses in Groundwater. Acacatech Materialien, Munchen., 3:10-13

[27] LeChevallier M. and Au K. 2004. Water Treatment and Pathogen Control, achieving Safe Drinking Water. IWA Publishing, Alliance House, London

[28] Leclerc, H., S. Edberg, V. Pierzo, and J. M. Delattre. 2000. Bacteriophages as indicators of enteric viruses and public health risk in groundwaters. Journal of Applied Microbiology, 88: 5-21.

[29] Le Minor P. 2003. The genus Salmonella. In: The Prokaryotes: an evolving electronic resource for the microbiological community, electronic release 3.14, 3th ed.; Dworkin, M., Falkow, S.,

[30] Rey D., Taylor JC, Lass A, van Rensburg L and Vosloo O. 2002. Determining the possible application of diatoms as indicators of general water quality: a comparison with South Africa. Water SA 30(3): 1-11

[31] Michelle K. 2013. Effects of water pollution on human health. Available at: http://filteredhealth.com

[32] Nasser, AM, Tchorch Y and Fattal B. 1993. Comparative survival of E. coli, F+ bacteriophages, HAV and poliovirus 1 in wastewaters and groundwaters. Water Science and Technology, 27: 401-407

[33] National Health and Medical Research Council (2003). Review of coliforms as microbial indicators of drinking water quality. Australian Government Publication. Available at http://www.nhmrc.gov.au/_files_nhmrc/publications/attachments/eh32.pdf

[34] OzoneFac. 2013. Ozone technology: ozone generator summary. Available at: http://www.ozonefac.com

[35] Plutzer J. and Torokne A. 2012. Free-living microscopic organisms as indicators of changes in drinking water quality. Water Practice and Technology, 7(3): 1-13

[36] Promolife. 2011. Ozone water purifiers: ozone water purification an effective water cleaning methods available to remove toxins, and FDA approved for sanitation and water purification. Trustwave's trusted commerce program. Promolife, Inc: Fayatteville

[37] Rosario K, Symonds EM, Sinigalliano C, Stewart J and Breitbart M. 2009. Pepper mild mottle virus as an indicator of faecal pollution. Applied and Environmental Microbiology, 75(22): 7261-7270

[38] San Francisco Public Utilities Commission. 2013. Chloramines. Available at: http://sfwater.org/index.aspx?page=357

[39] Scheutz F and Strockbine NA. 2005. Genus Escherichia. In Bergey's Manual of Systematic Bacteriology, 2nd ed.; Brenner, D.J., Krieg, N.R., Staley, J.T., Eds.; Springer: New York, NY, USA, Volume 2, Part B, pp. 607–623

[40] Sharma N. 2013. The health implications of reusing sewage. Available at www.cseindia.org/userfiles/nandini.pdf

[41] Strockbine NA and Maurelli AT. 2005. Genus Shigella. In: Bergey's Manual of Systematic Bacteriology, Volume 2, Part B. (Brenner DJ, Krieg NR, Staley JT, eds.). Springer: New York

[42] Tech Brief. 1999. Ozone: a national drinking water clearing house fact sheet, ultraviolet disinfection.

[43] Toze S. 1997. Literature review for urban water systems multi-divisional research program: Microbial pathogens in wastewater. CSIRO, Australia, 3: 5-10

[44] Trojan Technologies. 2013. UV Disinfection for Drinking Water .Trojan UV- water confidence. Available at: http://trojanUV.com/applications/drinkingwater

[45] US National Library of Medicine National Institute of Health. 2010. Pathogenic human viruses in coastal waters. Available at: http://www.nih.gov.

[46] Warrington P. 2011. Fungi: Aquatic pathogens. Available at: http://www-micro.msb.le.ac.uk/MBChB/6a.html

[47] WHO. 2001. Water Quality: Guidelines, Standards and Health (Fewtrell L and Bartram J), Eds.). IWA Publishing: London

[48] WHO. 2004. Guidelines for Drinking water quality, vol. 1. World Health Organization Press, Geneva

[49] WHO. 2006. Guidelines for drinking water quality: incorporating first addendum. World Health Organisation Press, Geneva

[50] WHO. 2008. Guidelines for Drinking-water Quality, Incorporating 1st and 2nd Addenda, Volume 1, Recommendations, 3rd ed.; World Health Organization (WHO) Press, Geneva.

[51] WHO. 2010. Enterotoxigenic Escherichia coli (ETEC). In: Diarrhoeal Diseases. World Health Organization (WHO). Available at: http://www.who.int/vaccine_research/diseases/diarrhoeal/en/index4.html

[52] Yahya MT and Yanko WA. 1992. Comparison of a long-term enteric virus monitoring data base with bacteriophage reduction in full scale reclamation plants. Paper Presented at the International Water Pollution Research Symposium, Washington, DC, May 26-29, 1992.

[53] Yates MV. 2013. Pathogens in reclaimed water. Available at: http://www.geoflow.com/wastewater/pathogens.htm

2

Dynamics of Land Cover and Impact on Stream flow in the Modder River Basin of South Africa

Onalenna Gwate[1], Yali E. Woyessa[2], David Wiberg[3]

[1]Department of Geography and Population Studies, Lupane State University, Zimbabwe
[2]Department of Civil Engineering, Central University of Technology, South Africa
[3]Water Programme, International Institute of Applied Systems Analysis, Laxenburg, Vienna, Austria

Email address:
onalennag37@gmail.com (O. Gwate)

Abstract: Understanding how dynamics in individual land use types influence changes in streamflow is vital. Most hydrological studies are based on paired catchment and modelling approaches. These are data intensive and require a long period of monitoring. To determine the hydrological effects of tropical vegetation conversion over large areas, it is manageable to study the same basin over different time periods. The study analysed changes in historical stream flow patterns with reference to dynamics in land cover in C52A quaternary catchment of South Africa. Landsat images for years 1993, 2004 and 2013 were used for the development of land cover maps. Subsequently, step-change (median point change, Mann-Whitney and Kruskall Willis) and trend detection tests (Spearman's rho and Kindall's tau) were applied to average annual discharge and rainfall data for the catchment between 1984 and 2013.Tukey's honestly significant difference (HSD) test was also used to compare the means. Results revealed that huge land cover changes coincided with significant ($p<0.05$) changes in streamflow although rainfall remained homogenous over the same period. This suggests that land cover change is intricately coupled to increases in streamflow. In addition, increased runoff is usually accompanied by increased rates of erosion and siltation. To ensure sustainable management of the catchment, therefore, soil and water conservation measures are critical within the broader context of integrated water resources management.

Keywords: Change Detection, Land Cover Classification, Landsat Image, Streamflow, Rainfall

1. Introduction

Scarcity and misuse of freshwater resources pose serious and growing threats to sustainable development and protection of the environment [1]. The quality and quantity of water heavily impact ecosystem services and human health. River flows are essential for the health of aquatic ecosystems and water supply [2]. Streamflow is also crucial for both ground and surface water resources. Rainfall-runoff relationships within a watershed are the result of the interplay of many factors, but are driven primarily by the interaction of climate, land cover, and soils [3]. Man as a keystone species has been modifying the environment for his own selfish needs. Increasing population necessitates further vegetation clearance to accommodate the farming and settlement needs of the communities which have significantly altered the magnitude and timing of river flows [4, 5, 6, 7].As a result, the health of aquatic ecosystems has declined and some water supplies have become stressed [8, 9, 10]. Therefore, it is imperative to understand the coupling between land use change and the generation of streamflow in a catchment as this influences water balance.

Land use and land cover play a crucial role in driving hydrological processes within watersheds[11].These include changes in water demands e.g. irrigation and urbanization, changes in water supply from altered hydrological processes of infiltration, groundwater recharge and runoff, and changes in water quality from agricultural runoff and suburban development [12]. Sound water resource management is central to the achievement of sustainable development. South Africa has adopted a total catchment approach to the management of water resources. As such the country has been divided into water management areas [13]. Land cover change is a primary concern in watershed management as it may also lead to increased flooding, soil degradation and decreased recharge of aquifers. Some studies have suggested that the consequences of land cover change may outweigh those from climate change [14, 8].Understanding the consequences of land-use change for hydrologic processes, and integrating this understanding into the emerging focus on land-change science are major needs for the future [15]. Woyessa et al., focused on the possible impacts on streamflow of scaling up rainwater harvesting techniques by farmers in the C52A catchment of South Africa [16]. However, for a complete understanding of the flow regimes of the system, it is prudent to look at a whole range of land use activities that could possibly pose impacts.

Studies on the effect of land cover changes on streamflow have yielded different and contrasting results [17, 18, 19, 20]. Bewket and Sterk noted that such contrasting findings suggest that the impacts of land cover changes on water resource systems vary from place to place depending on site-specific factors [21]. In addition, [22] believe that changes in a series of hydrological data can occur either gradually or abruptly or in a more complex form. Investigating stream flow patterns with respect to land cover dynamics enables assessment of sustainability of land use systems, because the stream flows are reflections of the ecological state of the entire watershed [21].This may also form a basis for forecasting the likely effects of any potential changes in land cover on water resource systems within this catchment. Therefore, this study analysed changes in annual historical stream flow and rainfall patterns with reference to dynamics in land cover in the C52A catchment in South Africa. The study findings are important in the development of responsive decisions and policy implementations with respect to catchment utilization in South Africa.

2. Material and Methods

2.1. Study Site

The study site lies between 26° 01`5 0 E and 27 ° 0 0` 0 E and 29° 0` 0 S and 29 ° 45` 0 S. The Modder river originates from the eastern hills of the quaternary catchment C52Abetween Dewetsdorp town and Rustfontein dam located at the outlet of the catchment C52A. The C52A catchment covers about 937km^{2}and lies in the semi-arid central South Africa (Figure 1). The soil texture sandy clay loam to sandy clay and is susceptible to crusting. The quaternary catchment C52A is located in the undulating and high altitude part of the basin between altitudes of 1400 and 1500 m above sea level. Mean annual rainfall for the last 30 years was 542mm and the maximum mean daily temperature is 29^{0}C. Vegetation physiognomic type is predominantly savanna grassland.

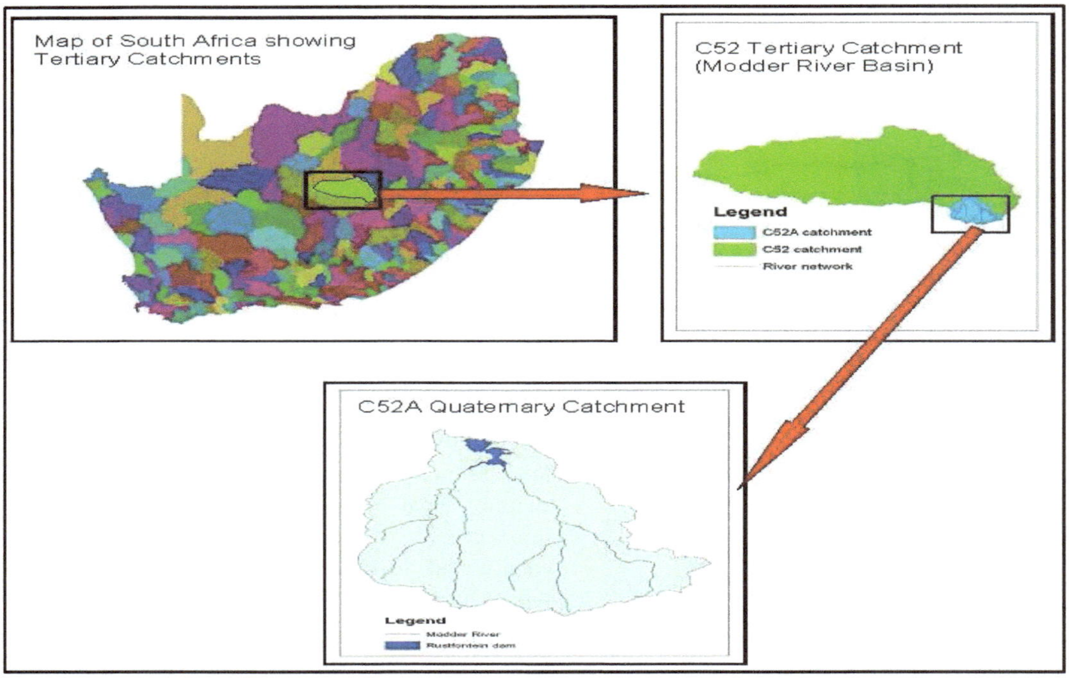

Figure 1. Location of the study area.

2.2. Method of Analysis

It has been observed that it is more manageable to study the same basin over different time periods of time rather than adopt a paired catchment approach when determining hydrological effects of tropical vegetation over large areas [23]. Unlike the conventional methods of the paired-catchment and modelling approaches which require a long period of monitoring and are data intensive, this study linked breaks or changes in streamflow to land cover dynamics in the catchment to ascertain the coupling between the two. To achieve this, dynamics in land cover between 1993 and 2013 were assessed using landsat images. Furthermore, average annual streamflow and rainfall data from 1984 to 2013 were subjected to a number of shift and trend detection methods. To this end, classified land cover maps for 1993, 2004 and 2013 were generated for the catchment. Throughout the text, significant difference means that $p < 0.05$ for means comparison.

Statistical shift detection methods applied

Test for step change

Rainfall and streamflow data were divided into three periods (1984-1993, 1994-2003 and 2004-2013), taking into consideration the availability of land cover maps The Median change, Kruskal-Wallisand the Mann -Whitney tests were used to detect any changes in the distribution of discharge and rainfall over the years. It is important to use more than one change detection method when dealing with hydrological data in order to ascertain the change [22].The median test compares medians across groups. These are essentially rank-based non parametric tests [23,22].

Test for trend

The Kindall's tauand Spearman's rhotests were applied to establish the presence of any trend in the data. These are non-parametric tests and particularly useful for eliminating influence of extreme or "outlier" values in hydrological data as well as for avoiding the requirements of normal distribution [24]. It should be noted that outliers are common in most hydrological data series. Two tailed test of significance for the correlation between time and hydrological data (discharge and rainfall) was performed.

Analysis of Variance (ANOVA)

Finally, ANOVA was computed to confirm changes in streamflow and rainfall patterns. As such to test for the homogeneity of variances, Levene's test was used. Subsequently, equal variances were assumed and post hoc Tukey's HSD test was applied to enable multiple comparisons of means across sub periods.

2.3. Land Cover Data and Image Analysis

The materials used in the study include Landsat TM, ETM and Landsat 8 satellite images for 1993, and 2004 and 2013. GIS and remote sensing software that were used include ArcGIS version® 10, ArcView® 3.2 and ENVI® 4.3 All the data were downloaded into the workstation for analysis using the GIS software. Image analysis was done using ENVI® 4.3 software as shown in Figure 2.

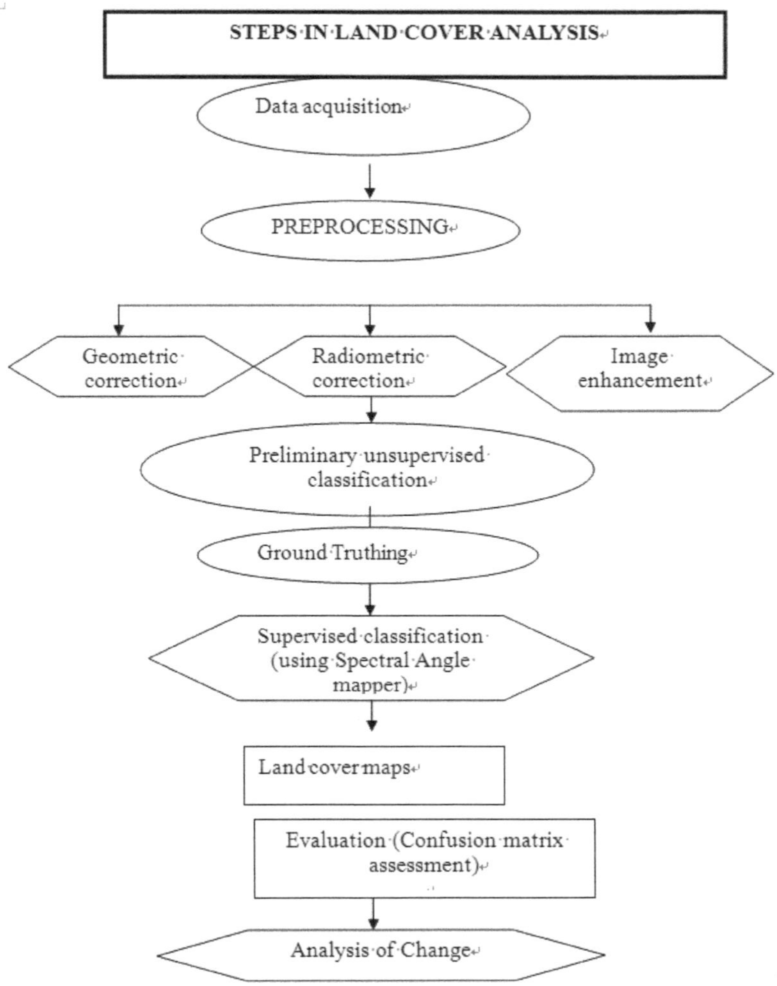

Figure 2. *Steps in vegetation analysis.*

2.3.1. Image Acquisition

A series of Landsat TM/ETM+ and Landsat 8 images (path 170 row 080) covering the watershed were acquired from NASA portal (GLOVIS). The images were selected for cloud free days during April, May, and June for the years 1993-03-18, 2004-06-04, and 2013-04-26. The acquired images were georeferenced and co-registered to one another to allow time line analysis.

2.3.2. Image Enhancement

Image enhancement techniques used were contrast stretching at 2.5%, and false colour composites to facilitate the identification of features. The near infrared composite was combined with visible bands (band combination NIR, Red, Green) to produce a false colour composite. Vegetation in the Near Infra-red (NIR) band was highly reflective due to chlorophyll pigment and false colour composite vividly showed vegetation in various shades of red . Built up areas were displayed in blue in the false colour composites while soil colours varied from dark to light brown. Deep red hues indicated broad leaf healthier vegetation while lighter reds signified grasslands or sparse vegetation. Water appeared dark and black depending on depth and clarity, due to absorption of energy in the visible red and NIR bands. False colour composites helped in differentiating water, soil and vegetated areas.

2.3.3. Normalization

Most sensors including LANDSAT record reflected electromagnetic radiation by earth features in the form of Digital Numbers (DN). These pose difficulties when comparing multi-temporal images because of differences in sun angle, sensor angle and flight height among other reasons. This problem was circumvented by changing DN values to radiance and then radiance to reflectance, a process called normalization. The following formula was used:

Digital Numbers to Radiance

$$\text{rad} = \frac{\text{LMAX}-\text{LMIN}}{255*\text{DN}+\text{LMIN}} \tag{1}$$

Radiance to reflectance:

$$\rho p = \frac{\pi * L\lambda * d2}{\text{ESUN}\lambda * \cos(\Theta s)} \tag{2}$$

where:

ρp = planetary reflectance,

L_λ=spectral radiance at sensor's aperture,

$ESUN_\lambda$ = band dependant mean solar exo-atmospheric irradiance,

Θs = solar zenith angle, and

d = earth-sun distance, in astronomical units

2.4. Image Analysis

The images were classified based on four techniques, namely unsupervised K-means, supervised Spectral Angle Mapper (SAM) and visual interpretation.

2.4.1 K-Means

K-means is a form of unsupervised classification technique which calculates initial class means evenly distributed in the data space. It then iteratively clusters the pixels into the nearest class using a minimum distance technique. All pixels were classified to the nearest class. It generated a general view of the area, creating a classified image with many clusters that aided in training site selection.

2.4.2. Spectral Angle Mapper (SAM)

The Spectral Angle Mapper (SAM) is a physically-based spectral classification that uses an n-dimensional angle to match pixels to reference spectra. It worked by determining the spectral similarity between two spectra, calculating the angle between the spectra, treating them as vectors in a space with dimensionality equal to the number of bands [25]. This technique was used on calibrated reflectance data and was more effective because it was relatively insensitive to illumination and albedo effects compared to other methods like maximum likelihood and minimum distance methods. The end member spectra were collected from false colour images through Region of Interest (ROI) average spectra. The spectra for different land cover classes were compared to typical spectra of known land features for the purpose of separating ambiguity features.

2.4.3. Visual Analysis

After the SAM analysis, visual aids were employed for further feature identity based on the features' site, location, arrangement and their shape. Land cover maps were produced for all the satellite image dates and the earlier date image was compared against the later date image in order to establish percentage changes. Image subtraction was done for each class to assess whether there were positive or negative changes on identified corresponding land cover classes.

2.4.4. Accuracy Assessment of Supervised Classification

The standard method of Confusion Matrix was used to assess classification accuracy for each image date by comparing classification results with ground truth region of interest (ROIs). The methods of accuracy assessment used included the Kappa statistic and Google Earth. The Kappa statistic is a statistical method of assessing the accuracy that took into account the chance of random agreement. Average accuracy of classification was 78%, 80 and 93% for 1993, 2004 and 2013 images respectively.

2.5. Streamflow and Rainfall Data

Stream flow data was obtained from the Department of Water Affairs and rainfall data was obtained from the South Africa Weather Services websites.

3. Results and Discussion

3.1. Land Cover Changes

Great land transformations have taken place within the catchment during the selected period. Figure 3 and Table 1 show results of land cover classification of C52A quaternary catchment for 1993, 2004 and 2013.

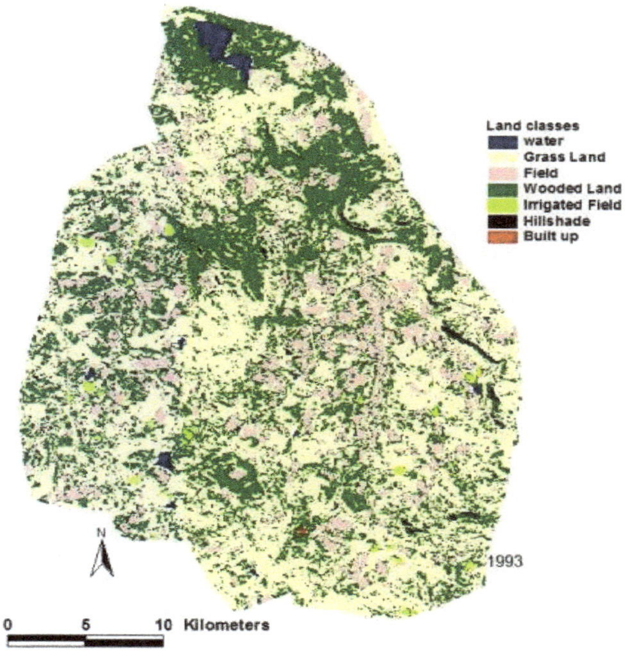

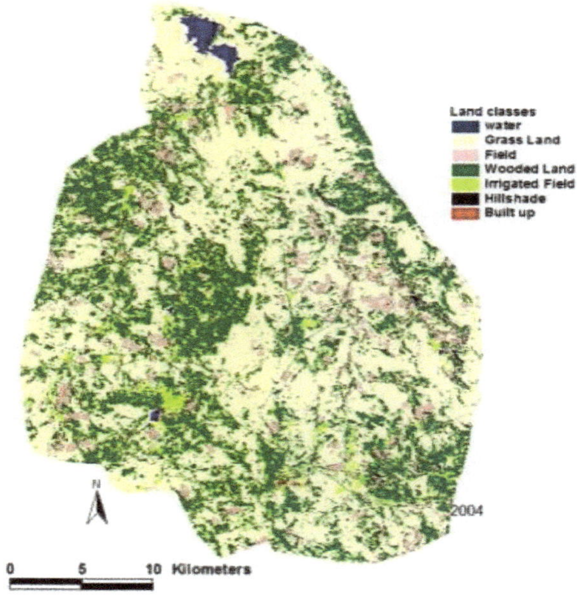

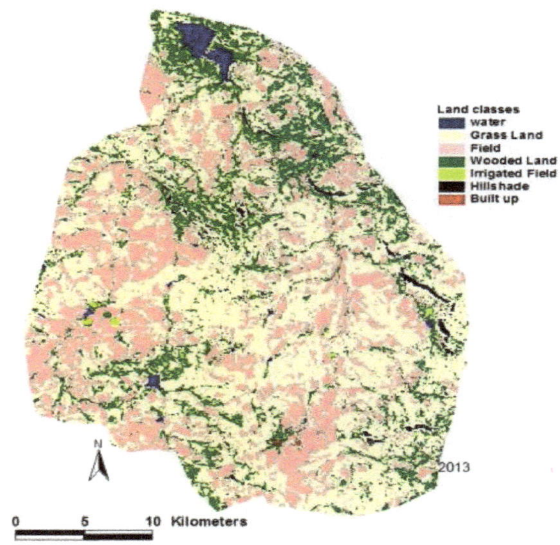

Figure 3. *Land cover maps of C52A quaternary catchment for 1993, 2004 and 2013.*

Table 1. *Land cover changes between 1993 and 2013 in C52A quaternary catchment.*

Land cover	%age cover	%age cover	%age cover	%age changes	%age change
	1993	2004	2013	1993-2004	2004-2013
Water	0.95	0.89	0.99	-6.7	12
Grassland	57	54.6	49.3	-4.3	-9.8
Cultivated area	15	16.6	31.8	14.9	92
Irrigated field	0.6	1.8	0.18	202	-89
Wooded land	25.8	25.4	16.5	-1.6	-35
Hill shade	0.5	0.3	0.9	-40	200
Built up	0.2	0.4	0.6	100	47
Total	100	100	100		

In 1993 most of the land was occupied by grasslands (57%), wooded land (26%) and cultivated area covered about 15 % of the total area. The irrigated land, built up area and water surface contributed 0.6%, 0.2% and 1% respectively to the total land cover. Between 1993 and 2004, water and grassland covers decreased by about 6.7% and 4.3% respectively. Wooded land slightly decreased by 1.6% owing to human encroachment. The area occupied by irrigated fields increased by 202% and the built up area increased by 100% (Table 1). The cultivated area also increased by about

15%. The decrease in grassland cover suggests that other land use options such as cultivation were encroaching into grasslands. By 2004 grassland continued to dominate land cover at 55% of the total area and declined by about 10% to contribute 49% of total catchment coverage in 2013. The wooded area covered about 25% in 2004, declining by about 35% in 2013. The cultivated area occupied about 32% of the total watershed in 2004 and declined by 92% to contribute15% of total area of the catchment in 2013.

3.2. Change Detection

3.2.1. Step-Change Detection

The results from the Median change point test/ Pettit's test for change reveal that the medians of discharge are not the same across the categories (p = 0.006).This suggests a shift in stream flow between 1984 and 2013. Mann-Whitney test revealed that streamflow was not significantly different between the first period (1984-1993) and the second period (1994-2003); but statistically different (p = 0.004) between the second period and the third period (2004-2013). The stream flow pattern was also significantly different (p = 0.003) when the first and third periods were compared. The Kruskal-Wallis test also confirmed that the distribution of streamflow across these three sub-periods was not the same (p = 0.003) and the mean has been increasing over the selected periods (Figure 4). The three tests above were also applied to rainfall data over the same periods and the differences were statistically insignificant.

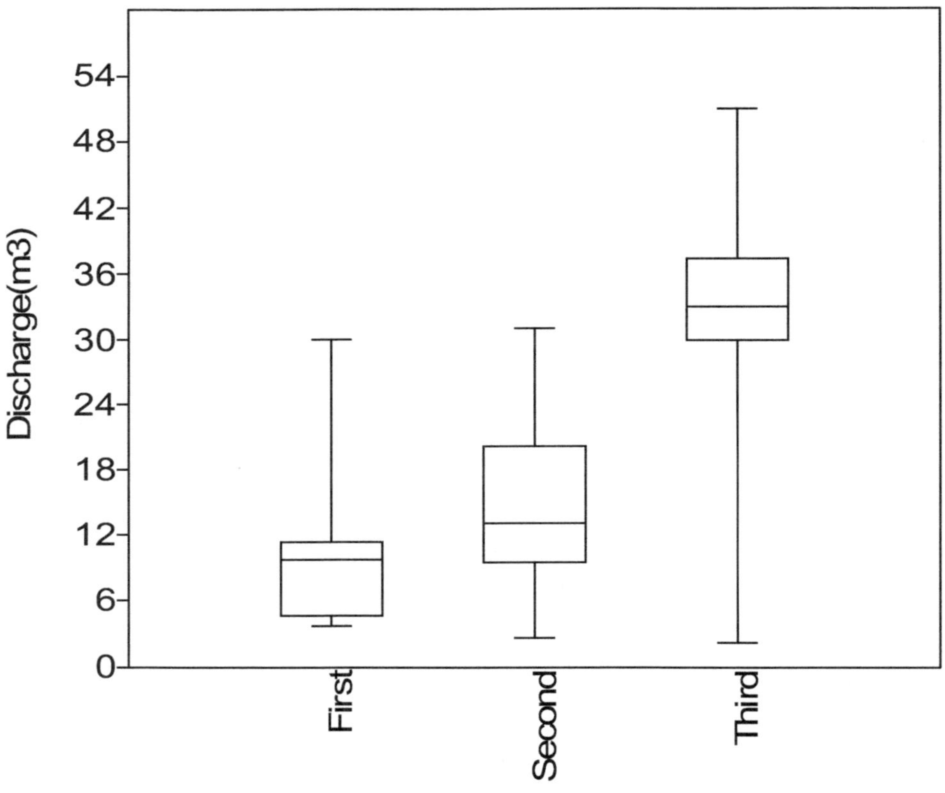

Figure 4. *Box plot showing changes in discharge over the first (1984-1993, second (1994-2003) and third periods (2004-2013).*

3.2.2. Trend Detection

Two trend detection tests in the form of the f Spearman's rho and Kendall's tau were applied to both discharge and rainfall data. The null hypothesis that there is no change in the mean of the data overtime is rejected since results revealed that there was a significant difference (p = 0.000001), suggesting that streamflow is either increasing or decreasing. Figure 4 confirms that mean discharge has been increasing progressively. However, the null hypothesis with respect to rainfall was accepted suggesting that the rainfall was homogenous over these time period.

The Levene test of the homogeneity of variance revealed that the variances were homogenous. Subsequently, the post hoc Tukey's HSD test was performed assuming equal variances.. Results confirmed that the means of the first and second periods were not significantly different (p = 0.74) while those of the first and third were significantly different (p = 0.00001). The second and third were also significantly different (p = 0.002).

3.3. Impacts of Land Use Change on Hydrologic Response

Results suggest that land use has greatly changed in the catchment and rainfall has remained essentially homogenous over the selected period. The observed changes in streamflow are to a large extent, not related to climate variability or climate change. The changes in land use coincided with shifts in streamflow patterns. For example during the first period, statistical tests could not detect significant changes in streamflow probably due to smaller land transformations that took place. During this period, cultivated area increased by 14.9% while grasslands and woodlands decreased by 4.3 and 1.6% respectively. These land use changes did not result in significant changes in streamflow and this was not surprising because, paired catchment experiments have shown that changes in annual water yield from forest cover reductions of less than 30% of the catchment could not be detected by streamflow measurement [26]. However, between the second and third period, statistical tests highlighted significant changes and these were accompanied by huge land cover changes. Cultivation increased by about 92% while wooded land decreased by about 35% and the grasslands decreased by about 9.8%. Therefore, the increase in streamflow is consistent with common knowledge on the prospective role of tropical forests, that reducing forest cover in forested catchments increases the water yield from the catchment with the majority of the increase occurring in the base flow component [21]. It is also consistent with results from paired catcthment studies[27, 17, 28, 29, 30, 31] that reduction in forest cover increased water yield.Also the area covered by the built up area grew by about 47% providing ample opportunities for runoff generation because of hard surfaces.

4. Conclusions

The contribution of land cover change to streamflow was evaluated by analyzing historical land cover dynamics with reference to streamflow and rainfall patterns. Results suggest that land cover change is one of the major factors influencing streamflow in the C52A catchment. Increase in streamflow coincided with dramatic changes in land cover within the catchment. Management of land use change will be vital for streamflow generation. Future decisions on land use in the catchment will be vital in determining the flow regime and ultimately the water balance in the catchment. Water balance at a catchment level is important since it informs water resource allocation for different user groups including the environment. The increase in streamflow over time suggests that there is increased runoff, usually accompanied by increased rates of erosion and siltation. It can also be discerned from the results that long-term monitoring of land use changes is important for determining trends in hydrology. This information is also important for the development of decision support tools for catchment farmers and managers and to inform policy. In order to ensure sustainable management of the catchment, soil and water conservation measures will be critical within the broader context of integrated water resources management.

Acknowledgements

We thank the SA YSSP for giving us the opportunity to be part of the 2014 program and for providing the requisite logistics.

References

[1] ICWE, The Dublin Statement and report of the conference. International Conference on Water and the Environment: Development issues for the 21st century. 26-31 January, Dublin, Ireland (1992).

[2] P.V. Caldwell, G. Sun, S.G, McNulty, E.C. Cohen, M. Myers, Impacts of impervious cover, water withdrawals, and climate change on river flows in the Conterminous US. Hydrological Earth System. Science Discussion 9 (2012) 4263 – 4304.

[3] M. Hernandez, SN.MillerC.D. Goodrich F.B.Goff, W.G. Kepner, C.M. Edmonds, J.K. Bruce, Modeling runoff response to land cover and rainfall spatial variability in semi-arid watersheds.Environmental Monitoringand Assessment64 (2000) 285 - 298.

[4] R.B. Jackson, S.R. Carpenter, C.N.Dahm, D.M. McKnight., R. J. Naiman, S.L. Postel, S.W. Running, Water in a changing world.Ecological applications11 (2001) 1027–1045.

[5] N.L. Poff, J.D. Olden, D.M. Merritt, D.M. Pepin, Homogenization of regional river dynamics by dams and global biodiversity implications, Proceedings of the National Academy of Sciences, USA, 104 (2007) 5732– 5737.

[6] S. Piao, P. Friedlingstein, P.Ciasis, N. de Noblet-Ducoudre, D. Labata, S.Zaehle, Changes in climate and land use have a larger direct impact than rising CO2 on global river 25 runoff trends, Proceedings of the National Academy of Sciences, USA 104 (2007), 15242–15247.

[7] D. Gerten, S. Rost, W. von Blo, W. Lucht, Causes of change in 20th century global river discharge, Geophysical Research Letters., 35(2008)L20405 doi:10.1029/2008GL035258.

[8] C.J. Vörösmarty, P. Green, J. Salisbury, R.B. Lammers, Global water resources: vulnerability from climate change and population growth. Science 289 (2000) 284 - 288.

[9] J. Alcamo, T. Henrichs, T. Ro¨sch., World water in 2025 - global modelling and scenario analysis for the World Commission on Water for the 21st Century. Kassel World Water Series 2, Center for Environmental Systems Research, University of Kassel, Germany (2000).

[10] D.M. Carlisle, D.M. Wolock, M.R. Meador, Alteration of streamflow magnitudes, and potential ecological consequences: a multiregional assessment, Frontiers in Ecology and Environment 9 (2011) 264–270, doi:10.1890/100053.

[11] D. Gerten, S.Schaphoff, W. Haberlandt, W.Luht, S. Sitch, Terrestrial vegetation and water balance—hydrological evaluation of a dynamic global vegetation model.Journal of Hydrology. 286(2004) 249–270

[12] R. DeFries, K.N.Eshleman, Land-use change and hydrologic processes: a major focus for the future, Hydrological Processes Journal 18(2004) 2183 – 2186.

[13] Department for Water Affairs and Forestry (DWAF),National Water Resource Strategy - First Edition: September 2004. Government of South Africa

[14] O.E. Sala, F.S. Chapin, J.J. Armesto, E. Berlow, J. Bloomfield, R. Dirzo, E. Huber-Sanwald, L.F.Huenneke, R.B. Jackson, A. Kinzig, R.Leemans, D.M. Lodge, H.A. Mooney, M. Oesterheld, N.L.Poff, M.T. Sykes, B.M. Walker, M.Walker, D.H. Wall, Biodiversity:global biodiversity scenarios for the year 2100. Science 287 (2000) 1770 - 1774.

[15] D.P. Turner, W.D. Ritts, W.B. Cohen, S.T. Gower, M. Zhao, S.W. Running, S.C. Wofsy, S.Urbanski, A.L. Dunn, J.W. Munger, Scaling gross primary production (GPP) over boreal and deciduous forest landscapes in support of MODIS GPP product validation. Remote Sensing of Environment, 88 (2003) 25 -270.

[16] Y.E. Woyessa, E. Pretorius, P.S. van Heerden, M. Hensley, L.D. van Rensburg, Impact of land use on river basin water balance: a case study of the Modder river basin, South Africa, Colombo, Sri Lanka: Comprehensive Assessment Secretariat. Comprehensive Assessment Research Report 12 (2006).

[17] J.M. Bosch, J.D. Hewlett, A review of catchment experiments to determine the effect of vegetation changes on water yield and evapotranspiration. Journal of Hydrology 55(1/4)(1982) 3–23.

[18] D.F. Scott, R.E. Smith, Preliminary empirical models to predict reductions in total and low flows resulting from afforestation. Water SA 23 (2) (1997) 135–140.

[19] W. K.B. Elkaduwa, R. Sakthivadivel, Use of historical data as a decision support tool in watershed management: A case study of the Upper Nilwala basin in SriLanka. Research Report 26. Colombo, Sri Lanka: International Water Management Institute (1998).

[20] J.K. Lørup, J.C.Refsgaard, D.Mazvimavi, Assessing the effect of land use change on catchment runoff by combined use of statistical tests and hydrological modelling: case studies from Zimbabwe. Journal of Hydrology 205 (1998) 147 - 163.

[21] W. Bewket, G. Sterk, Dynamics in land cover and its effect on stream flow in the Chemoga watershed, Blue Nile basin, Ethiopia, Journal of Hydrological Processes.19(2005) 445 - 458.

[22] Z.W. Kundzewicz, A. J. Robson, Change detection in hydrological records—a review of the methodology / Revue méthodologique de la détection de changements dans les chroniques hydrologiques.Hydrological Sciences Journal49(1)((2004)) 7 - 19. doi:10.1623/hysj.49.1.7.53993

[23] M. Costa, A. Botta, A.J. Cardile, Effects of large scale changes in land cover on the discharge of the Tocantins river, Southeastern Amazon.Journal of Hydrology 283 (2003), 206 - 217.

[24] A.N. Pettitt, A non-parametric approach to the change point problem. Applied Statistics 28(1979) 126 - 135

[25] Y. Lin, X. Wei, The impact of large-scale forest harvesting on hydrology in the Willow watershed of Central British Columbia .Journal of Hydrol0gy 359(1-2) (2008) 141-149. doi:10.1016/j.jhydrol.2008.06.023

[26] ENVI EX User's Guide, December, 2009 Edition, ITT Visual Information Solutions.

[27] A.E.Brown, L. Zhang, T.A. McMahon, A.W. Western, R.A.Vertessy,A review of paired catchment studies for determining changes in water yield resulting from alterations in vegetation. Journal of Hydrology 310(1-4) (2005) 28–61. doi:10.1016/j.jhydrol.2004.12.010

[28] A.R Hibbert, Forest Treatment effects on water yield, Reprint from Proceedings of International Symposium on Forest Hydrology, Pennyslavia State University 1965 (1967) 527 - 543.

[29] F. Gallart, P. Llorens, Catchment management under environmental change: impact of land cover change on water resources. Water International, 28 (3) (2003) 334 - 340.

[30] W. Xiaohu G. Sun, S. Liu, H. Jiang, G. Zhou, L. Dai, The Forest-Streamflow Relationship in China: A 40 year Retrospect. Journal of the American Water Resources Association 44 (5) (2008) 1076 - 1085.

[31] G. Zhou, X. Wei, X. Chen, P. Zhou, Liu, G. Sun, D.F. Scott, S. Zhou, L. Han, Y. Su, Global pattern for the effect of climate and land cover on water yield. Nature Communications 6 (2015) 5918.

3

African Indigenous Knowledge System and Environmental Sustainability

Alokwu Cyprian Obiora[1], Ezenwaji Emma Emeka[2]

[1]Department of Religion and Philosophy, Paul University, Awka, Nigeria
[2]Department of Geography and Meteorology, Nnamdi Azikiwe University, Awka, Nigeria

Email address:
obickol@gmail.com (A. C. Obiora), emmaezenwaji@gmail.com (E. E. Emeka)

Abstract: Observers in international political arena have brought to public knowledge the fact that environmentalism has gradually crept into international politics. This is because of the commonalities between environment and political economy. Governments at various levels, environmental policy makers, have therefore sought solutions from scientific, economic and political dimensions to address the problems while neglecting the important role indigenous knowledge system could play in this much needed synergy. After careful research of some related literatures, the paper emerged with the findings that non-inclusion of African indigenous knowledge in environmental discourse over the years was simply based on the assumption that scientific investigation and findings was the only authentic method and acceptable means of providing information/solution to myriad of problems of the environmental sustainability. It, therefore, strongly argues that solution to these problems particularly in Africa requires a multi and interdisciplinary approach and not just scientific inquiry. It calls attention to the fact that African indigenous knowledge system (which still remains untapped resource) could beuseful, in fostering the much needed culture of ecological responsibility. To achieve this, it recommends a new paradigm to environmental discourse where natural science integrates with African indigenous knowledge to achieve environmental sustainability.

Keywords: International, Politics, African, Indigenous, Ecological

1. Introduction

There is no doubt that what is on the lips of environmentally concerned people, including the international community and Nigeria in particular, is the issue of safeguarding our gradual but increasingly dilapidating environment. It has been mentioned earlier that confronting the environmental challenges of our time requires an interdisciplinary and even multi-dimensional approach. One of such approaches is the use of African indigenous knowledge system.

For us to understand the importance of African indigenous or traditional knowledge system in contributing to ecological sustainability, we need to have an understanding of the African world views, its cultural heritage, values and ecological construct. We also need to understand African myths that explain the various cultural practices of the African people. Understanding the traditional ecological knowledge of the Africans is important for the consideration of a broad range of questions regarding their nature-human relations. This is because different groups of people in various parts of the world perceive and interact with nature differently, and have different traditions underpinning their ecological ethics. Their perceptions and knowledge of the universe are in part shaped by their values, worldviews, and religion in the broader sense. In other words, when people attempt to answer fundamental questions about the place and relationship of humanity with the universe, their ideas about the universe accumulate and definite views and systems begin to emerge (Ibe, 2003). As a thought process, there can be no end to the development of people's views about the universe. This world view can be defined as "the complex of their beliefs and attitudes concerning the origin, structure of the universe and the interaction of its beings with particular reference to man (sic)" (Metu,, 1999). World views are embedded in the people's experience and, then, expressed or re-enacted in their cultures (Kalu, 2000). Nature is full of mystery and such mystery poses a challenge to people's understanding of the society in which they live. Society, therefore, constructs a world view in an attempt to unravel the mystery of the universe. The society's world view underpins their culture. The society constructs culture in their encounter with and effort to understand and relate with nature, in order to harness its resources for the nurture of the society.

From the above, it is clear that environmental discourse is heavily loaded with statements about the way indigenous and traditional societies relate to their environment (Cooper and Palmer, 1998). For example, the report of the World Commission on Environment and Development (the Brundtl and Report), referred to the harmony with nature and the environmental awareness which characterized the traditional way of life of many indigenous people (World Commission on Environment and Development, 1987). Similarly, Margaret Thatcher, the former British Prime Minister, praised the primitive people of the rain forest for having a "one-ness" with their environment that has been lost in the urban jungle (Cooper and Palmer, 1998). Furthermore, in his opening address to the United Nations Conference on Environment and Development in 1992, (the Rio Earth Summit), Maurice Strong advised as follows: "we must reinstate in our lives the ethic of love and respect for the earth which the traditional peoples have retained as central to their value system" (United Nations, 1993). The most important distinction to be drawn from the above assertions, is between beliefs about the way indigenous and traditional peoples act towards their environment (that they do not dominate or destroy it, that they live sustainably within it), and beliefs about what they think, feel or know about their environment-that they respect it and possess ecological wisdom (Cooper and Palmer, 1998).

The above description of indigenous people and their relationship with their environment is evidently true about indigenous peoples of Africa. Before the coming of Christianity and the introduction of Western world views, African peoples lived and understood their existence in terms of the belief system and cultural constructs of the society to which each belonged. In Africa, there are different ideas and world views about the world just like in other parts of the world. These views are expressed in myths, legends, proverbs, rituals, symbols, beliefs and wise sayings (Metu, 1985). Although each ethnic community has its own beliefs about humanity and the world around it, anthropological and ethnological studies have made it possible to observe some general characteristic features of African world view (Mugambi, 1988). The generalisation of these characteristic features has in no small way helped us to have a common understanding of the African world view since the task of describing a world view shared by all Africans or even a large section of African societies would almost be an impossible task in this work. We can only attempt to encapsulate some of these general characteristic features.

2. Concepts of Traditional Ecological Knowledge (TEK)

Scholars are not in agreement over the use of the term "traditional". Some consider it derogatory, backwardness, and instead favour "indigenous" or "local." Others point out that many indigenous peoples themselves see "tradition" in a positive light. They do not take it to mean inflexible adherence to the past but rather to mean time-tested and wise (Warren, 1992: 1). This paper uses both traditional and indigenous interchangeably in a non-derogatory sense.

In the search for an environmental ethics towards an ecologically sustainable society, indigenous peoples and traditional ecological knowledge have attracted considerable attention from both scholars and environmental movements as a viable option to be considered. Traditional ecological knowledge may be defined as "a cumulative body of knowledge, practice and belief evolving by adaptive processes and handed down through generations by cultural transmission, about the relationship of living beings (including humans) with one another and with their environment" (Possy and Dutfield, 1996). As a body of knowledge system, traditional ecological knowledge includes the worldview or religious traditions of a society. It is both cumulative and dynamic, building on experience and adapting to change, as societies constantly redefine what is considered "traditional." (Grenier, 1998). It is an attribute of societies with historical continuity in making a living in a particular place. African Traditional ecological knowledge may therefore be considered as a sub-set of African indigenous knowledge, defined as local knowledge held by African peoples or local knowledge unique to an African cultural society (Berkes, 1999).

African traditional ecological knowledge is comprised of a wide range of understandings of ecosystems and this includes traditional modes of agriculture especially in the form of shifting cultivation. Essentially, in this agricultural system, a family clears or slashes and bums a piece of wooded area. The ashes act as fertilizer for the next two years or so. Different types of crops are grown concurrently on the same plot as a simulation of biodiversity of a natural rainforest ecosystem. This protects the soil from turning into laterite, bricklikesoils. After three years of cultivation, one moves on to another piece of land while the former stays fallow to regenerate. Regeneration may take up to ten years or more. However, such asystem entails a reasonably low population density. The system of shifting cultivation was practiced in most African countries until recently when it became no longer possible due to high population densities (Kapileni, 1994). Other areas where traditional environmental knowledge is projected includes agro forestry, animal raising, hunting, fishing and gathering (Wolfgram, 2006).

3. Methodology

The research method adopted is qualitative which according to Merriam (1998) is an approach which recognises that meaning emerges through interaction and is not standardized from person to person as in quantitative research and thus allowing the researcher to study issues in detail without predetermined categorised analysis. This method was used because it assures an internal data collection which is achieved through generation of funs that encourages the subjects to participate in the research. The sampling method adopted was purposive sampling. We employed this method because we had to select respondents based on some defining characteristics that makes them holders of specific data needed for the study (Maree, 2007).

4. Result

Traditional environmental knowledge is not stored in books; rather it is retained and transmitted verbally, through subsistence practices and through cultural myths and taboos.

It is interesting to observe that the significance of traditional ecological knowledge-(TEK) has been recognized in the international arena and that the number of publications on this subject has grown rapidly since the 1980s (Werner, and Begishe, 1980). However, its relevance in environmental management and development discourse in general, including the relationship between it and Western science has remained controversial (Wolfgram. 2006). This controversy emanates from the fact that scientists tend to be sceptical, demanding evidence when confronted with traditional knowledge that may not easily lend itself to scientific verification (Wolfgram, 2006). The problem of credibility is compounded because some traditional knowledge which may include elements, such as the religious dimensions of the environment, (these could not be easily verified in a scientific manner), do not make sense to science. For example, many of the African societies believe that some non-living parts of the environment (including rivers and mountains), as well as all living beings, have spirit or at least is connected to one spirit or the other. Asante vividly captures this point when he observed that:

Their (i.e.) African's cosmologies were based on the existence of a Supreme Being who was spirit and creator, and the source of all power and energy. He granted existence to all created things. He animated the gods, the spirits and human beings, animals, rivers, trees, rocks and all matter. Their metaphysics conceived the creator as the Universal Vital Force that animated and energized all things whose real essence consisted not in matter but in the energy and power infused into them by the Creator (Asante, 1985).

In the face of such a belief grounded purely on religious conviction, it becomes difficult for science to make any scientific verification because it has no tools for the study of the spiritual dimensions of the environment. Nevertheless such beliefs are important for our understanding of traditional ecological knowledge.

On their own part, traditional knowledge holders are also sceptical of scientists, and tend to dismiss scientists who do not have extensive first-hand knowledge of a specific land area. As a result, they are often baffled by the preoccupation of scientists to measure and quantify almost everything and dismiss any that does not hold valid scientific proof.

Power relations concerning authority over knowledge therefore becomes a big issue here. This tension becomes more complex because not many "educated people" are willing to accept the fact that Western scientific methodology is merely one way, and not the only way, to acquire knowledge and provide solutions to humanity's multifarious problems. Unfortunately, the superiority of scientific method has gained more acceptance as the dominant knowledge system at the negligence of traditional knowledge. As a result of this perceived "superiority" of Western science over traditional knowledge, Western science is seen as having a monopoly on truth and regards knowledge and insights that originate outside institutionalized Western science as unacceptable fallacy. Practitioners in the field of Western science, therefore tend to dismiss understandings that do not fit their own which come from other bodies of knowledge such as TEK. As a result, it has become the sole basis of environmental decision-making by many environmental policy makers and development agencies throughout the world including Africa.

However, in his defence of TEK, Hunn articulates the following reasons and insists that traditional environmental knowledge should be considered as another method of acquiring knowledge alongside Western method and therefore argues that it should be preserved because:

- It is a locally grounded science that supplements Western science.
- It is an intellectual monument to our common humanity.
- It provides a kind of human alternative or human diversity option to the globalization of the Western market culture.
- The use of TEK contributes to conceptual pluralism, and expands the range of approaches and information needed to solve environmental problems (Hunn, 1999).

Furthermore, Wolfman sees the potential of TEK to build upon ecologically sustainable local knowledge and belief systems as an important asset for environmental conservation. This potential for him, justifies its significance (Wolfgram, 2006) . He argues that there are tangible and practical reasons why TEK is important for the rest of the world's people by outlining the following points:

It is a source of biological knowledge and ecological insights.

It is important for the sustainability of difficult-to-manage ecosystems such as tropical forests.

It is important for community-based conservation by connecting human values with conservation values.

In-depth local environmental knowledge and trends over time for a given site are important for environmental assessment.

Traditional knowledge is essential for development, especially for "bottom-up" (as opposed to top-down) development planning with people.

Traditional ecological wisdom is a source of inspiration for environmental ethics. Belief systems of many indigenous groups incorporate the idea that humans are part of the natural environment, and their relationship with nature may be characterized as peaceful coexistence.

Based on the inestimable importance of TEK, Posey and Graham caution that Traditional knowledge and Western science need not be thought of as opposites, rather, it may be useful to emphasize the potential complementarities of the two, and to look for points of agreement rather than disagreement (Darrell and Balee1989: 76). These observations are significantly instructive. This significance reflects the explosion of interest in TEK in recent years and therefore justifies the need to derive ecological insights from indigenous practice, and the need to develop a new ecological ethic based in part on indigenous wisdom.

Our understanding of the need to appreciate African traditional knowledge and the important role it could play in the search for a synergistic approach to the current global ecological crisis when it is integrated with other ethics that relate to ecological sustainability is necessary.

5. Discussion

How we think about the world affects the way we live in and relate with it. This is very important for our discussion because our understanding of nature-our cosmology, affects the way we understand ourselves, the way we relate to other people, and, of course, the way we relate to the earth and other forms of life (Sindima, 1990). From her past colonial experience, the people of Africa have been influenced by a cosmology inherited from the West. This is the whole point made in "Things Fall Apart" by Chinua Achebe and other "pan Africanists" about the British colonial exploitation of Africa. This Western cosmology is mechanistic in nature. It is a perspective that views all things as lifeless, commodities to be understood scientifically and to be used for human ends. This is contrary to the African understanding of the cosmos.

The African conceives of the universe beyond the diversity of its forms as a fundamentally one cosmic reality in its visible and invisible forms. To some extent, both the visible and the invisible elements of nature are interlinked, human life is inseparably bound to nature, and both human life and the life of other creatures are one with the divine. It is therefore unthinkable for Africans to objectify nature as the other' or to see nature as having only instrumental value. The African notion of bondedness of all things fosters a sense of care for all of creation. This sense of bondedness is embedded in individual personality and this guides the individual's behaviour from birth to death, through puberty and marriage. The processes of acquiring the sense ofbondedness, involve some levels of socialization through which the individual as it were is implanted with the sense of bondedness. This therefore explains why when African people die outside their ancestral home, they would be brought back to their place of birth to be laid to rest so as to be gathered with the rest of their kindred (Pobee, 1979). Africans do not allow their own to be buried outside their birth place.

According to Ikenga-Metu, "the dichotomy which is so characteristic of the West, especially the Greco-Christian world view is strikingly absent in African world view. There is no clear cut distinction or opposition between the visible and the invisible, the material and the spiritual, the temporal and the non-temporal, the sacred and the profane" (Metu. 1985). This is against the Western cosmology which is static, objective, dichotomised and in fact dualistic in its distinction between body and soul, matter and spirit (Hord, 1995). The African is, of course, sensitive to his or her external world, to the material aspect of beings and things. For him or her, matter in the sense of the European understanding is quite different. Matter in the African understanding serves only as a system of signs which translates the single reality of the universe (Metu, 1985). In this sense, the whole universe appears as an infinitely small and at the same time an infinitely large, network of life forces which emanate from God and end in God who is the source of all life forces. It is he who vitalizes and devitalizes all other beings, all the other life forces (Hord,199). Ibe, quoting Egbujie succinctly summarises this African understanding of cosmic oneness thus: "Beyond and above the visible tactile physical world, there is a non-visible, non-tactile world which envelops the former. It permeates the former through and through; it is simultaneously within an outside of the earth and the seas" (Ibe, 2003). Consequently, there is not only "less separation between subject and object, between self and non-self, but fundamentally all things share the same nature and the same interaction one upon another ... the living, the dead and the first ancestors, from the stone to the divinities a hierarchy of power but not of being, for all are one, all are here, all are now" (Metu, 1985). As a result of this cosmic oneness, it is generally believed that the deities inhabit the natural phenomena with which they are associated like rivers, forests, sky or sun. The ancestors are believed to be around their homes and hearths and take part in all important family affairs. This idea of cosmic oneness or corporate identity, as Mugambi calls it, is central in African cosmological thinking. According to him, thecommunity to which a person belongs extends to include past and future generations. The family extends beyond the conjugal unit to include the whole ethnic clan, and more extensively, the whole ethnic group (Mugambi, 1988).

In terms of creation story, there abound almost irreconcilable versions of creation stories in Africa. These differences could be as a result of different patterns of observation used by different societies to explain their existential reality. While some societies use models drawn from the visible physical world (the order of the planetary bodies), others draw their models from ecology-the sea, desert or fauna. Despite these differences in creation stories, the African understanding of creation is unanimously religious in nature. According to Mbiti, when people in Africa say that the universe was created by God, they are automatically looking at the universe religiously (Mbiti, 1980). Based on Mbiti's observation, we can say, therefore, that the African view of the universe is profoundly religious as opposed to the Western mechanistic view. Africans see the universe as a religious entity and treat it as such (Ibe, 2003). The African traditional cosmology views their existence as their religion and their tradition as their existence. The two cannot be separated. "They eat religiously, drink religiously and for them religious activities invigorate the person, supplying him (sic) with help from deities and unseen external forces and also protecting the person from dangerous activities of the spirits" (Madu, 1996).

The African understanding of the universe is two-fold: the visible world and the invisible world. While the visible world is populated by men (sic) and all material surroundings like sky, mountains, earth, rivers, forests and so on, the invisible world consists of the heavenly part (sky) inhabited by the creator and deities, and "the spirit land" of the ancestors, the spirits, disembodied spirits, all located somewhere in the ground (Ibe, 2003) . Following this classification, many African people hold that the universe is in the form of a three-tier creation, namely: the heavens, the earth and the underworld, which lies below it.

The heavenly part of the universe is the home of the stars, sun, moon, meteorites, sky, the windand the rain, with all the phenomena connected with them such as thunder and lightning, storms, eclipses and so forth. It is also thought to be the home of God, although people cannot locate where he dwells. The sky', heaven', or beyond the clouds' are used to describe his dwelling place. While some people believe that God has other beings living with or close to him-his messengers or ministers, others thought he is completely living alone. The earth, too, is full of created things. It is symbolically looked upon as the mother of the universe hence the expression- "mother earth or the goddess earth" (Metu, 1985).

From the foregoing, it is evident enough that the significance of the African world view in relation to ecological consideration is *sine qua non* to understanding African indigenous ecological ethics in order to be able to offer the synergistic solution to the current ecological problems in Africa. This is because the African world view is responsible to a very large extent for the profound communion Africans have with the universe. Ikenga-Metu agrees with this assertion when he argues that:

The main object of an African is to live a life in harmony with humanity and with nature. Man strives to be in harmony with God, the deities, and his fellow men (sic) both living and dead. He feels himself in intimate rapport and tries to maintain harmonious relationship with the animal, vegetation and other elements and phenomena in the universe. For him, the first evil is disintegration for this would spell disaster for himself and his immediate world. The ideal thing is integration, communion and harmony (Metu, 1985).

The above quotation about the African people's sense of intimacy and harmonious relationship with nature justifies the belief that Africans are ecologically responsible even before the advent of Christianity. This is evidently seen in the African belief that the deities have the responsibility to police people as it were, to ensure integration and harmony of the ontological order. This consciousness compels people to always ensure that their relationship with nature is not in any way impaired. Any impaired relationship with nature, they believed, would attract unfavourable consequences to them.

The consciousness mentioned above shows that the importance of moral order in African societyis of fundamental significance. It is understood that God gave moral order to people so that they might live happily and in harmony with one another and with nature. This moral order is instrumental to the establishment of customs and various institutions in all societies, to safeguard the life of individuals and the community of creatures which humans are part of. Each society is able to formulate its values because of this moral order. These values deal with relationship among people, and between people and God and other spiritual beings, and humanity's relationship with the world of nature (Ibe, 2003). To act contrary to this moral order is considered an abomination against nature and humanity. The functionality of moral order ensured that Africans maintained and lived in close relationship and harmony with nature before the introduction of the Western culture and Christian faith which dichotomizes and distinguishes between spirit and matter on one hand and between man and nature on the other hand.

6. Africa's Past Ecological Experience

Before the penetration of Africa by the Europeans, Africans had developed a unique system of preserving their environment (Kalipeni, 1994). This is in spite of the fact that traditionally, Africans had no written literature regarding their relationship with nature. However, this does not imply that they were environmentally bankrupt. On the contrary, they expressed their beliefs and attitudes to environment through myths, stories, proverbs, songs, taboos and rituals-oral history. Samson Gitau in his book, The Environmental Crisis, argues that Africa has a rich environmental culture and that if one wishes to understand the traditional African attitudes towards environment, he or she has to listen to their myths, taboos, stories, proverbs, and beliefs. One will also have to observe their symbols and rituals (Gitau, 2000). For example, among the Igbos of Nigeria, some ethical prescriptions-taboos-are associated with some rivers and forests to safe guard them from pollution, abuse and exploitation. In other African societies, certain animals and plants are associated with spirits in a particularly close way that ensured environmental sustainability (Forde, 1998).

In this manner, African culture had been able to successfully control the proper use of the earth over the years until the culture of the industrial market became a global process-attempting whether by seduction or coercion to destroy all other existing cultural forms (Alokwu, 2009). In doing so, it has created its own contradictions, giving humanity the power to extinguish not only itself but also many of the world's species. We would like to argue strongly on the basis of the above assertion, that the loss of African way of life, and with it, of cultural identity, lies at the root of our environmental problems, socio-economic and political underdevelopment. Before the advent of Western Missionaries and industrial market culture, Africans were noted for their rich cultural heritage. This had enormous influence on their life.

In those good old days when moral order was strictly

observed in traditional Igbo society, children were taught to observe some ecological ethics. They were taught not to defecate near the village stream where drinking water was fetched. They were taught that violation of this prohibition-the moral order, would attract some severe sanctions from the deity who owns the stream. Children were taught to observe the moral code which the Igbos called omenani. This omenani was believed to have been handed down from the earth goddess from time immemorial. The elements of omenani are made up of purely, religious, social, economic and political restrictions (Allen, 1951). These prohibitions known as *nso ani* (what the earth-goddess abhors) are meant to enforce moral order in the society and people strictly adhered to them. Ilogu articulates some of the components of *nso ani* which could attract severe penalties from both the society and the gods to include but not limited to the following;

- Stealing yam either from the barn or from farm
- Homicide
- Incest
- Suicide by hanging
- Poisoning someone
- Theft of domestic fowls especially a hen in its hatching pot
- Killing of a pregnant goat (Ilogu, 1974).

A careful look at the above list of what is considered as "nso ani" shows that most of the prohibitions (which if violated could attract severe punishments) have some ecological implications which are meant to safeguard nature as a whole. Generally speaking, traditional ecological ethics was based on the maxim; "do not do to others what you would not want them to do to you". This maxim, however, is not peculiar to Africans. Indigenous peoples all over theworld-Pacific Islanders, American Indians, and Aboriginal Australians-have similar mechanisms informed by their beliefs and philosophies regarding the land and their relationships to the land and the rest of the environment (Sowunmi, 1995). For example, it is widely believed among Africans that the land is "mother"-the source of nourishment, survival and indeed life. It shouldbe reverenced. In some communities, it is believed that land cannot be owned by the people since the understanding is that the people are owned by it. A typical example of this is the case of the Anam people of Anambra West of Anambra state-Nigeria. The traditional land ownership is communal in nature. No single individual owns a land. The entire community owns the land in trust for the earth goddess.

It has been observed that indigenous people have an excellent knowledge of their environment. This afforded them the wisdom of not attempting to over exploit its wide range of resources. Many communities are known to use more than a hundred different plant species for dietary, medicinal and other domestic purposes. The indication of their harmonious existence with the environment, and judicious exploitation of it, is that they can remain for generations, over thousands of years, in the same area; hunting was solely for food and other basic needs, not for sport or fun. Communal spirit and kinship bond were very strong, negative individualism, selfishness and greed were discouraged.

Traditional ecological knowledge promoted harmony between humankind and the environment and a healthy concept of how to relate to it. Hitherto what has been known about African cultures has come largely from academic studies in ethno-archaeology and anthropology, usually by foreign scholars who often regarded the people as "primitive". Colonizing powers imposed on these people their own cultures, which they in their arrogance and ignorance considered to be superior and "civilized" (Wolfgram, 2006). The fact that the so-called superior cultures have been the ones largely responsible for the degradation of the environment as well as the economic crisis of the impoverished ex-colonies is a clear indication of the failure of their ideologies and life styles (Sowunmi, 1995).

A reinvention of past African ecological ethics and the sets of moral orders referred to earlier in African society would be necessary in the face of the current ecological crisis bedevilling the entire universe. This is because the implication of the current global earth crisis is the destruction and distortion of our hitherto beautiful environment in many ways. The extent of this distortion is such that human existence on the surface of the earth is endangered. The threat to human existence and bio depletion' reminds us of Fidon Mwombeki's caution that "humanity is dangerously capable of destroying itself and the creation" (Mwombeki, 2001). The current rate of the earth crisis emanating from the quest for development and other human activities also increases the fear expressed by McDough that "when the earth is disfigured and its fruitfulness impaired, then every succeeding generation of people is automatically condemned to poverty" and the possible extinction of the whole universe (McDough, Sean. 1990). The consequences of focusing only on immediate ends and not thinking intergenerational occasioned by the Western mechanistic view of the world are a contributory factor to the environmental degradation and other problems of our time. This calls for a behavioural or attitudinal change in order to avert the impending doom described above. This change could come from TEK which emphasizes the essence of being human-"my humanity is caught up and inextricably bound up in yours ... I am because I belong. This fosters a strong sense of community. The African concept of community recognizes that we live in one household, that we need one another to survive and thrive. The word "community" refers to more than a mere association of atomic individuals. The term suggests bondedness; it refers to the act of sharing and living in communion and communication with each other and with nature. To communicate is to stay in a relationship and inculcate a sense of sharing. This is what Sindima meant by saying that "in community we share and commune with selves who are other than ourselves and yet united to us." Sindima goes further to suggest that in Africa "persons are not individual entities or strangers to one another". It is the community which "defines the person as person, not some isolated static quality of rationality, will or memory" (Shutte, 1993). They are nature itself seeking fullness in the actuality of present life. Since people belong to the fabric of life, their life-like nature-must be respected. This call for respect of persons is also a charge to the community to create possibilities for persons to realize full personhood in a community of life where all are bonded together, everyone is responsible for everyone else".

Setiloan's biocentric ethics sheds more light in our understanding of community in relation to African context. He uses the term biocentric ethics to emphasize the inclusive nature of community life in all its forms (Setiloane, 1995). The term "biocentric" was originally formulated by Leopold in his notion of land ethic which implies an expansive notion of community. This term was later broadened as "ecocentric" or "biocentric" ethics by different philosophers over the decades to mean that we are all part of a biotic community (Cock, 2007). In a community there is the structure as well as mutuality of care, assurance of belonging, sustenance and support. The African notion of "earth community" refers to relations of interrelatedness and interdependence, without rivalry and oppression. It is about community and reconciliation, wholeness, connectedness, love, sacrifice, generosity, and welcoming the It maintains that all living things have intrinsic worth-value in and of themselves-not just instrumental or utilitarian value. He concludes that land ethic involves the extension of our human ethics to include the other species with which we share the land. In this perspective, all ethics ... rests upon a single premise: that the individual is a member of a community with interdependent parts and the land ethic simply enlarges the boundaries of thecommunity to include soil, waters, plants and animals). Therefore, "biodiversityis necessary not only for utilitarian and humanitarian reasons (for maintaining the present and future health of the entire biosphere, for enhancing the quality of life, and for aesthetic enjoyment), but its own sake. Ecocentrism expands the goods of the human community to embrace and include within it the good of the biotic community" (Merchant, Caroline. 2004).

Setiloane's ecological ethics is also expressed in the African principle of live and let live. In recognition of this principle, the Igbos of Nigeria strongly believes in their notion of "*Egbe belu, ugo belu. Nke si ibe ya ebena, nku kwaa ya*". When loosely translated, it reads; let the kite perch as well as the eagle. If anyone refuses the other, let its wing be cut off. The rationalization of egbe belu principle tries to enforce respect for all life forms. It is a concept which recognizes the right of existence of all life and insists that none has the right of refusal to existence to other life forms. The *egbe belu* principle guarantees protection to all the community members including the weaker ones who would not have been able to protect themselves. Alokwu (2009) observesthat the application of this principle ensures safety in the whole community of life as none is marginalized for any reason. This healthy relationship which the *egbe belu* principle promotes eliminates greed in its various dimensions as only one's legitimate needs are sufficiently satisfied. Therefore, it is sinful for one not to recognize the right of existence of others or to disturb the constant cycle of existence which affects the corporate existence of the community. In order to maintain this *egbe belu* principle, the Igbos (and most African societies) developed the idea of taboos and totemism where certain groups of species-animals, plants, rivers etc. are prohibited from being killed or destroyed. It is generally believed that non adherence to this prohibitive order would attract the wrath of the gods on humanity. In a home there is the structure as well as mutuality of care, assurance of belonging, sustenance and support.

7. Conclusion/Recommendation

From the discourses so far, the paper has succinctly dealt on the need for the inclusion of TEK as environmental model for entrenching environmental responsibility and sustainability in Africa. In the search for collaborative solution to the current environmental crisis the paper suggests that African scientists and teachers should evolve a new approach to the study of natural sciences that integrates indigenous principles of natural resource conservation and management through TEK to ensure a holistic approach to the study of natural organisms and systems.

It further recommends that African countries should mount deliberate programmes of training to produce top level experts in the various fields of natural resources and the environment who are also interested in African tradition. The objective should be to create a pyramidal structure of expertise, with those in the top echelon constituting the critical mass of training for innovation in environmental sustainability.

African governments and wealthy individuals should be encouraged to assist researchers to conduct competitive and innovative research in African environmentalism. Locally designed educational programmes on the value of natural resources management should be integrated into educational curricula at all levels.

References

[1] Andrew, F. Walls. (1976). "Towards Understanding Africa's Place in Christian History" in John Pobee. Religion in Pluralistic Society. New York; E. J. Brill.

[2] Asanie, E.(1985). "Ecology: Untapped Resources of Pan-Vitalism. in Africa" African Ecclesial Review 27.

[3] Berkes, F.(1999). Sacred Ecology: Traditional Ecological Knowledge and Resource Management.

[4] Bums, A. (1951). History of Nigeria. London: George Allen and Unwin Ltd.

[5] Cooper, F. and Palmer, J. A. (1998). Spirit of the Environment. Religion, Value and the Environmental Concern. London: Routledge.

[6] Darby, K. R. (2006). Ecology, Economy and God, Theology that Matters. Minneapolis: Fortress Press.

[7] Dary II Forde (ed) (1998). African World. Studies in the Cosmological Ideas and Social Values of African Peoples. Hamberg: LIT James Currey with the IAI.

[8] Emeife, I. M. (1999). God and man in African Religion. A Case Study of the Igbo of Nigeria. Enugu: Snaap Press Ltd.

[9] Emeife, I. M. (1985). Comparative Studies of African Traditional Religions. Onitsha: Imico Publishers.

[10] Ezekiel, K. (ed) (1994). Population Growth and Environmental Degradation in Southern Africa. United States of America: Lynne Reinner Publisher.

[11] Fred, L. H. (Mzee Lasana Okpara) and Jonathan, S. L. (1995).(eds) "I am Because We Are".

[12] Grenier, L. (1998). Working with Indigenous Knowledge. A Guide for Researchers. Ottawa: International Development Research Centre.

[13] Gwinyia H. M. (1995). The origins of Development of the African Theology. Maryknoll: Orbis Books

[14] Hunn, E. (1999). The Value of Subsistence for the Future of the World. Ethnoecology, Situated Knowledge/Located Lives. Tucson: University of Arizona Press.

[15] Ibe, M. J. (2003). Environmental Ethics and Politics in Developing Countries.

[16] Ilesanmi, S. (1997). Religious Pluralism and the Nigerian State. Athens: Centre for International Studies, Ohio University.

[17] Ilogu, E.C. (1974). Christianity and Igbo Culture. Onitsha: University Publishing Company.

[18] Jacklyn, C. (2007). The War against Ourselves. Nature, Power and Justice. Johannesburg: Wits University Press.

[19] Jenkins, P. (2002). The Next Christendom. New York; Oxford University Press.

[20] JNK, M.(1988). African Heritage and Contemporary Christianity. Nairobi-Kenya: Longhorn Publishers.

[21] John, S. P. (1979). Toward an African Theology. United States of America. Parthenon Press.

[22] Kalu, O. U. (2000). Power, Poverty and Prayer. The Challenges of poverty and pluralism in

[23] African Christianity, 1960 - 1996. New York: Peter lang.

[24] Leal, R. (2004). The Environment and Christian Faith. Australia: St. Paul's Publications.

[25] Louw, J. P. and Eugene A. N. (eds) (1989). Greek-English Lexicon of the New Testament based on semantic domains. Vol. 1. Cape Town: Bible Society of South Africa.

[26] Madu, R. O. (1996). African Symbols, Proverbs and Myths: The hermeneutics of Destiny. Frankfurt am main. Peter lang.

[27] Mbiti, J. (1980). African Religion and Philosophy. Ibadan: Heinemann

[28] McDough, S. (1990). The Greening of the Church. New York: Orbis.

[29] Menkiti, I. A. (1979). 'Person and Community in African Traditional Thought". In Wright, R.A.(ed) African Philosophy. New York: University Press of America.

[30] Merchant, C. (2004). Reinventing Eden: The Fate of Nature in Western Culture. New York: Routledge.

[31] Mwombeki, F.R. "Ecology in the New Testament"

[32] Okadigbo,C. (1975). The Philosophical Foundations of the African Personality. Unpublished Notes. Enugu: Bigard memorial Seminary.

[33] Patrick, H.(2006).Unsustainable. A primer for Global Environmental and Social Justice (New York: Zed Books)

[34] Posey, D. A. and Graham, D. (1996). Beyond Intellectual Property: Towards Traditional Resource Rights for Indigenous Peoples and Local Communities. Ottawa: International Development Research Centre

[35] Posey, D. A. and William I. B.(1989).(eds). Resource Management in Amazonia: Indigenous and Folk Strategies. New York: New York Botanical Gardens.

[36] Samson, G.(2000).The Environmental Crisis. A Challenge to African Christianity (Nairobi: Action Publications)

[37] Setiloane, G. (1995). "Towards a Biocentric Theology and Ethic Via Africa". Journal of Black Theology in South Africa 9: (1) May.

[38] Shutte, A. (1993). Philosophy for Africa. South Africa: University of Cape Town. p 46

[39] Simone, A.M. (1999). Thinking About African Urban Management in an Era of Globalization. African Sociological Review: 3(2), 69-98,

[40] Sindima, H. (1990)."Community of Life: Ecological Theology in African Perspective" in B.C. Birch, W. Eakin and J.B. McDaniel, Liberating Life: Contemporary Approaches to Ecological Theology. Maryknoll: Orbis

[41] Sowunmi, A.A. (1995).Giver of Life -"Sustain Your Creation" in Hailman, David, G. (ed) Ecotheology.Voices from South and North. New York:Maryknoll.

[42] Steven A. W.(2006).Global Development and Remote African Villages. Environmental Conservation and Cultural Survival in Cameroon. United States of America: The Edwin Mellen Press.

[43] Story Earth (1993). Nature Voices on the Environment. A Compilation of Inter Press Service (California: Mercury House,)

[44] United Nations(1993). Report of the United Nations Conference on Environment and Development, Rio de Janeiro, 3-14 June 1992, Vol.11: Proceedings of the Conference: New York: United Nations.

[45] Wamback, A. E. Constructing an oikotheology: The Environment, Poverty and the Church in South Africa.

[46] Wambutda, D.(1986). An African Christian Looks at Christian Missions in Africa. Readings in Missionary Anthropology 11. California: William Carey Library

[47] Warren, D. M. (1992).Indigenous Knowledge, Biodiversity, Conservation and Development. International Conference on Conservation of Biodiversity in Africa: Local Initiatives and Institutional Roles. Nairobi: Kenya.

[48] Werner, and Begishe, K. Y. (1980). Ethno science and Applied Anthropology. Indigenous Knowledge Systems and Development. Washington, D.C.: University of America Press.

[49] White, R.(2004). (ed) Controversies in Environmental Sociology. London: Cambridge University Press.

[50] Wolfgram, S. A. (2006).Global Development and Remote African Villages. Environmental Conservation and Cultural Survival in Cameroon.

[51] World commission on Environment and Development(1987). Our Common Future. Oxford: Oxford University Press.

4

Local Authorities' Views and Attitudes on Sustainable Energy Policy

Anastasia Dimitriou[1], Dimitris Pimenidis[2]

[1]Department of Education Sciences in Early Childhood, School of Education Sciences, Laboratory of Research and Communication for the Environment and Environmental Education, Democritus University of Thrace, Alexandroupolis, Greece
[2]Department of Pre-school education and Design, School of Humanities, Msc Environmental Education, University of Aegean, Rhodes, Greece

Email address:
anadim@otenet.gr (A. Dimitriou), poimenidis@aegean.gr (D. Pimenidis)

Abstract: To create effective climate change mitigation energy policies that contribute to greenhouse gases reduction are required. To this direction, local authorities play a crucial role as they are responsible for the planning and implementation of energy policies in their territories. This study aims to investigate the local authorities' knowledge and attitudes on energy production, energy management, energy consumption and renewable energy sources (RES), as well as their energy practices and their intentions. The study was carried out in the island of Rhodes, Greece where 12 elective representatives of local authorities (municipalities and prefecture) were asked to fill in a questionnaire on the above issues. Almost all of the participants recognized energy saving as an important dimension of energy policy. They indicated their intentions for energy saving applications in the future, but the measures they use to apply in their area of responsibility are extremely limited. Despite respondents' positive attitudes towards the use of RES, they seem to hesitate to develop an energy model based mainly on RES, as they believe they cannot cover the energy needs of the island. They also, believe that they play a secondary role in the planning and development of energy policy at a local level, as they estimate that energy policy is mainly a matter of the central government. Findings show that there is a demand for a more specialized and focused energy policy. They further show the need for specialized environmental education programs addressed to local authorities aiming at increasing motivation as well as supporting their efforts for the implementation of sustainable energy policies.

Keywords: Local Authorities, Mayors, Attitudes, Views, Energy Policy, Sustainability

1. Introduction

Energy issue is one of the most important environmental issues as it is closely connected to a large number of serious environmental problems. Production and consumption of energy result air pollutants emissions that reinforce greenhouse effect with consequence climate change. These air pollutants are known as greenhouse gases (carbon dioxide, methane, nitrous oxide and fluorinated gases are the most important). These gases absorb and emit radiation within the thermal infrared range. This process is the fundamental cause of the greenhouse effect and climate change as well [1].

Taking measures against climate change is one of the outstanding aims of European Union, in the frame of which are formed energy strategies for the reducing of greenhouse gas emission. For the achievement of this aim, improvements in the field of energy policy are required because the use of energy contributed up to 80% of the greenhouse air pollutants. Actions such as the more rationally use of energy, energy saving, the implementation of renewable energy sources in the generation of electricity and other energy needs, are recognized as very important [2-4].

In this context, the role of local authorities is recognized as crucial, because they are responsible for the planning and implementation of energy policies; they shape citizens' environmental attitudes as well as public opinion on environmental issues and they also raising citizens active participation for confronting environmental problems. Local authorities are also significant energy consumers through their operations in services, in heating, cooling and lighting of public buildings (schools, hospitals, and offices) and street lighting as well.

European Commission has a crucial role to play in helping local and regional authorities to form sustainable energy policy and supporting their efforts to implement. Its decisive contribution reflects in the development of the 'Covenant of Mayors':

> The Covenant of Mayors is the mainstream European movement involving local and regional authorities, voluntarily committing to increasing energy efficiency and use of renewable energy sources on their territories. By their commitment, Covenant signatories aim to meet and exceed the European Union 20% CO_2 reduction objective by 2020 [5].

In Greece, to date, 89 Municipalities are participated in this movement, among which 11 island municipalities [6], that is very limited number given the total number of Greek municipalities comes up to 325 [7].

Local authorities have a leading role in the function of their communities. According to the national legislation local authorities in Greece, take responsibilities which permit them the planning and implementation policies for the environment, the development and the quality of life.

Given the environmental problems facing modern societies and the necessity to implement measures for their confrontation in the perspective of sustainability, the issue is the policies which implement in local level to immunize environmental quality and sustainable development. Local authorities' role for the achievement of sustainable development is vital while:

> Local authorities construct, operate and maintain economic, social and environmental infrastructure, oversee planning processes, establish local environmental policies and regulations, and assist in implementing national and sub national environmental policies. As the level of governance closest to the people, they play a vital role in educating, mobilizing and responding to the public to promote sustainable development [8].

A crucial question is whether local authorities' representatives, as policy makers, make their decisions in an environmental responsible way that contributes towards environmental preservation and protection. In other words, the issue is whether the policy makers act as environmental responsible citizens; as citizens with awareness, environmental knowledge and interest to understand the environmental problems and with the intention and the capability to solve them [9-10].

One of the most important contemporary environmental issues come under the responsibility of local authority is the energy management. Therefore, mayors' conceptions of energy issue and their attitudes towards energy management are key factors for implementing energy policy ensuring environmental quality and sustainable development.

While research has extensively studied citizens', educators' and pupils' conceptions of energy issues, [11-16] only a small number of relevant studies have focused on local authorities. In particular, Mayors' conceptions of energy issues and attitudes towards energy policies are scarce [17-18]. Other studies, however, have recorded the influence of local authorities in energy governance but do not focuses on mayor's conceptions and attitudes towards energy issues [19].

The above mentioned suggest the importance of investigating the mayor's conceptions and attitudes towards the production, management and consumption of energy, as well as the degree of implementation of RES technologies and technological solutions at a local level.

This research aims to explore the knowledge and attitudes of the elected representatives of local and regional authorities of Rhodes, Greece, towards the production, management and consumption of energy and RES; and the degree of implementation of energy technologies and technological solutions at a local level. In particularly, the research aims to investigate: a) the knowledge of the participants about energy issue, b) determine the extent to which they implement sustainable energy policy, c) their attitudes towards the implementation of measures for the rationale use and energy saving, d) determine the extent to which they understand the use of RES in confronting energy problem and their attitudes towards RES, e) the measures they take or they tent to confront the energy problem. Additional, the level of their environmental awareness and the kind of the resources they use for their information, are investigated. Findings of this research may show several dimensions of energy issues that may exploit to supporting the efforts of local and regional authorities for the implementation of sustainable energy policies, and for the planning and development of educational programs concerning energy issues. The results of this study are presented and discussed in the present paper.

2. Energy Policy and Sustainability

2.1. Energy Policy Towards Sustainability

The energy model of contemporary societies seems not to promote sustainable development and this, because it is connected with the overconsumption of energy, the exhaustion of natural resources, environmental degradation and climate change. In addition, political crisis between countries that produce and provide energy might lead to energy crisis in the countries which import energy. These conditions make it more and more important to implement sustainable energy policies at local and regional level. Due to the environmental problems connected with energy production and consumption, there has been an urge to change the model of energy production, which is currently based on fossil fuels (oil, lignite). The development of a sustainable energy model should definitely involve the use of renewable energy sources (RES), as they mainly contribute to climate change mitigation through the reduction of greenhouse gas emissions. Renewable energy sources continue renew by the cycle of nature and are never exhausted such as sun, wind, rivers, oceans and organic materials. Solar, wind, geothermal, bioenergy and hydropower, are considered to be important RES that contribute to the reduction of greenhouse gas [20 -22].

The greatest challenge for the planners of energy policy is the design, the development and the management of appropriate, economically affordable and reliable energy services in a sustainable way, in order to reinforce and support sustainable development [3, 22]. The implementation of renewable energy technologies, as well as, actions for energy saving, and improvement of energy efficiency contribute to a sustainable energy future [4, 22]. To this direction, the European energy policy is being formed mainly aiming at energy saving, the use of RES and the reduction of greenhouse gases emissions. For the application of this energy policy several national action plans have been designed which Member States are asked to implement [2, 24]. In particular, for the total of Member States, until 2020, main targets are as follows: a) reduce of greenhouse gas by at least 20% compared to 1990 levels, b) increase the share of RES in final energy consumption and c) an a 20% increase in energy efficiency [2, 24].

For Greece, the protection of climate through the promotion of electrical energy from RES constitutes an environmental and energy priority of the highest importance. The main national target, is the reduction of greenhouse gases by 4% by the end of 2020, in relation to 2005 levels, and achieves an 18% RES infiltration into gross final consumption. To this direction, the national energy policy for RES, aims at: a) contribution of energy produce from RES to the gross final energy consumption by a share of 20 % and b) contribution of electrical energy produce by RES to the gross energy consumption to a share of at least 40 % [25]. Energy needs are expected to be met with the production of approximately 13300MW electricity from RES (currently producing 4000MW) with the contribution of the available renewable energy technologies (7500MW from wind fans, 3000MW from hydro-electric plants and 2500MW from solar systems) as well as the improvement of buildings energy efficiency [25].

2.2. Area of Study and Energy Situation

The island of Rhodes is located in the prefecture of the Dodecanese, in the south-east of Athens and it is the fourth largest island of Greece with a population of 115.290 (figure 1). Rhodes Island is one of the most popular touristic destinations, especially in summer period. Over than 2.000.000 people visit Rhodes annually that means that its population increase dramatically. As a result, Rhodes faces an increased energy demands that, in turn, pressure the operation of the electricity productive units pushing them onto borderline levels [26].

A steam-electric power plant situated in Soroni, at the South of the island, about 23 km from the city of Rhodes, is the basic electric production unit in the island. The steam-electric power plant of Soroni is an independent and non-interconnected energy production system, which autonomously produces electrical energy for Rhodes and the neighboring island of Halki (figure 1).

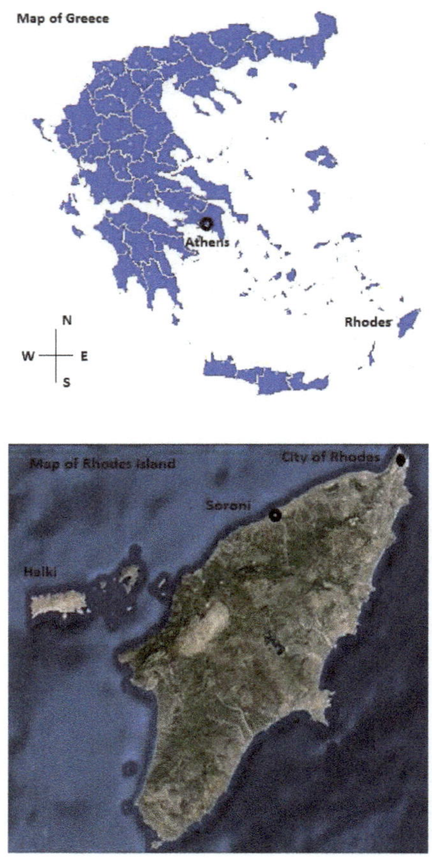

Figure 1. *The area of study*

The electric power plant operates by the use of crude oil and diesel that are transported by tankers. Fuels via undersea and underground pipelines are transported from the tanker to tanks situated in the area of the plant.

During last decades, it is observed a huge growth in the amounts of electrical energy consumption in Rhodes. Particularly, in 1995 the total electrical energy consumption was about 393.205 MWh. In 2010, it almost duplicated, reaching up to 764.593 MWh. During the years 2010-2014 the average annually electrical energy consumption was about 780.000 MWh [26-29].

The installed capacity of Rhodes power plant is currently 206 MW. Its marginally covers the amounts of electrical energy required especially during summer, because of the increasing touristic activity. Official data state as maximum power demand for 2012 being 202 MW and for 2013 as high as 210 MW [27-29].

In the recent years, since 2008, RES systems have been installed to support the local electrical energy production. The RES systems included wind generators and photovoltaic systems with installed capacity 49,15 MW and 16,92 MW respectively. These RES systems are owned by individuals-investors, except wind generators of 4.95 MW owned by the Public Power Corporation, of Greece [29]. The contribution of the electrical energy produce by RES seems to increase during last years. Especially, in 2010 up to 7.43 % of the total electrical energy production was coming from RES technologies (7.18 % and 0.25 % of wind and solar energy respectively) [26-28]. By the end of 2013, the contribution of RES into the local electric power system was increased reaching up to 15.5 % of the total energy production while by the end of the year before was 9.8 % [27, 29]. During the year 2014, the inflow of energy from RES into the local system reached up to 17.6 % of the total energy production [30].

As official data recorded, for the production of approximately 298.070 MWh from the steam-electric power station of Rhodes, 65.665 metric tons of crude oil and 10.723 Klit of diesel were used. Data, also estimated that carbon dioxide emissions reached up to 221.965,942 metric tons [30].

From the above it is clear that the electrical energy consumption in Rhodes Island increase over the years, mainly due to touristic activity. The existing electrical power plant seems no to respond to the required electrical energy demands. Thus, the issue is the planning and implementation of energy policy that ensure energy efficiency in a sustainable way.

It is astonishing that, despite the national action plan for the development of sustainable energy policy, at this time, it is scheduled the construction of one additional steam-electric plant station in Rhodes to support the existing one. The construction of a new conventional power plant in Soroni or in another area of the island triggers serious controversy between local community authorities and the residents of Rhodes, who demand an energy model towards environmental sustainability [31].

3. Methodology

3.1. Participants

The participants were twelve (12), elected representatives of local and regional authorities of Rhodes, Greece, for the period 2010-2014. Ten (10) of them were mayors of equal in number municipalities, and two (2) were representatives of Regional authority. Until 2014, there were 10 municipalities in Rhodes. Today, (2015) after the new administrative division that occurs in Greece, there is one union municipally in the island of Rhodes that consists of 10 municipal units, similar to the municipalities of the period 2010-2014. At this time (2015) the majority of the participants are also elected representatives of the new municipal units.

All participants are married and almost all of them are parents (11/12). Researchers communicated with the participants at a personal level for investigating their wiliness to participate in the survey. Participants were told the aim of the survey and that the survey results would be published. Participation in the study was voluntary and anonymous, and participants were also told that if they did not wish to participate they could simply not answer the questionnaire. Participants were asked to answer the questions during a face-to face interviews.

3.2. Instrument

For the purpose of the study, a questionnaire was designed by the authors that consisted of open and closed questions. The questionnaire aims to elicit information about mayors' views and attitudes regarding energy policy in a sustainable way. The questionnaire apart from the initial questions concerning participant's demographic characteristics consisted of four parts. In the first part the respondents were asked to estimate the level of their environmental awareness, their sources of environmental information, the degree of their knowledge and education on environmental issues and problems. The second part of the questionnaire consisted of open and closed questions (multiple-choice and Linker type) concerning the knowledge and beliefs of the respondents regarding the ways of production and use of electric energy, the forms of RES and their contribution to the energy production locally and nationally; the level of the application of RES technologies in energy production. In the third part of the questionnaire the respondents were asked to answer questions regarding RES and the extent to which RES technologies use in their municipality and the way they use them; as well as their attitudes towards RES. The forth part of the questionnaire focuses on the conceptions of the participants of the designing of energy policy including RES, the existing way of using and implementing measures towards energy saving in their municipalities, issues concerning personnel education on energy management and the possible existence of factors affecting their attempt to establish sustainable energy policy.

For developing and testing the adequacy of the questionnaire, a pilot study was conducted for overcoming any misunderstandings and assessing whether the research protocol is realistic and workable.

4. Results

4.1. Participant's Demographic Characteristics, Environmental Awareness

All participants are male that reflects the situation of the limited number of female mayors in Greece; and in terms of age are divided equally into two age group, 30-45 and 46-60 years old. Their educational level is very high, while the majority had graduated studies. During their studies, almost half of the respondents received education on environmental issues in contrast to the majority who state that they have not received relevant education (Table 1).

Table 1. *The participants' demographic characteristics*

Characteristics	Frequency (f)
Gender	12
Male	12
Female	0
Age	
30-45	6
46-60	6
Level of Education	
Higher	10
Secondary	2
Education on environmental issues	
Yes	5
No	7

The majority of the respondents (10/12) regard themselves as well-informed about the environment and related issues while two of them (2/12) say that they feel moderate informed; as they say, they receive environmental information very often and often respectively. Respondents were given a list of several information sources and were asked to identify their main sources of information about environmental issues (Table 2). Television, press, internet and local government environmental bodies are the most important information source for almost all the participants. Few of the mayors cite professional associations or research-training institutes at University level as one of their main information sources.

Table 2. *Participants' environmental information sources*

Sources	Frequency (f)
Television	11
Press	11
Internet	11
Local government environmental bodies	11
Professional associations	8
Research-training institutes	2

When the participants were asked to consider how important protecting environment is to them personally, all of them cite that is very important.

Overall, the majority of the respondents consider as most pressure contemporary environmental problems facing in Greece those are connected with the energy issue. Specifically, the energy production, the energy adequacy and the exhaustion of natural resources seems to be the most pressure environmental problems for the majority of the respondents (Table 3).

Table 3. *Participants' views about most pressure environmental problems in Greece*

Environmental problems	Frequency (f)
Energy production	10
Energy adequacy	10
Exhaustion of natural resources	10
Ecosystems degradation	10
Global warming	9
Acid rain	9
Water pollution	9
Use of toxic substances	2
Depletion of ozone layer	2
Urbanization	2

4.2. Knowledge Concerning Energy Issue and RES

For the purposes of the recording of the participants' knowledge on energy issues certain questions were formed related to the sources of energy production both at a national and local level, the forms of RES and the technologies applied for their utilization, the sufficiency of the existing power plant and its potential extension, the building of an additional one and the use of RES for the energy needs of the island and the most important domains of energy consumption. From the answers analysis it is clear that the majority of the respondents are highly informed on the ways of energy production both at a national and a local level. They refer to the burning of lignite and the use of oil as exclusive sources of energy production nationally and locally respectively. It is worth pointing out that three of the participants refer to lignite burning as a means of energy production locally, which is not applied. The RES are referred to by all the participants as sources partially applied nationally and locally.

Specifically, regarding RES forms used for energy production respondents focus on solar and wind power. Almost everybody (10 out of 12) is aware of the simultaneous existence of energy production systems, which at a local level involve a combination of solar and wind power. As to the ownership of RES on the island almost all locally elected representatives (11 out of 12) know that it is primarily private, while only one refers to mixed ownership, that is a partnership between private investors and public bodies, a situation that exists in Rhodes, as already mentioned, a partnership between Public Power Corporation and individuals.

In respect with the domains of large energy consumption, nearly all of the respondents refer to the hotel industry (11 out of 12). Instead, local authorities are considered a large energy consumer only by half of the participants (6 out of 12).

From the analysis of the answers given by participants concerning the energy production system on the island it seems that all of them agree with the fact that the existing system is insufficient. Nearly everyone (11 out of 12) seems to believe that apart from the existing power plant there is a need for a second one in the same area (Southern Rhodes) and a simultaneous broad application of RES. The closing down of the existing power plant and the building of a new more powerful one with the simultaneous energy production with the use of RES is an idea that most of the participants agree upon (10 out of twelve). Only two of the respondents seem to agree with the closing down of the existing system and the exclusive production of energy through RES. The rest are completely negative.

Regarding the degree to which the participants appreciate the information they receive on energy saving issues, the rational use of energy and the RES applications there are variations. In particular, the majority of the participants (9 out of 12) estimate that they are adequately informed on energy saving issues, fewer believe the same about issues related to RES applications (7 out of twelve) and even fewer (6 out of 12) about issues concerning the rational use of energy. As far as the above issues are concerned, basic sources of information for the majority of the participants are the web (11 out of 12) and the press (9 out of 12), while for half of them (6 out of 12) are relevant workshops or seminars organized by universities and institutes.

4.3. Views and Attitudes on RES

From the answers given it seems that most of the respondents (9 out of 12) do not believe that RES can exclusively meet the energy needs of the island. Only a small minority (3 out of 12) absolutely believe in such a possibility.

Investigating the attitude towards energy production exclusively with the use of RES, in case this is practically feasible, it seems that almost all of the participants are positive.

However, it is evident that most of them (7 out of 12) would opt for RES for the production of energy, on the condition that the cost would not exceed the current one of conventional energy production (use of oil). What is worth mentioning is that one out of three (4 out of 12) is willing to choose exclusively the use of RES for energy production affording even a double cost, while only one would afford a triple cost. The choice of RES as an exclusive means of energy production would be desirable by everyone, as they believe that RES are environmentally friendly and their use reduces greenhouse gases emissions, nevertheless they think they are ineffective.

The participants were asked to refer to the forms of RES they would prefer to be broadly applied at a local level (Table 4). The answers analysis showed that solar energy and wind power installations on land are highly preferable by the participants. The majority of the participants are negative towards RES installations in the sea (wind power systems in the sea) or the exploitation of its energy (wave/ tidal, small hydroelectric systems). Also, almost all of the participants do not choose at all natural gas or biomass burning to cover energy needs at a broader local level.

Table 4. *Preferable RES implementation on a local level*

Forms of RES	Frequency (f)
Which form of energy would you prefer to be used at a local level?	
Solar energy	12
Wind plants on land	11
Wind power systems in the sea	3
Small hydroelectric systems	3
Wave/tidal systems	3
Natural gas	2
Biomass	2

In order to investigate the participants' views on RES technology applications in public services, facilities and public buildings in their area of responsibility, they were asked to rank the available RES technologies depending on what they believe about the degree of applicability (options: 1: can be absolutely applied, 2: can be considerably applied, 3: cannot possibly applied) (Table 5).

Table 5. *Participants' views on RES technology applications in their territories*

RES technologies	Frequency (f)		
	1	2	3
Bioclimatic architecture	8	3	1
Small range solar systems	7	5	-
Wide range solar systems	6	6	-
Photovoltaic systems in buildings	6	6	-
Photovoltaic systems in several areas of the city	5	5	2
Wind power systems	9	3	-
Biomass	2	2	4
Use of urban waste	3	5	4
Geothermal systems	2	4	4
Natural gas	2	8	2

From the data of Table 5, it is clear that the available RES technologies, such as solar and wind power systems as well as bioclimatic architecture are highly preferable among the participants, who consider them absolutely applicable in practice. Relative to bioclimatic architecture they point out that it could be perfectly applied only in new constructions as for the older ones there are certain urban restructuring restrictions due to the historical value of the buildings. More than half of the respondents believe that small scale solar energy systems are absolutely applicable, while half of them believe the same for large scale solar energy systems, although there is the issue of the location of the installations locally. Typically, large scale systems are those used in factory units, above 10KW and small scale are household systems up to 10KW. What is more, half of them choose to install photovoltaic systems in buildings, while fewer refer to their broader application in several areas of the urban structure. Respondents' views on RES technologies that have to do with utilization of urban waste, biomass, geothermic and natural gas for energy purposes seem to be divided, as some of them think that they are highly applicable, while others believe that they cannot be applied at all.

4.4. Views on Policy Planning and Its Application

In order to study the views of the locally elected representatives on the planning and application of energy policy, certain questions were posed regarding the people in charge of policy planning, the obstacles and difficulties they think are involved in the application of energy measures aiming at energy saving, rational use and RES use and actions to be taken to reduce or overcome such obstacles.

Answers concerning energy policy planners, those in charge of applying techniques and practices related to energy issues showed that all of the participants believe that the State is mainly responsible (government, parliament, ministries). Local government seems to be highly responsible according to the majority of the respondents (8 out of 12), however opinions are divided because quite a few (4 out of 12) do not agree. Participants recognize partial responsibility to bodies such as the Technical Chamber and the Chamber of Commerce (9 out of 12). Less responsibility is attributed to factory and hotel units by the majority of the participants (8 out of 12), while some do not attribute any responsibility to them at all (4 out 12).

Almost all of the respondents state that they come across obstacles and difficulties when applying energy measures for energy saving, rational use of energy and RES use. Answers to a question posed on the types of obstacles and the degree to which they affect the application of measures showed that almost all of the participants (11 out of 12) state that institutional and procedural issues as well as insufficient financing and financial incentives make it impossible to implement measures for a sustainable energy practice. Respondents seem to be divided as to the high cost of application as an obstacle to the implementation of energy efficient practices. Almost half of them (7 out of 12) consider that high cost impedes sustainable energy practices application in contrast to the rest (5out of 12), who believe that the cost does affect application but to a smaller degree. For the majority of the participants (9 out of 12) technical difficulties, ignorance of those in charge, of the public or employees in the companies involved, as well as indifference towards the application of energy measures are quite a serious obstacle to their implementation. Half of the people asked (6 out of 12) believe that another important obstacle is the lack of specialized professionals, while the rest (6 out of 12) believe that this factor does not affect the implementation of the measures. Nearly one out of three (4 out of 12) states that the fact of non-acceptance of such measures by citizens or company employees is not a serious problem, in contrast to the majority (8 out of 12), who think that this is an important factor.

In order to investigate ways to overcome difficulties, which participants themselves refer to as obstacles to the implementation of sustainable energy practices, they were asked to rank a series of actions to be taken according to their importance. From the answers analysis it seems that almost all of the Mayors believe that obstacles could be removed with some absolutely necessary actions concerning institutional issues, such as the simplification of the institutional context, the application of relevant rules and regulations of the State, the adoption of regulations, the financing by the State and the integration of issues of energy saving, rational use of energy and RES use in schools. Their viewpoints on the importance of the information the public should receive from the Mass Media, the State or the local government are divided, as more than half of them (8 out of 12) claim that these actions are absolutely necessary, while the rest (4 out of 12) believe them to be of less importance.

All of the respondents totally agree with the need to plan and apply educational programs both for the citizens and certain social groups, such as politicians, businessmen and employees aiming at the sustainable development on the island of Rhodes and the Dodecanese. For the planning and application of the above programs all of the participants refer to the responsibility of the regional energy agency of the Dodecanese and the Municipalities themselves. Almost everyone believes that the Technical Chamber and the University of the Aegean (11 out of 12) and the Public Power Corporation department in Rhodes (10 out of 12) are responsible. Fewer (9 out of 12) attribute responsibility to the various professionals associations on the island and half of them (6 out of 12) to the institutional primary and secondary education.

4.5. Existent Energy Policy and Energy Applications for Energy Saving and RES Implementation in Local Government Services

For the investigation of the aspects of energy policy which is practically applied, the representatives of the local government were asked if there is a person in charge of energy management, if energy application, energy management and action plan studies are being carried out and by whom, if there has ever been set up special support and financial motives policy for the production, transportation and use of electric energy and if there is an attempt to design and apply energy educational programs for the employees in their area of concern.

From the answers analysis it is evident that only in two out of the twelve bodies of the local government of Rhodes there is a designated energy administrator, the first one being in charge of the energy facilities and the second being in charge of the planning of energy management in the services under his jurisdiction. Half of the local government bodies that took part in the present study (6 out of 12) state that there are energy management studies carried out and relevant action plans prepared in the services of their concern. Three of them use the Municipality's technical services and two of them the regional energy agency of the Dodecanese, while one is cooperating with individuals.

Half of the local government bodies (6 out of 12) claim that they are designing educational programs for their staff on issues regarding energy saving, rational energy use and RES application, while only two are already implementing such programs.

For the recording of the existing energy practices respondents were asked to choose among several applications that aim at energy saving and rational energy use, as well as RES use and suggest additional ones. The results are presented in Table 6.

Table 6. Energy-saving and RES implementation techniques in buildings and services

Statements	Frequency (f)
Which of the follow you have applied in the buildings and services in your territory	
Turning off the lights when not needed	11
Turn off electric appliances when not needed)	9
Appropriate adjustments in cooling and heating systems according to the season	8
Control the amount of sunlight that is admitted into buildings (use of shelters, curtains)	8
Interventions in the shell of the buildings (use of insulation materials)	2
Insulation interventions in the heating systems	2
Using energy efficient light bulbs	5
Use of ceiling fans instead of air conditioners	3
Use of solar water heating systems	10
RES implementation (photovoltaic system)	1
RES implementation (geothermal heating system)	1
Replace energy consuming appliances	3

From the answers analysis it is clear that almost everyone is taking action against overconsumption in their services such as turning off the lights when not needed, and closing the windows when the air conditioners are on. Most of them (9 out of 12) turn off electric appliances when not needed. Two out of three representatives (8 out of 12) apply practices that have to do with the cooling and heating of the buildings making the appropriate adjustments according to the season and practices to control the amount of sunlight that is admitted into buildings, such as shelters, curtains and ventilation systems. Interventions in the shell of the buildings that reduce energy waste, such as the use of insulation materials and windows are put into practice by only two of the participants. The same number of representatives, (2 out of 12), make insulation interventions in the heating system. Fewer than half of the participants (5 out of 12) have been using energy efficient light bulbs and even fewer (3 out of 12) have replaced energy consuming appliances. The use of ceiling fans instead of air conditioners is limited and only by the minority of the representatives (3 out of 12), while most of them have installed solar water heaters to cover hot water needs. Applications relevant to RES are extremely limited with only one of the municipalities having installed a photovoltaic system in its central building for the production of energy and applied a geothermal heating system in one of its other buildings.

5. Discussion

Attitudes of the local authorities are recognized as a crucial factor for energy saving, RES application for energy production locally, as well as for the promotion of the European Union aims regarding the energy issue on the prospect of energy sustainability [3, 17-19]. On the one hand, they are themselves significant energy consumers, in their territories, in the case of public lighting, building and services lighting and transportations. On the other hand, they are responsible for the planning and application of energy policy locally and in addition, they inform and educate citizens affecting their behavior. The present study was carried out in this context and the findings are to be discussed below.

The sustainable management of the environment and environmental issues requires the existence of several factors among which the degree of information and education of the people in charge. Regarding their education on environmental issues, fewer than half of the representatives of the municipalities and the prefecture that took part in the study, have been in some ways relevantly educated during their studies, however not particularly on the energy matter. They show a great interest on the environment and related issues because as they mention they are often informed by several sources, mainly the Media (TV and the press) and the web. The Media and the web are the most important sources of information on the environment for all citizens and especially some social groups (teachers, students) both in Greece and internationally [16, 32].

As far as the information on energy issues is concerned it seems that the degree to which they believe they are informed on certain aspects varies. In particular, it seems that the majority of the people asked think that they are adequately informed on energy saving matters, not adequately informed on RES applications and less informed on the rational use of energy. The majority of the respondents derive information on these issues mainly from the web and the press and less from workshops or seminars organized by universities or institutes.

What they consider as the most serious environmental issues are those relevant to energy, such as the production, the sufficiency of electric energy and the exhaustion of natural resources, in contrast to European citizens, who, in their majority, refer to air and water pollution [32].

Even though participants consider the energy issue and especially energy consumption as a modern environmental issue, it seems that they do not recognize the local government as a large energy consumer. On the contrary, they think that hotel industry is energy consuming. There is no doubt that hotels consume large amounts of energy [33-34]. However, the local government is also energy consuming, as its services and its buildings use large amounts of energy for lighting, heating and cooling. There is also energy demand for public lighting and transportations. It should be pointed out that at a European level the main target is to reduce CO_2 emissions (related to the use of energy) by 20% in local communities use and the improvement of energy efficiency by 20% or more by 2020 [1-3].

The participants are highly aware of energy production in the area of study, Rhodes, Greece, where energy is produced with the use of oil, although some of them believe that lignite is used, a practice applied nationally but not locally [21-22]. Almost everybody knows about the limited use of RES in the production of energy on the island which is a combination of solar and wind power systems.

All of the participants state that the existent energy production system is inefficient, as they claim that it cannot cover the needs on the island, especially during highly touristic seasons. They seem to agree that a second power plant is necessary, along with a broader application of RES. The above findings show that despite they believe that power plants cause environmental problems it is a priority for them to cover energy needs on the island even if that entails the building of a new power unit. Naturally, they refer to the simultaneous broad use of RES for energy production but they have reservations about their efficiency, as they think that they cannot meet the energy needs of the island. They believe that RES could only act as subsidiary means of energy production and that is their main thesis. This particular attitude of the locally elected representatives is contrasted to the European energy policy, which acknowledges the efficiency of RES and aims at raising the energy share of RES by 20% in the gross final consumption of energy by 2020 [2-4].

The cost seems to influence to a great degree their attitude towards RES applications to cover energy needs. Most of them set the condition that the cost should not exceed the conventional energy production cost that currently exists and few are willing to afford a greater cost. The economic factor seems to act negatively in RES application for local government representatives in other countries as well, like for example in the Czech Republic and America [18]. On the other hand, citizens in countries such as Germany [35] or in other parts of Greece [14] insist on the clean energy from RES regardless of the cost of its production.

The choice of RES as an exclusive means of energy production would be desirable by all of the participants, as they claim that they are environmentally friendly and they reduce greenhouse gases emissions, nevertheless, as already mentioned, they do not believe they are sufficient for the energy needs of the island. In order to deal with energy needs in Rhodes all participants primarily opt for solar energy as it is in abundance. In Rhodes sunshine reaches up to 115 hours monthly and the annual solar radiation is 2000 KWh/m^2

[36-37]. In order to take advantage of the solar energy they agree upon the installation of small and large scale energy systems, photovoltaic systems in buildings and the town for its lighting.

The option that follows is the creation of wind farms on land, while the idea of wind systems in the sea is turned down because it could devalue the area, which is of major touristic importance. In general, as relevant studies have recorded, citizens' views on wind farms appear divided. Both in Greece and internationally it is observed that a number of citizens support the installation of wind farms, while a number worth mentioning are negative towards it because of several factors, such as socio- economic, or the devaluation of landscape aesthetics [11-12, 14-15, 35, 38].

Respondents' opinions on RES technologies concerning the utilization of urban waste, biomass, geothermic energy and natural gas for energy purposes also seem to be divided. This happens because some of them believe those technologies are quite applicable and others partially or not applicable at all.

Concerning energy policy that elected representatives apply or are willing to apply locally, results are quite interesting. It should be pointed out that the majority of the participants claim that the State (government, parliament, ministries) is mainly responsible for designing the energy policy and applying techniques and practices relative to energy matters. Their views on the responsibility of the local government are divided, as most of them believe that it is crucial, while one out of three does not attribute any responsibility at all.

As far as the implementation of measures on energy saving, rational energy use and RES application is concerned, they state that they come across obstacles and difficulties. As most important they name institutional and procedural issues, insufficient financing and insufficient financial incentives. Technical difficulties, lack of specialized professionals, ignorance, the indifference of those in charge, company employees and citizens are thought to be a great obstacle to the implementation of energy practices. In order to overcome these problems, all of the representatives claim that some actions regarding institutional issues are absolutely necessary, for example the simplification of the institutional context, the application of relevant rules and regulations of the state, the adoption of regulations, State funding and the implementation of issues on energy saving, rational energy use and RES in schools for the education of future citizens. The provision of financial incentives and the reinforcement of the local government by the central state, the governments, are also considered significant by the locally elected representatives in other countries as well [19].

All participants absolutely agree on the need for planning and implementing educational programs both for the citizens and for special social groups, such as politicians, businessmen and company employees aiming at the sustainable energy management in accordance with the objectives of sustainable development on the island. To this direction, they believe that the role of the University of the Aegean, as well as the energy management bodies (Public Power Corporation, Hellas, chambers) is crucial. Despite the great importance they attach to information and education provided on energy issues the realization of such programs for the education of the staff under their administration is not a common practice but a future plan.

In relation to energy saving and the rational use of energy respondents seem to be taking energy austerity measures regarding the reduction of energy consumption by equipment used in their services mainly through actions and practices by the staff on a daily basis (turning off the lights, closing the windows when air conditioners are on, turning off appliances when not needed). What is more, there are certain cooling and heating adjustments made in the buildings according to each season's standards.

Despite their strong support to bioclimatic architecture as a means to deal with the energy issue, there are not any important actions taken, only some practices relative to the control of the amount of sunlight that is admitted into buildings, with shelters, curtains and ventilation systems all of them to a limited extent.

RES applications are almost inexistent, with only one of the municipalities having installed a photovoltaic system in its central building for energy production and used a geothermal heating system in another building it owns. Finally, only a few of the elected representatives have installed solar water heaters to meet hot water needs.

The island of Rhodes is one of the most popular touristic destinations, with more than 2,000,000 visitors annually. Therefore energy demands are high. However, it is necessary to reinforce local authorities for the development of sustainable energy policies. As it becomes evident from the present study the cooperation of institutions and specialized scientists on energy issue with the local government is crucial in order to design and apply an energy policy in a sustainable way. The Mayors' predominant view on the inability to support local energy needs with an energy model based primarily (or even exclusively) on RES should be thoroughly examined and suitable educational tools should be used to reduce resistance to RES implementation in order to support the local system. Education on issues related to RES could influence their overall attitude towards RES at the level of practical possibilities and applications far from theoretical environmentally friendly speculations, as what it is all about is the broad application of RES aiming at the reduction of the serious effects of conventional energy production process.

Education for the environmental issues, including, energy seems to be the most effective tool to promote sustainability. As in [39] quoted:

> In order to achieve sustainability, an enormous co-ordination and integration of efforts is required in a number of crucial sectors and rapid and radical change of behaviors and lifestyles, including changing consumption and production patterns. For this, appropriate education and public awareness should be recognized as one of the pillars of sustainability together with legislation, economy and technology.

6. Conclusions

The present study investigates the knowledge and views of the locally elected representatives of Rhodes on production, management and consumption of energy at a local level. Results show that locally elected representatives of Rhodes have the following tendencies:

- They are highly informed on energy production both on the island and in Greece and its effects on the environment.
- They believe that they play a secondary role in the planning and development of energy policy at a local level, as they estimate that energy policy is mainly a matter of the central government.
- They recognize the importance of RES for the production of energy; however they are skeptical about RES efficiency, as they believe they cannot cover the energy needs of the island.
- No RES technologies are being applied in buildings, services or other local facilities in their territories.
- Institutional and procedural issues, as well as insufficient financing incentives constitute serious obstacles to the implementation of energy saving practices and RES applications in their territories.

Acknowledgment

The authors are thankful to the Mayors of local authorities of Rhodes, Island, Greece, of the period 2010-2014 for their participation in this survey.

References

[1] European Environmental Agency (2006). Greenhouse gas emission trends and projections in Europe. Report No 9. Copenhagen: EEA

[2] European Commission (2013a). GREEN PAPER. A 2030 framework for climate and energy policies. Brussels: European Commission,

[3] European Commission (2013b). Empowering local and regional authorities to deliver the EU climate and energy objectives. Brussels: European Commission.

[4] European Commission (2010a). LIFE and local authorities *Helping regions and municipalities tackle environmental challenges.* Belgium: European Union.

[5] Covenant of Mayors. Available at http://www.eumayors.eu/IMG/pdf/covenantofmayors_text_en.pdf

[6] Covenant of Mayors. Available at http://www.simfonodimarxon.eu/ (in Greek).

[7] Skolarikos, M. (2013). (eds). Municipalities in numbers. Athens: The Central Union of Municipalities and Communities (in Greek).

[8] UNESCO (1992). United Nations Conference on Environmental and Development: Agenda 21. Switzerland: Unesco, Chapter 28.

[9] Hungerford, H., Volk T. (1990). Changing learner behavior through environmental education. *Journal of Environmental Education*, 21 (3): 8–21.

[10] Kollmuss, A., Agyeman, J. (2002). Mind the Gap: why do people act environmentally and what are the barriers to pro-environmental behavior? *Environmental Education Research*, 8 (3): 239-260.

[11] Ek, K. (2005) Public and private attitudes towards "green" electricity: the case of Swedish wind power. *Energy Policy* 33(13):1677–1689.

[12] Kaldellis, J.K. (2005). Social attitude towards wind energy applications in Greece. *Energy Policy,* 33(5): 595–602.

[13] Kaldellis, J.K., Kapsali, M., Kaldelli, E. & Katsanou, E. (2013). Comparing Recent Views of Public Attitude on Wind Energy, Photovoltaic and Small Hydro Applications. *Renewable Energy*, 2013 (52): 197–208

[14] Katsaprakakis, D. A. (2012). A review of the environmental and human impacts from wind parks. A case study for the Prefecture of Lasithi, Crete. *Renewable and Sustainable Energy Reviews*, 16(5): 2850–2863.

[15] Kontogianni, A., Tourkolias, C., Skourtos, M., & Damigos, D. (2014). Planning globally, protesting locally: Patterns in community perceptions towards the installation of wind farms. *Renewable Energy*, 66: 170–177.

[16] Liarakou, G., Gavrilakis, C., & Flouri E. (2009). Secondary School Teachers' Knowledge and Attitudes Towards Renewable Energy Sources. *Journal of Science Education and Technology*, 2009 (18): 120–129

[17] Busch, H,, McCormick, K.(2014). Local power: exploring the motivations of mayors and key success factors for local municipalities to go 100% renewable energy. *Sustainability and Society*, 4(5): 1-15.

[18] Frantál, B. (2014). Have Local Government and Public Expectations of Wind Energy Project Benefits Been Met? Implications for Repowering Schemes, *Journal of Environmental Policy & Planning*, DOI: 10.1080/1523908X.2014.936583.

[19] Fudge, S., Peters, M., Wade, J. (2012). Locating the agency and influence of local authorities in UK energy governance. Centre for Environmental Strategy, University of Surrey: United Kingdom

[20] Menanteau, F., Finon, D., Lamy, M. L. (2003). Prices versus quantities: Choosing policies for promoting the development of renewable energy. *Energy Policy*, 31:799–812.

[21] IEA (International Energy Agency), (2007). Key world energy statistics, 1–82.

[22] Roosa, A. S., (2008), *Sustainable Development Handbook*, Fairmont Press, Inc.

[23] Rogelj, J., McCollum, D. L, Riah,i K. (2013). The UN's 'Sustainable Energy for All' initiative is Compatible with a Warming Limit of 2°C. Nature Climate Change, 3: 545-551, doi:10.1038/nclimate1806.

[24] EC, European Commission (2010b). EUROPE 2020. *A strategy for smart, sustainable and inclusive growth.* Belgium: European Union.

[25] YPEKA, Ministry of Environment, Energy and Climate Change (2014). National action plan on renewable energy sources. Athens: YPEKA (in Greek).

[26] PPCH, Public Power Corporation, Hellas, (2012). Re newable energy sources in non interconnected islands. *Fact Sheet for the year 2012.* (In Greek).

[27] PPCH, Public Power Corporation, Hellas (2013). Renewable energy sources in non-interconnected islands. *Fact Sheet for the year 2013.* (In Greek).

[28] HEDNO, Hellenic Electricity Distribution Network Operator (2012). *Fact Sheet: electricity power production in non-interconnected islands for the year 2012.* (In Greek).

[29] HEDNO, Hellenic Electricity Distribution Network Operator (2013). *Fact Sheet: electricity power production in non-interconnected Islands for the year 2013.* (In Greek).

[30] HEDNO, Hellenic Electricity Distribution Network Operator (2014). *Fact Sheet: electricity power production in non-interconnected Islands for the year 2014.* (In Greek.)

[31] Pimenidis, D., Dimitriou, A. & Fokiali, P. (2013). Local society and energy: Citizens conceptions about the operation of the electrical power plant in Soroni, Rhodes, Greece. In Papavasileiou, V. (Eds). Sustainable local societies: a reality or a Vision (353-374). Rhodes: University of Aegean. Electronic edition. ISBN 978-618-81027-0-5. (In Greek).

[32] EC, European Commission European Commission, (2014). *Special Eurobarometer 2014. Attitudes of European citizens towards the environment.* EC: European Union

[33] Priyadarsini, R., Xuchao, W., Eang, LS, (2009). A study on energy performance of hotel buildings in Singapore. *Energy and Buildings*, 41: 1319–1324

[34] Xuchao, W., Priyadarsini, R., Eang, LS, (2010). Benchmarking energy use and greenhouse gas emissions in Singapore's hotel industry. *Energy Policy*, 38: 4520–4527

[35] Musall, F. D., & Kuik, O. (2011). Local acceptance of renewable energy – a case study from southeast Germany. Energy Policy, 39(6): 3252–3260.

[36] HNMS, Hellenic National Meteorological Service (2009). Climate report (In Greek).

[37] Huld T., Müller R., Gambardella A., (2012). A new solar radiation database for estimating PV performance in Europe and Africa. *Solar Energy*, 86: 1803-1815.

[38] SEI, Sustainable Energy Ireland (2005). *Attitudes towards the development of wind farms in Ireland.* Renewable Energy, Information Office: Ireland.

[39] UNESCO- EPD, (1997). *Declaration of Thessaloniki. UNESCO-EPD97/CONF.401/CLD.2,* (article 6). Paris: UNESCO.

5

Optimisation of Diesel Polluted Soil Using Response Surface Methodology

Olawale O.[2, *], Oyawale F. A.[1], Adepoju T. F.[2], Aikulolu S.[2], Akinmoladun A. I.[3]

[1]Mechanical Engineering Department, Covenant University, Otta, Nigeria
[2]Chemical Engineering Department, Landmark University, Omu Aran, Kwara State, Nigeria
[3]Department of Physics and Technology, Bronx Community College of the City University of New York, New York, USA

Email address:
Lamstock2@yahoo.com (Olawale O.), Olawale.olamide@lmu.edu.ng (Olawale O.)

Abstract: This research is based on the investigation of the bioremediation of diesel polluted soil using biostimulation strategy. The study was carried out on a diesel contaminated soil and concentrations: Tween 80 (5-15ml), Poultry droppings (20-60g), Hydrogen Peroxide (0.5-1.5ml) were added and incubated for 28 days of remediation period. Response Surface Methodology with Box-Behnken Design (BBD) was used with three factors and three levels of Tween 80, Poultry droppings and Hydrogen Peroxide as independent variables while diesel oil (Total Petroleum Hydrocarbon) removal was the dependent variable (response). The result showed disparities in the diesel oil biodegradation pattern with respect to Tween 80, Poultry droppings and Hydrogen Peroxide. The statistical analysis, via ANOVA showed coefficient of determination R^2 to be 99.89% and $P < 0.05$. The predicted optimum parameter of Tween 80: 10.10ml, Poultry droppings: 41.46g and Hydrogen Peroxide: 1.10ml were gotten while 56.565% was gotten as the optimal diesel oil removal. At this optimum condition, the observed diesel oil removal was found to be 56.568%. It can be concluded that bioremediation of diesel-contaminated soil resulted in petroleum hydrocarbon degradation.

Keywords: Bioremediation, Diesel, Tween 80, Soil

1. Introduction

Diesel is a fuel obtained from petroleum distillation and it is used in diesel engines. It is a lightweight mixture of liquid hydrocarbons and has a relatively low ignition temperature (540°C) and is ignited by the heat of compression. It is produced from the fractional distillation of crude oil between 200°C (392°F) and 350°C (662°F) at atmospheric pressure. When the soil becomes contaminated, the ecosystem is altered and agricultural activities are affected seriously due to this pollution. [5]; stated that used or improperly disposed diesel fuel contains potentially toxic substances, such as: benzene (carcinogens), lead, arsenic, zinc and cadmium, which can seep into the ground and contaminate ground water. Furthermore, it is common to observe diesel spills from the cleaning of oil tankers and also via the disposal of used oil from diesel cars by the road side mechanics. Physicochemical technologies have been previously used for the remediation of polluted soils [8]. According to [3]; bioremediation technology through the mechanism of biodegradation has been recognized to be a valuable alternative for the detoxification and disposal of toxic substances. This is because it is environment-friendly, financially viable, and technologically simple and it also conserves soil texture and characteristics.

Furthermore, bioremediation is an ecologically acceptable technology that is used for the removal of a pollutant from the biosphere through the use of micro-organisms [13]. It relies on biological processes to minimize unwanted environment impact of the pollutants. Nevertheless, this Research is focused on biostimulation method. According to [11]; biostimulation is the addition of nutrients to a polluted site in order to encourage the growth of naturally occurring chemical degrading microorganisms. This research is based on optimising diesel polluted soil.

2. Materials and Method

2.1. Method

Diesel-contaminated soil was excavated from the Generator House at Landmark University Cafeteria, Omu-Aran, Kwara State, Nigeria between 0-30cm from the soil surface. The soil was sieved with 5mm sieve. Organic wastes was used in this study instead of NPK fertilizer, to facilitate aeration through small pores to increase the water-holding capacity of the soil; thus enhancing bioremediation. The organic wastes used in this study was collected from Landmark University Research Poultry Farm. Physicochemical properties of organic wastes and soil employed were determined using standard methods.

2.2. Physicochemical Property Determination for Soil

Physicochemical properties determined for soil included pH, water holding capacity, textural class as well as Nitrogen, Potassium, Carbon and Phosphorus content.

2.3. Soil pH Determination

10 g of sieved (≤ 5 mm) and air-dried soil was weighed poured into a 50ml beaker and 25 ml of distilled water added. The content was stirred manually for 30 minutes with a glass rod and left to stand for 1 hour. The electrode of the pH meter (Kent EIL 7055) was then inserted and the pH determined.

2.4. Determination of Water Holding Capacity (WHC)

Six (6) inches of soil was placed in percolation tube and compacted by gentle bouncing. Water was added until the water level reached 2 inches above soil level. The tube was covered and left to stand for 2 days. After this period, the top half-inch of soil was discarded and wet soil was weighed into a pre-weighed evaporating dish. The sample was placed in oven at 110°C for 24 h. It was then removed and weighed to obtain bone dry weight of the sample. Percentage water holding capacity was calculated as [12]:

$$\%waterholdingwater = \frac{wetweight - bonedryweight}{bonedryweight} \quad (1)$$

2.5. C, N, P and K Content

Soil textural class, as well as soil carbon, nitrogen, phosphorus and phosphorus contents were analyzed at the Civil Laboratory of Landmark University, Omu-Aran, Kwara state. Soil textural class was determined using the hydrometer method [7]. The organic carbon, phosphorus and nitrogen contents were determined using an element detector.

2.6. Bioremediation Setup

200g of diesel polluted soil was placed in plastic bottles, labeled 1 to 17. Organic wastes (poultry droppings), H_2O_2 and Tween 80 were added to each of the diesel polluted soil in appropriate conditions. The soil was then mixed daily to provide sufficient aeration and moistened by the addition of water every other day to adjust the water holding capacity at 60% throughput throughout the experimental period. The plastic bottles were then incubated at room temperature. The controls contained only diesel polluted soil.

2.7. Experimental Procedure

Polluted soil samples (200g) were placed in plastic bottles (microcosm). The diesel-contaminated soil in each plastic bottle was amended with different amounts of poultry dung (20 - 60g), Tween 80 (5 - 15 mg/l) and hydrogen peroxide (0.5 - 1.5 g/l), respectively. Soil used as control was not amended with any biostimulating agents. In total, 17 microcosms were settled and incubated for 28 days. All microcosms were mixed manually once per week to enhance oxygenation, and kept moist during the 28 day experimental period. Samples were withdrawn after the incubation period for percentage diesel removal analysis.

2.8. Experimental Design and Data Analysis

Table 1. *Experimental range and the levels of the variables.*

Factors	High level (+1)	Medium level (0)	Low level (-1)
Tween 80 (A) ml	5	10	15
Poultry Droppings (B)	20	40	60
Hydrogen Peroxide (C) ml	0.5	1.0	1.5

Table 2. *Full-factorial Box-Behnken design for the three independent variables.*

Run	Factor 1 A: Tween 80 (ml)		Factor 2 B: Poultry Droppings (g)		Factor C C: Hydrogen Peroxide (ml)	
	Coded	Uncoded	Coded	Uncoded	Coded	Uncoded
1	0	10.00	+1	60.00	-1	0.50
2	0	10.00	0	40.00	0	1.00
3	+1	15.00	0	40.00	-1	0.50
4	+1	15.00	0	40.00	+1	1.50
5	0	10.00	-1	20.00	+1	1.50
6	0	10.00	0	40.00	0	1.00
7	+1	15.00	-1	20.00	0	1.00
8	0	10.00	0	40.00	0	1.00
9	-1	5.00	-1	20.00	0	1.00
10	-1	5.00	0	40.00	-1	0.50
11	0	10.00	+1	60.00	+1	1.50
12	-1	5.00	0	40.00	+1	1.50
13	+1	15.00	+1	60.00	0	1.00
14	-1	5.00	+1	60.00	0	1.00
15	0	10.00	0	40.00	0	1.00
16	0	10.00	0	40.00	0	1.00
17	0	10.00	-1	20.00	-1	0.50
18 control		-		-		-

The Box-Behnken factorial experimental design employed had three independent variables: Tween 80 (surfactant), Poultry droppings (nutrient) and hydrogen peroxide. Each of the independent variables was studied at three levels (1, 0, +1), with 17 experimental runs and one control. The levels were selected based on preliminary study results and literature. The variables optimized were Tween 80 (5 – 15 mg/l), Poultry droppings (20 – 60 g) and hydrogen peroxide (0.5 – 1.5 g/l) at three levels, respectively were shown in Table 1. Efficiency of diesel removal was assessed after 28 days. Table 2 shows the coded and actual values of factors and levels used in the experimental design. Diesel contaminated soil without biostimulation was also analyzed as a control. The statistical software Design Expert 6.0.8, (Stat-Ease Inc., Minneapolis, USA) was used to evaluate the analysis of variance ($P < 0.05$) to determine the significance of each term in the fitted equations and to estimate the goodness of fit in each case.

2.9. Total Petroleum Hydrocarbon Analysis

The TPH analysis was carried to determine the percentage diesel removed using the gravimetric method. 10g of each treatment was collected and dried at room temperature for 72 hours. 5g of the soil was placed in 200 milliliters beaker and 150ml of toluene was added. The mixture was stirred continuously for 30 min, left to stand in a fume cupboard for 2 hours and then filtered using Whatman No 42 filter paper. The residue, (soil), was allowed to dry in an oven at 50°C. The TPH was calculated according to [1]; [2]:

$$\% \text{ Diesel Removal } \left(\frac{g}{kg}\right) = \frac{(\text{initial weight of soil} - \text{weight of soil after soil extraction}) \text{in g}}{\text{initial weight of soil in kg}} \quad (2)$$

3. Results and Discussions

3.1. Soil Parameters

Table 3 showed the values for the soil parameters that were tested.

Table 3. Soil Parameters for the Diesel-polluted soil.

Parameters	Values
Soil Ph	7.29
Water holding capacity	16 (±2) %
Nitrogen content	0.34 mg/l
Potassium content	6.0 mg/l
Calcium content	26 mg/l
Phosphorus content	0.15mg/l

3.2. Bioremediation and Natural Attenuation

The experiment was designed using Response Surface Methodology, and after performing 17 experimental runs of the Box-Behnken design (BBD) and one control (see Table 2), the results of the statistical experiments were analyzed with reverence to the coded design matrix. The regression equation showed that Diesel degradation rate was an experimental function of test variables in coded units. The result from Table 4 showed that on day 28 (4th week), diesel content had decreased in all the soil microcosms.

In control, natural Bioattenuation removed 21.15% of petroleum hydrocarbons from the soil. It was observed that the respective reduction in petroleum hydrocarbon (TPH) of soil microcosms with amendments was much higher when compared to control in the same period in Table 4. This observation showed that the addition of biostimulants increased the rate of TPH degradation in the soil. This is in agreement with the report of [3] that an increase SEO biodegradation with the addition of biostimulants such as NPK, Tween 80 and Pig Manure as supplements. However, [9]; stated that one of the major factors limiting degradation of hydrocarbons is their low availability to the microbial cells [9]. In addition, hydrocarbon-oxidizing potential has also been shown to increase with hydrocarbon exposure. Thus, in Table 4, run numbers 12 and 13 (at lower concentration of Poultry droppings and H_2O_2), and run numbers 8 and 9 (at higher concentration of Poultry droppings and H_2O_2) had same remediation conditions but with different surfactant (Tween 80) concentration, results showed that addition of surfactant can enhance diesel degradation. Similar observations was made by [6] concerning the impact of biosurfactants on soil bioremediation.

The effect of different concentrations of hydrogen peroxide supplementation were investigated at the same condition of Poultry droppings and Tween 80 (run numbers 4 and 8, and run numbers 5 and 14) and the findings demonstrated that addition of hydrogen peroxide (H_2O_2) can enhance the bioremediation process of soil contaminated with diesel. This is in agreement with [4]. On the other hand; run numbers 11 and 12 and run numbers 5 and 7 through similar condition but with different amount of Poultry droppings were tested and the results showed that extra amount of Poultry droppings can improve diesel removal from contaminated soil.

Table 4. *Experimental design and Results for bioremediation of Diesel polluted soil.*

Run	Percentage of Diesel removal	
	Observed value	**Predicted value**
1	56.260	56.446
2	41.670	41.554
3	56.780	56.446
4	42.470	42.685
5	47.760	47.709
6	56.240	56.446
7	46.580	46.249
8	44.930	44.763
9	45.700	45.485
10	48.210	48.261
11	41.570	41.406
12	39.190	39.306
13	40.660	40.824
14	49.300	49.631
15	43.460	43.627
16	56.320	56.446
17	56.630	56.446
18 (control)	21.15	-

3.3. Second Order Polynomial Regression Model and Statistical Analysis

The experimental data were fitted to a second order polynomial regression model containing 3 linear, 3 quadratic and 3 interaction terms [10] using the same experimental design software to derive the Regression equation for diesel removal from polluted soil as stated in eqn. 3:

$$Y = \beta 0 + \beta 1 A + \beta 2 B + \beta 3 C + \beta 11 A2 + \beta 2^2 B2 + \beta 33 C2 + \beta 12 AB + \beta 13 AC + \beta 23 BC \quad (3)$$

where β0 is the value of the fixed response at the center point of the design; β1, β2, β3 are linear coefficients; β12, β13, β23 are quadratic coefficients; are the interaction effect coefficients regression terms, respectively; A, B and C are the levels of independent variables. The significance of each coefficient in the equation was determined by F-test and P-values. F-test showed that all the factors and interactions considered in the experimental design are statistically significant i.e. $P < 0.05$, at 95% confidence level. The regression equation obtained after analysis of variance gave the level of diesel removal as a function of the different biostimulation variables: Tween 80, Poultry droppings, and Hydrogen peroxide.

The response (Y) generated is:

$$Y = 56.45 + 0.42A + 0.71B + 0.98C - 9.75A2 - 5.93B^2 - 2.56C^2 - 0.34AB - 0.055AC - 0.023BC \quad (4)$$

Where A is Tween 80 concentration, B is Poultry Droppings concentration; C is Hydrogen peroxide concentrations. To test the fit of the model, the regression equation and determination coefficient (R^2) were evaluated. The model F-value of 717.44 infers the model is significant. The low probability value (<0.0001) indicates that the model is significant. Coefficient determination ($R^2 = 0.9989$) being a measure of goodness of fit to the model indicated a high degree of correlation between the observed value and predicted values. The determination coefficient ($R^2 = 0.9989$), suggested that more than 99.89% of the variance is attributable to the variables and indicated a high significance of the model. Thus, 0.11% of the total variance cannot be explained by the model. The fitted model was considered adequate if the F-test is significant ($P < 0.05$) [4]. The Analysis Of Variance (ANOVA) as shown in Table 5 demonstrated that the model was highly significant, as was apparent from the very low probability ($P < 0.0001$) of the F - test and insignificant result from the Lack of Fit model ($P = 0.1948$). The lack of fit test was implemented by comparing the variability of the current model residuals to the variability between observations at replicate settings of the factors. The Lack of Fit F-value of 2.54 implied the Lack of Fit is not significant relative to the pure error.

The Lack of Fit was designed to determine whether the selected model is adequate to describe the observed data, or whether a more complicated model should be used. The Predicted R-Squared value of 0.9881 is in reasonable agreement with the Adjusted R Squared value of 0.9975. Adequate Precision measures the signal to noise ratio. A ratio > 4 is desirable. The ratio of 71.064 obtained in this research indicates an adequate signal. This model can be used to navigate the design space.

The coefficient of variation (CV) as the ratio of the standard error of estimate to the mean value of the observed response is a measure of reproducibility of the model, generally a model can be considered reasonably reproducible if its CV is not greater than 10 per cent. Hence, the low variation Coefficient value (CV = 0.66 per cent) obtained indicates a high precision and reliability of the experiments. The coefficient of the model (parameter estimation) and the corresponding P-values are presented in Table 6.

The significant regression coefficients was considered, ignoring those with an insignificant effect on the response at a significance level of 95%. The P-values of the regression coefficients suggested that among the test variables, linear, quadratic and interaction effects of Tween 80, Poultry Droppings and hydrogen peroxide are highly significant. The insignificant effects (factors and interactions) with P-values higher than 0.05, were ignored. In this study, A, B, C, A^2, B^2, C^2, AB, AC and BC are significant model terms. Thus, statistical analysis of all the experimental data showed that NPK fertilizer, Tween 80 and hydrogen peroxide concentration had a substantial effect on diesel removal in this study.

***Table 5.** Analysis of variance (ANOVA) for the quadratic response surface model fitting to the biodegradation data of diesel.*

Source	Sum of Squares	DF	Mean Square	F- value	P–value
Model	638.54	9	70.95	717.44	< 0.0001
Residual	0.69	7	0.099	-	-
Lack of Fit	0.45	3	0.15	2.54	0.1948
Pure Error	0.24	4	0.060	-	-
Correlation Total	639.23	16	-	-	-

Standard deviation = 0.31 C. V=0.66 R^2=0.9989.

Adjusted R^2 = 0.9975, Predicted R^2= 0.9881, Adequate Precision = 71.064.

***Table 6.** Coefficient of the model for diesel biodegradation.*

Variables	Coefficient Estimate	Standard Error	F-value	P-value	Remarks
β0	56.45	0.14	717.44	< 0.0001	Significant
β1	0.42	0.11	14.02	0.0072	Significant
β2	0.71	0.11	40.49	0.0004	Significant
β3	0.98	0.11	78.29	< 0.0001	Significant
β11	-9.75	0.15	4045.81	< 0.0001	Significant
β22	-5.93	0.15	1494.94	< 0.0001	Significant
β33	-2.56	0.15	278.60	< 0.0001	Significant
β12	-0.34	0.16	4.74	0.0658	Significant
β13	-0.055	0.16	0.12	0.7368	Significant
β23	-0.023	0.16	0.020	0.8902	Significant

Figure 1 showed the studentized residuals and normal per cent probability plot. Residual showed the difference between the observed value of a response measurement and the value that is fitted under the theorized model. Small residual values indicated that model prediction is accurate.

The predicted versus actual plot of diesel oil biodegradation were shown in Figure 2. Actual values were determined for a particular run and the predicted values were calculated from the approximating function used for the model. The Cooks distance and studentized residuals illustrated the normal distribution and constant variance of the residuals, the goodness of fit, linearity of the fitted model, and the independence. Figure 2c showed Cook's distance plot; according to this plot there were no points that were potentially powerful due to their location in the factor.

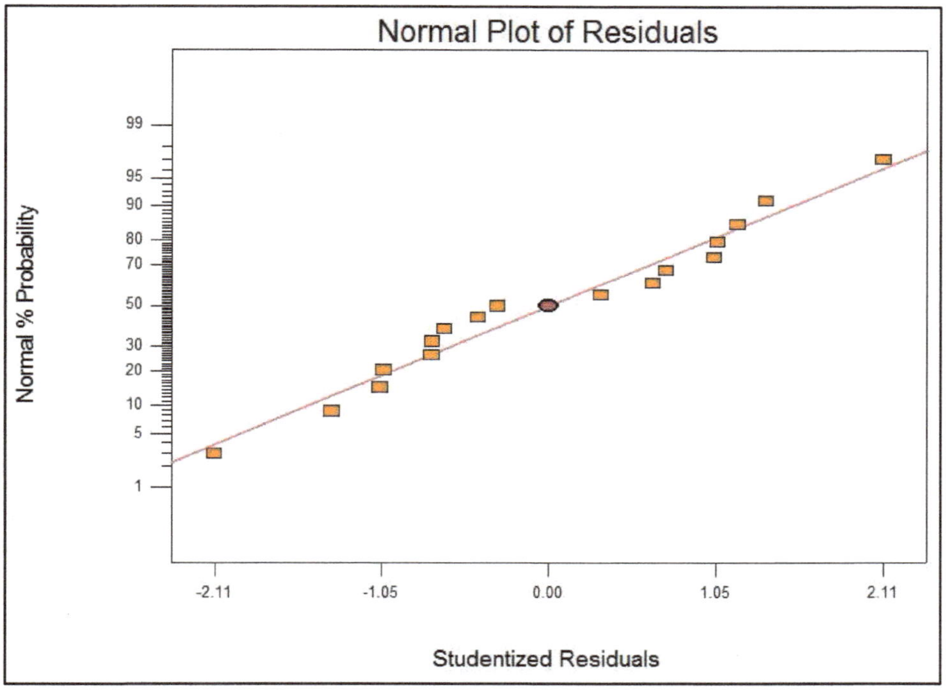

Figure 1. *Normal plot of residuals plot of soil diesel bioremediation.*

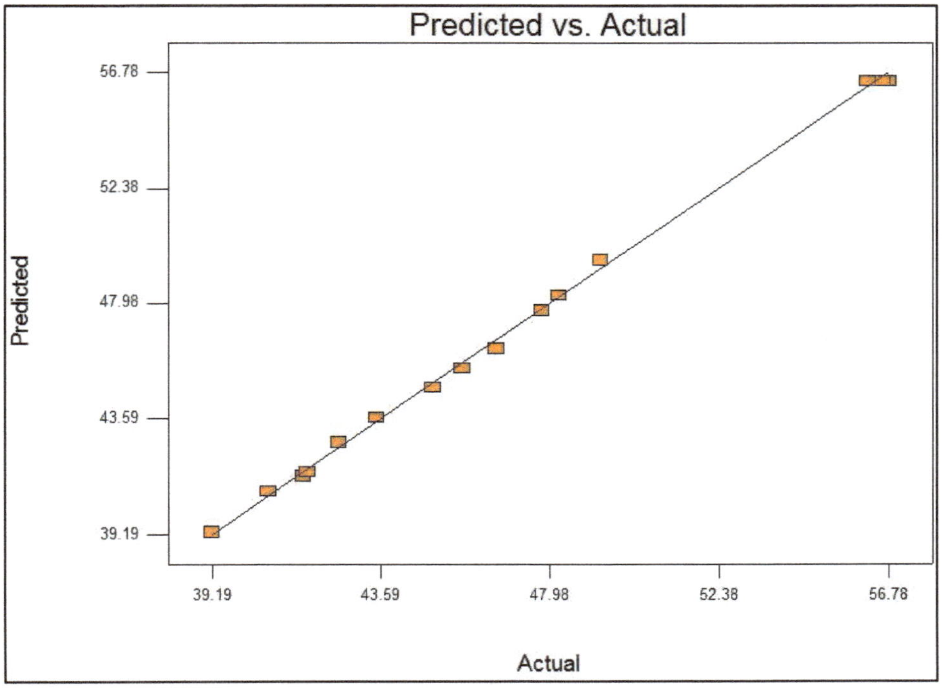

Figure 2. *Predicted versus actual plot of soil diesel bioremediation.*

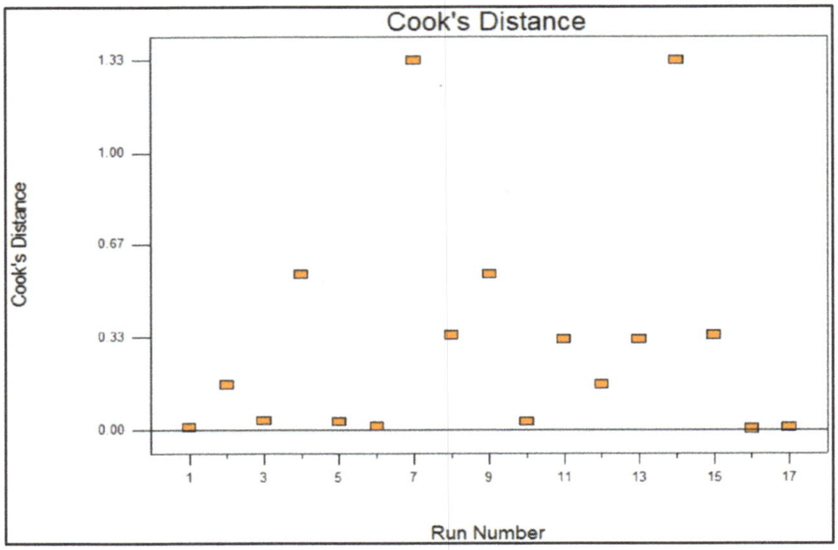

Figure 3. Cook's distance plot of soil diesel bioremediation.

3.4. Interaction Among Factors That Influence Diesel Oil Removal

The graphical representation of the response shown in Figures 4– 6 showed the effect of Tween 80 (A), Poultry droppings (B) and hydrogen peroxide (C) on removal of diesel. The effect of the interaction of poultry droppings and Tween 80 on diesel bioremediation is illustrated in Figure 4. It was observed in this study that; higher rate of diesel removal was attained with higher surfactant (Tween 80) concentration and relatively high amount of Poultry droppings. The maximum degradation yield of diesel (56.5652%) was obtained with 10.10ml of Tween 80 surfactant and 41.47g of Poultry droppings at a fixed hydrogen peroxide concentration of 1.10 ml. This was because of better bioavailability of substrate for the inherent microorganisms.

Figure 5 shows the 3D response surface plot of the interaction effect between Tween 80 and hydrogen peroxide (H_2O_2) concentrations. This plot demonstrated that both Tween 80 and hydrogen peroxide have the best interaction performance at optimum concentrations. A higher percent diesel oil removal was obtained at a higher hydrogen peroxide concentration with relatively high amount of tween 80. This three dimensional plot explained that both tween 80 and hydrogen peroxide have individual impact on diesel oil removal as the individual coefficient of both Tween 80 and hydrogen peroxide is positive and their interaction effect is positive. Figure 4.6 showed the response surface 3D plot of the effect of interaction between Poultry droppings and hydrogen peroxide concentrations. Higher rate of diesel oil removal was observed with increase in hydrogen peroxide and poultry droppings concentration due to positive interaction effect. Due to dominating interaction effects of hydrogen peroxide, higher levels of this variable gave higher yields of diesel oil removal.

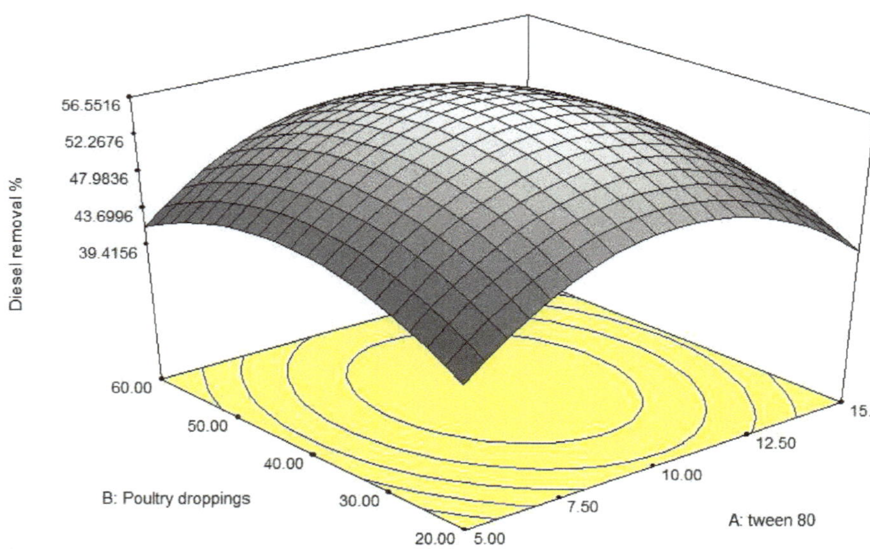

Figure 4. Response surface 3D plots indicating interaction effects of factors Tween 80 and Poultry droppings.

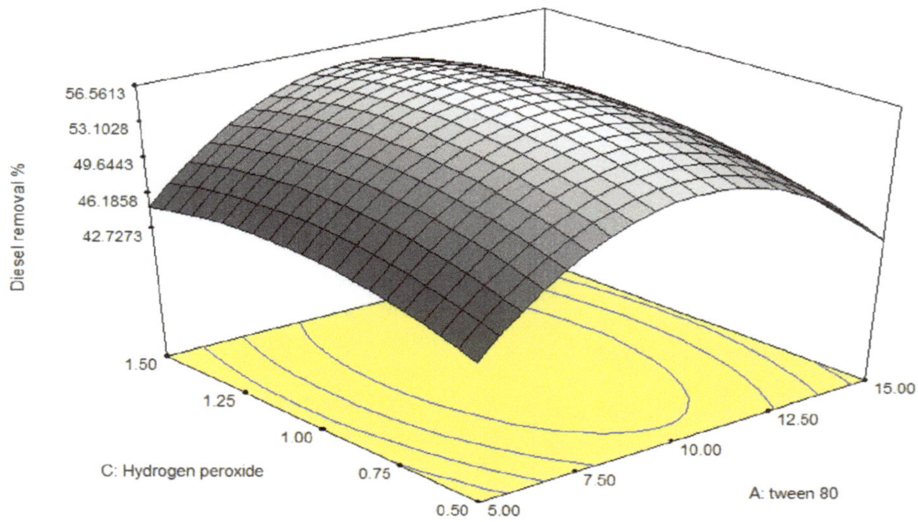

Figure 5. *Response surface 3D plots indicating interaction effects of factors Tween 80 and hydrogen peroxide.*

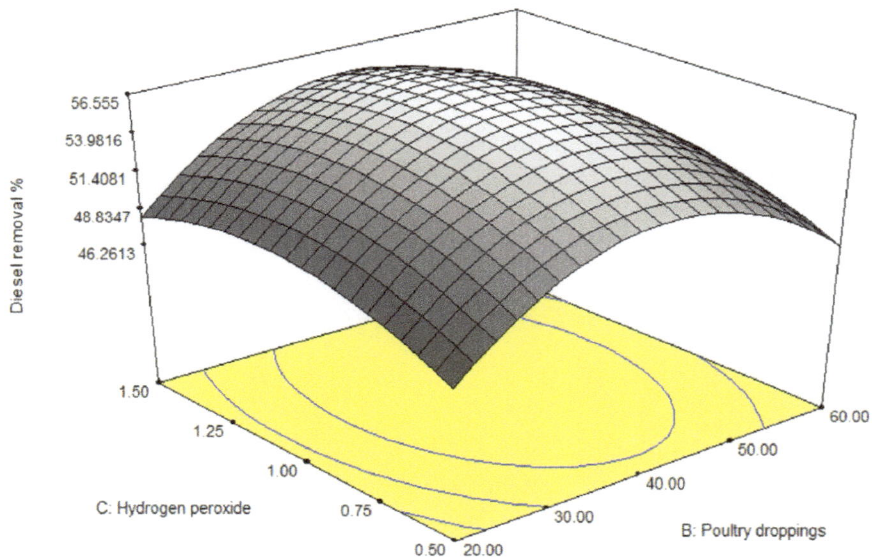

Figure 6. *Response surface 3D plots indicating interaction effects of Poultry droppings and hydrogen Peroxide.*

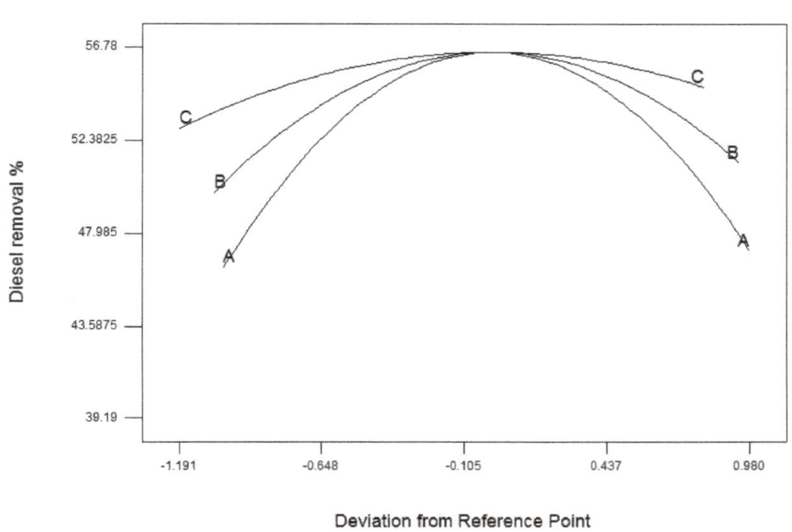

Figure 7. *Factor plot representing the individual variable effect on diesel bioremediation (A= Tween 80, B = Poultry droppings and C = Hydrogen peroxide).*

3.5. Factor Plot

The factor effect function plot was used to assess the effect of each factor graphically. From the trace plot as shown in Figure 7, it showed that each of the three variables used in the present study has its distinct effect on diesel removal by the inherent microbial populations in the soil. Gradual increase in Tween 80, Poultry droppings and hydrogen peroxide concentrations from low level (coded value –1) to a higher level (coded value +1) resulted in both increase and decrease of diesel oil degradation. Moreover, it was also observed from Figure 7 that over the range of -1 to +1 of Tween 80, the diesel degradation change in a wide range, which was also the case for Poultry droppings. However, for hydrogen peroxide the diesel oil removal did not change much over a wide range. This clearly indicates that keeping hydrogen peroxide at the optimum level, a change in Poultry droppings and Tween 80 concentrations will respectively affect the process more severely than done otherwise.

3.6. Optimization and Validation

Numerical optimization technique based on desirability function was used to determine the workable optimum conditions for the diesel oil bioremediation process. In order to provide an ideal case for biodegradation, the goal for Tween 80, Poultry droppings and hydrogen peroxide was set in range based upon the requirements of the diesel bioremediation and diesel oil removal was set on maximize. The predicted optimum (uncoded) values of Tween 80, Poultry droppings and hydrogen peroxide were found to be: 10.10ml, 41.46g and 1.10 ml, respectively, to achieve 56.565% maximum diesel oil removal; while desirability for the predicted optimum values was 0.988 (Figure 8).

Nevertheless, validation experiment was conducted to determine the optimum diesel oil removal when the biostimulation factors were set at the favorable optimum levels established above, through BBD and RSM. Standard deviation and percent error were investigated for validation of experiments. Errors between predicted and actual values were calculated according to the formula below:

$$Error = \frac{Actual - PredictedError}{ActualError} X100 \quad (5)$$

At the optimized condition for diesel removal, 56.586% diesel removal was obtained. The percentage error between the predicted and actual values was found to be 0.037%. The results clearly showed that no substantial difference was observed.

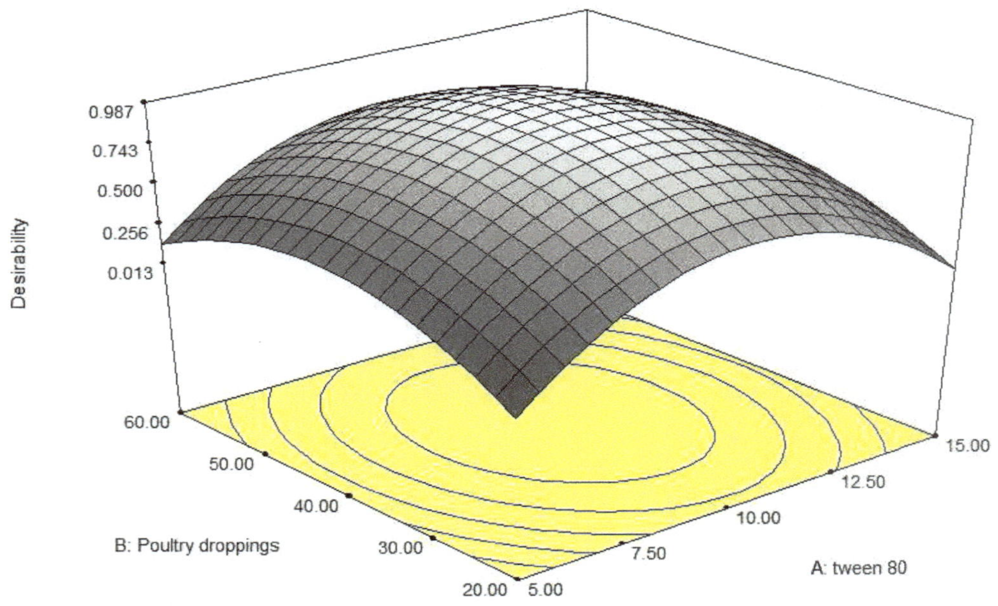

Figure 8. Desirability plot to optimize the bioremediation of diesel-polluted soil process.

4. Conclusion

This study investigated the bioremediation of diesel polluted soil and its optimization using Response Surface Methodology. The diesel contaminated soil was incubated for 28 days bioremediation period and statistical analysis was carried out. The predicted optimum parameters were Tween 80: 10.10ml, Poultry droppings: 41.46g and hydrogen peroxide: 1.10 ml. The optimal diesel oil removal was gotten to be 56.565%. At this optimum condition, the observed diesel oil removal was found to be 56.586%. It can be concluded that bioremediation resulted in petroleum hydrocarbon degradation.

Acknowledgement

The authors thank the Chancellor and also all the staff of Chemical Engineering Department of Landmark University, Omu-Aran, kwara State; Nigeria.

References

[1] Ameh A. O, Mohammed Dabo I. A, Ibrahim S., Ameh J. B, (2013). Earthworm-assisted bioremediation of Petroleum Hydrocarbon contaminated soil from Mechanic Workshop". Afr. J. Environ. Sci. and Technol: 7(6): 531-539.

[2] Ameh A. O, Mohammed Dabo I. A, Ibrahim S., Ameh J. B, T. K Bello, (2012). "The effect of Surface Area and Periodic Tilling on the Bioremediation of Used Engine Oil contaminated soil". Niger. J. Sci: 46: 9-11.nter. J. Ener. & Envrion, Eng: 1-14.

[3] Agarry, S. E. and Ogunleye, O. O (2012) Box-Behnken Design Application to study enhanced bioremediation of soil artificially contaminated with spent engine oil suing biostimulation strategy.

[4] Agarry, S. E, Owabor, C. N, Yusuf, RO, (2012): Enhanced bioremediation of soil artificially contaminated with kerosene: optimization of biostimulation agents through statistical experimental design. J. Pet. Environ. Biotechnol: 3: 120.

[5] Chaineau, C., Setier, J. and Morillon, A. (2002), Is bioremediation a solution for the treatment of oily waste? Society of Petroleum Engineers Inc. SPE78548: 1-10.

[6] Franzetti, A; Gandolfi, I; Bestetti, G; Smyth, T.PJ and Banat, M. (2010) Production and applications of trehalose lipid biosurfantant. Eur. J. Lipids Sci. and Technol: 112: 617-627.

[7] Gee, G. W and Bauder, J. W (1986) Particle Size Analysis by hudrometer. Asimplified method for routine analysis and a sensitivity test of measurement parameters. Soil. Sci. Soc. Am. J.: 43: 1004-1007.

[8] Hamdi H, Benzarti S., Manusadzianas L, Aoyama I and Jedidi N 2007: Bioaugmentation and Biostimulation effects on PAH dissipation and soil ecotoxicity under controlled conditions. Soil Biol Biochem; 39: 1926–35.

[9] Mohajeri, L; AbduAziz, H; Isa, M. H; and Zaheed, M. A (2010) A statistical experimental design application for optimizing biodegration of weathered crude oil in coastal sediaments. Bioresour. Technol: 101: 893-900.

[10] Montgomery DC (2008) Design and Analysis of Experiments. (Seventh Ed.) John Wiley, New York.

[11] Mozik, A. and Seget, Z. F (2010). "Bioaugmentation as a strategy for cleaning up of Soils contaminated with Aromatic Compounds". Microbial. Res. 165: 363-375.

[12] Srivastava, S and Thakur, I. S (2006) Bio sorption potency of *Aspergillusniger* for removal of chromium (VI). Curren. Microbio: 153: 232-237.

[13] Vidali, M. (2001), Bioremediation: An overview. J. Appl. Chem. 73 (7): 1163-1172.

Patterns for Arsenate and Trace Element Relationships in Surface Water and Groundwater of Obuasi Municipality of Ghana

Alexander Kwaku Banson*, **Richard Kwasi Amankwah**, **Samuel Agyarko Ndur**

Dept. of Minerals Engineering, University of Mines and Technology, Tarkwa, Ghana

Email address:
alexkbanson@gmail.com (A. K. Banson), ramankwah@gmail.com (R.K. Amankwah), samndur@gmail.com (S.A. Ndur)

Abstract: Obuasi Municipality is a historic gold mining tropical community that has a massive cocktail of old and active anthropogenic arsenic and trace metal storage structures and activities. Given the potential health impact of elevated arsenic and trace metals in surface and ground water of Obuasi Municipality, it is necessary to assess the distribution and possible link between arsenic and trace metals found in surface water and ground water of the study area. Arsenic is predominant as arsenate in the Obuasi Municipality. In response to this need, the study used statistical and geochemical analytical tools to identify patterns of relationships between arsenate distributions and trace metals in surface water and groundwater of the study area.

Keywords: Arsenic, Arsenate, Trace Metals, Obuasi, Ghana, Surface Water, Groundwater

1. Introduction

Obuasi Municipality is located in the western part of the main Ashanti-Prestea gold belt mineralization and lies in the southern part of Ashanti Region of Ghana. The study area has a long history of artisanal mining predating large scale mining. Artisanal mining was recorded in the area as early as 1471[1, 2, 3, 9]. Currently, active artisanal mining is intense, but conducted by illegal miners, since almost the entire area is covered by the concession of Anglogold Ashanti Limited, Obuasi Mine [1, 9]. Arsenic which is often associated with gold in the study area occurs predominantly as arsenopyrite and as a trace element in the pyrite, pyrrhotite, bournonite, tetrahedrite, sphalerite and aurostibnite [4, 9]. The gold particles are generally fine-grained but some are visible to the naked eye [4]. Gold sulfide ores occur as disseminated assemblage of arsenopyrite, pyrite, pyrrhotite with minor chalcopyrite, stibnite, sphalerite and galena in metasediments and metavolcanics in which carbonate and sericite are the prominent alteration minerals [4,5]. Important gangue minerals in the study area are quartz, carbonate, chlorite, and sericite, along with carbonaceous matter [4,5]. Generally, the finer arsenopyrite grains are found to contain high gold concentrations occurring as sub-microscopic gold particles, and supergene concentration of gold from the sulfides extends from depth of between 30 m to 80 m [2]. Ore deposits of the study area occur in a major shear zone with a proven lateral extent of 8km and depth of 1600m (Amanor and Gyapong, 1988). Figure 1 shows a geological map of the study area.

Previous mining operations in the Obuasi Municipality has left behind open mined out pits which are filled with water serving as habitat for fishes and other aquatic organisms. Liquid discharges at AGA from the past to present which are likely to contribute to release of arsenic and other trace metals into surface and ground water are effluent from the processing plants, seepages and discharges from tailings dams, run-off pits and exploration activities. AGA has operated the Sulphide Treatment Plant (STP), Tailings Treatment Plant (TTP) commissioned in 1988, Oxide Treatment Plant (OTP), Pompora Treatment Plant (PTP) commissioned in 1947, and the Heap Leach Plant (HLP) commissioned in 1990. Liquid effluent from STP which includes precipitated arsenic as ferric arsenate, sulphur as calcium sulphate, metals as insoluble hydroxides and flotation tails CIL residue from the BIOX leach process is pumped to the Sansu Tailings Dam (STD). The TTP tailings network of AGA, include the Pompora Tailings Dam, Kokoteasua East Dam and the Sansu Tailings Dam. Seepages from finger drains and run-offs from the Pompora Dam area flow into coffer dams, whereas decant water from the Kokoteasua East Dam flows into the Pompora valley. Decant from STD reports to a holding pond and is subsequently discharged into the Nyam River. Liquid effluent from the OTP is mainly from CIL residue circuit which is discharged into the STD. Process plant water at the OTP is recycled through the STD. Liquid effluent from the PTP area which include rainfall runoffs, decant water from Kokoteasua East Tailings Dam, Pompora Dam, underground water from North Ramp Mine, excess water from the Pompora sump is discharged into the Kwabrafo stream. The HLP has been decommisioned. There are various pits and waste dumps scattered with most of them concentrated at the central portion of the study area. Most of these pits were backfilled with material that potentially contain gold and arsenic and are likely to release arsenic when rehabilitation delays or the rehabilitated area is further disturbed. Arsenic occurs as a major constituent in more than 200 minerals, including elemental arsenic, arsenides, sulphides, oxides, arsenates and arsenites [17]. Most are ore minerals or their alteration products. However, these minerals are relatively rare in the natural environment [11, 17]. The greatest concentrations of these minerals occur in mineralised areas and are found in close association with the transition metals as well as Cd, Pb, Ag, Au, Sb, P, W and Mo [6, 13, 17]. Arsenic is found primarily in the earth's crust with natural concentrations ranging between 0.1 to 40.0 mg/kg and average concentrations from 1.5 to 5.0 mg/kg and is released naturally through volcanic, forest fires and geothermal, activities [14, 20]. It is also found naturally in rocks, soils, sediments, water and biota around the surface of the earth [12]. Arsenic is often associated with iron and manganese in some sedimentary and igneous rocks [12, 18,19]. Relevant and common arsenic containing minerals are arsenopyrite (FeAsS), realgar (AsS), and orpiment (As_2S_3). In the Obuasi Municipality, arsenic has been found to be associated with Ferrisymplesite ($Fe_3(AsO_4)_2.6H_2O$), Symplesite ($Fe_3(AsO_4)_2.8H_2O$), Liskerdite ($(AlFe)_3AsO_4(OH)_6.5H_2O$ and Arseniosiderite ($Ca_3Fe_4(AsO4)_3(OH)_8$ [4]. Arsenic and trace metals may be released naturally to ground water or surface water through weathering, erosion and dissolution. Release of arsenic into the environment often results in significant environmental consequences [7, 9, 10, 11] as it is released along with other toxic metals. Table 1 shows a list of common minerals that contain arsenic. Arsenic has been observed to be prevalent in the Obuasi Municipality [1, 4, 15, 17]. Currently no pattern has been established as a relationship between arsenate and trace metals in surface or groundwater of the Obuasi Municipality. This study seeks to identify any available pattern for management decisions.

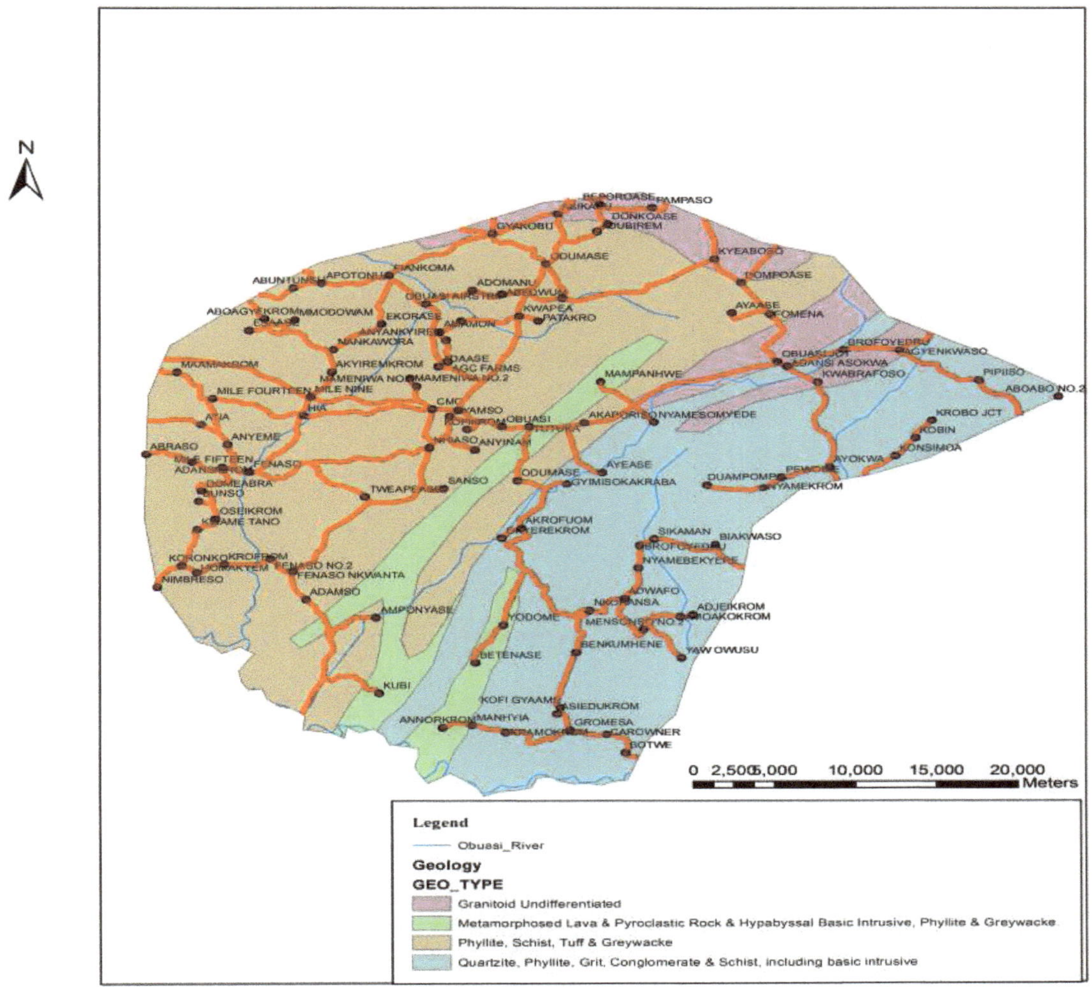

Figure 1. *Geological Map of Obuasi Municipality (Source: Extracted from Geological Map of Ghana).*

Table 1. Common Minerals of Arsenic.

Name	Chemical Formula
Arsenopyrite	$FeAsS$
Smalite	$CoAs_2$
Lollingite,	$FeAs_2$
Cobaltite	$CoAsS$
Orpiment	As_2S_3
Gersdorffite	$NiAsS$
Realgar	As_4S_4
Tennantite	$4Cu_2SAs_2SiS_3$
Chloanthite	$NiAs_2$
Proustite	$3Ag_2SAs_2S_3$
Niciolite	$NiAs$
Enargite	$3Cu_2SAs_2S_5$

[Source: Ferguson, 1990]

2. Materials and Methods

Water Sampling: In total, 230 surface water samples were collected from 25 river sampling points, 160 stream sampling points and 45 ponds. As well, 142 ground water samples were collected from 108 community boreholes, 18 monitoring boreholes, 11 hand-dug wells and 5 mechanised borehole systems in 96 settlements in the study area (Figure 2).

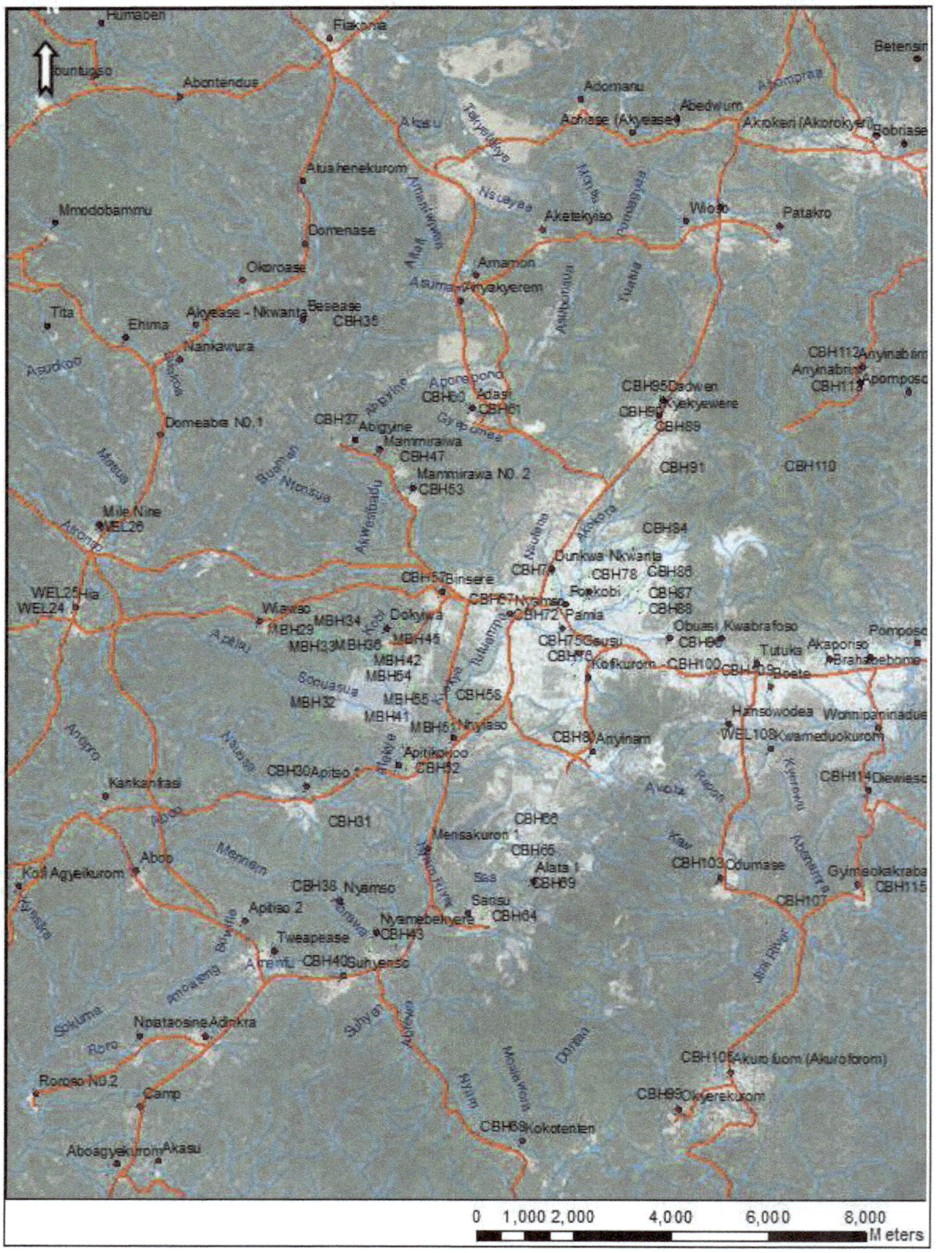

Figure 2. Surface and Groudwater Sampling Points on Surface Water Map Superimposed on Rapid Eye Satellite Image of Obuasi Municipality.

The groundwater and surface water samples were transported in ice chests with ice packs stored at approximately 4°C and analyzed at the Societe General Surveillance (SGS) MASLAB Environmental Laboratory in Ghana. Samples which were analysed for heavy and base metals were preserved with 5ml of ultrapure grade concentrated HCl acid in the laboratory. All samples collected for arsenic analysis were preserved using ultrapure grade HCl acid. Approximately 90 μl of acid was added per 60 ml sample that resulted in a pH slightly below 2.0. A Varian SpectrAA 55 atomic absorption spectrometer with a Varian Vapor Generating Accessory (VGA) 77 was used for analyzing the samples for dissolved arsenic. The instrument was calibrated using freshly prepared arsenic standards (2, 5, 10, 20, 30 μg/L) with a 0.1% HCl acid concentration. Agilent Spectra AA55 Atomic Absorption Spectrophotometer (AAS) was used to analyse metals whiles the Agilent VGA 77 Vapour Generating Accessories along with the Agilent Spectra AA 55 was used for analysis of total As. The Agilent VGA 77 unit was operated with 99.99% Argon in a dust-free environment at the required pressure range of 300 to 400 kPa (43 to 57 psi) and normal flow rate of 100 mL/minute. For quality assessment/quality control standards including NIST Standard Reference materials and a dionised water (as a reference sample blank) were analyzed together with the water samples. Data quality was assessed using charge balance calculations, replicate samples and sample blanks. Charge balance errors are typically <5%. Replicate arsenic analysis samples and standards showed a precision within5%. Blank assays were found to be negligible. The limit of detection for As(III) and total arsenic with the VGA/AAS technique is 0.001 mg/l with an error of ± 70% at the 95% confidence interval. Concentrations between 0.001 and 0.010 mg/l have an error of ± 25% at the 95% confidence interval. Concentrations between 0.010 and 0.050 mg/l have an error of ± 7% at the 95% confidence interval, and concentrations greater than 0.050 mg/l have an error of ± 3% at the 95% confidence interval. Inorganic As(V) was determined by calculating the difference between total inorganic As and As(III). This assumes there is no organic arsenic present in the water.

Table 2. *Arsenic Guiding Standard for Ghana Groundwater and Surface Water.*

ARSENIC GUIDING STANDARDS		
Guiding Standards	Risk Class	Arsenic Conc. (mg/l)
Greater than EPA Akoben Effluent Limit	Very High	>0.100
Greater than ECLimit to EPA Akoben Limit	High	>0.050 – 0.100
Greater than WHO/GWC Limit to EC Limit	Medium	>0.010 – 0.050
Greater than Detection Limit to WHO Limit	Low	>0.001 – 0.010
Less than or Equal to Detection Limit	Very Low	≤0.001
EPA – Ghana Environmental Protection Agency		
EC – European Commission		
WHO – World Health Organisation		
GWC – Ghana Water Company		

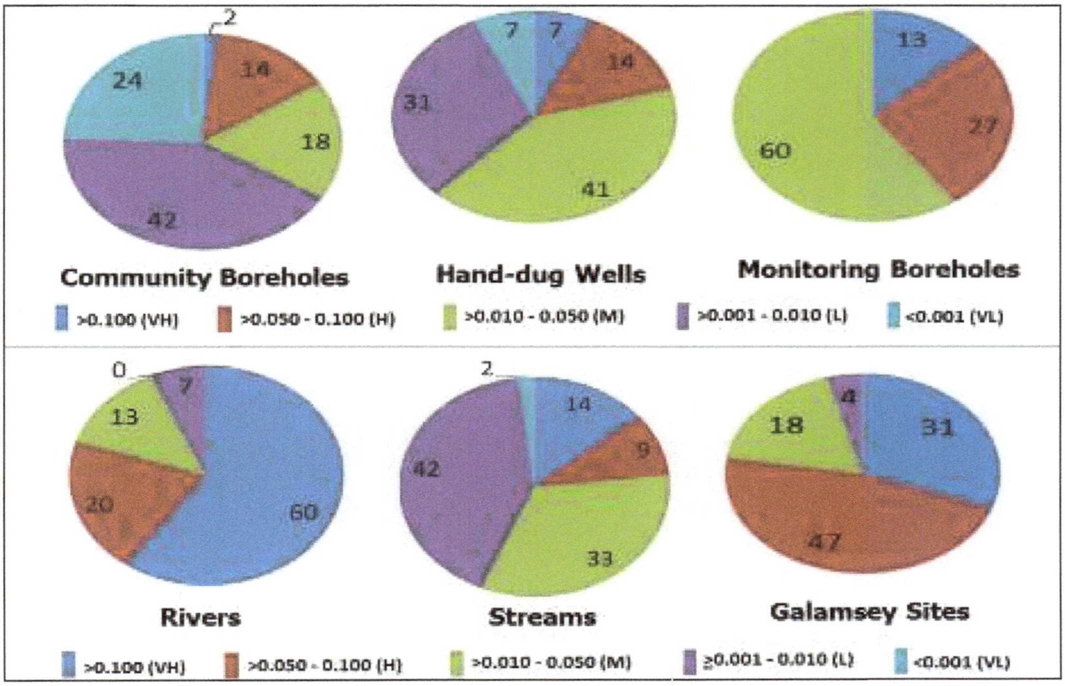

Figure 3. *Percentage Arsenate in Surface Water and Groundwater of Obuasi Municipality (As concentrations in mg/l).*

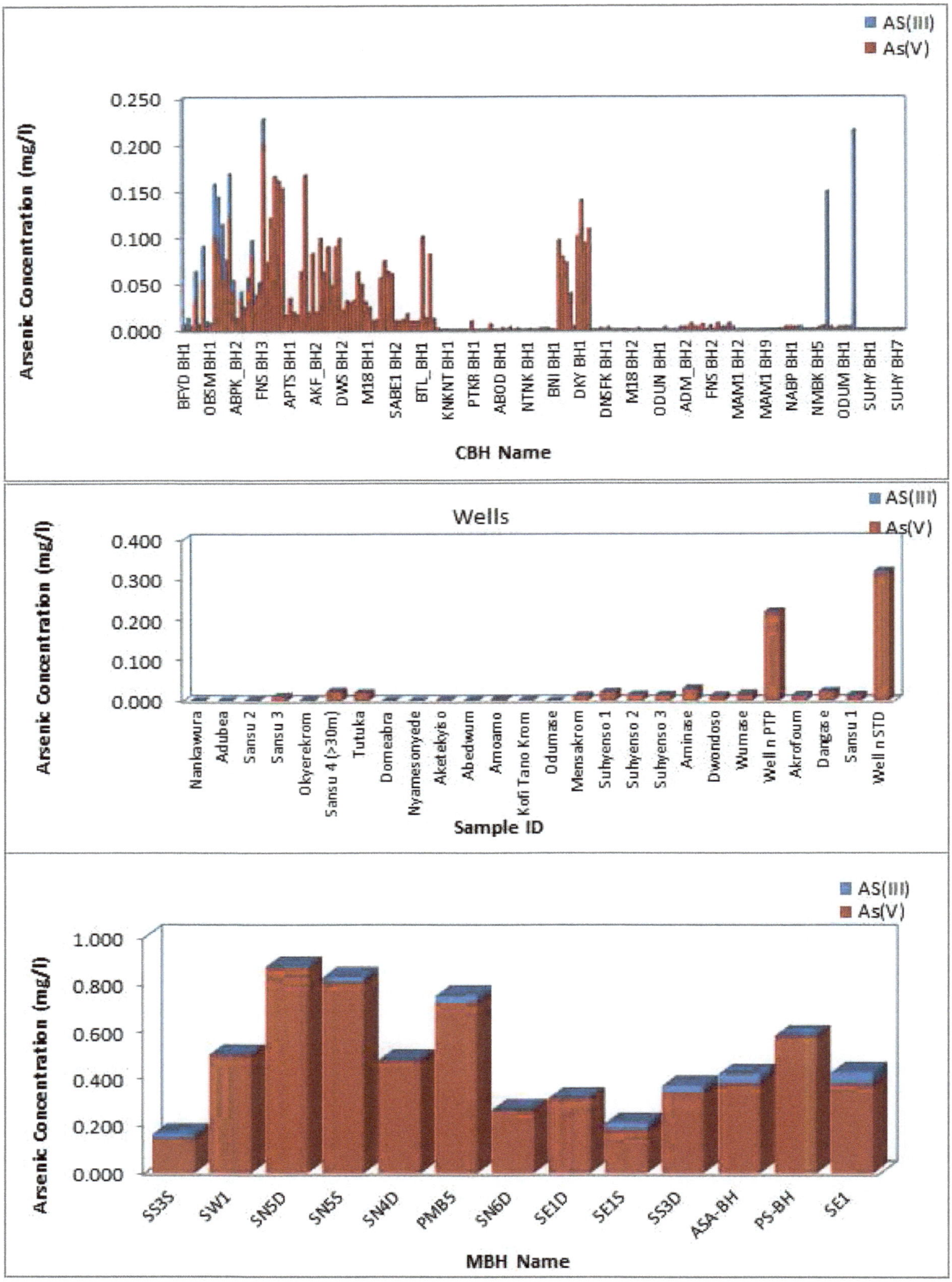
AS(III)
As(V)
Arsenic Concentration (mg/l)
0.250
0.200
0.150
0.100
0.050
0.000
BFYD BH1
OBSM BH1
ABPK_BH2
FNS BH3
APTS BH1
AKF_BH2
DWS BH2
M18 BH1
SABE1 BH2
BTL_BH1
KNKNT BH1
PTKR BH1
ABOD BH1
NTNK BH1
BNI BH1
DKY BH1
DNSFK BH1
M18 BH2
ODUN BH1
ADM_BH2
FNS BH2
MAM1 BH2
MAM1 BH9
NABP BH1
NMBK BH5
ODUM BH1
SUHY BH1
SUHY BH7
CBH Name
Wells
AS(III)
As(V)
Arsenic Concentration (mg/l)
0.400
0.300
0.200
0.100
0.000
Nankawura
Adubea
Sansu 2
Sansu 3
Okyerekrom
Sansu 4 (>30m)
Tutuka
Domeabra
Nyamesonyede
Aketekyiso
Abedwum
Amoamo
Kofi Tano Krom
Odumase
Mensakrom
Suhyenso 1
Suhyenso 2
Suhyenso 3
Aminase
Dwondoso
Wumase
Well n PTP
Akrofoum
Dargase
Sansu 1
Well n STD
Sample ID
AS(III)
As(V)
Arsenic Concentration (mg/l)
1.000
0.800
0.600
0.400
0.200
0.000
SS3S
SW1
SN5D
SN5S
SN4D
PMB5
SN6D
SE1D
SE1S
SS3D
ASA-BH
PS-BH
SE1
MBH Name

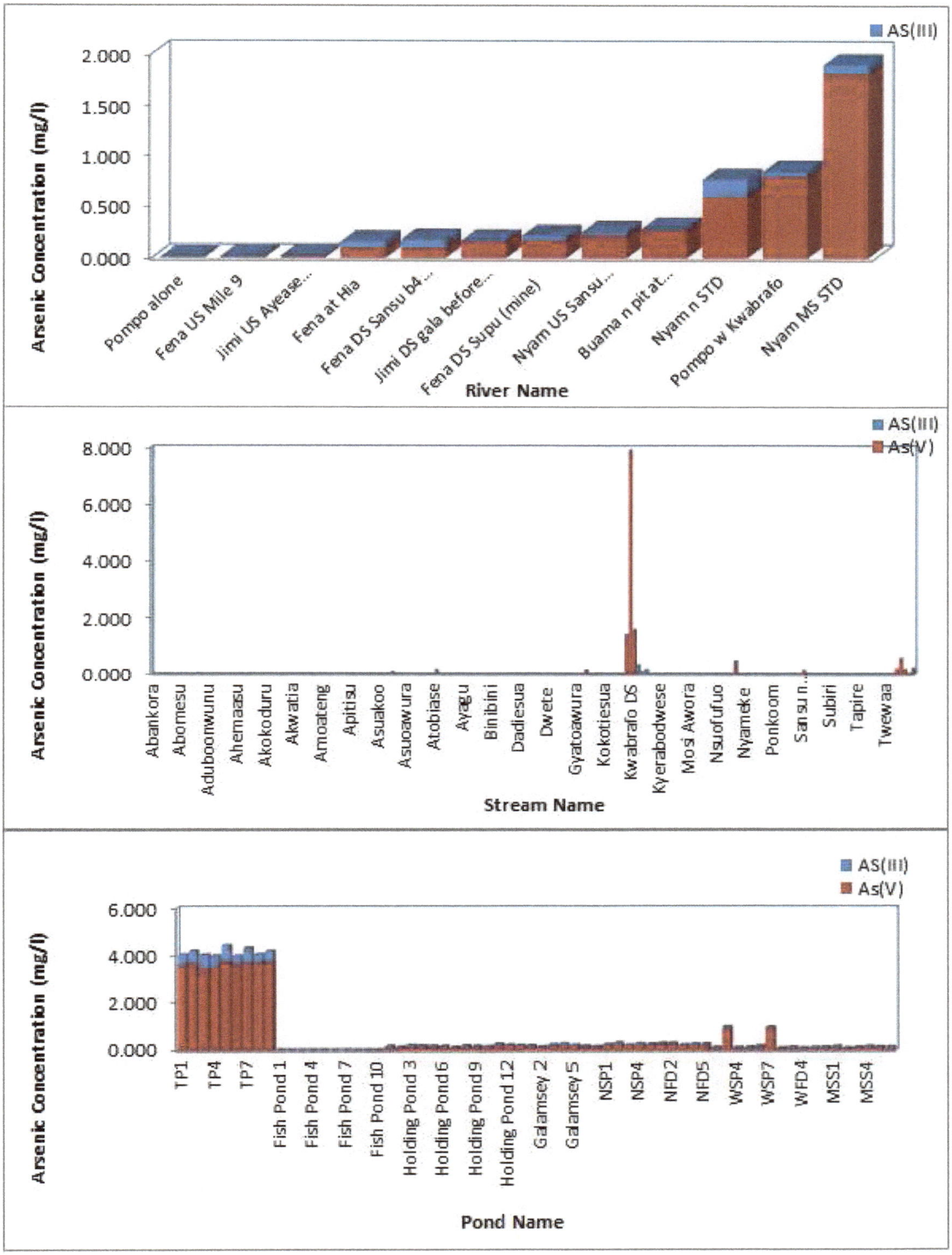

Figure 4. *Arsenate Distributions in Surface Water and Groundwater of Obuasi Municipality.*

3. Results and Discussion

3.1. Statistical Summary of Arsenic in Surface and Groundwater of Obuasi Municipality

Percentage distribution of arsenate in the study area is shown in Figure 3 whiles arsenate content of surface and groundwater sites is shown in Figure 4. Generally, except for monitoring boreholes which are not used for domestic activity, groundwater (CBH, 66%; Wells, 40%; MBH, 0%) has relatively higher percentage of sampling sites than surface water (streams, 44%; rivers, 7%; and ponds 4%) with arsenate concentration below the WHO standard of 0.010 mg/l) in the study area. It is also observed that except for monitoring boreholes, groundwater (CBH, 98%; Wells, 93%; MBH, 87%) has relatively higher percentage of sampling sites than surface water (streams, 86%; rivers, 40%; and ponds, 69%) with arsenate concentration below the EU standard of 0.050 mg/l) in the study area. Generally, except for monitoring boreholes, groundwater (CBH, 84%; Wells, 81%; and MBH, 60%) has relatively higher percentage of sampling sites than surface water (streams, 77%; rivers, 20%; and ponds, 22%) with arsenate concentration below the EU standard of 0.050 mg/l in the study area.

3.2. Influence of Trace Elements on Arsenate Concentration

To evaluate the distribution of arsenic in the studied water, concentrations of various trace elements including Al, Ag, B, Ba, Be, Ca, Ce, Co, Cr, Cs, Cu, Eu, Dy, Fe, Ga, Gd, Ge, Ho, K, Li, Lu, Mg, Mn, Mo, Na, Nd, Ni, Pb, Pr, Rb, Sb, Sm, Sr, Tb, Tl, Tm, U, Y, Tb, Zn and Zr were determined in selected water samples and statistical analysis performed on the dataset to identify trend in the analytical results. Generally, trace elements such as B, Be, Ce, Co, Cr, Cs, Eu, Dy, Ga, Gd, Ge, Ho, Lu, Mo, Nd, Ni, Pr, Pb, Rb, Sb, Sm, Sr, Tb, Tl, Tm, U, Y, Tb and Zr were not detected in statistically significant quantities for analysis. However, trace elements such as Al, Ba, Cu, Fe, Li, Mn, Sr, and Zn showed varying trends and were used in the analysis to determine their relationship with total arsenic or any of its species. Arsenic appears to show some relationship with Sr and to a lesser extent Ba in both surface and groundwater. The associations between As and Sr in groundwaters of the study area supports the fact that arsenic has been derived from arsenic bearing ore minerals.

3.3. Influence of Trace Elements on Arsenate Concentration in CBH

Three classes of CBH were identified per their trace element relationship with As(V) in the study area as follows (Figure 5):

1. CBH with very low to medium arsenate (0.001 mg/l to 0.010 mg/l) showed positive linear correlation with Fe, Sr, Li, Mn, Zn whilst, Ba, Cu showed negative linear correlation; and
2. CBH with medium to high arsenate (0.010 mg/l to 0.100 mg/l) showed positive linear correlation with Sr and Al whilst, Fe, Ba, Cu, Zn, Li showed negative linear correlation.

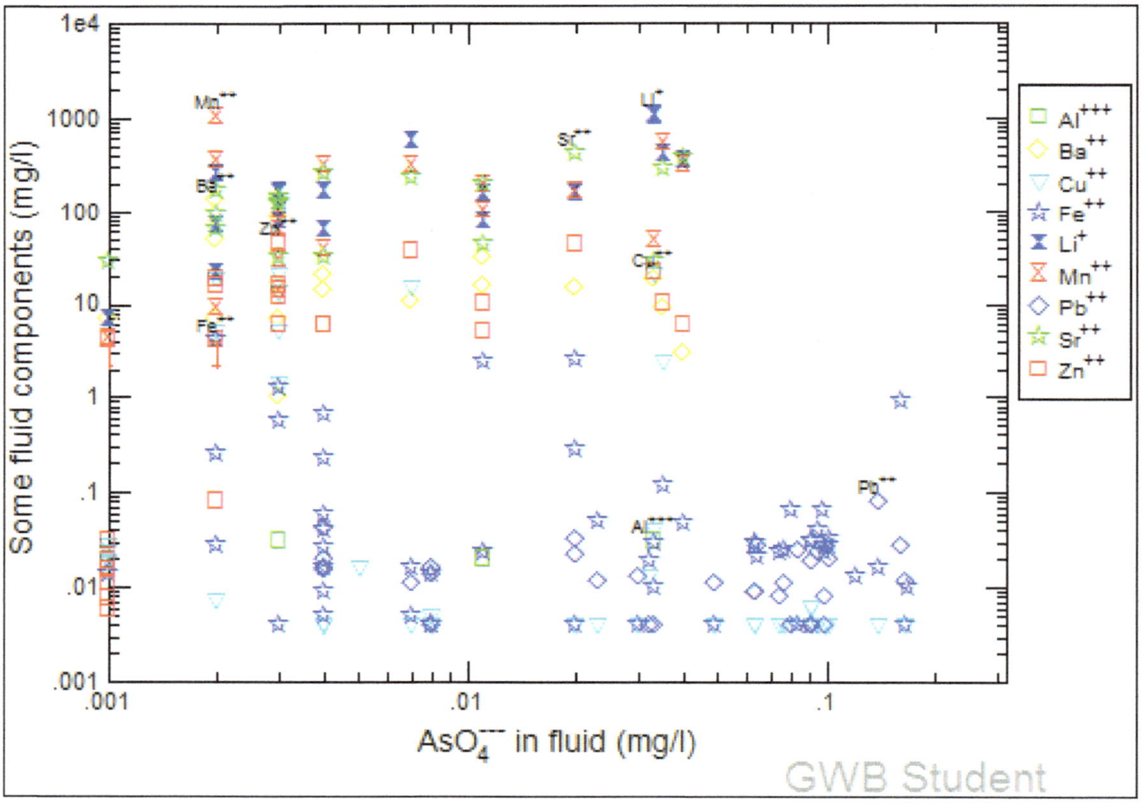

Figure 5. *Log-Log Relationship between As(V) and Trace Elements in CBH.*

3.4. Influence of Trace Elements on Arsenate Concentration in Wells

Two classes of wells were identified per their trace element relationship with As(V) in the study area as follows (Figure 6):

1. Wells with very low to medium arsenate (<0.010 mg/l) showing negative correlation between As(V) and Fe, Zn, Al and positive correlation for Sr, Li, Ba, and Mn; and
2. Wells with medium to very high arsenate (>0.010 mg/l) showing negative correlation between As(V) and Cu, Ni, Sr, Zn, Mn and positive correlation for Fe, Ba, Al.

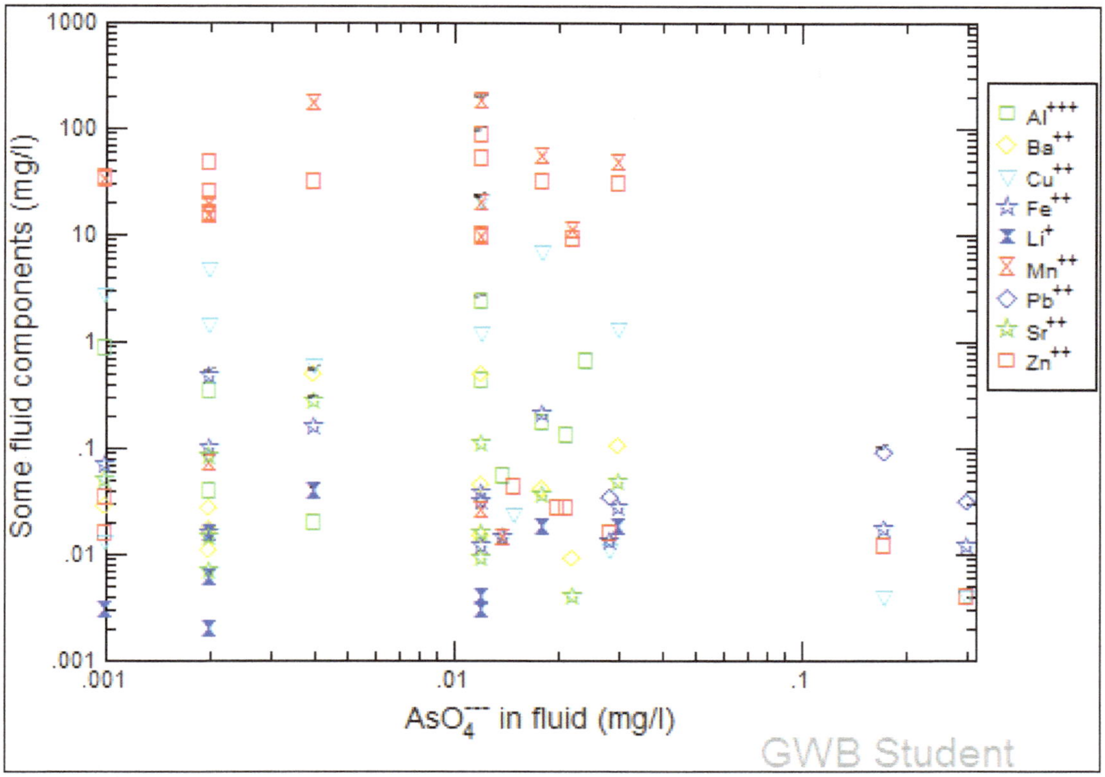

Figure 6. *Log-Log Relationship between As(V) and Trace Elements in Wells.*

3.5. Influence of Trace Elements on Arsenate Concentration in MBH

Arsenate concentration in MBH showed positive correlation with Cu and negative correlation with Fe, Zn and Mn (Figure 7).

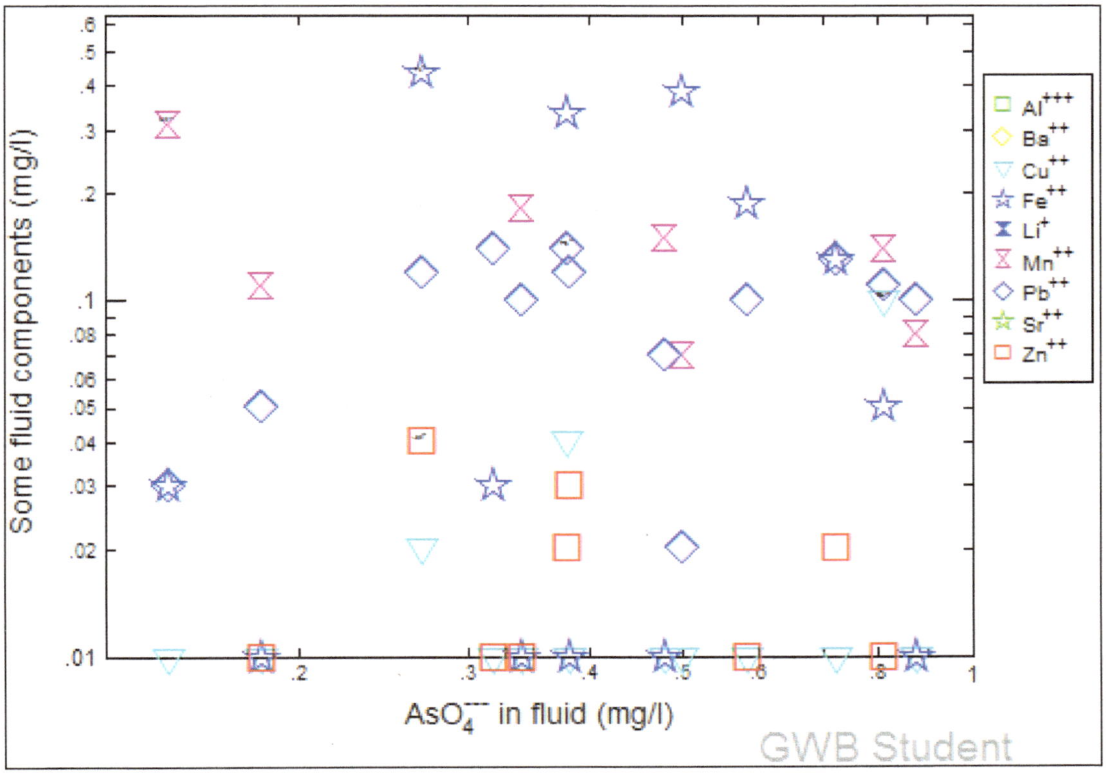

Figure 7. *Log-Log Relationship between As(V) and Trace Elements in MBH.*

3.6. *Influence of Trace Elements on Arsenate Concentration in Streams*

Three classes of streams were identified per their trace element relationship with As(V) in the study area as follows (Figure 8):

1. Very Low to low arsenate (0.001 mg/l to 0.010 mg/l): Negative linear correlation between Mn, Cu and positive linear correlation for Fe, Al, Ba, Sr, Li and Zn; and
2. Medium arsenate (0.010 mg/l to >0.050 mg/l): Negative linear correlation between Mn, Cu and positive linear correlation for Sr, Li, Fe, Zn, Ba, and Al.

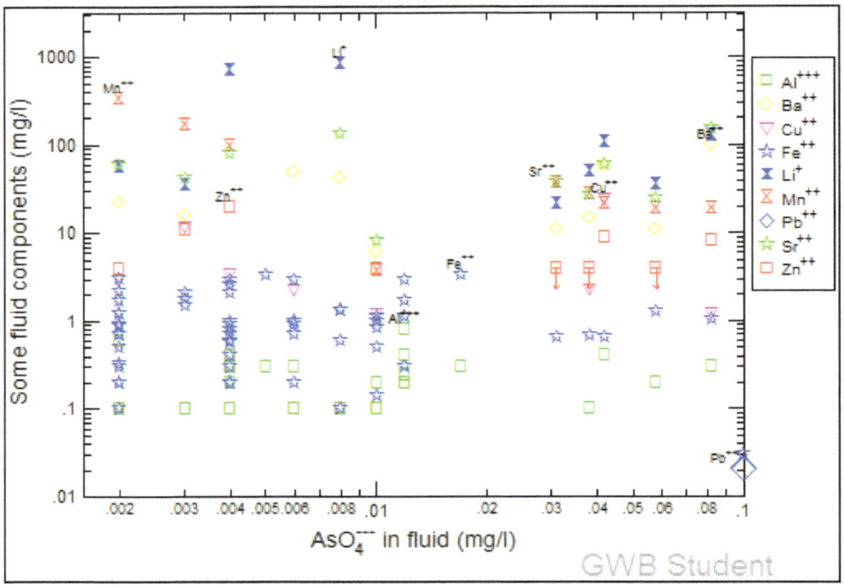

Figure 8. *Log-Log Relationship between As(V) and Trace Elements in Streams.*

3.7. *Influence of Trace Elements on Arsenate Concentration in Rivers*

Rivers were identified per their trace element relationship with As(V) in the study area as follows (Figure 9):

1. Very low to high arsenate (<0.100 mg/l): negative linear correlation with Sr and Cu; and positive linear correlation with Sr, Mn, Li, Al, Fe, Zn and Ba.
2. Very high arsenate (>0.100 mg/l) negative linear correlation with Zn, Fe and Mn; and positive linear correlation with Al, Cu, Li, Sr and Ba

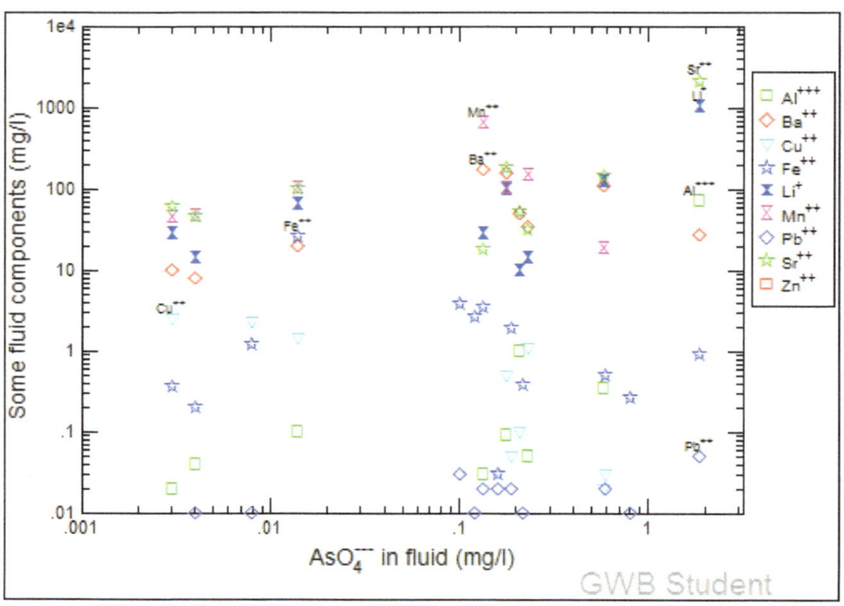

Figure 9. *Log-Log Relationship between As(V) and Trace Elements in Rivers.*

3.8. Influence of Trace Elements on Arsenate Concentration in Ponds

1. No clear trend between As(V) and the trace elements was identified. However, three main clusters were observed on the Semi-log plot of As(V) *vs.* log of selected trace elements (Figure 10).

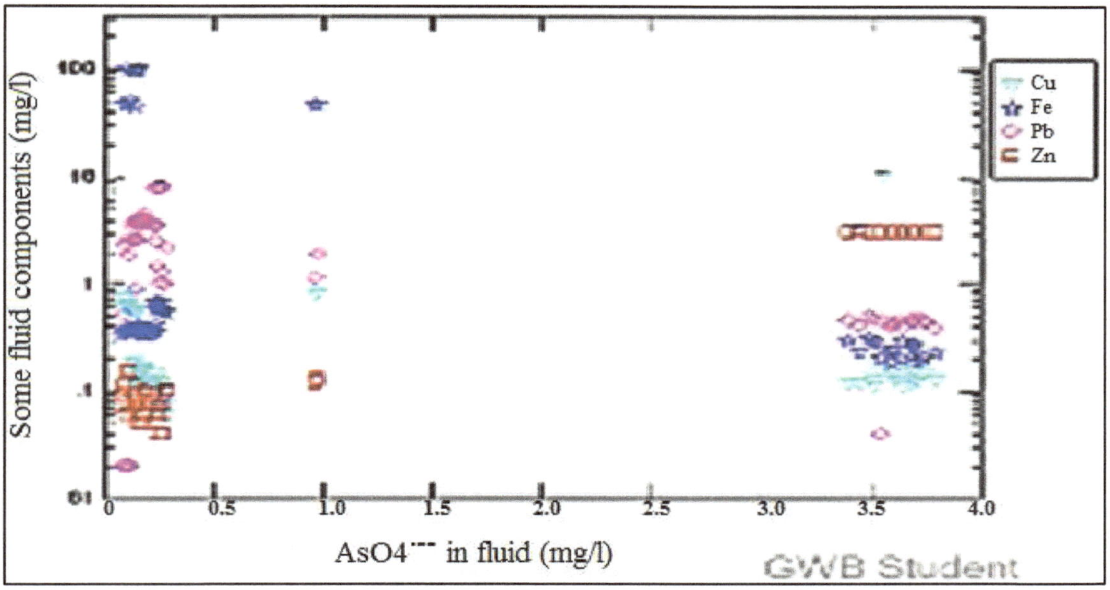

Figure 10. *Semi-Log Relationship between As(V) and Trace Elements in Ponds.*

3.9. Summary of Trace Elements and Arsenate Concentration in Surface and Groundwater

The relationship between arsenate and trace elements in groundwater of the study area can be summarized as follows:

CBH: It was observed that generally for CBH at low to medium As(V) concentration, as As(V) increase Fe, Sr, Li, Mn, Al and Zn increase. The positive trend changes at medium to high As(V) concentration when Fe, Mn and Zn begin to decrease whilst Sr, Li, and Al continue to increase with arsenate increase. Ba and Cu however consistently decrease with arsenate increase.

Wells: Interestingly for wells at low to medium As(V), Sr, Li, Mn, Ba begin to increase whilst Fe, Al and Zn, begin to decrease. As arsenate concentration increases from medium to very high, Fe and Al begin to increase and Cu, Li, Sr, Zn, Mn begin to decrease. Ba however is consistently increasing. The following is a breakdown of the relationship between the individual trace elements and As(V) concentration in groundwater:

1. Cu is constantly reducing in groundwater with increasing As(V);
2. Li and Sr are constantly increasing in wells and constantly decreasing in CBH whilst Ba and Li are constantly increasing in CBH and constantly decreasing in wells;
3. At low arsenate: Fe increase in CBH and decrease in wells, but at high As(V) Fe decrease in CBH and increase in wells;
4. At low As(V): Al decrease in CBH and wells, but at high As(V), Al increase in CBH and wells; and
5. At low As(V): Mn, Zn, Li increase in CBH and wells, but at high As(V), Mn, Zn, Li decrease in CBH and wells.

Figure 11 presents patterns for As(V) and trace elements relationship in wells and CBHs for the study area.

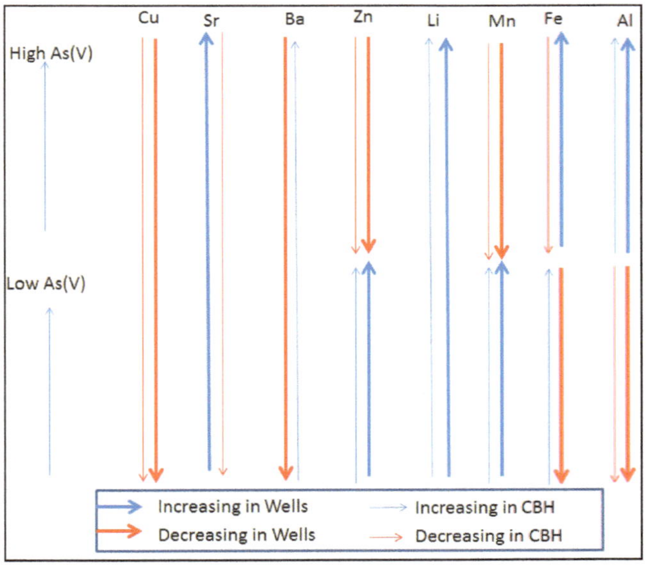

Figure 11. *Patterns of Arsenate and Trace Elements Relationships in Wells and CBHs.*

The generally high concentration of Mn, Cu, Li, Sr, Ba, Zn, Fe and Al in wells and boreholes suggests they are products of dissolution of host minerals occurring together. The relatively low pH of the water and concentration of these trace elements reflect the elements' abundance in the geological units being drained. It can also be observed that as Sr and Li are being leached from the host rock in wells at all concentrations, the process is reversed with the removal of Ba, and Cu from solution. Likewise, Zn and Mn are leached into solution at low arsenate concentration with the removal of Fe and Al. The process is reversed at high arsenate where Zn and Mn are removed, whilst Fe and Al are leached back into solution. Iron (Fe) and Mn may be released microbiologically and sometimes concurrently by the reductive dissolution of their metal oxides. At the circum-neutral pH of the study area (low arsenate), Mn is much more slowly oxidized than Fe and explains the low Fe and high Mn content of the groundwater. As the redox potential of the groundwater is lowered oxygen and nitrate will be reduced followed by Mn, Fe and SO_4^{2-} at high arsenate concentrations. Repeated redox cycles can increase the concentration of secondary oxides of Fe, Mn and Al in the sediments of the groundwater which can be demobilized later upon agitation, microbial action or temperature changes. This suggest two arsenic mobilization processes are taking place in wells where in one instance there is infiltration from an external source (probably close arsenic storage structures) whilst the second process is leaching of the mineralized host rock. The processes taking place in CBH are similar to that in wells, except that in CBH, Zn, Mn and Fe are leached into solution at low arsenate concentration with a concurrent removal of only Al. The process is also reversed at high arsenate where Zn, Mn and Fe are removed, whilst Al is leached back into solution. It would therefore be misleading to attempt to predict the concentration of arsenic in groundwater on the basis of their Fe contents alone. However, it can be noted that wells (not CBH) with low Fe will tend to have low arsenate concentration. All others cannot be clearly predicted. Even though it is not sufficient to predict the actual concentration of arsenic from the Fe concentration of the groundwater, the presence of high Fe in any well or borehole is a definite indication of high arsenic in groundwater. The relationship between arsenate and trace elements in surface water of the study area can be summarized as follows:

Streams: At very low to low increasing arsenate, Fe, Ba, Sr, Li, Al, Zn increase and Mn, Cu decrease. As arsenate increase to medium Fe, Ba, Sr, Li, Al, Zn continue to increase whilst Cu and Mn consistently decrease.

Rivers: At very low to high increasing As(V), Fe, Al, Zn, Mn, Ba, Sr, and Li begin to increase and Cu begin to decrease till very high arsenate when Al, Cu, Li, Sr increases and Ba, Fe, Mn, Zn begin to decrease. The following is a summary of the relationship between the individual trace elements and As(V) concentration in surface water:

1. Sr, Ba, Li and Al consistently increase in Streams and rivers;
2. Fe and Zn consistently increase in streams, but increase in rivers at low As(V) and decreases at very high As(V);
3. Mn consistently decrease in streams, but increase in rivers at low As(V) and decrease at very high As(V); and
4. Cu consistently decrease in rivers, but decrease in rivers at low As(V) and increase at very high As(V).

Figure 12 presents patterns for As(V) and trace elements relationship in surface water for the study area.

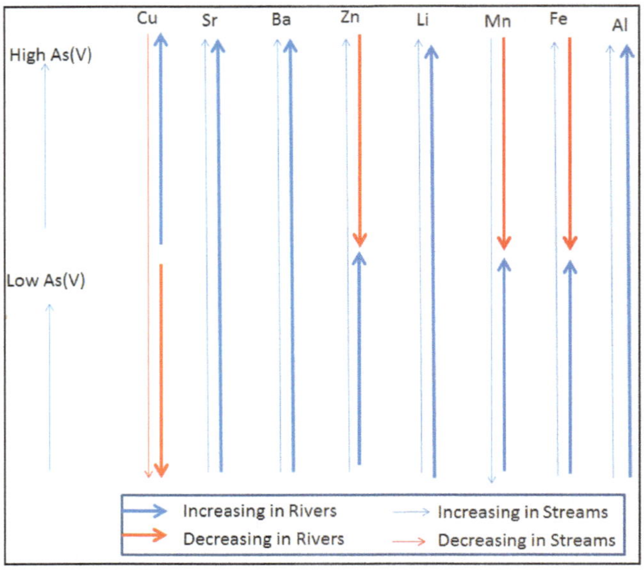

Figure 12. *Patterns of Arsenate and Trace Elements Relationships in Streams and Rivers.*

The positive correlation between arsenate and Sr, Ba, Li, Al at all concentrations of arsenate suggest that arsenic and these trace elements are likely derived from leaching of the localized mineral [16]. Low pH and high concentrations of Mn, Zn and Fe suggests oxidation of the host minerals is taking place in streams to oxidise pyrite and release SO_4^{2-} and Fe into solution. The oxidation process is likely to mobilise arsenic into streams. The reverse process of reduction is taking place at high arsenate concentrations in rivers where pyrite is being formed with the cementation of zinc and formation of minerals of manganese. The relatively low pH and highly reduced rivers is likely to initiate reduction of As(V) and mobilise As(III) into rivers. Figure 13 shows the relationship between As(III) content and Eh with pH of rivers.

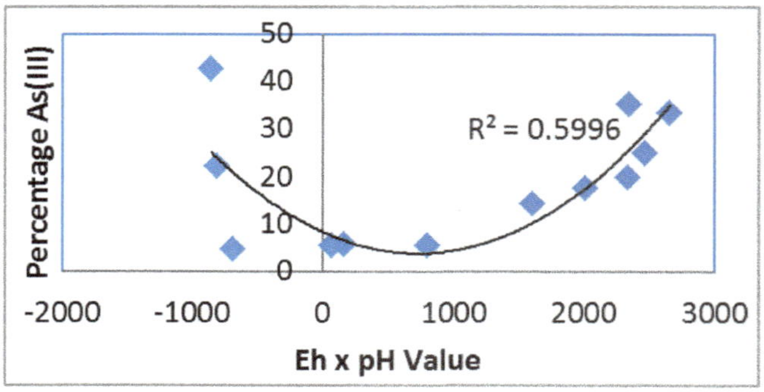

Figure 13. *Relationships between Percentage As(III) and Eh x pH of Rivers.*

4. Conclusion

The study has confirmed the distribution of arsenate and other trace metals in the Obuasi Municipality and identified patterns between some of the trace metals and arsenate concentration of surface water and groundwater in the study area. In particular metals such as Cu, Sr, Ba, Zn Li, Mn, Fe and Al have been found to have definite patterns with low or high levels of arsenate in the surface and groundwater of the study area. It is not advisable to attempt to predict the concentration of arsenic in groundwater on the basis of their Fe contents alone. However, it can be noted that wells (not community boreholes) with low Fe will tend to have low arsenate concentration. All others cannot be clearly predicted. Even though it is not sufficient to predict the actual concentration of arsenic from the Fe concentration of the groundwater, the presence of high Fe in any well or borehole is a definite indication of high arsenic in groundwater of Obuasi Municipality.

References

[1] Akabzaa, T.M., Banoeng-Yakubu, B., and Seyire, J.S. (2005). Heavy metal contamination in some mining communities within Jimi river basin in Ashanti Region. *Journal of the Ghana Science Association*, 7 (1).

[2] Amanor, J. A. and Gyapong, W. A. (1988). The Geology of Ashanti Goldfields. Report of the Ashanti Goldfields Corporation, Obuasi, Ghana.

[3] Amponsah-Tawiah, K., and Dartey-Baah, K. (2011). The Mining Industry in Ghana: A Blessing or a Curse, *International Journal of Business and Social Science* Vol. 2 No. 12; July 2011. 62p.

[4] Anon. (1998). Environmental Impact Assessment Report for Backfilling of Dokyiwa Pits, Obuasi, Ashanti Region, Ghana (Contract No C1419), Ashanti Goldfields Company Limited, Maisa Consul Ventures, UST, Kumasi, p.1-41.

[5] Bowell, R. J. (1993). Mineralogy and geochemistry of tropical rain forest soils: Ashanti, Ghana. *Chem. Geol.* 106, p.345-358.

[6] Clifford, D., Zhang, Z. (1994). Modifying ion exchange for combined removal of uranium and radium, *Journal of the American Water Works Association*, 86(4):pp. 214–227.

[7] Courtin-Nomade, A., Neel, C., Bril, H., and Davranche, M. (2002). Trapping and mobilisation of arsenic and lead in former mine tailings – Environmental conditions effects, *Bull. Soc. Geol. Fr.* 173, p.479–485.

[8] Ferguson, J. E., and Gavis, J. (1972). A review of the arsenic cycle in natural waters, *Water Res* 6:pp. 1259-1274.

[9] Foli, G., Apea, O. B., and Amedjoe, C. G. (2011). Pre-mining water quality prediction from non-weathered sulphide ores along the Ashanti metallogenic belt in Ghana using Acid-Base accounting procedure *Am. J. Sci. Ind. Res.*, 2(5): pp.827-833.

[10] Irgolic, K. J. (1994). Determination of total arsenic and arsenic compounds in drinking water. In: Arsenic: Exposure and Health, Chappell, W. R., Abernathy, C. O. and Cothern, C. R. (eds.) Science and Technology Letters. Northwood, England, pp. 51 - 60.

[11] Jordan, T. E., Correll, D. L., and Weller, D. E. (1997). Effects of agriculture on discharges of nutrients from Coastal Plain watersheds of Chesapeake Bay, J. *Environ. Qual.*, 26, pp. 836–848.

[12] Moreno-Jiménez, E., Clemente, R., Mestrot, A., Meharg, A. A. (2013). Arsenic and selenium mobilisation from organic matter treated mine spoil with and without inorganic fertilization, *Environmental Pollution* 173, pp.238-244.

[13] National Research Council (NRC). (1977). *Arsenic: Medical and Biological Effects of Environmental Pollutions*. National Academy of Sciences. Washington D.C., 332 p.

[14] Navarro Torres, V. F., Aduvire, O., Singh R. N. (2012). Assessment of natural attenuation of acid mine drainage pollutants in El Bierzo and Odiel basins: A case study, *Journal of Mining & Environment* Vol.2, No.2, 2011, pp.78-85.

[15] Perrouty, S., Aillères, L., Jessell, M. W., Baratoux, L., Bourassa, Y., and Crawford, B. (2012). Revised Eburnean geodynamic evolution of the gold-rich southern Ashanti Belt, Ghana, with new field and geophysical evidence of pre-Tarkwaian deformations, *Precambrian Research* 204– 205 pp. 12– 39.

[16] Plumlee, G. S., Smith, K. S., Montour, M. R., Ficklin, W. H., and Mosier, E. L. (1999). *Geologic controls on the composition of natural waters and mine waters draining diverse mineral deposits*, in Plumlee, G. S., and Filipek, L. H., eds., The Environmental Geochemistry of Mineral Deposits Part B: Case Studies and Research Topics, Reviews in Econ. Geology, vol. 6B: Littleton, CO, Soc. Econ. Geologists, p.373-409.

[17] Smedley, P. L. (1996). Arsenic in rural groundwater in Ghana. J. Afr. Earth Sci. 22, pp. 459–470.

[18] Stollenwerk, K. G. (2003). *Geochemical Processes Controlling Transport of Arsenic in Groundwater*: A review of Adsorption. In Arsenic in Groundwater, Alan H. Welch & Kenneth G. Stollenwerk Editors. Kluwert Academic Publishers, pp. 67-125.

[19] Yidana, S. M., Banoeng-Yakubo, B., and Sakyi P. A. (2012). Identifying key processes in the hydrochemistry of a basin through the combined use of factor and regression models, *J. Earth Syst. Sci.* 121, No. 2, April 2012, pp. 491–507.

[20] Zhang, Xuexia; et al. (2008). "Phylogenetic analysis and arsenate reduction effect of the arsenic-reducing bacteria enriched from contaminated soils at an abandoned smelter site". Journal of Environmental Sciences 20 (12).

Soil Erosion Is Influenced by Grain for Green Policy in Loess Plateau Area of Northern Shaanxi, China

Tao Wang

College of Geomatics, Xi'an University of Science and Technology, Xi'an, China

Email address:
wht432@163.com

Abstract: Grain for Green Policy (GGP), one of China's important ecological construction and protection policies, have been made remarkable achievements since implementation in 1999. The purpose of this work is to assessment of soil erosion influenced by GGP, which implementation is lead to great change of land use and vegetation cover, using RUSLE model on the basis of land use, MODIS NDVI, DEM and precipitation data in Loess Plateau area of northern Shaanxi, northwest of China. Result showed: (1) GGP implementation in northern Shaanxi was mainly during 1999-2003 and the area of returning farmland in the period accounted for 81.28% of the total area from1999 to 2008. (2) Farmland in northern Shaanxi and slope >25° was decrease 4.94% and 4.84%, respective ly, while grassland was increase 4.04% and 4.05%, respectively. The change of decreasing farmland and increasing grassland was mainly from 2000 to 2005 in northern Shaanxi and slope >25°. (3) Farmland was convert into woodland and grassland, which leaded to increasing of vegetation NDVI. Especially, the vegetation NDVI value of farmland in slope >25° was increase larger than whole area of northern Shaanxi from 2000 to 2010. (4) The change area of soil erosion in northern Shaanxi and slope >25° was decrease in 1 degree, while was increase in other degrees from 2000 to 2010. But when recalculated by rainfall erosivity in 2000, the change area of soil erosion in northern Shaanxi and >25° was increase in 1 degree, while was decrease in other degrees from 2000 to 2010. Vegetation cover change induced by GGP was reduce soil erosion, but due to rainfall erosivity increased, soil erosion was show increase trend in northern Shaanxi and slope >25°.

Keywords: Land Use Change, MODIS NDVI, Rainfall Erosivity, Loess Plateau

1. Introduction

Grain for Green Policy (GGP) is one of the major ecological protection and construction policies in China. GGP pilot implementation involve 1897 county-level administrative units since 1999, and those counties included in loess plateau area of northern Shaanxi province are key counties of construction. By 2010, the area of returning farmland to forest was $906.30\times10^4 hm^2$, afforestation of barren hills was $1533.97\times10^4 hm^2$, new forest conservation was $246.81\times10^4 hm^2$, total food subsidies was $\$322.79\times10^8$ ($1=￥6), and subsidies of living cash amounting was $\$43.74\times10^8$ [1]. With the implementation of GGP, forest cover rate has been improved, and GGP has achieved significant environment benefits [2].

Loess Plateau area in northern Shaanxi province is a serious soil erosion area, and the key area of GGP implementation. Affected by GGP, ecological environment has been improved obviously, vegetation cover increased, land use structure adjusted, a large area of arable land on steep slopes has been converted into forest or grassland, soil nutrients has been recovered and soil erosion decreased in northern Shaanxi [3,4]. Implementation of GGP was lead to reduction of arable land, promotion of rural surplus labor to non-agricultural industries, and improved of rural household's economic structure [5]. By the impact of GGP, total arable land of grain output was declined but grain yield of per unit area was increased [6,7]. Overall, ecological environment – human social system had been changing in many aspects by GGP. By the continued implementation of GGP, issues of implementation were highlighted by scholars [8].

Vegetation cover are important factors affecting soil erosion [9, 10, 11]. The goal of GGP implementation in northern Shaanxi is to reduce soil erosion. Relevant regulations and laws required the farmland which has serious soil erosion, important ecological status but low grain production, should be included in GGP plans and is prohibited reclamation in slope >25° area. Therefore, even though the policy does not explicit propose, but the arable land especially in slope >25° should be the main target of GGP [4, 12]. The purpose of this work is to assessment of soil erosion influenced by GGP, which implementation is lead to great change of land use and vegetation cover, using RUSLE model on the basis of land use, MODIS NDVI, DEM and precipitation data in Loess Plateau area of northern Shaanxi, northwest of China.

2. Materials and Methods

2.1. Study Area

Loess plateau area in northern Shaanxi include Yulin and Yan'an, which contain 25 countries and locate in 107°15′56″E-111°13′10″E and 35°21′56″N-39°35′06″N (Figure.1), while Mu Us Sand land along the edge of northern Yulin. By the end of 2011, the total population of the area is 554.64×10^4, and the total land area is 80 624km^2. The climate is temperate continental monsoon climate, and the average temperature is 9.5℃. Annual average rainfall and the average sunshine are 506.5mm and 2540h/a, respectively, and frost-free period is 194d. The whole territory belongs to the Yellow River watershed, and the main tributary is Wuding River and Yanhe River, etc [13].

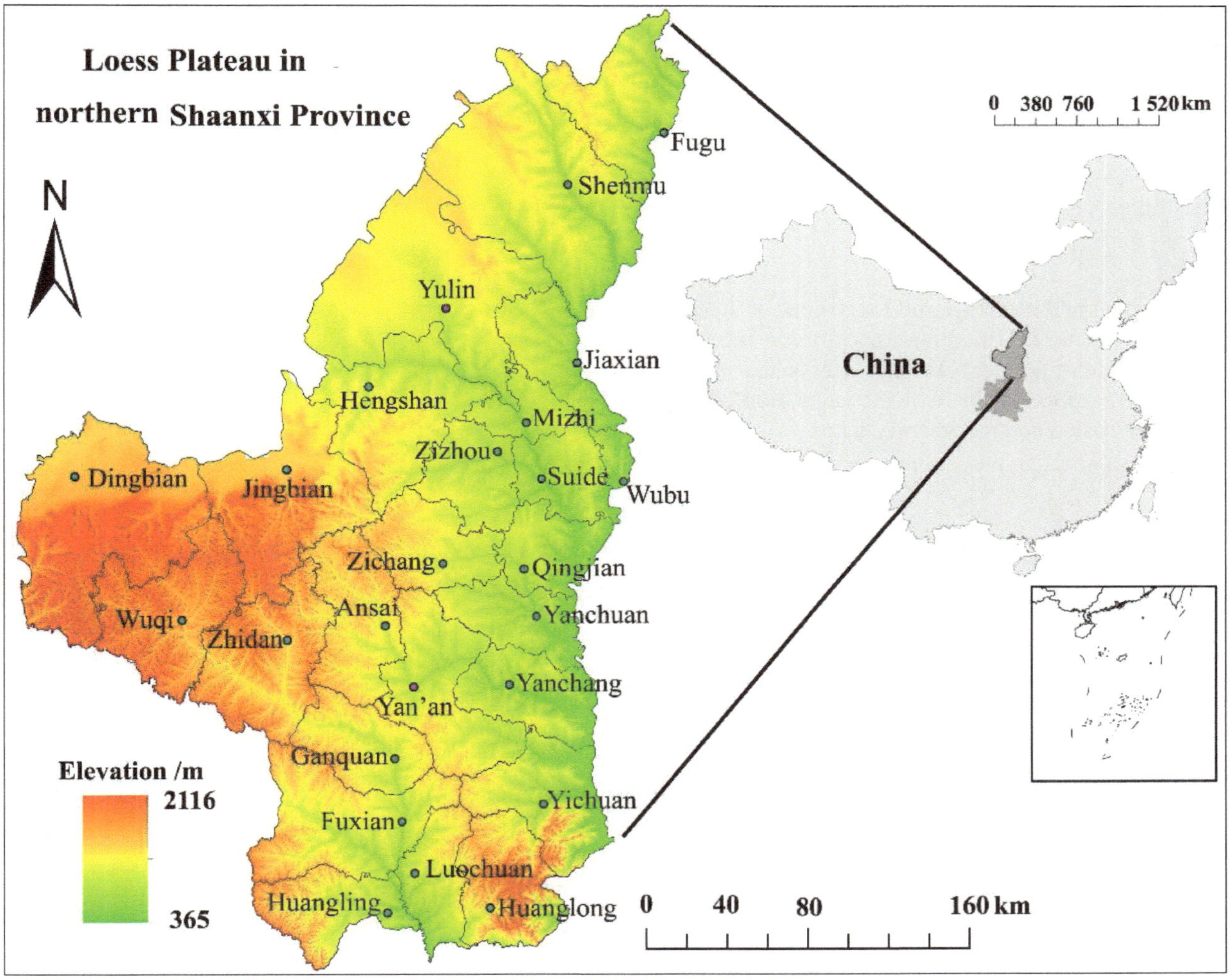

Figure 1. *Study area.*

Precipitation in study area is focus on the summer and autumn, which account for 80% of annual precipitation, and which also the important reasons of soil erosion in northern Shaanxi, China. Yulin and Yan'an are the mainly area of GGP implementation in Shaanxi province. According to the statistical yearbook of Shaanxi province (1999-2008), the area of returning farmland to forest in northern Shaanxi is 472.1×10^3hm^2 from 1999 to 2008, accounting for 41.14% of total area of returning farmland to forest in Shaanxi province [13].

2.2. Materials

The data used in this research include: (1) 1:100 000 vector land use map of northern Shaanxi in 2000, 2005 and 2010 year, comes from the project "Remote sensing survey and evaluation of ecological environment change in China (2000-2010)". Land use map in 2010 was classified by environmental satellite images, while TM images were used in 2000 and 2005. Geographic coordinates of land use map is GCS_WGS84, and projection is Albers ellipsoid with central longitude 105°. Classification system of land use map include

7 primary classifications and 30 secondary classifications. (2) Vector county administrative boundary map of northern Shaanxi. (3) ASTER GDEM data with 30m resolution of northern Shaanxi. The dataset was provided by Geospatial Data Cloud, Computer Network Information Center, Chinese Academy of Sciences (http://www.gscloud.cn). (4) Daily rainfall data of 7 stations from 1991 to 2010 is collected from China Meteorological Data Sharing Service System (http://cdc.nmic.cn). The station is Dingbian, Yulin, Hengshan, Suide, Wuqi, Yan'an and Luochuan. (5) Soil map of northern Shaanxi with scale of 1:500 000. (6) MODIS NDVI 13Q1-Level 3 images data of vegetation index with 250m resolution of northern Shaanxi in 2000, 2005 and 2010. (7) Shaanxi statistical yearbook from 2001 to 2011.

2.3. Methods

The Revised Universal Soil Loss Equation (RUSLE) is used to calculate soil erosion of northern Shaanxi. RUSLE is represented as follows [14]

$$A = R \times K \times L \times S \times C \times P \quad (1)$$

where A is the amount of the average soil loss ($t \cdot hm^{-2} \cdot a^{-1}$); R is the rainfall erosivity factor ($MJ \cdot mm \cdot hm^{-2} \cdot h^{-1} \cdot a^{-1}$); K is the soil erodibility factor ($t \cdot hm^{-2} \cdot h \cdot hm^{-2} \cdot MJ^{-1} \cdot mm^{-1}$); L is the slope length factor; S is the slope factor; C is the vegetation cover factor and P is the erosion control practice factor.

2.3.1. Rainfall Erosivity Factor (R)

R factor is the indicator for evaluating the impact of rainfall on soil separation and transportation, and is used to estimate by rainfall parameters. Annual rainfall erosivity is calculated by daily rainfall data proposed by [15,16]:

$$R_i = \alpha \sum_{j=1}^{k} P_j^{\beta} \quad (2)$$

$$\alpha = 21.586\beta^{-7.1891} \quad (3)$$

$$\beta = 0.8363 + 18.144P_{d12}^{-1} + 24.455P_{y12}^{-1} \quad (4)$$

where R_i is the rainfall erosivity of ith year, $MJ \cdot mm \cdot hm^{-2} \cdot h^{-1} \cdot a^{-1}$; k is the number of days with erosive rainfall in one year, d; and P_j is the erosive rainfall of jth day, mm. The standard rainfall leading to soil erosion is ≥12mm/d, otherwise is calculated by 0 value [17]. α and β are characteristic parameters, and P_{dl2} is annual average daily rainfall with daily rainfall ≥12mm, mm, while P_{y12} is average annual rainfall of rainfall ≥12mm, mm.

2.3.2. Soil Erodibility Factor (K)

K factor affected by many factors, is the indicator to evaluate soil erosion probability when rainfall erosivity is occurred [18, 19]. K value is calculated by EPIC model [20] as follows:

$$K = \left(0.2 + 0.3\exp\left(0.0256M\left(1 - \frac{F}{100}\right)\right)\right) \cdot \left(\frac{F}{N+F}\right)^{0.3} \left(1.0 - \frac{0.25T}{T + \exp(3.72 - 2.95T)}\right) \cdot \left(1.0 - \frac{0.7\delta}{\delta + \exp(-5.51 + 22.9\delta)}\right) \quad (5)$$

where, M, F, N is mass fraction of sand, silt and clay, respectively, %; T is mass fraction of soil organic carbon, %; $\delta = 1 - M/100$. The K value is need to change into the international system of units through multiplied by 0.1317, $t \cdot hm^2 \cdot h/hm^2 \cdot MJ \cdot mm$.

2.3.3. Topographical Factor (LS)

Slope and slope length (LS) are topographical factor, which mainly reflecting the effects of topography on soil erosion. L is calculated by DEM according to [21] proposed formula and S is calculated by the formula proposed by [22] and [23]. The formula is expressed as follows:

$$L = (\lambda/22.13)^{\alpha} \quad (6)$$

$$\alpha = \beta(\beta + 1),\ \beta = (\sin\theta/0.0896)/(3.0(\sin\theta)^{0.8} + 0.56) \quad (7)$$

$$S = \begin{cases} 10.8\sin\theta + 0.03,\ \theta < 9\% \\ 16.8 sin\theta - 0.50,\ 9\% \le \theta < 14\% \\ 21.9 sin\theta - 0.96,\ \theta \ge 14\% \end{cases} \quad (8)$$

where λ is the slope length; α is the slope length index; θ is the slope (%) extract by DEM.

2.3.4. Vegetation Cover Factor (C)

C factor is vegetation index, which is calculated based on MODIS NDVI data, the formula is expressed as follows [24,25]:

$$AVHRR\ NDVI = 0.18 MODIS\ NDVI + 0.131 \quad (9)$$

$$C = \exp\left(-\alpha \cdot \frac{AVHRR\ NDVI}{\beta - AVHRR\ NDVI}\right) \quad (10)$$

where α and β are parameters with the value of 2 and 1, respectively. C value is calculated by AVHRR NDVI data, which need to covert MODIS NDVI into AVHRR NDVI.

2.3.5. Erosion Control Practice Factor (P)

P factor is calculated by the following formula proposed by [26]:

$$P = 0.2 + 0.03S \quad (11)$$

where S is the slope extracted by DEM (%).

3. Results

3.1. GGP on Land Use Change

3.1.1. Decreased Area of GGP Implementation in Northern Shaanxi

The area of returning farmland to forest or grassland of Shaanxi province and northern Shaanxi region including Yan'an and Yulin city was obtained by statistical yearbook of Shaanxi province from 2000 to 2011, and the result showed in Figure.2.

GGP implementation was occurred mainly during 1999-2008, and statistical yearbook of Shaanxi province was not recorded about the area of returning farmland to forest or grassland after 2009. From 1999 to 2008 of GGP implementation, the area of returning farmland was about 11476.2km^2 in Shaanxi province, where the area of returning farmland in northern Shaanxi (Yulin and Yan'an) was 4721km^2, accounting for 41.14%.

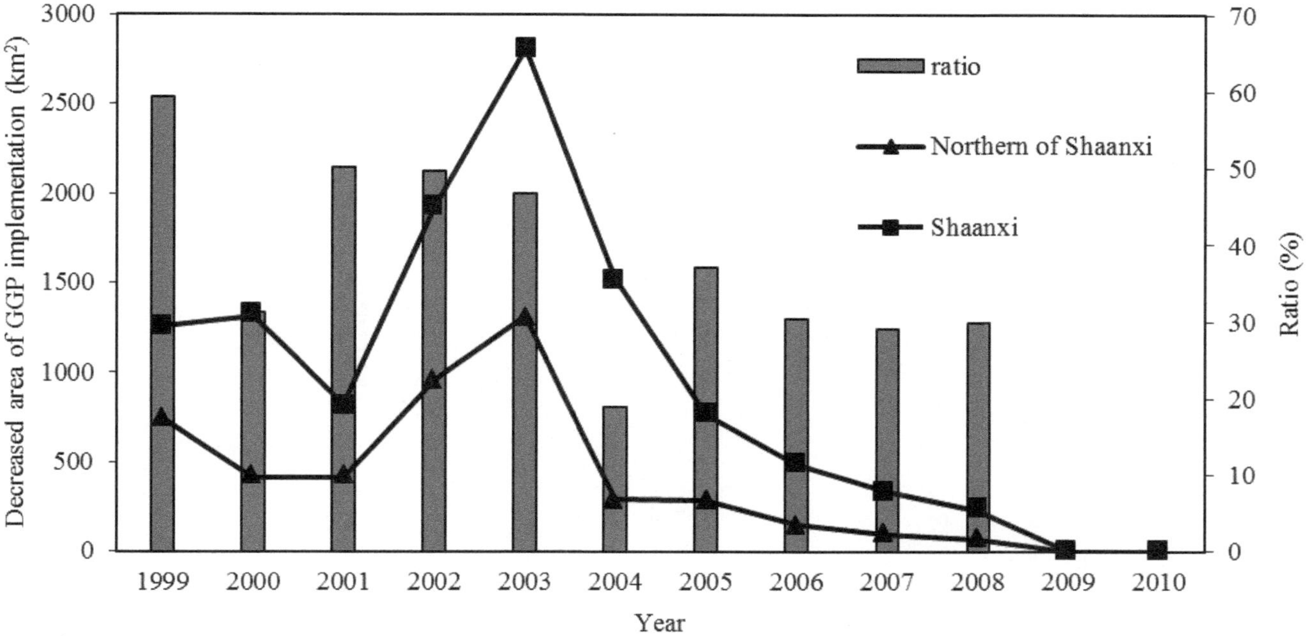

Figure 2. *Decreased area of GGP implementation in Shaanxi province and northern Shaanxi.*

The process of area change can be divided into 2 stages. The first stage was from 1999 to 2003 with increasing trend of area of returning farmland in Shaanxi province and northern Shaanxi, and the ratio of returning farmland area in northern Shaanxi accounting for Shaanxi province was higher than 40% (except in 2000). The second stage was from 2003 to 2008 with decreasing trend of area of returning farmland in Shaanxi province and northern Shaanxi, and the ratio of returning farmland area in northern Shaanxi accounting for Shaanxi province was about 30% (except in 2004).

Overall, GGP implementation in northern Shaanxi was mainly during 1999-2003, and the area of returning farmland in the period accounting for 81.28% of the total area during 1999-2008.

3.1.2. Land Use Change

Land use type in northern Shaanxi was woodland, grassland and farmland, which three land types accounting for more than 90% of the total land area. Woodland and grassland were show an increasing trend and the percentage increased from 20.21% to 20.73%, from 44.74% to 48.78%, respectively, during 2000-2010, while farmland showed decreasing trend and the percentage decreased from 25.16% to 20.22% (Figure.3a).

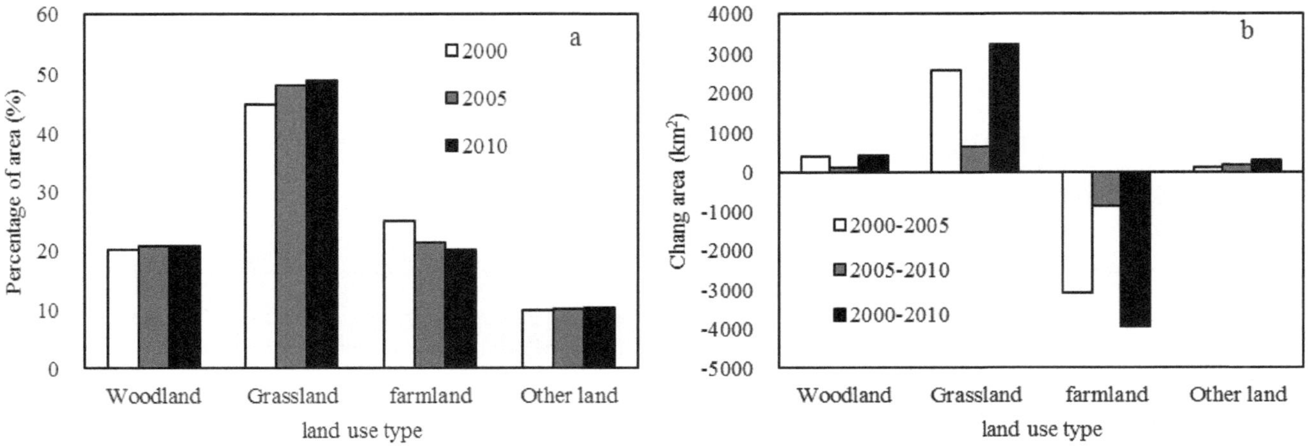

Figure 3. *Land use structure (a) and change area (b) in different stage of northern Shaanxi.*

Woodland, grassland and farmland were the largest types of land use change. The area of woodland and grassland were increase 412.35km^2 and 3229.71km^2, respectively, from 2000 to 2010, while farmland was decrease 3948.16km^2 (Figure.3b). The change rate of woodland, grassland and farmland were 0.52%, 4.04% and -4.94%, respectively, during 2000-2010.

The most intense period of land use change was from 2000 to 2005 in northern Shaanxi. In the period of 2000-2005, the change area of woodland, grassland and farmland were 377.49km^2, 2583.13km^2 and -3081.71km^2, respectively, and accounting for 91.55%, 79.98% and 78.05% of the total change area of 2000-2010, respectively.

3.1.3. Land Use Change in Slope >25°

Land use type in slope >25° of northern Shaanxi was woodland, grassland and farmland, which three land types accounting for more than 99% of the total land area in slope >25 °. Woodland and grassland were show an increasing trend and the percentage increased from 36.47% to 37.17%, from 50.50% to 54.55%, respectively, during 2000-2010, while farmland showed decreasing trend and the percentage decreased from 12.59% to 7.75% (Figure.4a).

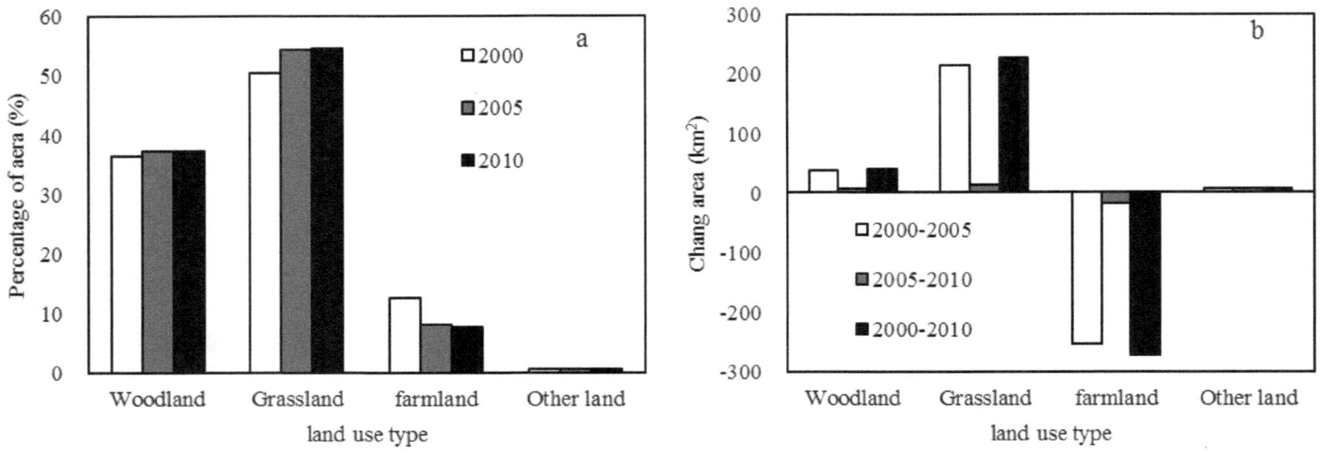

Figure 4. *Land use structure (a) and change area (b) in different stage of slope >25° in northern Shaanxi.*

Woodland, grassland and farmland were the largest types of land use change in slope >25°. The area of woodland and grassland were increase 39.51km^2 and 226.90km^2, respectively, from 2000 to 2010, while farmland was decrease 270.67km^2 (Figure.4b). The change rate of woodland, grassland and farmland were 0.70%, 4.05% and -4.84%, respectively, during 2000-2010.

The most intense period of land use change in slope >25° was from 2000 to 2005 in northern Shaanxi. In the period of 2000-2005, the change area of woodland, grassland and farmland were 38.41km^2, 213.65km^2 and -254.04km^2, respectively, and accounting for 97.21%, 94.16% and 93.86% of the total change area of 2000-2010, respectively.

3.2. GGP on Vegetation Cover Change

Large changes of land use were happened in northern Shaanxi by GGP implementation, which also lead to changes of vegetation cover. Land use and vegetation NDVI data were used to extract vegetation NDVI value corresponding to 2000, 2005 and 2010 of farmland area in 2000 in northern Shaanxi and slope >25°, and the value of vegetation NDVI of changed and unchanged area of farmland in 2000. Results show in Table 1.

Table 1. *Vegetation NDVI value in different area.*

	Type	2000	2005	2010	2000-2005	2000-2010	2000-2010(%)
	2000	0.4111	0.5053	0.5623	0.0942	0.1512	36.78
Farmland area	Unchanged during 2000-2010	0.4152	0.5034	0.5565	0.0882	0.1413	34.03
	Changed during 2000-2010	0.3947	0.5129	0.5859	0.1182	0.1912	48.44
	2000	0.4074	0.5197	0.5856	0.1123	0.1782	43.74
Farmland area in slope >25°	Unchanged during 2000-2010	0.4035	0.5024	0.5630	0.0989	0.1595	39.53
	Changed during 2000-2010	0.4135	0.5468	0.6210	0.1333	0.2075	50.18

The direct effect of GGP implementation was lead to a significant reduction of farmland converting to woodland and grassland, thereby vegetation NDVI value increased. The vegetation NDVI value of farmland area in 2000 showed increasing trend from 2000 to 2010, with an increase of 0.1512 which accounting for 36.78% of 2000. In the farmland area of 2000, the vegetation NDVI of farmland unchanged area was increase 0.1413 with 34.03% while of farmland changed area increased 0.1912 with 48.44% (Table 1). Therefore, the vegetation NDVI of changed area of farmland in 2000 was increase larger than the vegetation value of unchanged area, which was influenced by GGP implementation.

The vegetation NDVI value of farmland area in slope >25° in 2000 showed increasing trend from 2000 to 2010, with an increase of 0.1782 which accounting for 43.74% of 2000. In the farmland area of slope >25° in 2000, the vegetation NDVI of farmland unchanged area was increase 0.1595 with 39.53% while of farmland changed area increased 0.2075 with 50.18% (Table 1).

Overall, farmland was convert into woodland and grassland

from 2000 to 2010, which leaded to vegetation NDVI increase by GGP implementation. Especially, the vegetation NDVI value of farmland in slope >25° was increase more than whole area of northern Shaanxi.

3.3. GGP on Soil Erosion

The value of *R*, *K*, *L*, *S*, *P* and *C* was calculated by the formula mentioned in the context. *R* and *C* had different value in 2000, 2005 and 2010, so the amount of the average soil loss in northern Shaanxi and the area of slope >25° were obtained in 2000, 2005 and 2010. While for evaluating GGP on soil erosion, the rainfall erosivity should be not considered. Therefore, the amount of the average soil loss in northern Shaanxi and the area of slope >25° were also obtained in 2000, 2005 and 2010 using the rainfall erosivity of 2000. Six soil erosion degree was divided according to the classify standard of soil erosion [27], that was 1, 2, 3, 4, 5, 6 degree with the amount of the average soil loss <1000 $t\cdot hm^{-2}\cdot a^{-1}$, 1000-2500 $t\cdot hm^{-2}\cdot a^{-1}$, 2500-5000 $t\cdot hm^{-2}\cdot a^{-1}$, 5000-8000 $t\cdot hm^{-2}\cdot a^{-1}$, 8000-15000 $t\cdot hm^{-2}\cdot a^{-1}$, >15000 $t\cdot hm^{-2}\cdot a^{-1}$, respectively. The percentage of change area in different degree during 2000-2005, 2005-2010 and 2000-2010 showed in Figure.5.

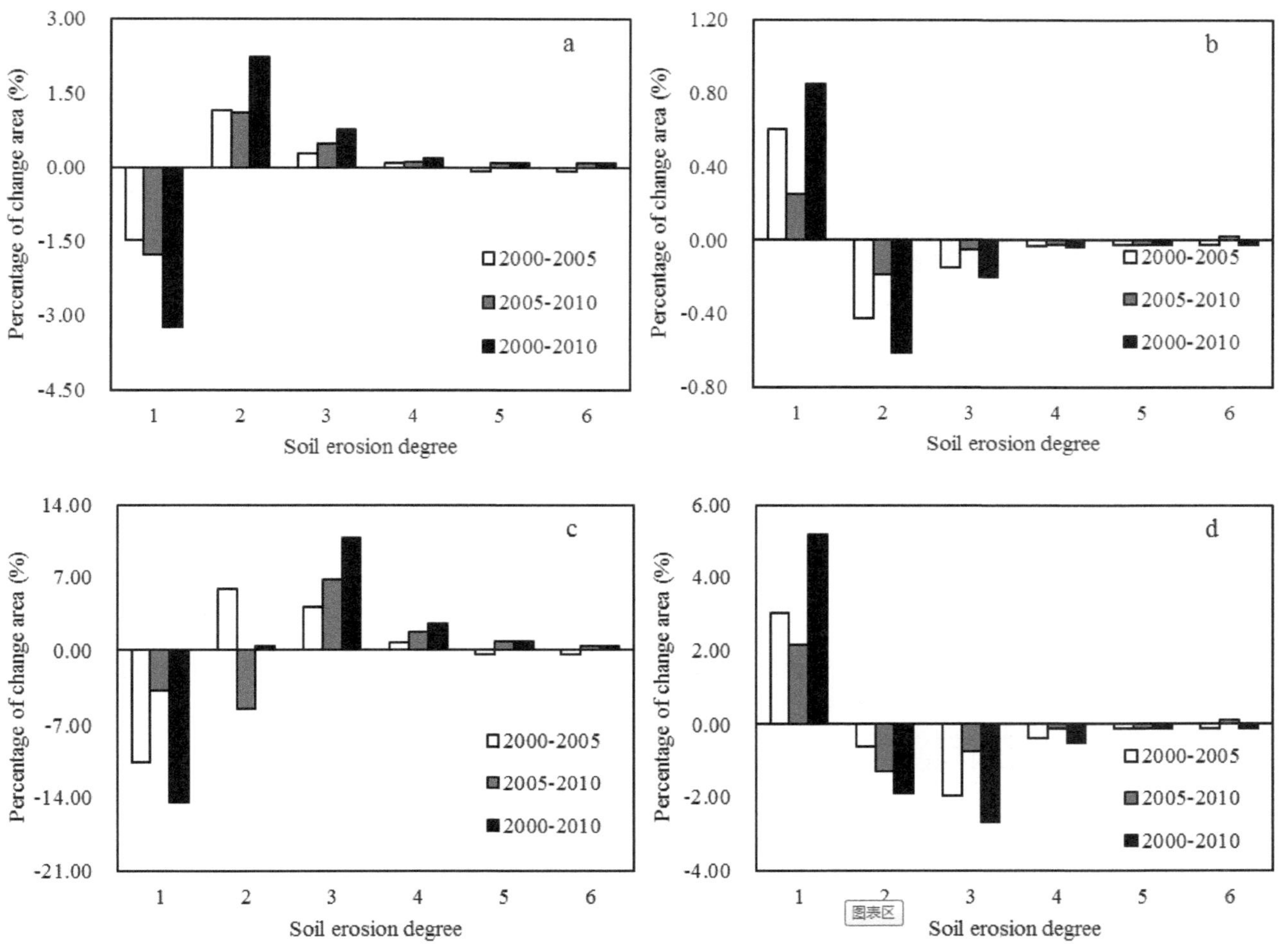

1, <1000 $t\cdot hm^{-2}\cdot a^{-1}$; 2, 1000-2500 $t\cdot hm^{-2}\cdot a^{-1}$; 3, 2500-5000 $t\cdot hm^{-2}\cdot a^{-1}$; 4, 5000-8000 $t\cdot hm^{-2}\cdot a^{-1}$; 5, 8000-15000 $t\cdot hm^{-2}\cdot a^{-1}$; 6, >15000 $t\cdot hm^{-2}\cdot a^{-1}$. a and c were calculated by rainfall erosivity in 2000, 2005 and 2010, while b and d were recalculated by rainfall erosivity in 2000.

Figure 5. *Change area of different soil erosion degree in the area of northern Shaanxi (a, b) and the area of slope >25° (c, d).*

The percentage of change area of northern Shaanxi in 1 degree was decrease with 3.23%, while other degrees were increase with 2.24%, 0.76%, 0.18%, and 0.06% in 2, 3, 4, 5 degree, respectively, and the percentage of change area in 6 degree was not obviously from 2000 to 2010 (Figure.5a). When recalculated by rainfall erosivity in 2000, the percentage of change area of northern Shaanxi in 1 degree was increase with 0.85%, while other degrees were decrease with 0.61%, 0.20%, 0.04% and 0.01%, respectively, and the percentage of change area in 6 degree was also not obviously from 2000 to 2010 (Figure.5b).

The percentage of change area in slope >25° of 1 degree was decrease with 14.46%, while other degrees were increase with 0.31%, 10.83%, 2.52% and 0.80%, respectively, and the percentage of change area in 6 degree was not obviously from 2000 to 2010 (Figure.5c). When recalculated by rainfall erosivity in 2000, the percentage of change area in 1 degree was increase with 5.17%, while other degrees were decrease with 1.89%, 2.70%, 0.51% and 0.07%, respectively, and the percentage of change area in 6 degree was also not obviously from 2000 to 2010 (Figure.5d).

Overall, the change area of soil erosion in northern Shaanxi and slope >25° was decrease in 1 degree, while was increase in other degrees from 2000 to 2010. When recalculated by rainfall erosivity in 2000, the change area of soil erosion in northern Shaanxi and slope >25° was increase in 1 degree, while was decrease in other degrees from 2000 to 2010. Vegetation cover change induced by GGP was reduce soil erosion, but due to rainfall erosivity increased, soil erosion was show increase trend in northern Shaanxi and slope >25° area.

4. Conclusions

As an important ecological construction and protection policy, the implementation of GGP has played important role in ecological environment. In this paper, the impact of GGP on soil erosion was evaluated by land use, MODIS NDVI, DEM, precipitation, and vegetation cover data in northern Shaanxi and the area of slope >25°. Result showed:

(1) GGP implementation in northern Shaanxi was mainly during 1999-2003 and the area of returning farmland in the period accounted for 81.28% of the total area from1999 to 2008.

(2) Farmland in northern Shaanxi was decrease largely with 3948.16km^2, while grassland was increase largely with 3229.71km^2. The percentage of change area of farmland and grassland was 4.94% and 4.04%, respectively. The process of decreasing farmland and increasing grassland was mainly from 2000 to 2005. Farmland in the area of slope >25° was decrease largely with 270.67km^2, while grassland was increase largely with 226.90km^2. The percentage of change area of farmland and grassland was 4.84% and 4.05%, respectively. The process of decreasing farmland and increasing grassland was mainly from 2000 to 2005, too.

(3) Farmland was convert into woodland and grassland from 2000 to 2010, which leaded to vegetation NDVI increase. Especially, the vegetation NDVI value of farmland in slope >25° was increase larger than whole area of northern Shaanxi.

(4) The area of soil erosion in northern Shaanxi and slope >25° was decrease in 1 degree, while was increase in other degrees from 2000 to 2010. When recalculated by rainfall erosivity in 2000, the area of soil erosion in northern Shaanxi and slope >25° was increase in 1 degree, while was decrease in other degrees from 2000 to 2010. Vegetation cover change induced by GGP was reduce soil erosion, but due to rainfall erosivity increased, soil erosion showed increasing trend in northern Shaanxi and the area of slope >25°.

This paper has many limitations. One of the limitations is soil erosion model with not detailed analysis in the process of research, but it is an important element. There are many methods and models to calculate rainfall erosivity, soil erodibility, topographical factor, vegetation cover factor and erosion control practice factor. Different choice of methods and models would produce different result in soil erosion. Moreover, the paper only estimated macro-scale soil erosion influenced by GGP in northern Shaanxi, but the results lack of validation by hydrological sediment monitoring data. So soil erosion model and results validation need to further be improved.

Acknowledgements

This work was supported by the National Natural Science Foundation of China (41271103), Cultivation Foundation of Xi'an University of Science and Technology (2014007), Provincial Natural Science Foundation of Science and Technology Department in Shaanxi (2015JQ4110), and Provincial Science Foundation of Department of Education in Shaanxi (14JK1479).

References

[1] State Forestry Administration of PRC, "2011 Statistical Yearbook of National Forestry Analysis Report." China Forestry Network, http://www.forestry.gov.cn. 2011.

[2] M. Persson, J. Moberg, M. Ostwald, J. Xu, "The Chinese Grain for Green programme: assessing the carbon sequestered via land reform." Journal of Environment Management Vol. 126, pp. 142-146, 2013.

[3] H. Zhou, A. Van Rompaey, J. Wang, "Detecting the impact of the "Grain for Green" program on the mean annual vegetation cover in the Shaanxi province, China using SPOT-VGT NDVI data." Land Use Policy Vol. 26, pp. 954-960, 2009.

[4] J. J. Zhang, F. J. Chen, J. Q. Bai, X. P. Zhang, Y. N. Lei, E. J. Liu, "Evolution analysis of cultivated land structure characteristics on gully area of loess plateau in 1983-2009." Transactions of the CSAE Vol. 28(16), pp. 232-239, 2012.

[5] R. Chen, C. Ye, Y. Cai, X. Xing, Q. Chen, "The impact of rural out-migration on land use transition in China: past, present and trend." Land Use Policy Vol. 40, pp. 101-110, 2014.

[6] D. Li, J. Zhuo, Z. Sun, "Monitoring the effects of ecosystem restructuring project after returning farmland to forest based on RS and GIS." Transactions of the CSAE Vol. 24(12), pp. 120-126, 2008.

[7] Z. Li, B. Li, "Spatial and temporal changes in grain production before and after implementation of Grain for Green project in Loess Plateau region." Transactions of the CSAE Vol. 28(11), pp. 1-8, 2012.

[8] S. Cao, C. Xu, L. Chen, X. Wang, "Attitudes of farmers in China's northern Shaanxi province towards the land-use changes required under the Grain for Green Project, and implication for the project's success." Land Use Policy Vol.26, pp. 1182-1194, 2009.

[9] L. Chen, Z. Huang, J. Gong, B. Fu, Y. Huang, "The effect of land cover/vegetation on soil water dynamic in the hilly area of the loess plateau, China." Catena Vol. 70, pp. 200-208, 2007.

[10] P. Zhou, O. Luukkanen, T. Tookola, J. Nieminen, "Effect of vegetation cover on soil erosion in a mountainous watershed." Catena Vol. 75, pp. 319-325, 2008.

[11] Z. M. Wen, B. G. Less, F. Jiao, W. N. Lei, H. J. Shi, "Stratified vegetation cover index: a new way to assess vegetation impact on soil erosion." Catena Vol. 83, pp. 87-93, 2010.

[12] Y. N. Zeng, W. P. Jin, H. M. Li, L. X. Tan, "Analysis and evaluation of cultivated land decrease in eastern part of Qinghai Plateau." Transactions of the CSAE Vol. 29(21), pp. 214-222, 2013.

[13] Shaanxi Bureau of Statistics, NBS Survey Office in Shaanxi, "Shaanxi statistical yearbook." China Statistics Press, Beijing. 2000-2009.

[14] D. Yoder, J. Lown, "The future of RUSLE: inside new revised universal soil loss equation." Journal of Soil and Water Conservation Vol. 50(5), pp. 484-489, 1995.

[15] W. B. Zhang, Y. Xie, B. Y. Liu, "Rainfall erosivity estimation using daily rainfall amounts." Scientia Geographica Sinica Vol. 22(6), pp. 705-711, 2002.

[16] W. B. Zhang, J. S. Fu, "Rainfall erosivity estimation under different rainfall amount." Resources Science Vol. 25(1), pp. 35-41, 2003.

[17] Y. Xie, B. Y. Liu, W. B. Zhang, "Sutdy on standard of erosion rainfall." Journal of Soil and Water Conservation Vol. 14(4), pp. 6-11, 2000.

[18] K. Auerswald, P Fiener, W. Martin, D. Elhaus, "Use and misuse of the k factor equation in soil erosion modelling: an alternative equation for determining USLE nomograph soil erodibility values." Catena Vol. 118, pp. 220-225, 2014.

[19] L. S. Zhao, X. L. Liang, F. Q. Wu, "Soil surface roughness change and its effect on runoff and erosion on the loess plateau of China." Journal of Arid Land Vol. 6(4), pp. 400-409, 2014.

[20] United States Department of Agriculture, "EPIC-erosion/productivity impact calculator: 1. Model Documentation." Technical Bulletin Number 1768. Washington, D C: USDA-ARS, 1990.

[21] W. H. Wischmeier, D. D. Smith, "Predicting rainfall erosion losses: a guide to conservation planning with universal soil loss equation (USLE)." Agriculture Handbook 537. Washington, D C: USDA-ARS, 1978.

[22] D. K. McCool, L. C. Brown, G. R. Foster, L. D. Meyer, "Revised slope steepness factor for the universal soil loss equation." Transaction of the ASAE Vol. 30(5), pp. 1387-1396, 1987.

[23] B. Liu, M. A. Nearing, L. M. Risse, "Slope gradient effects on soil loss for steep slopes." Transactions of the ASAE Vol. 37(6), pp.1835-1840, 1994.

[24] S. M. DeJong, "Applications of reflective remote sensing for land degradation studies in a Mediterranean environment." Koninklijk Nederlandse Aardrijksunding Genootschap, 1994.

[25] T. Li, L. Zheng, "Soil erosion changes in the Yanhe watershed from 2001 to 2010 based on RUSLE model." Journal of Natural Resources Vol. 27(7), pp. 1164-1175, 2012.

[26] A. Lufafa, M. M. Tenywa, M. Isabirye, "Prediction of soil erosion in a Lake Victoria basin catchment using GIS based universal soil loss model." Agricultural Systems Vol.76, pp. 883-984, 2003.

[27] The Ministry of Water Resources of PRC (MWR of PRC), "SL190-2007 Standards of classification of soil erosion." Beijing: China Water & Power Press, 2007.

Analysing the Effects of Physical Conditions of the Workplace on Employee's Productivity

Mohammad Taghipour[1, *], Maryam Mahboobi[2], Arash Nikoeifar[2], Ebrahim Soofi Mowloodi[2]

[1]Industrial Engineering, Science & Research Branch of Islamic Azad University, Tehran, Iran
[2]Industrial Engineering, non-profit institution of higher education, Aba, Abyek Qazvin, Iran

Email address:
mohamad.taghipour@srbiau.ac.ir (M. Taghipour), mahboobi64@gmail.com (M. Mahboobi), arash2076@gmail.com (A. Nikoeifar), ebrahimmowloodi@gmail.com (E. S. Mowloodi)

Abstract: One of the issues that today will improve the productivity of any organization is attention to the human factors engineering. Experience in worldwide has shown that modifying of work conditions, improving the quality and quantity of production, loss of illnesses due to the work, cost reduction and thereby enhancing the productivity, all are the positive consequences of suitable usage of Ergonomics in organizations which may be feasible under the shadow of following correction system of work condition, paying more attention to the workplaces and applying macro-ergonomics approach. This study was carried out to analyze the environmental and organizational condition of workers in the assembly unit of an automotive company with ergonomics optimization approach. In fact, the aim of this study was to find the amount of employee's satisfaction from environmental and organizational factors of their jobs, thus providing guidelines for improving the identified problems which eventually will lead to increase the productivity. To do this descriptive research, after providing a questionnaire which it is valid and approved final, 50 out of 200 workers of the second shift of assembly unit have been randomly selected and were asked about the condition of the workplace and also disorders while working. After collecting and observing responses, frequency was obtained using diagram. Meanwhile, using mask had the least attention and environment noise had the most frequently in disorder categories. In addition, workers were complaining about the shortage of labor force, high speed of production lines and lack of break time.

Keywords: Physical Condition, Micro and Macro Ergonomics, Productivity, Workplace Condition

1. Introduction

Today, ergonomics is one of the least known applicable knowledge in country's industrial and services community, while ergonomics paves the way for many management and engineering policies and systems. Of course, modern ergonomics is pretty vaster and developed compared with the past and is still under growth and pervasive in various fields of applied sciences for organizations' management.

Generally, purpose to applying ergonomic knowledge is to reach appropriate and logic relation between employees and environment, machine, job as well as organization. This is a condition in which production and productivity are desirable. Ergonomics can investigate and research employees' psychology and physiology in workplace which is a complicated system between human, machine and environment [1]. Experiences in the industrial counties show that macro-ergonomics is an important and effective factor to initiate suitable workplace. By applying it, more incentive is built in people to participate more and by increasing satisfaction; organization's productivity would be increased [2,3]. Regarding macro cases of a system such as management and organizational designing is not a new issue for ergonomists becauseexactly in Europe, there is a belief to resolve ergonomic problems holistically. Macro-ergonomics creates this belief using specific tools and techniques which are resulting in organizational massive scales [4].

Ergonomics plays a key role in identifying causes of incidents and productivity reduction and job quality reduction[5]. Human inadaptability with technology and organization can have negative impact not only on system outcome and productivity, but also on staffs' motivation, commitment and also job satisfaction. Experience in the world class has shown positive effect of applying ergonomics in different sectors for modifying job condition, production qualitative and quantitative enhancements, decreasing prevalence of musculoskeletal injuries derived from job, cost reduction and thereby increase in productivity and this issue is under the shadow of following appropriate correction system of working conditions and applying macro-ergonomics approach[6].

Helali articulates that scholars address importance of applying ergonomics in developing industrial countries for years as an attempt to modify job condition and productivity. Therefore, there is still lack of management awareness about ergonomics in industries of developing industrial countries [2,3]. Hendrick, H and Kleiner (2007) explain that macro-ergonomics based designing resolves many of designing principles of micro-ergonomics, because by applying macro-ergonomics, various principles of micro-ergonomics are spontaneously considered. Therefore, optimal ergonomics consistency between system components with overall system structure, reliability is obtained [8]. In socio-technical system literature, macro-ergonomics provides common optimization of technical subcategories and employees from up to down in the organization. Applying it will cause more guarantee of system optimal effectiveness and efficiency which includes worker's productivity, safety, comfort, inherent incentive and work life quality [7].

In the recent decades, especially about automobile, increase in customer demand has led to accelerate production lines of this product in the country. On the other hand, by intensification of production trend and due to the lack of reform program and system of working condition, health, safety and ergonomic problems in the workplaces especially in developing countries have increased. This issue will cause quality drop and ultimately productivity. This matter is one of the barriers of competition with other firms. Nevertheless, emphasize on job condition modifying in a few number of these firms is mostly emphasize on micro-economics interferences. Results obtained from this approach have been a few and inadequate. Thus, it is required to macro-economic approach is being employed [6]. Today, ergonomic experts have a high trend to apply strong macro-methodology for organization management. By these disorders, apparently, macro-ergonomics is more efficient than traditional ergonomics (which has the direct relation with organising job area).

Current study was carried out to analyze the environmental condition and the reform strategies of working conditions and increasing work life quality and thereby productivity increase in the assembly unit of an automobile company with macro-ergonomics approach. In fact, this study is using macro-ergonomic analysis method to identify workers' ergonomic demands using their participation throughout the process. Proposed macro-ergonomic study in some of the details expresses that how different aspects of micro-ergonomics, Total Quality Management and socio-technical systems can form a triangle with a common approach. In the framework of this method, quality and safety have been introduced as two important performance cases.

To conduct this analytical-descriptive study, after providing questionnaire, among three shifts of mornings, afternoon and night of assembly unit of automobile factory, afternoon shift has been chosen and among workers which are almost 200 persons, 50 persons were asked questions about workplace condition and disorders while working. Some of the macro-ergonomic problems were also proposed by workers in questionnaires. After collecting and checking responses, frequency of the results was obtained using diagram. Also, recommendedstrategies to modify were proposed.

2. Ergonomics

Emerging ergonomics dates back to the industrial revolution (at the end of 19th century and beginning of 20th century). Attempts which were simultaneously started during the Second World Wide in the USA and Britain caused ergonomic knowledge to start growth and development and in the most of European countries have been regarded. Unlike other sciences, official date of birth of ergonomic science dates back to 12th of January, 1949 in which for the first time in Britain, a group of scientists and researchers were interested in ergonomic issue and recognizing this new branch of science, but ergonomics was known as a science from the early 1950s by forming Ergonomic Research Community in Britain [9]. In 1980s, unfortunately, many technological disasters were happened and it caused people to get familiar with ergonomic knowledge more.Plant builders, manufacturers, workers, politicians, and consequently the public have turned to ergonomic designing [10]. Ergonomics has been formed from two parts: Ergo that means job and Nomos that means rules and principles and it is principles of human relations with the job. In one word, ergonomics is a science of optimal applying of tools and people in workplace so that the maximum efficiency would be obtained [11]. Underlying aims of ergonomic science are enhancing how to work, job mythology, job tools and their adaptability with human mental and physical features. Also, ergonomics trends to scientific conformity between job, condition, tool and work environment with human mental and physical features so that workplace and job are designed to be matched with physical features of people average (by considering respective standard deviation) [12]. Since appearing ergonomic knowledge, various definitions of ergonomics have been proposed and these definitions have been gradually got more optimal, accurate and comprehensible. One of them expresses that "Ergonomics is a study between human and job, equipment, environment and specially applying physiologic and psychological knowledge to solve problems". Brazilian Association of Ergonomics also believes that "Ergonomics is a study of human interaction in technology, organization and environment with the aim ofintegrated improvement of safety, comfort, welfare and productivity of human activity [9]. Also, International Ergonomic Association knows ergonomic as a scientific field which approaches to study human interactions with different elements of system and applying principles, data and techniques in the projects which are aimed to enhance human welfare and optimize systems' performance [9].

All of definitions have addressed more quality of employees' job, physical dimension, cognitive and organizational features and ergonomist (a person who is responsible for organization ergonomic principles) is attempting to, by minimizing the defective factors, provide more safety and satisfaction and thereby helping organization with enhancing productivity, but generally, in the primary definitions of experts in this filed, ergonomics is more concentrated on human physical factors. Then mental factors were added to this. By mentioning this point that these physical and mental features were used for a human in the past and are remembered as micro-ergonomics, therefore micro-ergonomics is called traditional ergonomics. But, improvements occurred in ergonomics caused traditional definitions were replaced by the modern definitions. In this new definition, "System" term was used as anything which is not human. Regardless of accurate definition proposed by International Ergonomics Association (IEA), modern ergonomics can be defined as the following: Human- System Interface Technology and optimizing human relation with the system. System sector in this definition also is divided into 4 subsystems which are: machine, environment, software, job and organization in which human relation with each of these subsystems is known Hardware-Ergonomics,Environment-Ergonomics, Cognitive-Ergonomics, and Work Design-Ergonomics, respectively. The four first subsystems are known as micro-ergonomics. Accordingly, ergonomics can be regarded in 2 parts: 1) micro-ergonomics which is concentrating on one human 2) macro-ergonomics which is concentrating on the organization (more than one human). In fact, micro-ergonomics considers physical designing of job environment and job tool. Macro-ergonomics wants to optimize organizational and job system designing via considering personnel, technologic and environmental variables and interactions between them, in terms of history, micro-ergonomics is a classic ergonomics which has 50 years history, while macro-ergonomics is a modern ergonomics and has one decade history. In fact, ergonomic objectives were developed since 1980s. This developed version is called macro-ergonomics which is technology application and development in human-machine relation (socio-technical system) in a macro class or in another hand in all over the organization. By this description, ergonomics is exited from micro area and it is more powerful to coordinate hundreds and maybe thousands of workers with systems, equipment and machineries. The main purpose of ergonomics is to evaluate and promote productivity through a specific method. Four primary subsystems include environmental, technological, organizational and individual variables which are interfering systematic inactions between people and job [13]. As it was mentioned macro-ergonomic based designing also meets many of micro-ergonomic needs and guarantees system as a result of ergonomic adaptability of system components with system overall structure. Macro-ergonomics is conceptually defined as a top-down socio-technical approach of systems for designing organizational structures of work and jobs related to human-machine, human-environment and user-system.Qualitymanagement and ergonomics were appropriately emerged during history after socio-technical systems theory which is a basis for integrated method required by organizations [13].

3. Productivity

In the literatures of scientific fields related to the organization such as management, economy and sociology, the best functional form of an organization has been defined synergy which has been interpreted as a high productivity. Today, productivity in organizations has been regarded by the majority of managers. Productivity can be interpreted as the sum of effectiveness and efficiency. In another document, productivity is to achieve the maximum possible profit using and taking advantage of job force, ability, human resource skill and talent in order to promote society's welfare. It can be articulated that the word "Productivity" was firstly developed by a person namedQuizni in 1776. After one century, that is 1833, a person named Litr defined productivity as a power and ability to produce.

Productivity has been considered equivalent to indices such as efficiency, effectiveness, performance, beneficiary, innovation, quality, culture, work life quality and so on. Due to this issue that in each of different scientific fields, special definitions and inferences of productivity have been proposed, in managers' viewpoint also, productivity due to roles, their knowledge and skills, coworkers, subordinates and condition in which they are working is different with others. In sum, it can be said that the most comprehensible definitions proposed about productivity belong to the managers. In mangers' viewpoint, productivity is including all of organizational aspects which are important in its function. In this viewpoint, productivity has been considered equivalent to efficiency, effectiveness, performance, products' quality and stakeholders' satisfaction and so on .

World modern science has solved majority parts of individual problems in the various job systems. In this regard, there are many scientific methods which are analyzing different angles of human health and hygiene and also their efficiency as a part of way to reachproductivity. One of them is ergonomics science. Also, less activity can be found that is free from any kind of threating factor. Accordingly, protection and preservation of human resources is arisen as an imperative mission. Paying more attention to human factors in each organization is undoubtedly one of the most important principles and criteria in the organization, because organizations without using human resource are in fact unreliable. Human work is the outcome of various and diverse variables. Sometimes, work is derived from human variables such as character, perception and attitudes and sometimes, it is influenced by variables which have physical nature. These variables are called work environment. Today, it was verified that condition of work environment such as noise and health all are affecting on individual performance.Marilyn Joyce, president of the Joyce Foundation in Seattle, America, believes that now this belief has been generalized that firms which are considering quality control and productivity, are using ergonomics as a commercialtalent. Successful firms have combined ergonomic programs with safety, quality control and production plans to reach the maximum profit [14].

4. Data Collection

Studied sector is assembly unit of automobile factory. This sector was chosen because of productions' qualitative and productivity drop and also inappropriatework condition of workers in this unit and complaint of inconsistencies between work, work environment and workers' job. Since, in assessments done by ergonomists, workers' participation seems pretty necessary to accelerate finding problems, applying appropriate changes and enhancing work condition. But, survey from workers has been considered as an assessment basis. Approximately, 630 male workers aged between 28-44 years old are working in three shifts in the assembly unit.Work shifts are from 7:30 to 15:30 and 15:30 to 23:30 and 23:30 to 7: 30. In each shift, around 210 workers are working. To survey workers, a questionnaire including questions about rate of workers satisfaction from workplace condition and barriers which causelack of progress in their work and thereby decreasing productivity was provided. Then, among the second shift workers, 50 persons were randomly chosen and they were asked questions about their work environment condition. Also, at the end of questionnaire, they were asked to mention other cases which haven't been brought in questions and cause interruption in their job. Due to this issue that our case study includes one fourth of workers of a work shift, therefore frequency diagram of outcomes has been drawn by Excel software.

5. Results

Results obtained from survey have been brought in table 1:

Table 1. *Data obtained from 50 compliment questionnaires of workers (Number of agreed workers with each item)*

Questions Response (Number of agreed)	High	Average	Low
Workplace space	3	36	11
Space allocated to your work	2	18	30
Noise rate in work environment	32	12	6
Environment temperature	0	34	16
Machineries' safety	2	32	16
Using mask while working	0	11	39
Surface inequality	28	16	6
Health and sanitation at work environment	5	19	26
Drinking water health	0	25	25
Work environment pollution (soldering smoke etc.)	28	17	5

Of course in this regard, in addition to the aforementioned cases, workers have mentioned other cases causing idsiorders in their work which are being investigated, but in sum, results showed that in spite of growth, development of assembly unit, environment and job condition provided for workers, positive opinion is not proposed and they are doing their duties in a condition which is far from ideal.

6. Discussion

Some of the environmental, organizational, physical and equipment factors which cause disorders in work trend and eventually decreasing quality and productivity will be evaluated using data in table 1:

6.1. Environmental Factors

6.1.1. Noise

Job performance can be harmed by the noise, it can be articulated that work environment noise is a very dangerous attacker, because it destroys a number of auditory cells every day. When a worker or employee finds an auditory loss, it is very late, because hardness of hearing or deafness caused by noise is irreversible. Damage derived from noise is not just limited to the ears, but it surrounds all over the body. Noise has its direct or indirect effect through bothersome or job dissatisfaction [15]. Although, many of primary studies showed that industrial workers can do their duties without effects derived from noise, but the recent studies have shown another results. For example, Bragdon showed that noise especially when is unexpected and unfamiliar, severely influences job performance [16]. Furthermore, according to the reports done by this researcher, noise above 90 dB causes increasing stress, fatigue and increasing rate of errors by physiological stimulus of harmful organism responses and quality and productivity are reducing. An unfamiliar and sever sound also will cause more anxiety and interruption in doing duties and even will cause disorder in verbal communications. Environment which has noise pollution creates reactions which all indicate jeopardizing physical health such as severe headaches, stomach discomfort, unusual fatigue, lowered resistance against damage, extreme vulnerability against the Cardiovascular disease. Work environment noise is also affecting mental health including impatience, irritability, stress, lack of concentration and difficulty. In work environment, noise pollution can threat workers and employees safety seriously. Also, high noise decreases concentration and obviously reduces job quality.

As it is clear in table 1, 32 persons out of 50 persons who are participating in the survey had consensus about high level of work environment noise and only 12 percentage of workers have commented about low level of environment noise. Therefore, the frequency of environment noise by 64% votes has been received the maximum votes. Diagram 1 shows the results obtained from this environmental factor. It is transparent that being more than 2 hours exposed by even a moderate amount of noise will cause work stoppage, tension, headache and ultimately feedback drop.

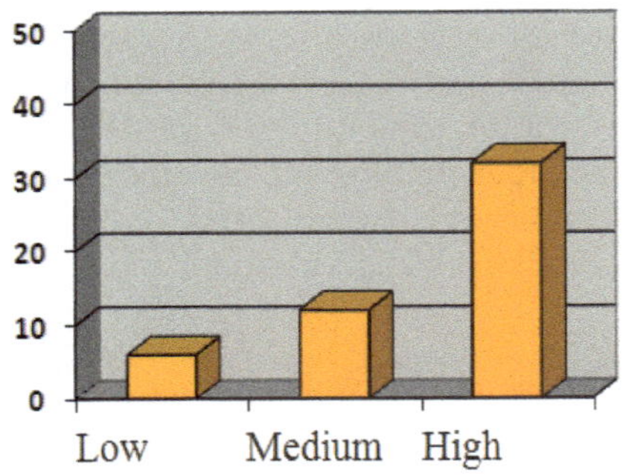

Diagram 1. *Results obtained from assessment of environmental factor, noise*

Now, if this noise was relevant to the current resources in industrial unit such as source of electricity or other equipment, it can be articulated that this noise can be considerably decreased by proposing strategies. Some of strategies can be referred like replacing these devices with other similar low noise equipment and also more update systems. Another method is to isolate these kinds of resources. To do this, by putting devices in the cabin or a closed container with pores for device ventilation, we can prevent from noise dissemination. Also, to prevent from noise derived from devices' vibration while working, we can fix device in its place using suitable flooring. We can change workers' work shift by performing an especial plan in order to travel between low and high noise places. We can also change workers' work environment by proposing overall plan in the factory and transfer them to the low noise place which is far from devices, but in a sense that none of aforementioned cases is performable, worker has to use ear protection. However, workers should be protected against noise in the best way. Accordingly, negative effects caused by noise on employees performances and also human errors will be decreased. It would be led to increase worker accuracy and thereby high quality productions and also increase in workers' and employers' satisfaction.

6.1.2. Temperature and Ventilation

Appropriate temperature condition for the effective work environment is usually 20-24 centigrade degree with 40-80 percentage relative humidity. So in this condition, temperature changes should not exceed 4 ° C [9]. Also, ventilation devices are pretty necessary especially for industrial units and they cause airflow exchange. To rebuild breathing air, each personrequires 12 to 15cubic meters of air per hour.In intensive work condition, this is double increasing[9].

In this industrial unit, among 50 comments, 34 workers (68 percent) have confessed the balance of their work environment temperature and 16 persons believe that work environment temperature is low and none of workers participated in the survey haven't believed that environment temperature is high. Of course, this issue is a little related to the season (autumn) in which survey has been done. Diagram 2 exhibits results of this assessment. To keep air quality in indoor environments; ventilation should be done at least equal to the volume of air per hour.

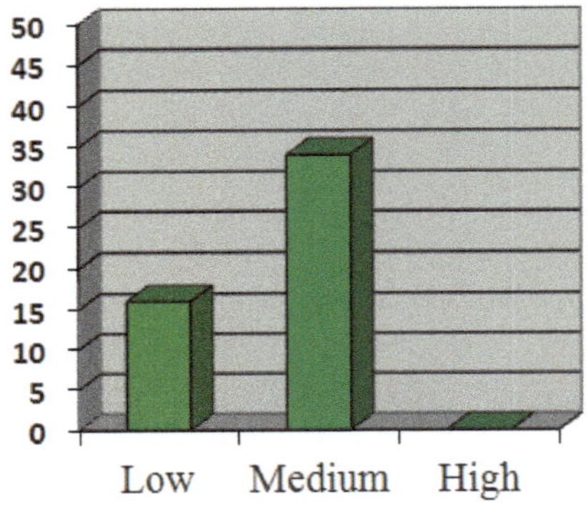

Diagram 2. *Results obtained from assessment of environmental factor, temperature*

But workplace temperature must be balanced, because very high temperature causes fatigue and drowsiness, physical decrease in performance and increase in danger of afflicting to the mistakes [17] and very low temperature will be job interruption. Work environment with suitable temperature and ventilation causes receiving better oxygen supply by workers and this enjoyable work environment will have positive effect on their performance and also organization's productivity.

6.1.3. Work Environment and Drinking Water Health

Environmental health is to control all of factors which have adverse impact on sustainability of human health. Environmental health manages all of individual physical and chemical aspects, apparent biological factors and behavioural effective factors. This matter includes assessing and controlling environmental factors which are potentially effective on health. Environmental health has targeted prevention from diseases and supporting healthy environments. In fact, the main goal of environmental health is to control all of factors which are potentially and practically imposing adverse effects on human survival and health. To reach this goal, it necessitates taking an advantage of environmental knowledge and engineering principles to control; amend and improve environmental physical, chemical and biological factors in order to retain and promote human health and welfare.

In the context of work environment and drinking water health, we have encountered interesting responses. As it is seen in table 1, 50 percent of respondents have referred to average drinking water health and half of them have referred to the lack of drinking water health. Diagram 3 exhibits resultsobtained from assessment of these organizational factors. Also, in the work environment health, 52 percent of workers have commented the lack of health and only 14 percent have considered work environment health and sanitation high.

We can assign programs instead of weekly cleaning;20 minutes are given to each worker before finishing his work. Applying this policycausesfurthermaintaining health and sanitation in the workplace and enhancement in work space will cause to increase workers' satisfaction, incentive and efficiency.

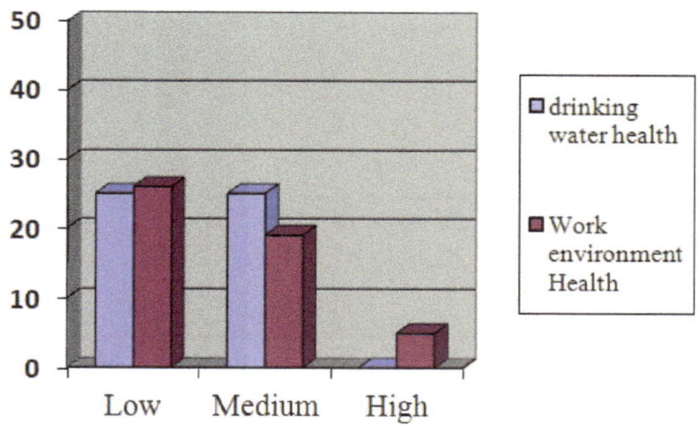

Diagram 3. *Results obtained from assessment of organizational factors, work environment and drinking water health*

6.1.4. Work Environment Pollution

Work environment pollution is one of the factors which have an impressive impact on work process. The less the pollution is, the more tangible the progress in productivity of their activities is. It is inferred from survey that 56 percent of workers have considered work environment pollution high and only 10 percent of them have considered this pollution low. Diagram 4 shows results obtained from survey about this organizational factor.

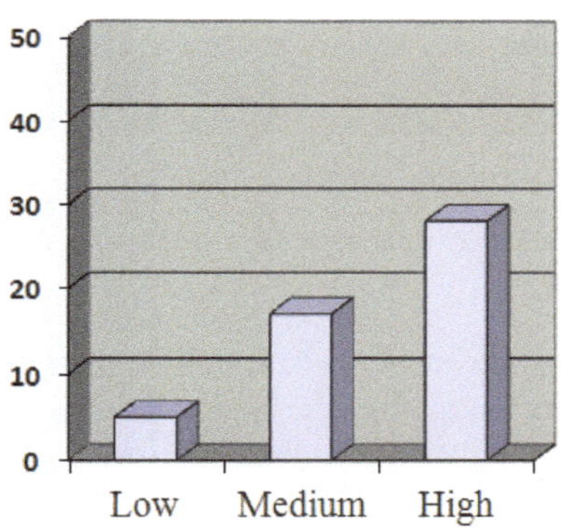

Diagram 4. *Results obtained from assessment of organizational factors, work environment pollution*

It is natural that in such polluted environment, workers are losing their incentives and their concentration and efficiency will be decreased and this would be along withphysical problems for the workers and organization doesn't gain respective feedback. To resolve this problem, we can use some techniques such as using mask, air purifiers or in a whole, changes in the workers' work environment. We will take this issue into the consideration in the work environment sector. Due to the normalization of the data, pearson parametric test can be used for data analysis. Pearson correlation test results showed in table 2.

As Table2 shows, according to Pearson correlation coefficient and significance level was between productivity and environmental factors, there is a significant positive correlation. The highest level is allocated to work environment and drinking water health.

The results of Table 3 shows that the coefficient of determination adjusted multiple linear combination of environmental factors and the productivity is equal to 0.326, which means that environmental factors account for 32 percent of the variance in productivity.

Table 2. *Pearson correlation test results between environmental factors and total productivity*

Variables	correlation coefficient	significant level
Noise	0.49	0.01
Temperature and ventilation	0.54	0.01
Work environment and drinking water health	0.63	0.01
Work environment pollution	0.44	0.01

Table 3. *results of multiple regression analysis of environmental factors and productivity*

Variables	Coefficient of Determination	Adjusted Coefficient of Determination	significant
Environmen tal factors	0.329	0.326	0.01

6.2. Equipment

6.2.1. Machineries Safety

Fundamentally, to develop projects' safety, prevention policy before action should be used. So that workers are allowed to use equipment only after preventive safety actions. In industrial units, due to this issue that there is a possibility of equipment errors because of high function, burnout and stuff like that, the possibility of dangerous incidences is not unexpected.These events will be led to the closure of all or part of the work.

As it was inferred from obtained data, 64 percent of participants in the survey have emphasised in medium machineries' safety and 16 persons have commented the low equipment safety. Diagram 5 shows results obtained from this assessment.

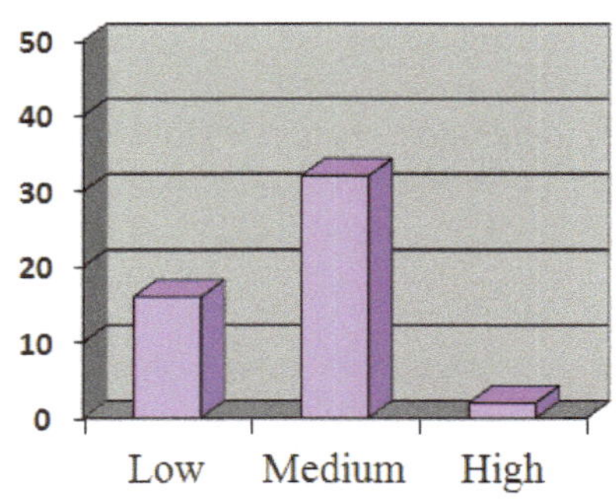

Diagram 5. *Results obtained from assessment of equipment factor, Machineries safety*

The simplest way to protect work with machineries is to remote human operator from dangerous equipment by putting a physical container between them, but if operator is directly connected to this equipment, it is better to separate all of dangerous parts of machine or at least parts which have direct contact with the user or they are protected well by establishment of protective devices in order not to provide contexts for human damages. Also, if unpredicted events are happened, device has to be ready for the sudden stoppage. This issue can be carried out through applying steel cables and leverages in order to reduce financial and lateral damages caused by incidents occurrences, but the best way to obtain incidents is to use Personal Protective Equipment (PPE). These kinds of safety instruments include gloves, hat, boots and glasses. However, periodic control of devices' operations should be considered as the first priority of each industrial unit, because it prevents from errors and delays in project implementation process.

6.2.2. Using Mask

Using Personal Protective Equipment (PPE) is generally bothersome. Mask also is one of these kinds of equipment which might harden breathing and cause performance disorder. Furthermore, most of filters will be inefficient against many of gases. As it is observed in survey data, 78 percent of workers haven't used mask and about 11 percent have used it a few. Diagram 6 displays results obtained from survey about this factor.

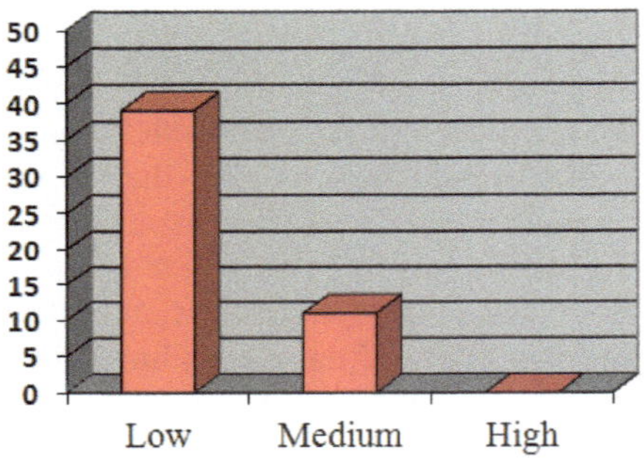

Diagram 6. *Results obtained from assessment of equipment factor, using mask*

So, Personal Protective Equipment (PPE) always must be employed as the last parapet to defend workers and only in the cases that other resources are inefficient and impractical. Nevertheless, while eliminating or remote worker from dangerous work environment is not possible, matter of using Personal Protective Equipment should be more addressed. Due to the normalization of the data, pearson parametric test can be used for data analysis. Pearson correlation test results showed in table 4.

Table 4. *Pearson correlation test results between equipment factors and total productivity*

Variables	correlation coefficient	significant level
Machineries safety	0.44	0.01
Using mask	0.57	0.01

As Table 4 shows, according to Pearson correlation coefficient and significance level was between productivity and equipment factors, there is a significant positive correlation. The highest level of is allocated to the use of a mask.

The results of Table 5 shows that the coefficient of determination adjusted multiple linear combination of factors, equipment productivity is equal to 0.354, which means that factors in equipment express 35% of the variance in productivity.

Table 5. *results of multiple regression analysis of equipment factors and productivity*

Variables	Coefficient of Determination	Adjusted Coefficient of Determination	significant
Equipment factors	0.366	0.354	0.01

6.3. Physical Factors

6.3.1. Surface Inequality

As data show, 56 percent of workers announced that surface inequality is high which this issue will be followed by waste of worker's useful time, fatigue and performance drop. Diagram 7 displays results obtained from assessment of workers' comments about this physical factor.

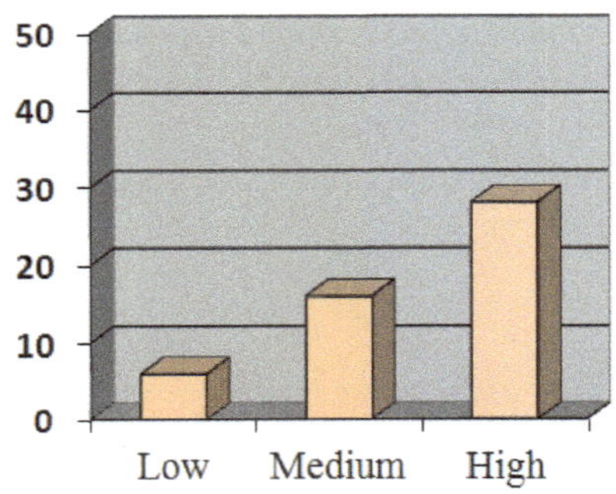

Diagram 7. *Results obtained from assessment of physical factor, surface inequality*

Enhancements in physical space should be carried out like amending surface inequality so thatparts' transport and especially commuting workers will be safer. This work causes preventing from job accidents and also provides a context for more health in environment.

6.3.2. Overall Space of Work Environment and Space Allocated to the Worker

By complicating organizations' structure and performance, organizing work environment either in terms of environmental or in terms of dimensions has to be done so that it would lead to the greater activation of human resources, increasing job incentive, workers' freshness, omitting probable dangers, increasing quality in the work and goods, decreasing depressions, growing productions, growing positive services and eventually achieving respective productivity. Work environment in organizations is one of the critical issues which is continuously proposed. Also, importance and how to create suitable and efficient environment for human resources arealways addressed. A proper environment can be effective on growing personnel values, increasing their ability and productivity. That is why, human management science or ergonomics is so critical for organization mangers and leaders [18].

According to the conducted survey, overall space allocated to the assembly unit of this automobile factory is average and low, so that 72 percent of workers (36 persons) have mentioned the average work space and 11 persons have emphasized on the lack of space, so that space allocated to the worker has the reverse results. Accordingly, 30 persons out of studied 50 persons, that is, 60 percent of workers have emphasized on the lack of space and 16 persons have emphasized on average space. Diagram 8 shows that assessment of results obtained from this survey.

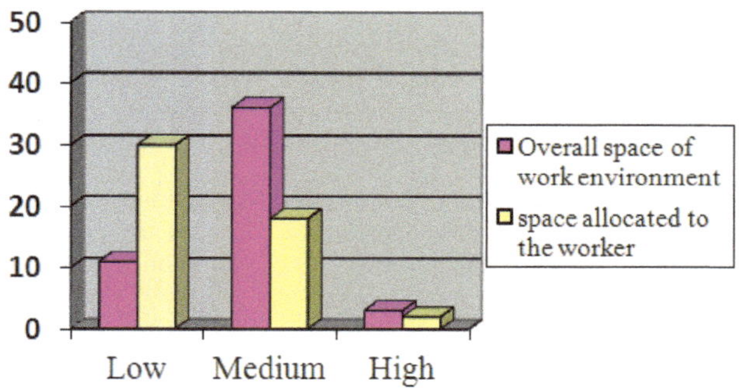

Diagram 8. *Results obtained from assessment of physical factor, work space*

Psychological researches and investigations indicate that physical condition of work environment has a great effect on mental health of organizations' employees. Appropriate lightening, type of windows, built green space, type of space layout, music, health as well as space color have a considerable effect on personnel performance. Since, employees are spending most of their work time in the work environment, so effect of internal factors on employees' performance is very high. That is why; type of layout and details of internal space of work environment based on type of organization activity and strategy, employees' viewpoint and organizational culture will be variable[19]. Psychological researches show that if physical condition was appropriate in the work environment, it would cause progress productivity and qualitative promotion of employees' work. Due to the normalization of the data, pearson parametric test can be used for data analysis. Pearson correlation test results showed in table 6.

Table 6. *Pearson correlation test results between physical factors and total productivity*

Variables	correlation coefficient	significant level
Surface inequality	0.42	0.01
Overall space of work environment and space allocated to the worker	0.68	0.01

As Table 6 shows, according to Pearson correlation coefficient and significance level was between productivity and physical factors, there is a significant positive correlation. The highest amount is allocated to Overall space of work environment and space allocated to the worker.

The results of Table 7 shows that the coefficient of determination adjusted multiple linear combination of physical factors with productivity equal to 0.452, which means that physical factors express 45% of the variance in productivity.

Table 7. *results of multiple regression analysis of physical factors and productivity*

Variables	Coefficient of Determination	Adjusted Coefficient of Determination	significant
Physical factors	0.434	0.452	0.01

7. Role of Other Factors

As it was previously mentioned, in the questionnaire, workers had been asked to mention other cases which cause disorder in their work process. We have obtained interesting statistics. Almost, most of them had complained about production line speed, shortage of people in production line, workers' stress, long work time and thereby short work time, erratic work process, lack of adequate time to do personal affairs which all are accounted as macro-ergonomic factors of work environment. These factors will cause to reduce quality of productions and by customer dissatisfaction; system's productivity is exposed by danger. They can be appropriate indices for the next macro-ergonomic studies.

8. Conclusion

Studying work environment in the organizations is one of the critical issues which is continuously proposed in the meetings and papers. Importance and how an efficient and proper work environment in terms of physical and mental aspects for human forces is created, is addressed. It is transparent that a good work environment, either in terms of physical aspect, by providing facilities to decrease human pollutants, continuous attempt to keep work environment safer and decreasing context for work accidents and initiating adornment culture in work environment or in terms of mental aspect, by creating credit and respect and considering justice, all cause emerging honour sense about the work and establishing friendship between employees. Sum of these factors can affect growing personnel values and increasing their ability and productivity. That is why, human management science or ergonomics for consistency between employees with job and work environment has the high importance. Since, factors affecting work are a lot and these factors are directly or indirectly affecting individual health and work, but to have real efficient ergonomics and to meet upcoming decades' needs, it is necessary to involve organizational and managerial factors in the ergonomic studies and programs. Macro-ergonomics is a human-oriented approach for overall fitness of organization designing and the system which is working with human. In fact, macro-ergonomics determines interactions and fitness between each of system's components by analyzing the whole system. The greater assurance of system optimal function and its effectiveness including productivity, safety, welfare, and incentive and work life quality can be known as the advantages of applying macro-ergonomic principles in the organizations.

Since, in each industry and production, worker plays an important and determining role in production, quality, and productivity and eventually organization place among competitors, our researches have been carried out in the context of environmental, organizational, physical and equipment factors by collecting data from questionnaires of workers in assembly unit of an Iranian automobile company. This dissertation showed that in the context of environmental factors, low temperature of work environment and high noise of work environment will cause workers' tension and dissatisfaction, while decreasing negative effects derived from noise on employees performance will cause decreasing human errors which are led to increase workers' accuracy and thereby productions with the high quality and also increasing workers' and employers' satisfaction. On the other hand, when activities done by job agent are ongoing, effective temperature also can have the maximum effect on individual job performance. In the context of physical factors, surface inequality in work environment in viewpoint of workers is high which creates some concerns for workers' safety. Also, in relation to the organizational factors, discussions such as drinking water health and work environment health have relatively less assessed. Also, pollution of work environment is high. In safety equipment discussion, job safety was in the moderate level and using mask was too low which asks for an especial attention, because despite high environment pollution, lack of using mask by the workers shows lack of managers' attention to the workers' health and subsequently creating unanticipated problems for the workers and employers.

All of these factors plus factors mentioned by the workers are macro-ergonomic factors. If they are correctly applied in the factory, they will lead to health, safety, workers' incentive and eventually promoting products' quality, customer satisfaction and increasing organization productivity. It has to be articulated that production line speed, shortage of people in production line, workers' stress, long work time and thereby short work time, erratic work process, lack of adequate time to do personal affairs all are the factors which have been proposed by the majority of workers. They can be appropriate indices for the next macro-ergonomic studies.

References

[1] Abarghoei, N. Sadra, HosseiniNasab, Hassan; Farokhzad, M, "Macro-ergonomics interventions and their impact on productivity and reduce musculoskeletal injuries (including case study)", 2012, Quarterly Journal of Iran's occupational Health.

[2] Helali F. "Developing an Ergonomics Intervention Technique Model to Support the Participatory Ergonomics Process for Improving Work Systems Organizations in an Industrially Developing Country and its 'Meta-Reflection'", 2008, Doctoral Thesis, Department of Human Work Sciences, Luleå University of Technology, Sweden.

[3] Helali, F,"The Ergonomics 'Know-how' Transfer Models to IDC's Industries (Concept, Theory, Methodology, Method, Technique)", 2009, U.K and the U.S.A: LAMBERT Academic Publishing AG & Co. KG.

[4] Kleiner, B.M.; "Macroergonomics as a Large Work-System, Applied Ergonomics"; 2005, Vol 14 (2): 81–89.

[5] Shahnavaz, H; "Making ergonomics a world – wide concept. Ergonomics". 1995, Vol 39 (12) 1391- 1402.

[6] Helali F, Lonnroth, E; Shahnavaz H; "Ergonomics intervention in industries of the industrially developing countries, Case study: Glucosan–Iran". In: Brown O Jr, HendrickHW, editors. Proceedings of Human Factors in Organizational Design and Management V. Amsterdam, the Netherlands: North- Holland. 2008, p. 141-6.

[7] HendrickHal.W, and Kleiner, B.M; "Macroergonomics Theory, Methods, and Applications"; book, 2002; Lawrence Erlbaum Associates. Publishers Mahwah, New Jersey London.

[8] Abdollahpour, N, "macro-ergonomics and its impact on productivity". 2011, M.S Thesis, Welfare and Rehabilitation Sciences University - Institute of Social Welfare.

[9] Iida, I. "Ergonomia: projeto e produçۥo". 2005, Vol 2. ed. Sۥo Paulo: EdgardBlücher.

[10] Sandrez, Mark, Mc Cormic, Ernest; "Ergonomics (human factors) in engineering design"; Translate by Mohammad Reza Afzali, Seventh Edition. 1998, Oloom Daneshgahi of Tehran Publication.

[11] Karzar Jedi, R., "The role of ergonomics and its impact on the production, production trends", MS Thesis, Industrial Management Field, Islamic Azad University of Tabriz, 2000.

[12] Forouzanfar, Behzad,"What is ergonomics", Industry and Safety Magazine, 1999, No 46.

[13] Kleiner, B. M.; "Macroergonomic analysis of formalization in a dynamic work system". Applied Ergonomics, Elsevier Science Ltda, 1998, v. 29, n. 4, p. 255-259.

[14] Taheri, Shahnam. N,. "Work and methods evaluation", 1997, Tehran: Arvin Pub.

[15] Abel, S.M., "The extra-auditory effects of noise and annoyance: An overview of research". J. Otolaryngol. 19, Suppl, 1990.

[16] Bragdon, Clifford R. "Noise pollution; the unquiet crisis". Philadelphia, Univ. of Pennsylvania Press [c1971], 1971.

[17] Grandjean, E.; Kroemer, K. H. E. "Manual de ergonomia: adaptando o trabalhoaohomem". 2005, Vol 5. ed. Porto Alegre: Bookman.

[18] Soltani, Iraj; book "productivity of human resources", Isfahan, Arkan Pub, 2005.

[19] Golabchi, M; Yousefi, S., Forouzanfar, M, "prioritizing the role of physical and architectural work in promoting personnel performance in project-based organization",Third International Conference on Construction Industry. 2012, Tehran.

9

Experience with Practical Environmental Education in Primary Schools

Manfred Fehr

Institute of Geography, Federal University, Uberlândia MG, Brazil

Email address:
mfehr@ufu.br

Abstract: This paper relates ongoing research on environmental quality in the Brazilian school universe, with analyses and results from six schools. The objective is to achieve environmental education of students through exposure to the impacts of their school on city infrastructure. In the first two schools, the work plan addressed composting of food residues from the canteen. The work in the third school included measurements of raw waste composition and destinations for sorted waste. In the fourth school, students composted biodegradable residues, sorted inert waste and delivered it to reverse logistics. In the fifth school, the scope included water and energy management. The work with this community resulted in the proposal of an environmental management system. In the sixth school, the students developed score sheets for sustainable behavior, which referred to water and energy consumption, waste production and degree of sorting and noise levels. In conclusion, this research established quantifiers to show the contribution of a school to a sustainable city.

Keywords: Composting in Schools, Environmental Behavior, Environmental Education, Environmental Management Systems, Environmental Protection, Score Sheets for Environmental Performance

1. Introduction

Enterprises of the production and service sectors routinely use environmental performance analyses, but its application in primary schools is uncommon.

Since 2004, the International Organization for Standardization (ISO) operates the Kid's ISO 14000 Program with the declared objectives "to develop environmental awareness among children, to teach them to implement environmental management in homes and communities and to open them to the value of networking with young people in other schools" [18]. By 2009 an estimated 210,000 children worldwide had participated and achieved a 70,000-ton reduction of CO_2 emissions.

Reports from Cambodia relate efforts to involve schoolchildren in the protection of the architectural, historical and cultural site of Angkor from deterioration by tourism [19]. The authors of [22], [21] and [6] describe the importance of environmental education in the school context, but the treatments remain general. No specific procedures for environmental performance analysis are given. Author [25] defends the idea that a school is not an isolated entity. It is part of the social context of the city and as such, cannot remain inert with respect to its interaction with the municipal environment. Exactly this idea forms the basis of the present study. The query is "Does the school community know its impact on city infrastructure and what does it do to minimize it?"

UNESCO declared the decade from 2005 to 2014 the "Decade of Education for Sustainable Development" [32]. This declaration attributes special responsibilities to educational institutions worldwide. The challenge is to stimulate new attitudes towards the environment that can lead the way to sustainable nations. Authors [16] challenge the success of the UNESCO initiative when they argue, "The UN Decade of Education for Sustainable Development failed to acknowledge or challenge neoliberalism as a hegemonic force blocking transitions towards genuine sustainability."

The present study introduces the idea of measuring the environmental performance of school communities and setting voluntary benchmarks. This idea is original in the region where the schools are located.

According to [12], Brazilian primary schools have not yet assimilated the challenge put forth by UNESCO, such that, generally speaking, the corresponding course content has not extrapolated the classrooms. They describe the implementation of water and energy saving initiatives, but admit that this type of procedure is restricted to less than one percent of Brazilian primary schools. Those authors also state that environmental impacts of schools have become more visible due to increasing consumption of water and energy, increasing waste generation and increasing use of private means of transportation by students.

This is a natural consequence of the growing importance of the service sector in general, and the school sector in particular, to national economic performance. Here is one more justification for the present study, which quantifies the environmental impact of schools in order to make it visible to the community. Authors [24] describe an experiment with eco-literacy activities, but do not provide any quantification. The Foundation for Sustainable Development provides "on-site workshops led by field staff to train community partners in different parts of the world to create economically and environmentally sustainable programs through community assessments, grant proposal development, project design and management, financial management, and monitoring and evaluation." [13].

The ECO-Schools initiative [8] recommends environmental reviews in schools. The reviews form the basis for action plans and typically contemplate the items "Litter, Waste Minimization, School Grounds, Biodiversity, Energy, Water, Transport, Health and Well-Being, Sustaining our World." The reference stipulates that the initiative of review and actions be in the hands of students. The United Kingdom section of ECO-Schools provides detailed review procedures for primary schools [9]. The following is an example of monitoring water consumption:

"Is there a water meter to record water use in the school?
Are there push-on or self-stopping taps in the toilets?
Are there water saving devices in the toilets?
How often does the school run water-saving campaigns?
Are pupils involved in taking and displaying readings?
In addition, if so, are the results recorded on graphs and shared with the rest of the school?"

The Australian Institute of Architects adopts criteria for an environmental review in schools from the Foundation Environment Education Europe. There are seven criteria as follows [3]:

"1 litter and water management
2 hot water, insulation, radiators, electricity and heating system
3 vehicle use
4 washroom taps, toilets and rain water use
5 use of recycled paper and responsible purchasing
6 landscape and wildlife features
7 environmental education, school interior and school involvement".

In addition, the same reference lists the items related to the Schools' Environmental Assessment Method as follows:

"1 recycling facilities
2 energy rating, low-NO_x combustion equipment
3 home-to-school transport policy
4 water savings and quality
5 caretaker training
6 use of recycled materials
7 site selection for new buildings
8 ventilation
9 integrated lighting controls".

The present study incorporates many of those criteria in its set of environmental performance parameters intended for targeting and competition in schools.

The project competition for schools run by The Guardian [30] specifically assigns value to projects that

"Measure the benefit to students.
Bring out ideas for improving school life.
Target long-term results.
Are transferable to other schools."

The Trust for Sustainable Living [31] runs a charity that organizes competitions for schools in a variety of environmental subjects. In 2014, it challenged students worldwide to answer the question "What does sustainable living mean to you?" and to produce a three-minute video on their school's best environmental project.

The method of key performance indicators (KPI) allows any school to choose the indicator considered most important by the community, such as the graduation rate of students and the success of finding employment after graduation [1].

Author [14] deals with Education for Sustainable Development as advocated by the UN. In quite general terms, "it should teach goals for conservation, social justice, appropriate development and democracy in order to build a society that is ecologically, socially and economically sustainable."

Authors [5] present proposals for school design that incorporates sustainability concepts. They argue that children spent most of their time in the school, and this environment ought to be inviting to them. They advocate concepts of good lighting, low energy use, good inside air quality, easy maintenance, water use efficiency, rain water harvesting and waste reduction as adamant to a sustainable school environment.

In a review paper, authors [20] state that theoretically, environmental knowledge leads to awareness and from there to behavior, which mimics the awareness. Their findings reveal that in practice, this process is not necessarily continuous. Knowledge does not always result in the pertinent behavior. In order to reach desired patterns of behavior, it is necessary to resort to direct experience. The present study uses this procedure in as much as it inverts the order of progress. It induces the community to measure the results of its behavior, which in turn will stimulate interest in acquiring knowledge. Exemplifying, the community measures its water consumption, and with this information starts asking questions about the availability of water in the city.

The Green Action Centre [15] addresses the specific subject of composting in schools. "Students can not only learn about how waste can be turned into a rich soil, but also the related environmental concerns." This experience from Canada provided motivation for the present study because it is perfectly applicable in Brazilian schools.

Authors [27] perform experiments in British schools about the effect of noise on classroom performance of students. They provide evidence that external as much as internal noise negatively affects memory, motivation and reading ability and that it influences the results of standardized tests and exams.

The Environmental Performance Index [10] is an initiative based on the statement "We are what we measure. It's time to measure what we want to be". It proposes to institutions that they measure their environmental impact in order to conceive plans for improvement.

The present study follows this line of thought and measures the impacts schools exert on their neighborhood and on the municipal environment generally. Students are not usually conscious of this type of impact, because school performance measurements up to now are restricted to academic achievement. The primary objective is to expose students to direct experience with identifying and measuring environmental impacts. The secondary objective is to induce environmental consciousness through the direct experience, and create the desire for acquiring the relevant knowledge. The tertiary objective is to develop environmental performance indicators in schools that guide students in the quest for improved behavior and allow for comparisons and target setting.

The foregoing references testify to the fact that competitions within the school universe are common events the world over, be it on academic or environmental topics. The present study pretends to prepare schools for this type of competition through diagnoses of their performances and the search for suitable indicators to set environmental targets.

Following personal contacts, the study proceeded successively in six schools located in Central Brazil in the municipalities of Araguari, Uberaba, Uberlândia and Ituiutaba. The study in itself represented a learning process. Starting with simple procedures like separating and composting food residues from the canteens, the scope steadily widened to include measurements of noise, water and energy consumption and degree of source separation of waste. The stipulation of dimensionless performance indicators agreed upon by the communities closed the learning cycle.

2. Methods

All school administrations granted permission for data collections and direct involvement of students in the various measurements and waste movement activities. On a voluntary basis, students gathered in various teams to tackle different impacts of the schools on the environment. The author's group supplied guidance, but the action remained with the students. The first two schools, located in Araguari, served as starting points of the study. Teams of students collected food scraps from the canteens and ran composting operations in open spaces of the school grounds [26]. In the third and fourth schools, located in Uberaba and Uberlândia, apart from running composting operations for food scraps, students measured the total waste of the school, determined its composition and identified correct destinations for the various components [28], [23]. In the fifth school, located in Uberlândia, the scope opened considerably. The procedure involved students as well as staff, included monitoring of water and energy consumption and established the base for an environmental management system [7]. The sixth and last school of this sequence, located in Ituiutaba, reached the highest level of quantification. Measurements taken by students included the following items:

Rate of solid waste production, degree of sorting at the source and delivery to pertinent destinations, by manual weighing of all waste items produced per day.

Water consumption and specific forms of use, by counts of valves and taps, showers, drinking fountains, hydraulic brooms and dish washing machines and times of use and analysis of water utility bills.

Energy consumption and specific forms of use through counts of lamps and appliances and times of use and analysis of energy utility bills.

Noise levels around the premises measured with decibel meter at different points and hours.

Constructed and open soil areas by manual measurement [4].

The items selected for performance evaluation closely follow those reported by [3], although no copy is attempted.

3. Results

Material balances for the composting operations appear in Table 1.

Table 1. *Material balances of composting experiences run by students in schools 1, 2 and 4, kg.*

Input:	food scraps	structuring material	water	Output:	sieved product	humidity	rejects
School 1	91.22	39.95	30.00		45.75	30%	41.00
School 2	197.85	65.67	42.00		111.07	36%	57.23
School 4	140.00	108.00	60.00		116.44	46%	97.56

Table 2. *Waste composition in school 3.*

	Biodegradable material	recyclable inert	rejects
School 3	38.2%	42.6%	19.2%

The students learned from this balance that approximately half of the input material escapes as biogas as result of microbial activity. An exception is the compost from school 4, the humidity of which is excessive.

In school 3, students separated and weighed all waste produced, with the result shown in Table 2.

The biodegradable material originated from the canteen and lunch counters. The students composted this part, and identified reverse logistics operators who took away the dry recyclable items. The direct involvement taught students that they could recycle 80.8% of all school waste.

In the fifth and sixth schools, the scope of the study opened to move beyond waste production and include other environmental parameters. Students measured water and energy consumption, noise levels and area available for rain harvesting.

Table 3 shows the populations of these two schools.

Table 3. *School population in schools 5 and 6.*

	students	teachers	service personnel	total
School 5	1924	175	40	2139
School 6	908	82	79	1069

The high number of service personnel in school 6 derives from the presence of a full-day nursery.

Table 4 summarizes the environmental situation in school 5.

Table 4. *Results of measurements in school 5.*

Water consumption	3131 m^3/ year
Energy consumption	46.1 Mwh / year
Solid waste production	8.1 tonnes / year
Average noise level on premises	65 decibels
Total area of school 6936.5 m^2 (permeable 2841.0 m^2, impermeable 4095.5 m^2)	

In consequence of this analysis, the author's team proposed to the school administrators, the participative development of an environmental management system (EMS) within one year by assigning specific tasks to all segments of the school community with physical and temporal targets and demands for result reporting. The proposal indicated the formation of teams consisting of teachers and groups of students for carrying out each task. The tasks initially contemplated were the following:

Water management. Scope: Monitor monthly water consumption, identify points of irresponsible use and compare to consumption at home. Time line: one year. Responsible persons: arts teachers and grade 6 students. Progress indicator: monthly reduction of utility bill.

Energy management. Scope: monitor energy consumption and identify unnecessary uses, compare with consumption at home. Time line: one year. Responsible persons: language teachers and grade 10 students. Progress indicator: monthly reduction of utility bill.

Solid waste management. Scope: Establish best practice handling procedures for all waste and compare to production at home. Time line: one year. Responsible persons: chemistry teachers and grade 5 students. Progress indicator: reduction of percentage of waste taken to landfill.

Noise control. Scope: monitor the noise levels inside and outside the school complex and determine the contribution of the school community to overall noise. Responsible persons: physics teachers and grade 8 students. Time line: one year. Progress indicator: separation of school noise from traffic noise.

Rainwater capture and use. Scope: determine to what extent rainwater could be collected and used. Time line: one year. Responsible persons: mathematics teachers and grade 9 students. Progress indicator: knowledge acquisition on the subject.

The work with students in school 6 produced the following analyses.

Solid waste production and destination.

The study identified three destinations for solid waste produced in the school.

1. biodegradable material from the canteen – destination pig farming.
2. dry packaging material – destination reverse logistics.
3. rubbish – destination landfill.

Manual measurement by weighing produced the following results:

1. biodegradable material 42.0 kg/day (5 days/week)
2. dry packaging material 8.0 kg/day (5 days/week)
3. rubbish 71.0 kg/day (5 days/week)

Total 121.0 kg/day (5 days/week)

The sorting of biodegradable and dry material was inadequate. Much of this type of material was present in the rubbish. The sorting ratio of (42+8)/121=0.413 was subjectively considered insufficient.

The production of 121.0/1,069 = 0.113 kg / (person*day) was considered acceptable for the full-day school context and additive to home production. No literature data are available on school waste production for comparison. The number 0.113 is an initial value open to improvement through environmental education activities. On a nation-wide basis, the solid waste production in Brazil in 2013 stood at 1.041 kg / (person*day) [2]. This indicates that production in the school (or elsewhere in the workplace) amounts to 10.9% of total waste produced by people. As for the sorting ratio, experiments reported from apartment buildings produced numbers in the range of 0.670 to 0.900 as consequence of intensive coaching [11]. They provide a reference for targeting in the school.

Energy consumption and types of use

The school community consumes energy with lighting and with appliances. The light bulbs present in the institution were the following:

441 bulbs of 20 w
819 bulbs of 40 w
1 bulb of 100 w
85 bulbs of 250 w
16 bulbs of 400 w
total bulbs 1,362, total power 69,330 w.

The appliances were shop equipment, computers, cloth and

dishwashers and dryers, pumps, electric showers and control circuits. The diagnosis estimated the times of use of all bulbs and appliances and arrived at the following monthly consumption: lighting 7,787 kwh + appliances 6,727 kwh = total 14,514 kwh.

This number coincided with the monthly energy bill received from the electric utility company for April 2011. The contracted power demand was 55 kw. In April 2011, the maximum demand actually reached 64 kw, which generated a fine of 28% on the energy bill. This fact demonstrated that the school administration did not exert control over the contracted power usage.

As energy consumption parameter, the study produced the number of 14,514 kwh/month * (1 month/20 working days) * (1 / 1,069 persons) = 0.679 kwh / (person*day) in the school, which is an initial value for discussion and future targeting by the school community.

Water consumption and points of use

Water is consumed in the canteen for food preparation and dishwashing, in drinking fountains, in toilets, in the nursery for hygiene and washing, in house cleaning and in irrigation of green spaces. The total consumption is provided by the main hydrometer, which serves the utility company for billing. From the utility bills for 2010 and 2011, the mean monthly consumption was 1,067 m^3 / month. This excludes the months without school activity. In terms of individual daily consumption, this reduces to the water use parameter of 1,067,000 liters/month * (1 month / 20 active days) * (1 / 1,069 persons) = 50 liters / (person*day) in the school.

This value is high due to the type of school with nursery and canteen. The average water consumption in homes of this city is 180 liters / (person*day) [33]. As the consumption is additive, the numbers indicate that persons who frequent the school consume 180 + 50 = 230 liters / day.

The utility company bills sewage collection and treatment as 70% of water consumption. This flow cannot be measured. Accepting the fraction indicated, the sewage parameter of the school would be 50 * 0.7 = 35 liters / (person*day).

Rain water harvesting

Storm water catchment had never been considered in this school. The present study quantified the possibilities and provided numbers for discussion. The mean annual precipitation in the city is 1,432 mm [17]. The school has 4,898 m^2 of roof area available for capture. Presently, all rainwater runs off to a nearby creek. The possible annual catchment from roofs is 4,898 m^2 * 1.432 m = 7,014 m^3/year. The water consumption as calculated above is 1,067 m^3/month or 8,536 m^3/year (8 months with school activities). This means that rain water could attend to more than 80% of water use, specifically to non-potable water uses, such as toilets, house cleaning and irrigation. The metering system does not allow for discrimination between potable and non-potable uses. Additional hydrometers would be required. The present rain contribution ratio is zero, but the target could be 7,014 / 8,536 = 0.822, which indicates the fraction of water use that rain harvesting could provide if the corresponding equipment were installed. The concept of sustainable school design of [5] supports and recommends this type of installation.

Noise levels

The school is located in what the law on noise defines as a mixed area, predominantly residential with a few commercial establishments. For this type of urban area, the law limits noise levels to 55 decibels (db.) during the day and 50 decibels (db.) at night [29]. This research verified noise levels in all parts of the school with a decibel meter. The noise was within the established limits in all sectors, except for the workshops with 67 db and the gym with 56 db. It will be relatively easy to correct those excesses. The noise excess ratio is 67 / 55 – 1 = 0.218.

Summary of parameters considered for evaluation

For ease of discussion and argumentation, the parameters experimentally determined are summarized in Table 5. They are divided into positive and negative impacts. Positive impacts refer to parameters to be maximized. Negative impacts refer to parameters to be minimized. As this type of measurement had never been done in the school, the community, students as much as staff, were not knowledgeable about their impact on the municipal infrastructure. Rainwater harvesting is not included because it is a design parameter not improvable with behavior patterns.

Table 5. *Results of measurements in school 6.*

Positive impacts:	
solid waste sorting ratio	0.413 sorted waste / total waste
Negative impacts:	
solid waste production	0.113 kg / (person*day)
energy consumption	0.679 kwh / (person*day)
water consumption	50 liters / (person*day)
noise excess ratio	0.218 noise fraction above legal limit

4. Discussion

As indicated in the Introduction, the study represented a learning process. Solid waste is the most visible environmental impact. It was the natural starting point for all measurements. The work in the first four schools demonstrated the possibility of recycling more than 80% of it through source separation, composting and reverse logistics. Table 2 shows that only 19.2% of all school waste is not recyclable. To the school community, this was an unexpected result. It stimulated new ways of thinking about waste. The composition in Table 2 is typical of schools where paper is the main consumption item. In households, the biodegradable fraction is in the order of 65% [11]. The contribution of the study to sustainable waste management practices lies in the source separation, which allowed for delivering each fraction to the correct destination. Composting on the school grounds provided to students the satisfaction of making a product out of material that previously went to the landfill. The separation of dry material provided the challenge of marketing the clean items, a procedure that required contacts with reverse logistics operators. The experience of direct involvement with measureable results induced students to adjust their behavior and learn about the schools' interactions with municipal waste management.

The measurements of water and energy consumption in schools 5 and 6 introduced the students to reading and interpreting bills from the utility companies. This was another completely new challenge, which proved to be a learning tool. Through this exercise, students became aware of their school's presence in the general municipal infrastructure. They started to reflect about the quantities of water and energy available in the city and about conscious use of these resources.

The proposal of an environmental management system in school 5 brought the community in contact not only with the environmental quality in the city, but also with best practices in enterprises and organizations worldwide. The permanent teams formed in the school produced the sense of responsibility. As an example, the Arts teacher and the class of grade 6 students, whoever they might be in any particular year, had to assume the responsibility of monitoring the use and waste of water and of checking the monthly utility bills in order to measure their progress. Omissions were out of the question.

The community of school 6 started to use Table 5 as reference for discussion and targeting improvements. The interaction with the municipal infrastructure is apparent again. The sorting ratio of waste measures the amount that is recycled through reverse logistics and does not go to the landfill. The Table has taken the guesswork out of environmental performance evaluation. Precisely defined numerical values are now available to school administrators for verification and steady improvement. The school community can set its own collective targets and document its numerical progress. As all parameters are defined clearly in terms of easily measurable values, they can be readily adopted by schools anywhere. This is the original contribution of the present study to environmental performance determination in schools. It is in line with the requirement for transferability of school projects supported by [30].

The noise excess index is the simplest candidate for correction. Adequate instructions to the workshop personnel solve the problem. Authors [27] stress the importance of this parameter and provide evidence of the negative impact of noise on classroom performance. The parameters referring to solid waste are controllable up to a certain limit. Administrative talent will provide instructions and examples for improvement. The sorting ratio is the first candidate for intervention. In city residences, [11] attained values in the range of 0.670 to 0.900. This range presents itself as reference for imitation.

Finally, the energy and water consumption parameters are the most visible performance indicators, because the utility bills remind the administration month after month. Design changes may help. Modern water taps and toilet flushing devices as well as solar panels for water heating and electricity generation are readily available. Savings will result from new technologies and from educational measures. These two items are of direct interest to the city administration, who might consider providing specific instructions to all schools. There are 77 schools in the city of Ituiutaba with a total student population of 7,172 [33]. Table 5 will show the city administration, e.g. that the daily water consumption in this universe reaches 7,172 * 50 = 358,600 liters. Depending on the supply situation in the municipality, coercive measures may come into play.

The parameters address all stakeholders of the environmental engineering and management community. Engineers will design and install solar panels, rain harvesting equipment and water saving devices. Managers, like school principals and municipal department heads will organize the solid waste movement. Table 5 offers to all of them data for comparison and extrapolation, and allows for the establishment of targets. As the community itself runs the regular diagnoses, it is exposed to what [20] refer to as direct experience, which is a more powerful tool for building consciousness and forming patterns of behavior than indirect experience like lectures.

The parameters in Table 5 have dimensions. This makes it complicated to perceive the overall performance improvement achievable by working individually on any one parameter. Therefore, the author proposes transforming the parameters into fractions of ideal values, such that one single performance index defines the situation of the school. The procedure is as follows.

A reference value is established for each parameter. For the positive impacts, the reference is placed in the denominator and the measured value in the numerator. For the negative impacts, the reference is placed in the numerator and the measured value in the denominator. Each fraction, then, aspires to unity as the positive impacts increase and the negative impacts decrease, and the average value of all fractions represents the overall performance index.

The reference values proposed for school 6 are shown in Table 6. They are guides for the school community in the quest for continuous improvement of its own index. Obviously, the choice of the references is subjective as long as no universal, national or municipal indicators are available. Every school has to exercise discretion when defining its reference values. If the municipal administration pretends applying the index method to all schools, then it has to stipulate municipal reference values or key performance indicators (KPI).

Table 6. *Reference values of environmental performance parameters. for school 6, established by debate within the community and subject to modification from experience.*

Positive impacts:	
solid waste sorting ratio	0.670 sorted waste / total waste
Negative impacts:	
solid waste production	0.100 kg / (person*day)
energy consumption	0.550 kwh / (person*day)
water consumption	40 liters / (person*day)
existing noise	55 highest db. measure in the school

Tables 5 and 6 allow for the calculation of the overall performance index of the school. As illustration of a positive impact, the solid waste sorting index is 0.413 / 0.670 = 0.616. The reference appears in the denominator. As illustration of a negative impact, the water consumption index is 40 / 50 = 0.800. The reference appears in the numerator.

Table 7 shows the complete calculation of a possible overall index based on the subjective reference values stipulated.

Table 7. Present environmental behavior index of school 6.

Positive impacts	
Solid waste sorting index	0.413 / 0.670 = 0.616
Total	= 0.616
Negative impacts	
Waste production index	0.100 / 0.113 = 0.885
Energy consumption index	0.550 / 0.679 = 0.810
Water consumption index	40 / 50 = 0.800
Noise index	55 / 67 = 0.821
Total	= 3.316
Overall performance index (0.616 + 3.316) / 5 = 0.786	

Table 7 is the typical result of a practical approach to measuring and influencing environmental behavior. The method has drawn on the experience of [8] and [3] and has adapted them to the specific local context.

It is an initial proposal subject to modification through discussions and debates within the school community. This is a direct consequence of the absence of universal environmental KPI for schools. The present study contributes its modest share of arguments to compose a common set of KPI at least at the municipal level. The index represents a reference towards which the community can work in preparation for national and international contests. It depends on the adoption of realistic long-term behavioral targets agreed upon by all members of the community as consequence of environmental education. This represents the most important outcome of the study. It fosters the idea that present-day school life is not only about classroom performance, and it puts into practice the teachings of the Environmental Performance Index [10], which state, "We are what we measure".

The following hypothetical example illustrates the type of progress that could be targeted for one year. The administration implements an environmental education program to involve students in pursuing the targets waste sorting ratio 0.600, energy consumption 0.590 kwh / (person*day), water consumption 45 liters / (person*day), excess noise level 0.000. Upon attaining those targets, the original environmental behavior index of Table 7 would change to that shown in Table 8.

Table 8. Hypothetical environmental behavior index after attaining yearly target.

Positive impacts	
Solid waste sorting index	0.600 / 0.670 = 0.896
Total	= 0.896
Negative impacts	
Waste production index	0.100 / 0.113 = 0.885
Energy consumption index	0.550 / 0.590 = 0.932
Water consumption index	40 / 45 = 0.889
Noise index	55 / 55 = 1.000
Total	= 3.706
Overall behavior index (0.896 + 3.706) / 5 = 0.920	

With the elaboration and dissemination of monthly progress reports, the school community follows the improvements achieved and can find stimulation to contribute. In case of the hypothesis at hand, and with reasonable collective effort, the environmental performance index could move from 0.786 (Table 7) to 0.920 (Table 8) within one year. This procedure represents a powerful educational tool in the hands of dedicated school administrators. Table 8 provides information on the main remaining bottlenecks. Water consumption and waste production have the lowest values. The index procedure brought those facts to the attention of the school community.

This hypothetical example illustrates how a school community can agree upon a set of targets in order to challenge itself and derive satisfaction from new patterns of behavior.

Although in an ideal case, KPI for schools are established at the municipal or regional level, nothing prevents individual schools from taking the initiative and providing benchmarks even in the absence of those KPI. In fact, the idea of the present study was exactly this: induce bottom-up procedures that might provoke actions by municipal or regional administrators. The method of index targeting is in accordance with the project competition parameters of The Guardian [30] in as much as it brings out ideas for improving school life, targets long-term results and is transferable to other schools. The relevance of the index resides in its simplicity. A single number with known significance conveys to the community continuous information on its progress in pursuing behavioral targets.

The practice of regular measurements to update the diagnosis is in accordance with [20] who concede that direct experience is more effective in producing behavior change than indirect experience like lectures on the environment.

5. Conclusions

Students are not conscious of the impact their school exerts on the municipal environment, because school performance measurements up to now are restricted to academic achievement.

Experiments have demonstrated to students that 80% of school waste is recyclable.

Students have been involved directly in waste sorting and composting.

Students learned how to make compost as a product and how to market recyclable items.

Through measurement and monitoring of water and energy consumption, students have become knowledgeable about availability and constraints.

An environmental performance index has been developed for purposes of targeting improvements in the school proper and of inducing competition between schools

Performance parameters on a "per person per day" basis have been transformed into fractions of ideal values in order to provide a framework for comparisons and target settings in the school universe. This is the theoretical contribution of the study to school performance analysis.

The definition of an environmental performance index as fractional approach to perfection simplifies comparative performance evaluation.

The general dimensionless nature of the proposed index gives it universal significance and applicability.

The index method represents a benchmarking experience in the region for evaluating environmental impacts of schools and is transferable to other schools.

The most appropriate way to close this report is to repeat the statement from the Environmental Performance Index [10]: "We are what we measure. It's time to measure what we want to be", and by extending an invitation to school communities worldwide to use the method proposed here to measure what they want to be.

Acknowledgements

This research received support from the Brazilian Research Council CNPq through grant 303480/2013-0-PQ.

References

[1] About Money http://management.about.com/cs/generalmanagement/a/keyperfindic.htm (accessed 2015 08 24).

[2] ABRELPE 2013, Eleventh Overview of Solid Waste in Brazil (in Portuguese) http://www.abrelpe.org.br/Panorama/panorama2013.pdf (accessed 2014 08 17).

[3] AIA Australian Institute of Architects, Environment Design Guide November 2003 http://www.yourbuilding.org/library/1_GEN23.pdf (accessed 2015 08 24).

[4] Andrade, V. S. C. S., Fehr, M. 2016, Search for objective environmental performance indicators of primary schools, *Benchmarking – An International Journal, Emerald Insight*, ISSN 1463 5771 Accepted for publication 2015 11 11, vol. 23 No. 2 pp. xxx.

[5] Brković, M., Milošević, P., 2012, Sustainable Schools as 3D Textbooks: Safeguards of Environmental Sustainability, *FACTA UNIVERSITATIS*, Series: Architecture and Civil Engineering Vol. 10, No 2, pp. 179 – 191, DOI: 10.2298/FUACE1202179B *UDC 727.1: 502.131.1 =111* http://facta.junis.ni.ac.rs/aace/aace201202/aace201202-06.pdf (accessed 2014 09 23).

[6] Coimbra, JA., 2004, A Linguagem e Percepção Ambiental (Language and environmental perception), in: PHILIPPI JUNIOR, A., ROMÉRO, MA., BRUNA, GC. (Org.). Curso de Gestão Ambiental (Course on Environmental Management). Manole Publishers, Barueri BR, pp. 527-570.

[7] Crosara, R., Fehr, M., 2011, Creating an environmental management system in a school community. *Journal of Research in International Business and Management, International Research Journals* vol. 1 no. 8 pp. 245-250, ISSN 2251 0028.

[8] ECO-Schools http://www.eco-schools.org/menu/process/seven-steps (accessed 2014 10 20).

[9] ECO-Schools review manual for UK schools http://www.eco-schools.org.uk/gettingstarted/environmentalreview (accessed 2014 10 20).

[10] Environmental Performance Index http://www.epi.yale.edu/indicators-in-practice/making-grade-indicators-sustainability-colleges-and-universities (accessed 2014 10 20).

[11] Fehr, M., 2014, The management challenge for household waste in emerging economies like Brazil: realistic source separation and activation of reverse logistics, *Waste Management & Research, SAGE Publications*, vol. 32 no. 9 pp. 32-39 http://www.dx.doi.org/10.1177/0734242X14541985.

[12] Feres, YN., Antunes, FZ., 2007, Gestão Ambiental em Instituições de Ensino (Environmental Management in Educational Institutions), Curitiba. Unpublished Research Report.

[13] FSD Foundation for Sustainable Development http://www.fsdinternational.org/about/mission (accessed 2016 01 23).

[14] Gough, A., 2006, Sustainable schools in the UN decade of education for sustainable development: Meeting the challenge? *Southern African Journal of Environmental Education,* vol. 23 pp. 48-63.

[15] GAC Green Action Centre http://greenactioncentre.ca/live/school-composting/ (accessed 2015 01 23).

[16] Huckle, J, Wals, AEJ, 2015, The UN Decade of Education for Sustainable Development: Business as usual in the end, *Environmental Education Research,* vol. 21 no. 3 pp. 491-503.

[17] INMET 2014 http://www.inmet.gov.br/portal/index.php?r=clima/faixaNormalPrecipitacaoTrimestral (accessed 2014 10 24).

[18] ISO 2004, Kid's ISO 14000 Program ISBN 92 67 10388 1 http://www.iso.org/iso/pressrelease.htm?refid=Ref895 (accessed 2014 06 20) http://www.artech.or.jp/english/kids/envedu/index.html (accessed 2014 06 20).

[19] ISO 2005, Cambodian schoolchildren put EMS basics into practice to preserve Angkor site, http://www.iso.org/iso/pressrelease.htm?refid=Ref952 (accessed 2014 06 20).

[20] Kollmuss, A., Agyeman, J., 2002, Mind the gap: Why do people act environmentally and what are the barriers to pro-environmental behavior, *Environmental Education Research* vol. 8 issue 3 pp. 239-260.

[21] Leite, ATA., Medina, NM., 2000, Educação Ambiental: Curso Básico a Distância (Environmental Education: Distance Learning Course) part 1 first edition, Ministry of the Environment, Brasília, DF, Brazil.

[22] Mamede, F., Leite, ALA., 1999, Environmental education for sustainable development (in Portuguese), *Ação Ambiental, Federal University at Viçosa*, Brazil, vol. 2, issue 8, pp. 18-20.

[23] Melo, M. R., Fehr, M., 2010, Management of institutional discards; how to eliminate the landfill option (in Portuguese), *Scientia Plena, Associação Sergipana de Ciência, Aracajú BR* vol. 6 no. 7 article 3, ISSN 1808 2793.

[24] Pires, JDTS., Fonseca, MMCS., Smith, RE, Philippi, LS., 2008, Integração entre gestão e educação ambiental na Escola de Ensino Básico Getúlio Vargas (Integration of school administration and environmental education). *EXTENSIO – Revista Eletrônica de Extensão. Número 6.*

[25] Oyafuso, A., 2004, Teaching Program (Plano Escolar) (in Portuguese), 4th edition, Biruta Publishers, São Paulo, Brazil.

[26] Santos, H. M. N., Fehr, M., 2008, Environmental education through composting of food residues in schools of Araguari (in Portuguese), *Caminhos de Geografia, IG-UFU, Uberlândia BR, vol. 9 no. 25 pp. 65-86,* ISSN 1678 6343.

[27] Shield, B. M., Dockrell, J. E., 2008, The effects of environmental and classroom noise on the academic attainments of primary school children, *The Journal of the Acoustical Society of America,* vol. 123, issue 1, pp. 133-144, http://dx.doi.org/10.1121.1.2812596.

[28] Silva, A. G., Fehr, M., 2013, How students can handle school waste: a case study (in Spanish), *Revista AIDIS de Ingeniería y Ciencias Ambientales: Investigación, Desarrollo y Práctica.* vol. 6 no. 2 pp. 65-71, ISSN 0718 378X.

[29] State of Minas Gerais, Brazil, 1990, Law N° 10100/1990 section 2.

[30] The Guardian http://www.theguardian.com/teacher-network/zurich-school-competition/schools-competition-overview (accessed 2015 08 24).

[31] Trust for Sustainable Living http://www.livingrainforest.org/charity (accessed 2015 11 03).

[32] UNESCO, 1999, Education and Population Dynamics: Mobilizing Minds for a Sustainable Future, EPD-99.

[33] Wikipedia, http://en.wikipedia.org/wiki/Ituiutaba (accessed 2014 10 21).

Assessing Farmers' Perception of Enclosures, Kewot District, Northeastern Ethiopia

Mengistu Asmamaw Mengesha[1, *], Mekuria Argaw Denoboba[2]

[1]Department of Biology, Debre Berhan University, Debre Berhan, Ethiopia
[2]Department of Environmental Science, Addis Ababa University, Addis Ababa, Ethiopia

Email address:
mengistilash99@gmail.com (M. A. Mengesha), mekuriaa@sc.aau.edu.et (M. A. Denoboba)

Abstract: The environmental and socioeconomic effects of land resources degradation are severe, especially in developing nations, where inappropriate land use and farming systems are practiced. Consequently, management options like enclosures are among rehabilitation strategies practiced in the degraded areas of Ethiopia. This study was conducted to assess the farmers' perception of enclosures in Kewot District, North Shoa, Ethiopia. Observation and questionnaire were used to collect data in the study sites. A total of 168 households were selected from two peasant associations using systematic sampling technique. Data were presented in descriptive statistics. The results indicated that the local farmers had perceived the existence of land degradation and its possible causes. Local farmers perceived enclosures positively and are optimistic to the performance of enclosures. However, the benefit sharing has satisfied the community moderately. As a result, participatory resource management strategy is recommended to ensure the sustainability of the enclosures.

Keywords: Enclosure, Local Community, Land Degradation, Rehabilitation

1. Introduction

Background

Land degradation is an indicator of under development resulted from amalgamation of social and economic factors such as poverty and inequitable distribution of land resources, inappropriate land use systems and farming practices in developing countries [31]. In developing countries, it causes a severe crisis to the livelihoods of the rural community and the environment [26].

In Ethiopia, due to the extreme dependence of the rural community on natural resources, particularly land, as a means of livelihood, the country is vulnerable for land resources degradation [10]. Land use land cover changes, mainly agricultural expansion in response to the demands of population growth, has caused accelerated erosion and loss of biodiversity in Ethiopia [15]. The country experiences a loss of nearly 1.9 billion tons of top soil per annum per year, especially from the highlands [11, 12]. Consequently, it has significantly declined agricultural production with an estimated cost ranging from 2 to 6.75% of the Agricultural Gross Domestic Productivity (AGDP) per annum [12, 29].

In response to the alarmingly degraded ecosystems, the practice of rehabilitation of degraded ecosystems is becoming an option to reclaim degraded sites globally [35, 18]. In Ethiopia, the trend of rehabilitation made in different watersheds has improved ecosystem health and land productivity [30]. However, the current rate and status of environmental degradation still calls for more extended and coordinated intervention actions to rehabilitate degraded lands [6, 20].

Enclosures (Area closures) are among various land management and rehabilitation strategies practiced to improve species diversity, soil quality and ecosystem productivity [22]. They are degraded lands that have been excluded from human and livestock interference and left to regenerate naturally [1]. The strategy has been instrumental to reclaim degraded lands in terms of cost, time of revival and the benefit it offers to the rural communities [20, 19].

Enclosures improve ecosystem conditions and enhance the provisioning services of ecosystem services that can improve the food access and economic wellbeing of the rural poor [13]. Enclosures in Ethiopia has been effective in maintaining vegetation resources for energy sources, on which 78-80% of the total household's energy supply of Ethiopian depend [7]. Rehabilitated mountains and fragile ecosystems reduce risk of flooding, improve biodiversity in a watershed. The practice safeguards down streams from flood and siltation hazards and ensures infiltration resulting in replenishing of ground waters [25].

Although land resources management strategy through enclosures is becoming a common trend in Ethiopia, especially in the highlands, researches documents that evaluate local community' perception on socioeconomic and environmental contributions is lacking [9]. Therefore, this study was designed to assess local community' perception on the role of enclosures in Kewot district, Northeastern Ethiopia.

Perception of the local communities' is a base mark for their choice of land resource management practices and its sustainability. Assessment evaluation is a pre-requisite for the actual implementation of the rehabilitation strategy. This evoked the researcher to choose the issue as a research title so that the level of farmers' perception will be assessed and documented. The research findings will have significances for stakeholders by providing concrete scientific evidences about the local communities' perception of land management practices (enclosures) and become a basis for future studies in the field area.

2. Materials and Methods

2.1. Description of the Study Area and Study Design

This study was conducted in Kewot District, 225 km northeast of Addis Ababa at Meriyena insirt (Merye) and Abomsana wuruba (Abomsa) peasant associations. These peasant associations were purposively selected due to long term experience in enclosure practices and availability of comparable adjacent grazing lands. The Meriye site has 23 years old enclosure and is located at 9^0 59′ 95″ N to 10^0 00′ 05″ N and 39^0 55′ 45″ E to 39^0 55′ 50″ E. The Abomsa site has 10 years old enclosure and is located at 10^0 00′ 98″ N to 10^0 01′ 475″ N and 39^0 55′ 54′ E to 39^0 56′ 15″ E. (Fig 1).The study sites receive a mean annual rain fall of 1071.4mm; with mean minimum and maximum temperature of 23.87 °C and 24.44°C, respectively.

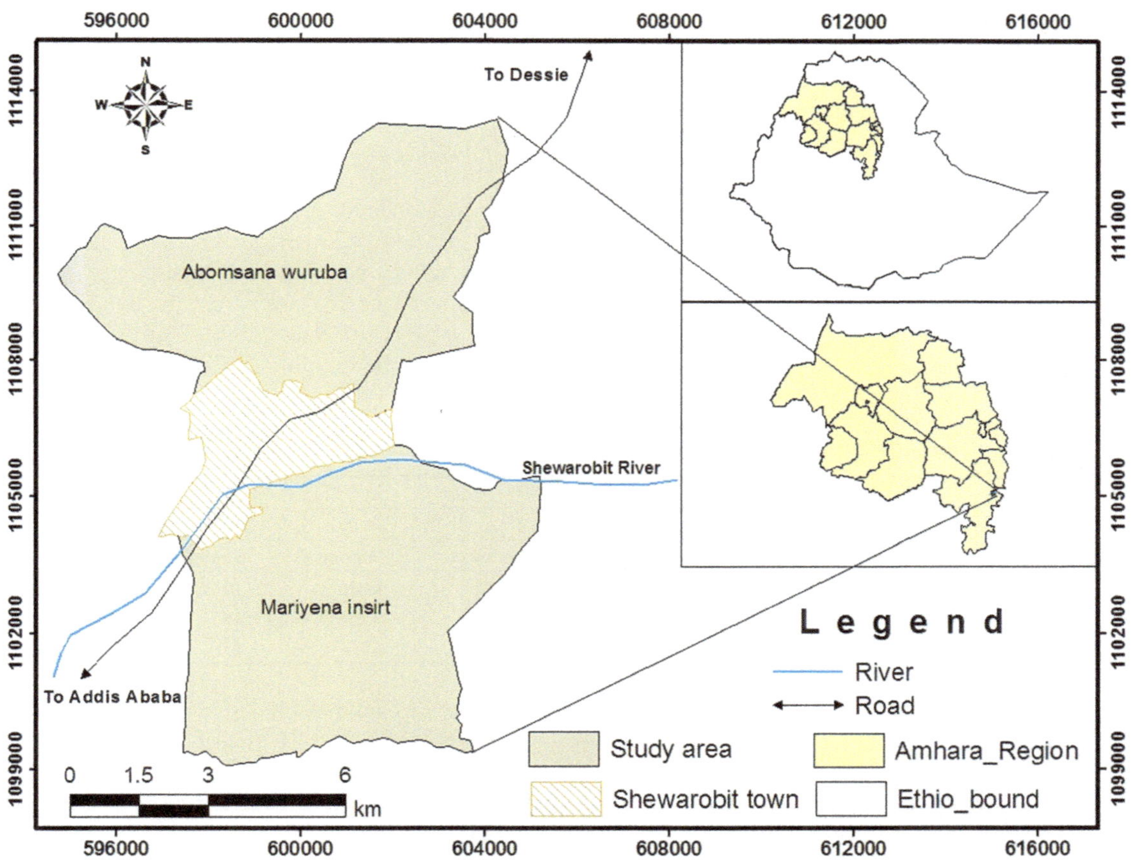

Figure 1. Map of study area.

Study Design

Reconnaissance field survey was made to obtain an overview of the study sites, followed by detailed preliminary survey, which was made between 4th weeks of January to the end of February, 2011. A cross-sectional survey research design was used to collect primary qualitative and quantitative data from the field.

2.2. Data Collection and Sampling Technique

Purposive and Systematic sampling technique methods were used to select study sites and households, respectively. Observation and semi-structured questionnaire were employed to collect information related to households' perception on the role of enclosures.

A total of 168 households were systematically selected from the list of local community members received from the Village administration. Semi-structured questionnaire was provided to these households with language translator to collect their perception on area closure and benefits gained from it. Respondents were included using a statistical formula as follows.

$$\text{i.e. } n = \frac{Z2*p*q*N}{Nd2+Z2pq} \tag{1}$$

Where, n= sample size
N= total population of households in both sites
Z^2= confidence interval (1.96, constant)
d^2= margin of error
p= proportion of population (0.5, constant)
q= 1- p
Assumption: let d= 0.05 and q= 0.5

$$n = \frac{(1.96)^2(0.5)(0.5)(298)}{(289)(0.05)^2 + (1.96)^2(0.5)(0.5)}$$

$$n = 168$$

Hence the data was stratified in to two groups, the number of households in each site was calculated as:

$$n1 = \frac{n*N1}{N} \tag{2}$$

Where, n_1= sample size in the first site
n= number of households in the first site
N1= total number of households included in the study
N= total number of households in both sites

$$n1 = \frac{186 * 168}{298}$$

$$n1 = 105$$

Similarly,

$$n2 = \frac{n*N1}{N} \tag{3}$$

Where, n_2= sample size in the second site
n= number of households in the second site
N1= total number of households included in the study
N= total number of households in both sites (Daniel, 1995)

$$n2 = \frac{112 * 168}{298}$$

$$n2 = 63$$

Besides, the performances of area closures on biophysical qualities of the areas were noticed and recorded by observation.

2.3. Household Survey Data Analysis

The data from household survey were fed on Excel spread sheet and described in descriptive statistics.

3. Results and Discussion

3.1. Local Community Perception on Land Degradation

All the respondents perceived that land degradation, mainly soil erosion was a major challenge in their locality and its impact was very severe. Accordingly, soil erosion, deforestation and lack of fodder were the most common land degradation types (105 households (63.6%), 95 households (57.58%) and 118 households (71.52%)) in the area by their degree of severity (Table 1).

Table 1. *The degree of severity of land degradation as perceived by respondents.*

	Degree of severity					
	1st		2nd		3rd	
Land degradation types	Frequency	Percent (%)	Frequency	Percent (%)	Frequency	Percent (%)
Soil degradation	105	63.6	25	15.15	12	7.27
Deforestation	27	16.4	95	57.58	35	21.2
Lack of fodder	33	20	45	27.27	118	71.5

Table 2. *Possible causes of land degradation as perceived by local communities.*

Possible causes	Value	Percent (%)
Deforestation	82	50
Poor land management	42	25
Over grazing	21	13
Poverty	15	9
Topography	5	3
Population growth	0	0
Soil characteristics	0	0

Respondents explained various possible causes of land degradation and stressed that deforestation, poor land management and overgrazing were believed to be the most possible causes of land degradation (82 households (50%), 42 households (25%) and 21 households (13%)) in the site. However, none of the households pointed that population growth, drought, and soil characteristics as possible causes of land degradation (Table 2).

Communities' awareness about the possible causes and existence of land degradation may be a basis to alleviate the problem. However, it might be due to the perception of local community that considering large family size as an asset and labor force for agricultural activities in rural Ethiopia, respondents did not recognize population growth as possible cause to land degradation. This calls for continuous education to raise peoples' attitudes. Similarly, studies conducted in northern Ethiopia reported that a rapid population growth causes a negative impact on agricultural land and in turn declines the carrying capacity of the environment [8].

3.2. Local Communities' Perception of Enclosures

Respondents stressed that enclosures are effective land management options that promote surface cover and mitigate soil degradation resulting in enhanced land value and productivity (Table 3). This positive attitude of local communities is fundamental for the sustainability of enclosures [17] and also for future rehabilitation projects [34].

Farmers' perception is supported by previous study that states irrespective of the increase in population size, soil and water conservation and land rehabilitation efforts such as enclosures resulted in decreased sheet and rill erosion [27] Besides, studies concluded that increased vegetation density in enclosures result in increased infiltration, that in turn triggers vegetation rehabilitation through superior biomass production thereby improved land productivity including spring discharge [28], [34].

Table 3. Local communities' perception on the contribution of enclosures.

Farmers' Opinion	Yes	Percent (%)	No	Percent (%)
Provide fodder access	137	83	28	17
Provide grass for sales	143	86.7	22	13.3
source of fuel wood	50	30.3	115	69.9
Means of finance	120	72.7	45	27.27
Ensure surface cover	154	93.3	11	6.67
Reduces land degradation	165	100	0	0
Reduces available grazing land	126	76.4	39	23.6
Limit free access of fuel wood	119	72.1	46	27.9
Restrict the use of common resources	117	70.91	48	29.09

The majorities of respondents are optimistic to the performance of enclosures and reported that enclosures are effective in rehabilitating degraded lands; hence they support vegetation growth on degraded lands. It was also noticed that the enclosed sites were completely covered with vegetation which were regenerated naturally; but open lands were bare sites. However, a few respondents explained that the performance of enclosures could be enhanced with integrated soil and water conservation mechanisms. This households' perception is in consistence with [33]) study that states enclosures are effective in controlling soil degradation [33] Similarly, [32] found that re-establishing natural vegetation is an option to reverse land degradation, rehabilitate landscape integrity, and realize the environmental and social benefits of natural resources; it is now widely practiced around the world [32].

Respondents reported that enclosures are among rehabilitation mechanisms mostly practiced in their locality to return degraded lands and improve agricultural productivity as well. It agrees with studies that state proper land management practice that maintains extensive ground cover is a guarantee to reduce soil erosion. Hence, run off and soil losses are inversely related to ground cover, vegetation cover increase surface roughness and reduce soil detachment and transport of soil particles [2]. As a result soil erosion rates in unprotected areas may be 100-1000 times higher than fields with permanent vegetation cover [4].

The majorities of respondents agreed that humans and livestock should be restricted from enclosed sites for effectiveness of enclosures. However, some disagree with complete restriction of human activities and livestock. According to the later assumption, selective grazing (goat and oxen) and cutting of construction materials (occasionally) could better be allowed to ensure the use of communal resources. However, studies reported that fencing involves restricting of damaging agents to maintain degraded lands [16] and it is a recommended practice to facilitate land rehabilitation [14]. Moreover, research result states that for overall vegetation rehabilitation, excluding of live stocks ensures the growth of woody vegetation which provides shelter for next generation of succession; as a result a better regeneration could be achieved by excluding of humans and animals from degraded sites [14].

3.3. Local Communities' Perception on Management of Enclosures

A total of 132 households (80%) had participated in onsite selection and the majorities of interviewed respondents agreed on selection criteria. As explained by respondents, the selection criteria were the extent of degradation (the more the degraded the area, it is likely to be abandoned for rehabilitation), which could be evaluated by its productivity history and sensitivity to hazards (like erosion); being marginal and communal lands to ensure common sharing of cost and benefits leading to sustainable management in the future.

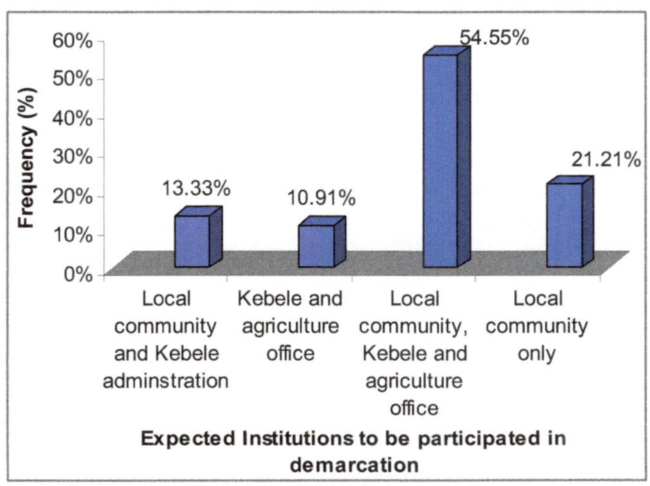

Figure 2. Institutions expected to be involved in demarcation of Enclosures as perceived by local communities.

Respondents pointed out that the local community should be involved in decision making and bottom up approach management of area closures develop sense of belongingness among the community. Consequently, a total of 90 households (54.55%) responded that for future sustainability of enclosures, demarcation should be collaborative (Local community, Kebele (Village) officials and Agricultural office) (Fig. 2), and the majorities of the respondents concluded that sense of belongingness could be developed if the management is designated to the local community after demarcation.

Households' conclusion is supported by studies that state although establishment of enclosures in Ethiopia have considerable economic and ecological significances, it might be due to top-to-down implementation and limited participation of local community in decision making and utilization of resources during the Derg regime, the community denied the contribution of enclosures [5]. This has significantly affected the sense of ownership and community's commitment for effective protection and sustainable management of land resources [26].

3.4. Local Communities' Perception on the Role of Enclosures

Respondents replied that cut-and-carry mode of using grass and beekeeping is activities allowed in enclosures. However, the use of common resources in the enclosures was negatively assessed by the majorities of the respondents (117 households (70.91%)) (Table 3). Besides, though half of the respondents replied that the befit sharing has moderately satisfied the community (84 households (51%)), a quarter of respondents (42 households (25.5%)) explained that the benefit sharing has satisfied the community poorly. This might be due to an assumption that enclosures will be permanently owned and no possibility to share benefits to youth in the future and those who were not accepted the establishments of enclosures at the beginning are not benefited from the resources. In line with assumption, the Abomsa Village administration has initiated to partition enclosures for youth associations which will ensure benefit sharing considering the landless youth.

The local communities disclose that enclosures provide three fundamental benefits to the local community. These include (i) Social value: accordingly, they were asking for resettlement due to accelerated soil erosion. However, nowadays the practices of enclosures have enabled them to control soil erosion and have increased land productivity. (ii) Economic value: respondents replied that enclosures provide considerable fodder access for live stocks and are efficient to increase financial income for households (137 households (83%) and 120 households (72.7%), respectively) (Table 3). This has helped them to save the cost of grass purchase for fodder and thatching. Besides, more than 1200 Ethiopian Birr could be gained from grass sales on annual basis.

This optimistic perception of the local community is supported by studies that state enclosed sites provide grass and wood access for local community beyond their aim of establishment [24] and are also attractive in financial terms, as people could had to purchase grass from other areas [14]. Respondents' view of economic benefits of enclosures also agrees with Lovejoy's (1985) research result that states resources from area closures contribute to the households' economy, suggesting that economic and social wellbeing is enhanced by focusing on rehabilitation of degraded lands [23]. Moreover, biophysical composition increased by 50% after enclosure leading to economic development of the community [24].

(iii) Environmental value: respondents explained that area closures are effective strategies in controlling accelerated soil erosion and agricultural lands below area closures become more productive than lands below grazing. This agrees with studies that state agricultural lands below free grazing were strongly affected by water erosion than below enclosed sites [33]. In addition, studies concluded that enclosures facilitate natural regeneration thereby reducing surface runoff. This will promote accumulation of soil organic matter and other plant nutrients that excel soil quality and capable of support diverse communities [14].

Although some claim enclosures due to their restriction of free access of resources, the majorities of the respondents are optimist towards enclosures as they promote surface cover, reduce surface erosion and provide grass access for fodder and sale. The majorities developed sense of belongingness and ready to expand the practice in the future, as they are involved in decision making and management of enclosures. Households' conclusion is supported by studies that state although establishment of enclosures in Ethiopia have considerable economic and ecological significances, it might be due to top-to-down implementation and limited participation of local community in decision making and utilization of resources during the Derg regime, the community denied the contribution of enclosures [5]. This has significantly affected the sense of ownership and community's commitment for effective protection and sustainable management of land resources [26].

4. Conclusion

The results in this study revealed that enclosures have significant contributions to the socioeconomic and ecological systems; as they generate ecological and socioeconomic benefits. The majorities of the local communities developed sense of belongingness and developed positive attitude to the performance of enclosures. Such perception is a base mark for future sustainability of the practice. Enclosures often restrict use of communal resources. Particularly in Ethiopia, having large number of cattle with free grazing habit is a common tradition in Ethiopia. As a result, respective governmental organizations are required to work on education. Thus, agroforestry strategies like animal fattening and apiculture activities need to be practiced to ensure the socioeconomic wellbeing of the society while maintaining natural resources.

Acknowledgements

I am grateful to the Kewote Ditrict and Shewa Robit town office of Agriculture staffs for their help during data collection. My deepest gratitude also goes to the Development Agents and Village Administrators of Abomsa and Meriye Villages for their cooperation during data collection and equally important was the contribution of households who provided information voluntarily.

References

[1] Betru Nedassa. 2003. Soil and water conservation program in the Amhara National Regional State. In: Tilahun Amede (ed.) Natural resources degradation and environmental concerns in the ANRS, Ethiopia: Impact on Food Security. Proceedings of the Natural Resources Management Conference, July 24-26, 2002, Bahir Dar, Ethiopia. pp. 109-125.

[2] Costin AB. 1980. Run off and soil nutrient losses from an improved pasture at Ginninderra, Southern Tablelands, New South Wales. Australia. *Journal of Agricultural Research*, 31: 533–546.

[3] Daniel W. 1995. Biostatics: A foundation for analysis in the health science, (6th Ed.). Georgia State University, New York.

[4] Deploey J. 1989. Erosion systems and perspectives for erosion control in European loess areas. *Soil Technology Series,* 1: 93–102.

[5] Dessalegn Rahmato 1994. Land Policy in Ethiopia at the crossroads. Working papers of Ethiopian development *No. 8.* University of Trondheim.

[6] Edward IN. 2000. Applied Ecology and Environmental Management. Blackwell Publishing Company, Australia.

[7] EFAP (Ethiopian Forestry Action Plan) 1994. Ministry of natural resources development and Environmental protection. Vol. II Final Report. Addis Ababa.

[8] Ejigu Belay 1999. Extension policy and its implementation in relation to sustainable land use and food security intervention. In: Proceedings of the 1st National Workshop. NOVIB partners Forum on sustainable land use, Ethiopia.

[9] Emiru Birhane, Demel Teketay, Barklund P. 2006. Actual and potential contribution of exclosures to enhance biodiversity of woody species in dry lands of Eastern Tigray. *Journal of the Dry lands,* 1(2): 134-147.

[10] EPA (Ethiopian Environmental Protection Authority) 1998. National action program to combat desertification. Federal Democratic Republic of Ethiopia Environmental Protection Authority, Addis Ababa.

[11] FAO (Food and Agriculture Organization) 1984. Ethiopian Highland Reclamation Study (EHRS). Final Report, Vol1: 2. Rome, Italy.

[12] FAO (Food and Agriculture Organization) 1986. Ethiopian Highlands Reclamation Study (EHRS). Final Report, Volume 1. FAO, Rome, Italy. 334 pp.

[13] FAO (Food and Agriculture Organization) 2001. News and highlights: Forestry forum spot lights poverty alleviation. Rome, Italy.

[14] Girma Kelboro 2009. The Role of fences for managing land resources. Interdisciplinary term paper. Bonn Interdisciplinary Graduate School for Development Research, the Netherlands.

[15] Hadgu Tesfa, Kirose Meles, Epema GF, Bruggen AC. 2008. Biodiversity and sustainability in agricultural landscapes in Tigray, Northern Ethiopia. Submitted to Landscape Urban Plan.

[16] Hayward MW, Kerley GI. 2009. Fencing for conservation: restriction of evolutionary potential. *Biological Conservation,* 142:1-13.

[17] Heitschmidt RK, Vermeire LT, Grings E. 2004. Is rangeland agriculture sustainable? *American Society of Animal Science,* 82: 138-146.

[18] Hobbs RJ, Cramer VA. 2008. Restoration Ecology: Interventionist approaches for restoring and maintaining ecosystem function in the face of rapid environmental change. Murdoc University, Murdoch, Australia.

[19] Kidane Giday 2002. Woody biomass estimation in community managed closure areas in Tigray: Implications to sustainable management and utilization. MSc Thesis, SLU, Sweden.

[20] Kindeya G/Hiwot 1997. Area enclosures as an approach in the management of dry land biodiversity: a case study in Tigray Region, Northern Ethiopia. Proceedings of the workshop on management of dry land biodiversity. 30 July-1August, 1997, Nairobi, Kenya.

[21] Kindeya G/Hiwot (2004). Dry land agro-forestry strategy for Ethiopia. Mekele, Tigray, Ethiopia, pp. 1-20.

[22] Lal R, Stewart BA. 1992. Need for land restoration. In: *Advances in Soil Science*, 17: 1 - 11.

[23] Lovejoy TE. 1985. Rehabilitation of degraded tropical forest lands. Commission on Ecology, Occasional Paper No. 5. Gland, Switzerland: International Union for Conservation of Nature and Natural Resources.

[24] Mengistu T, Teketay Demel, Hulten H, Yemishaw Y. 2005. The role of enclosures in the recovery of woody vegetation in degraded dry land hillsides of Central and Northern Ethiopia. *Journal of Arid Environment,* 60: 259 - 281.

[25] Mitiku Haile, Kindeya G/Hiwot 1997. Opportunities and constraints for afforestation in Tigray: the case study of Dega-Tembin. Paper presented at the 13th international Ethiopian Studies, 13-14 December, 1997, Kyoto, Japan.

[26] Nedessa Betru, Ali J, Nyborg I. 2005. Exploring ecological and socioeconomic issues for the improvement of area enclosures management: A case study from Ethiopia. Dry lands coordination group report No. 38 Oslo, Norway.

[27] Nyssen J, Poesen J, Desta G, Vancampenhout K, D'aes M, Gebremedhin Y, Govers G, Leirs H, Moeyersons J, Naudts J, Nigussie H, Mitiku H, Deckers J. 2007. Interdisciplinary on site evaluation of stone bunds to control soil erosion on cropland in Northern Ethiopia. *Soil and Tillage Resources*, 94: 151-163.

[28] Nyssen J, Poesen J, Moeyersons J, Luyten E, Veyret-Picot M, Deckers J, Mitiku Haile, Govers G. 2002. Impact of road building on gully erosion risk: a case study from the Northern Ethiopian Highlands. *Earth Surface Processes and Landforms,* 27 (12): 1267 – 1283.

[29] Sonneveld BG. 2002. Land under pressure: The impact of water erosion on food production in ethiopia. Shaker Publishing (PhD dissertation). The Netherlands.

[30] Sonneveld BG, Keyzer MA. 2003. Land under pressure: Soil conservation concerns and opportunities for Ethiopia. *Land Degradation and Development*, 14: 5-23.

[31] UNEP (United Nations Environmental Program) 1986. Sands of change: Why land becomes desert and what can be done about it? UNEP environmental brief No 2, United Nations Environment Program, Nairobi, Kenya.

[32] Vanandel J, Aronson J. (Eds.) 2006. Restoration Ecology. The new frontier. Malden: Blackwell Publishing.

[33] Wolde Mekuria, Mitiku Haile, Veldkamp E., Nyssen J Muys B, Kindeya Gebrehiwot, 2000. Effectiveness of exclosures to restore degraded soils as a result of overgrazing in Tigray, Ethiopia. *Journal of Arid Environment*, 69: 270-284.

[34] Wolde Mekuria, Veldkamp E., Mitiku Haile, Kindeya G/Hiwot, Muys B, Nyssen J. 2009. Effectiveness of exclosures to control soil erosion and Local community perception on soil erosion in Tigray, Ethiopia.

[35] Young TP. 2000. Restoration Ecology and Conservation Biology. *Biological Conservation,* 92: 73–83.

11

Environmental Management Compliance, Law and Policy Regimes in Developing Countries

Cuthbert Casey Makondo[1, *], Sydney Sichilima[1], Matthews Silondwa[1], Richard Sikazwe[1], Lombe Maiba[1], Chawezi Longwe[1], Yvonne Chiliboyi[2]

[1]Environmental Engineering Department, Copperbelt University, Kitwe, Zambia
[2]Institute for Water Research, Rhodes University, Grahamstown, Republic of South Africa

Email address:
SimplyCuthbert@live.co.uk (C. C. Makondo), nsichilima@gmail.com (S. Sichilima), msilondwa2@gmail.com (M. Silondwa), maibalombe@yahoo.com (L. Maiba), Richard.sikazwe@yahoo.com (R. Sikazwe), chalongwe@yahoo.co.uk (C. Longwe), Yvonne.chiliboyi@gmail.com (Y. Chiliboyi)

Abstract: This paper is a review of environmental management compliance of the extractive industry; particularly mining in Zambia and the government's commitment to environmental protection in view of sustainable development .The main objective of this work was to review industrial compliance against Zambia's environmental legal regime for the period between 2009 and 2014, track environmental policy development, its implementation within the institutional functional framework in view of precautionary principle, polluter pays principle and sustainability. The authors concluded that while the environmental legal framework may be considered relatively new compared to developed nations like United Kingdom, it dates back to the 1970s and is comparable in terms of strength, to that of developed nations. However, there are many challenges regarding effective implementation. Thus, although there are many stiffer and tougher regulations enacted in Zambia (to a point where some advocates of neo-classical frontiers economics have criticized Zambia as slowly introducing one-more among the most highly regulated environmental sectors in the world, deterring development) the authors concluded that implementation was a huge challenge: implementation of Zambia's environmental regulatory regime faces wide range of problems, from a highly centralized financial and decision-making system and budgetary allocation of negligible size, to lack of appropriate tools, equipment and personnel technical capacity. It was found that while the environmental regulations have been further stiffened around 2011-2013; compliance of mining companies was lagging behind. This was evidenced by failure of 8 in 10 mining firms to submit periodic reports as per regulatory conditions on their permits. Cases of discharging and disposing of hazardous waste to gain savings from the expenses that would otherwise arise through normal disposal procedures still exist. The authors concluded that while new and stronger measures have been put in place by government to ensure that the environmental degradation caused by mining and other extractive industry activities are adequately managed, the measures are not working effectively. These findings are consistent with the findings of the 2014 Auditor General's Report. It was clear that co-ordination within the government's regulatory institutional framework with interacting mandates was weak, while the selective nature on the application of the Environmental Impact Assessment (EIA) regulations was demeaning environmental protection and socio-economic justice.

Keywords: Extractive Industry, Triple-Bottom-Line, Sustainability, Environmental Management, Compliance, Law and Policy

1. Introduction

In developing countries where economic development is driven mostly by the extractive industry, terms such as "environmental management and sustainability" and "triple-bottom-line" are relatively new concepts [1, 2]. In these nations, economic growth concerns have tended to override environmental sustainability considerations [2, 3, 4]. However, with an increased global concern following

Rio-1992 (the first Earth Summit on Environment and Development) issues of sustainable development (the idea that development is threefold, encompassing-environmental, social and economic balance or triple-bottom-line) have gained understanding in resource management practices [4]. With a rapid growth of the third-sector organizations in the 1990s demanding environmental justice and green growth from governments through tough industrial regulation, stringent environmental regulatory regimes and reorganizations of institutional framework arrangements became inevitable in most nations. In Africa however, existing environmental regulatory regimes have been cited as weak in many studies [1, 3]. And because of this perception, some multinationals operating in the extractive industry have been forced by International Financing Corporations (IFC; e.g., Citi-group and World Bank) to implement initiatives considered 'good practice' such as equator principles when implementing IFC financed projects costing over US$4m [5]. However, equator principles have had their weaknesses which are not part of this paper.

After agriculture, mining in Zambia is probably the second oldest extractive industry. It has taken place for centuries; starting with rudimental activities recorded in 1880s to relatively larger scale artisanry 48 years later in 1902 around Kabwe-kamukuba, and then to the first commercialized operation of 1928. Mining has undeniably played a central role in Zambia's socio-economic development across all political regimes, albeit its associated environmental effects [6]. The mining of copper particularly, has been the most significant mineral resource with ubiquitous deposits in two main areas of the country, namely; Copperbelt and North-western Provinces. Other mining projects include coal, uranium and nickel in Southern Province and gemstones in many parts of the country. The sector has gone through three main phases of ownership: first, private ownership under a colonial administration prior to political independence in 1964, then national ownership post-independence in 1971under the Kaunda government, until the re-privatization of the 1990s under the Chiluba government [6].

During mines privatization in 1990s, with the demands of due-diligent audits, the burden of environmental liabilities was shouldered squarely by the government before new mine owners' formalized acquisitions of mine facilities across Copperbelt. With the help of the World Bank, and Nordic Development Fund, the Zambian government initiated the Copperbelt Environment Project (CEP) worth about US$62M, to address all former mining liabilities in the Country. By 2003, CEP was launched to address among others, strengthening the regulatory capacity of ECZ and Mine Safety Department (MSD). One of the most successive stories of CEP was the acquisition of various environmental monitoring equipments; and the creation of an Environmental Protection Fund (EPF) to ease future rehabilitation funding mechanisms by operationalizing environmental rehabilitation in view of polluter pays principle. Under this fund, mining liabilities of current and future operations would be addressed using these pool funds. Consequently, all mining firms were required to contribute certain amount of money annually commensurate to their foreseeable environmental liabilities. EPF is still operational today, albeit difficulties in costing environmental liabilities of ongoing mining activities.

While the positive side of mining has been greatly felt and more often openly discussed, the negative legacy; a much dark side of mining had remained secret and only known to few medical doctors and environmentalists. Thanks to the growth of the third sector organizations such as non-governmental organizations and investigative journalism: it is now public knowledge and clearly understood by many that mining in Zambia have had serious negative legacies; beginning with lead poisoning:-the silent killer in Kabwe, through sulphur-dioxide (locally known as *senta)* the "lung-eating" gas in Mufulira and Kitwe West, to the poorly managed radioactive materials present in Kitwe's Mindolo area, and from acidified surface water and ground water emanating from in-situ leaching and acid mine drainage challenges, to heavy metals contamination and siltation loading of streams, and deliberate hazardous chemical waste disposals on the Copperbelt due to an unethical business practices of some companies. All these practices have significant cumulative impacts not only on receptor environmental media quality and alterations of ecological functionalities, but also on public health and safety especially on the Copperbelt, and Kabwe. This is likely to be the same problem in the Northwestern Zambia where some of the world's largest copper smelting facilities have now been commissioned. Some cases of firms of multinational standing have been prosecuted by Zambia Environmental Management Agency (ZEMA), but with such multinationals, solicitation and lobbying usually results in reversal of judgments against conviction on their offences. Meanwhile, the lives of thousands of people remain threatened from toxic chemical discharge into local streams, to ambient air emissions of oxides of sulphur (SO_x) of concentrations ranging between 500 and 1000 μgm^{-3}.

2. Methodology and Limitations

This paper is a review of environmental management compliance of the mining sector in Zambia, and the government's commitment to environmental protection in view of sustainable development. It is based on policy tracking and subsequent field investigation works that spanned over 5 years from 2009 to 2014. The methods employed included regular sampling of stream water and analyzing for heavy metal levels using atomic absorption spectrometry method. Sampling at one specific point of the stream (for all the streams) investigated was done quarterly per year, for a five year period. At the end of each year, quarterly concentration readings were averaged to obtain an annual mean concentration. The water quality data (average data) was then compared with regulatory limits to measure level of compliance. Field observations and conducting simple interviews with 10 mines was part of the methodology to gauge commitment. Company documents in public domain like Environmental and Social Management Plans (ESMPs) in the EIA reports were also evaluated.

However, some of the limitations with this work include loss of individual sample signatures (actual figures) in the generalization or averaging process: averages are highly influenced by extreme values. Conversely, though sampling was done on a regular interval, results may not reflect an accurate picture. Other environmental media, e.g., air quality assessment and soil heavy metal analysis was not done due to lack of equipment for the former, and due to complexity of procedures for gaining access to mines facilities and the size of the study area for the latter.

3. The Environmental Legal Framework

Environmental regulation in Zambia dates back to 1970 when the Natural Resources Act was passed aiming at nature conservation. This was followed by Game Parks and Birds Act a year later whose objective was to conserve wildlife. In 1987, The Zambezi River Authority Act, targeting shared river systems bordering Zambia and Zimbabwe was enacted. Table 1 below shows specifically important pieces of environmental regulations. In terms of heritage conservation, this was recognized in 1989 when the Heritage Conservation Act was passed to preserve local culture, shrines, monuments and specific sites of national significance [7].

In spite of the National Conservation Strategy (NCS) being in place at this stage however, the cost of environmental liabilities and the negative legacy of mining were specifically felt at privatization after due-diligent audits were conducted in 1990s. It was at this time that the first need for clearer environmental legislation was recognized [7]. Consequently, with works already initiated by the Kaunda government from the recommendation of the NCS, the Environmental Protection and Pollution Control Bill (EPPCB) was passed and enacted by parliament in 1990 as Environmental Protection and Pollution Control Act- EPPCA (CAP 204).

The EPPCA provided for the establishment of the Environmental Council of Zambia (ECZ) in 1992 which assumed an advisory role to the government on environmental matters. Its other mandate was to issue environmental permits and monitor compliance of extractive and processing industries; draw up new regulations as necessary and enforce such on environmental media quality while educating the general public on matters of environmental and public health. With a view to building up on the EPPCA, recognizing that every citizen has a right to health and clean environment, government through ECZ formulated the National Conservation Action Plan to provide for mandatory appraisals for all new major projects and existing major ones that would need upgrading? This was going to be done through a precautionary principle tool provision - Environmental Impact Assessment (EIA), while providing for objective public participation in the process. In 1997, Statutory Instrument (SI) No. 28 (Environmental Impact Assessment) Regulations was passed to govern the EIA process. Under these regulations, it is an offence to implement any project categorized under first and second schedules of these regulations. Penalty for violating these regulations can be at very minimum US$45,000 or a three-year-jail term; even both depending on the gravity of the offence [7].

In 2011, the Environmental Management Act No. 12 repealed the EPPCA of 1990. By this Act, ECZ changed its name to Zambia Environmental Management Agency (ZEMA) but with exactly the same mandate, only with more powers to prosecute environmental offences. The Act also provided for the upgrading of all regulations that complemented EPPCA. In 2013, all the EPPCA regulations were upgraded by SI No.112-Environmental Management (Licensing) Regulations with yet stiffer penalties upon conviction of environmental violation. Table 1 shows all the most important regulations and acts and policies in Zambian [7].

Table 1. *Zambia's Key Pieces of Environmental Legislation and Regulations.*

Year	Environmental Component Target /Objective
1970	*Natural Resources Act/Nature Conservation.
1971	*Game Parks and Birds Act/Wildlife Conservation.
1974	National Fisheries Act/ Protection.
1978	*Public Health Act/ Waste Management & Environment Ratified the 1972 World Culture and Heritage/Eco-tourism
1982	Statutes of International Union for the Conservation of Nature and Natural Resources (IUCN).
1985	Petroleum Exploration & Production Act/Pollution Control National Conservation Strategy/Sustainable use of resources
1986	*Local Administration Act (Trade Effluents)/ Pollution Control.
1987	Zambezi River Authority Act/ Water Resources Management
1988	National Heritage Act/Conservation Tourism.
1990	*Enacted Environmental Protection and Pollution Control Act/Integrated Pollution Control and leads to the establishment of Environmental Council of Zambia in 1992. *Forestry Act (Amended in 1999) and currently under review Ratified Montreal Protocol and Vienna Convention.
1991	*Zambia Wildlife Act (Amended 1998).
1993	Ratified Ramsar and Bonn Convention; and the Endangered Species of Wild Fauna and Flora (CITES) Convention on Biodiversity. **Water Pollution Control Regulations (Effluent and Waste SI No. 72).
1994	Ratified Basel Convention/Transboundary hazardous waste; and Pesticides and toxic substances Regulations (SI No. 20)/Agriculture and Environment.
1995	Town and Country Planning Act 1995 (approval and revocation of development plans). The Energy Regulations Act (Cap 436; SI No. 16 of 1995)/Energy and Environment.
1996	**Air Pollution Control (Licensing and Emissions Standards) Regulations (SI No. 141). Ratified UN Framework Combating Desertification.
1997	Environmental Impact Assessment Regulations (SI. 28).
2000	**Hazardous Waste Management Regulations (SI No. 125).
2001	**Ozone Depleting Substances Regulations (SI No. 27)
2003	*Water Act: applies to water rights, impounds for irrigation.
2006	Ratified Kyoto Protocol-UNFCCC.
2008	Launched the first Environmental Policy
2011	Environmental Management Act No.12/ Repeals Environmental Protection and Pollution Act of 1990. More and stiffer penalties to violators. Water Resources Management Act/Repeals Water Act of 2003.
2013	Environmental Management (Licensing Regulations) SI. No. 112/ Repeals all Regulations except for regulations 1997, SI No 28. To make stringent all environmental regulations.

*Repealed Acts or regulations. ** Repealed specifically by Environmental Management (Licensing Regulations) SI. No. 112 of 2013.

4. Institutional Framework

From being Ministry of Environment, Natural Resources and Tourism in 1991, the ministry under which environmental protection falls has changed its name many times. At publication date of this paper, environmental protection fell under the Ministry of Lands, Natural Resources and Environmental Protection, from Ministry of Mines, Energy and Water Development which used to combine environment, mining, lands, energy and water development, a scenario that brought considerable administrative challenges making it the home-mark of work replication, bureaucracy and inefficiency. Essentially, policy synchronization proved difficult and cases of opposing objectives were not uncommon under this arrangement.

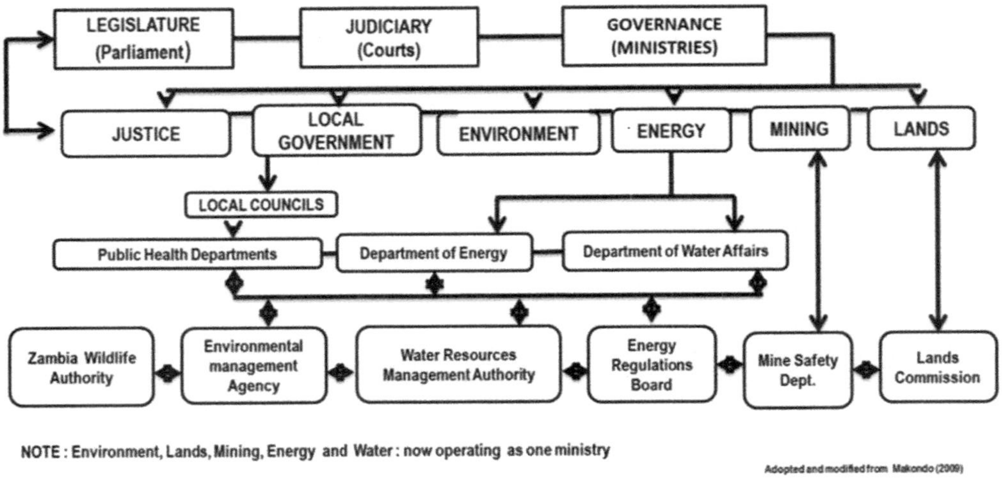

Figure 1. *Simplified Environmental Governance Institutional Framework*

Figure 1 (above) illustrates the simplified governance structure. At the top are the supreme organs of governance, parliament, judiciary and ministries. Ministries have different departments below them and under these ministries are various statutory organs with overarching mandates. The Zambia Environmental Agency (ZEMA) is one of these organs working in coordination with others e.g. Zambia Wildlife Authority on wildlife and natural resourced conservation, and with Water Resources Authority and Department of Water Affairs on water access rights, ground and surface water quality preservation. ZEMA also works with local councils on public health and safety while working with Lands on land matters for industrial project location approvals. It is this complex co-ordination that slows down information flows between organs of interacting mandates that was found to be weak. Lack of up-to-date information systems for easy decision making processes has made informed decision making difficult.

5. Findings/Results Discussion

Heavy metal concentrations results are shown in Table 2 and graphic presentation in Figure 3 below. These results indicate that cobalt (Co) concentrations were far too high in Uchi stream, followed by Mindolo and upper Kafue in Chingola ranging from 9 to 32 times more than the regulatory limits of $1mgL^{-1}$ (Table 2 and Figure 3). Manganese (Mn) was also high and above regulatory limits of $1mgL^{-1}$ in all the streams investigated and was found to be at least 1.5, and at most 4 times more than regulatory limits. However, cadmium (Cd) and arsenic (As) were low with the latter generally below detectable limits in all the streams investigated and when it was detectable; it was found to be below the regulatory limit of $0.05mgL^{-1}$ and thus raised no concerns. Cadmium though found in all the streams investigated, it was below regulatory limits of $0.5mgL^{-1}$ and thus raised no concerns as well. Lead (Pb) presence was generally detected but like arsenic, it was below the regulatory limit of $0.5mgL^{-1}$ and thus raised no concerns too. Copper was generally below regulatory limits of $1.5mgL^{-1}$. Potential of Hydrogen (pH) was within acceptable limits for Uchi, Mindolo, and down Kafue streams at Sabina, except for the upper Kafue in Chingola were it was averaging between 3.4 and 5.5.

Findings on institutional framework revealed a complex co-ordination system that completely slows down decision making-process, due to poor information flow between organs of interacting mandates. It was found that a lack of up-to-date information system for easy decision making process has made informed decision making process difficult if not impossible. Government is aware as evidenced by the continued relocation of the environmental protection component from one ministry to another. This has delayed implementation of environmental agendas as the department involved is continuously under restructuring with shifting reporting arrangements almost annually.

The capacity of statutory agencies (ZEMA and MSD particularly) is still lacking though slightly improved compared to 5 years ago. The technical capacity of personnel is still inadequate, though there are improvements. Highly trained personnel who include lawyers and inspectors with police training background in criminal investigation now handle some of the most sophisticated environmental pollution cases. However, due to poor salaries, most of the young ones leave for corporate environmental positions as soon as they get experienced. Conversely, while the Environmental Protection Fund (EPF) has been cited as one of the most successive stories of the Copperbelt Environment Project (CEP), findings revealed that there are still some challenges regarding its ways of valuing environmental liabilities of mining companies. For instance, it was found that the initial methods which were used at its inception were discarded soon after a year or two after the realization of gross weaknesses. New methods now use quantum based determinations of liabilities. However, even quantum based determinants have been criticized by some environmentalists of under-valuing the liabilities of mining companies.

With regards to the CEP, due to the procedures of the World Bank the project suffered massive bureaucracy financial disbursement delays. Instead of taking only 5 years to address environmental liabilities from former mining activities, the project took 8 years, and still never exhausted the funds, and never did land rehabilitation works in some locations like Kabwe for instance. Today, Kabwe is still ranked among the top-ten most polluted spots on the planet [10].

On EIA process and its implementation, it was found not to be effective. Challenges range from poor public consultation, weak stakeholder engagement to weak EIA out-put documents. The Environmental Impact Statements (EIS) lack scientific rigor. In worst case scenarios, some EIS reports were found to have "copy and paste" problem, from other EIS documents done by same consultants. These documents would have exactly the same errors, discussing the same locations, but were for separate project areas which were not only in different locations, but also for different developers. Other than suggesting that there is far too much mediocrity in the way EISs are being done, this also provides evidence that there are unqualified consultants doing these works. These findings are consistent with those of [10]. Apart from poor EIS works, there is further evidence that the EIA application is selective. Government funded, or government backed projects were found to be 3-4 times more likely to be approved without a thorough EIA processes. There are cases where EIAs for government owned projects have been done in retrospect, with the project already on commissioning phase. The other challenge identified on the EIA practice is the lack of post decision follow-ups by EIA approving body, ZEMA. There are no follow-ups to see whether the ESMP provided in the EIS is being followed by the project proponent, or whether the predicted impacts actually occurred. Post-decision monitoring audits are very useful in improving future EIA quality. Consequently, it can be argued that an EIA process in Zambia has simply become a weak academic exercise with little to offer on the precautionary and polluter pays principles, other than being merely used to seek stamps of approval from the authorities.

Table 2. *Water samples and associated levels of different parameters of tributaries of the Kafue River.*

Sampling period & name of stream	**Annual mean concentrations**						
Uchi Stream (Kitwe South)	**pH**	**Cu Conc. mgL-1**	**Pb Conc. mgL-1**	**Mn Conc. mgL-1**	**Cd Conc. mgL-1**	**Co Conc. mgL-1**	**As Conc. mgL-1**
Year 1	5.10	0.61	0.02	4.45	0.02	9.61	<0.01
Year 2	6.20	0.40	0.02	2.61	0.01	28.30	0.02
Year 3	5.57	0.43	0.07	2.10	0.03	12.00	0.01
Year 4	6.37	0.37	0.01	3.13	0.05	22.10	<0.01
Year 5	7.02	0.34	0.11	4.30	<0.01	32.00	<0.01
Mindolo Stream (Kitwe Northwest)							
Year 1	5.20	0.43	0.01	1.37	0.03	4.28	0.00
Year 2	6.00	0.61	0.03	3.21	0.01	1.96	0.00
Year 3	7.36	0.55	0.02	3.51	0.04	2.13	<0.01
Year 4	7.12	0.43	0.04	2.11	0.01	1.91	<0.01
Year 5	7.60	0.33	<0.01	2.72	0.02	3.90	0.00
Up Kafue River (Chingola Northwest)							
Year 1	3.40	0.52	0.13	1.78	0.04	1.87	0.02
Year 2	4.50	0.31	0.10	3.01	0.01	2.21	0.01
Year 3	4.02	0.32	0.01	1.98	0.03	1.99	0.01
Year 4	7.32	0.04	0.16	1.74	0.02	1.08	0.00
Year 5	5.50	0.05	0.11	1.21	<0.01	1.71	<0.01
Dwn Kafue River (Sabina-Kitwe North)							
Year 1	6.50	0.71	0.12	1.61	0.01	0.81	0.01
Year 2	7.20	0.57	0.08	1.33	<0.01	0.12	0.02
Year 3	7.10	0.98	0.06	1.62	0.02	0.13	0.01
Year 4	7.30	0.59	0.31	0.67	0.01	0.34	0.01
Year 5	6.96	0.65	0.01	0.92	0.01	0.51	<0.01

Figure 2. Low pH wastewater effluent discharging with heavy metal precipitates, silt and sedimentation, and erosion of un-vegetated dumps.

Table 3. Government Budgetary Allocation for Environmental Protection in the last decade. Source: Ministry of Finance, http://www.mofnp.gov.zm/.

Year	2006	2007	2008	2009	2010	2011	2012	2013	2014	2015
Amount ZMW'Million *	43.3	103.3	95.7	117.3	148.5	121.3	31.8	74.2	165.2	174.96
% of Total Budget	0.004	0.009	0.007	0.008	0.009	0.006	0.001	0.002	0.004	0.004
Total Budget ZMW'Million*	10,236.8	12,034.4	13,761.4	15,279	16,717.8	20, 537.4	27,698.3	32,212.2	42,682	46,666.56

* US$1=ZMW7.3 (Bank of Zambia. http://www.boz.zm)

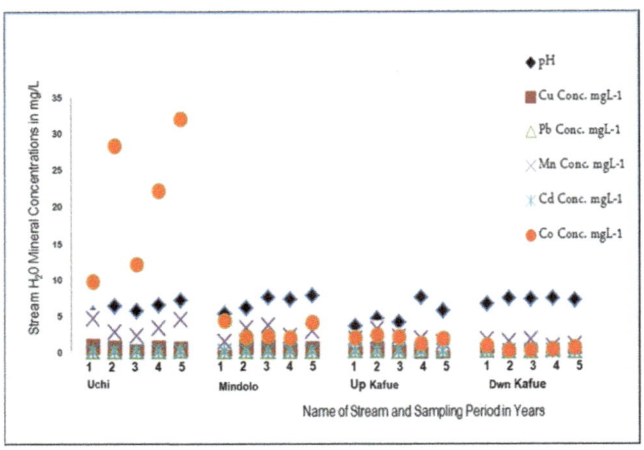

Figure 3. Graphic Presentation of stream water heavy metal concentrations.

Another interesting finding was the absence of environmental management systems certification: no mining company in Zambia is ISO 14001 certified, but most of them are in the planning phase to have systems put in place. With such lack of commitment to good practice environmental management systems, it is easy to argue that the mining industry in Zambia cannot comply voluntarily: it cannot drive the environmental protection agenda without being forced to, through regulation. In other words, the driving factor in this industry, it can be argued; is a single bottom-lime: - profit. This is not new: other studies have revealed a similar behavior [8, 9, 10, 11, 12, 13, 14, 15, 16, 17]. In Zambia however, there is further evidence of extreme profit loving, as exhibited by some companies' false profit declarations, and price transfer systems as a means through which to evade taxes. Some firms also threaten government of abandoning projects and use local nationals' job losses as a means to maintain status quo whenever there are proposed production output based tax changes in the sector. One can argue that there is no commitment to the triple-bottom-line, since the societal and environmental wellbeing has been receiving least attention. These findings are consistent with some of those by [10].This was evidenced by the fact that nearly all mining companies (80%) are still heavily relying on environmental management plans that were done more than 10 years ago, coupled with low impact "smoke screen" social corporate responsibility programs. Ideally, Environmental and Social Management Plans (ESMPs) must be revised every 3 years as loosely provided for in the EIA regulations. However, because this provision refers specifically to a non-developed project whose EIS may have been approved but just never got implemented within a three-year-period after approval, then a new EIA must be done with a new ESMP. And because it so unclear on the already existing ESMPs of operating firms, companies may almost always find ways to exploit such vague regulations.

Government commitment to environmental protection in monetary terms has been displayed in Table 3. As can be seen, the allocation is far less than 1% of the total national budget. It is the least funded sector in Zambia. Recreation, Culture and Religion is the second least funded sector and gets about 2.5 times more than the environmental sector budget. Much of the funds to the environmental sector come from donor countries to fund training and personnel capacity building programs, and for tools and equipment acquisition. The lack of proper funding obviously has a huge bearing on the monitoring and the economics of law enforcement in the sector.

6. Conclusion

Although the information collected over the last five years has shown that environmental compliance is improving generally except for a few incidences, such as the acid levels in the Kafue tributaries, and cobalt levels around Kitwe streams; mining companies are generally not complying with the environmental regulations and environmental licensing conditions set by Government. This was evidenced by failure of firms to submit periodical reports as required by ZEMA. Occasionally, some firms have dumped toxic waste materials into local streams to save disposal costs accruing contamination in the Kafue since all these streams eventually drain down the entire load into the Kafue main river which is a major source of agriculture and domestic water supply, from Chililabombwe through Chingola, Mufulira, Kitwe, and all the way to Lusaka.

Thus, although wastewater quality compliance has generally improved, there were still cases of hazardous chemical waste discharge by some firms to gain savings from the expenses that would otherwise have been incurred through normal disposal procedures, even if the savings could be far less than the penalties they would incur if convicted. Such unethical companies are motivated to engage in such illegal activities for profit, having known that the probability of being caught is negligible due to poor monitoring, while conviction is difficult due to problems facing the judicial system where cases drag-on for decades in court.

Secondary data and field observations (Figure 2) suggest that there is poor management of mine waste dumps and tailing dams. Most of the waste dumps are unstable and are the main sources of siltation on the Kafue River, reducing the water flow and the river channel. Mining and erosion of unstable waste dumps and tailing dams on the Copperbelt contribute about 135,000 tonnes per year to the siltation problems in the local streams-drainage-system on one hand, while on the other, the landforms where these dumps rest may be difficult to put to any alternative land-use, thereby potentially depriving the nation of future economic activities [10].

Conversely, records from the year 2000 onwards clearly indicated that copper smelting alone has been contributing about 350,000 tonnes of SO_x annually, with concentration ranging from 500-1000 μgm^{-3}[10]. These daily concentrations are 10 to 20 times more than the regulatory limits. Mining firms are also failing to submit periodic reports as required under their permits on air quality, probably due to lack of compliance. While the mining firms were found to be concerned about good corporate citizenship, the environmental protection commitment of most of them (80%) was found to be weak, and lacked specific goals. This was evidenced by the use of old and un-updated ESMPs, and lack of certification with Environmental Management Systems; i.e., ISO 14001 standards.

Although monitoring for compliance by the regulatory agencies was found to be poor, Zambia has probably one of the strongest environmental regulations in the region. Thus, although it only emerged in the 1970s, the environmental regimes in Zambia compare very well with that of developed nations like the United Kingdom [7] to the point where this has been seen as anti-development by some critics. For instance, some neo-classical frontiers economists now argue that Zambia is introducing anti-development environmental agendas, by slowly making the environmental area among the most highly regulated sectors in the world which may increase the cost of doing business. Because of such criticisms, the Zambian government reactions often end up in allowing projects in ecologically sensitive areas [25]. However, these arguments lack validity in view of causes of market failure: most environmental problems can be understood as resulting from market failures, where the most relevant type is negative externality. To overcome negative externality, polluters must pay for the pollution damage. It has been proven globally in many studies, that when companies comply and go greener in their operations further than what the regulations have prescribed, there is less regulation in return, and less policing by government as a result [4, 9, 18, 19, 20, 21, 22, 23, 24]. This does not only improve the company's reputation, but also reduce the costs of doing business [26, 27]. In Zambia, while the regulations can be considered as tough, the means for implementing these regulations is weak, as well as the means for measuring policy outcomes are unclear. A further challenge is the lack of resources and capacity to drive these environmental strategies for the desired outcomes. The institutional framework lacks co-ordination, which makes it difficult to implement strategies that cut across related sectors such as lands, environment, agriculture, water, forestry and energy. These findings are consistent with those by [10], who also found that co-ordination even within the same regulatory framework was weak. The conclusions are clear.

Selective application of the Environmental Impact Assessment (EIA) regulations exists. State owned and government backed projects are 3 to 4 times more likely to get approvals even with very poor quality Environmental Impact Statements (EIS). The EIA process and the EIA practitioners in Zambia do not adhere to professionalism and the demands of environmental justice. Abuse of office also does exist within ZEMA as some employees conduct environmental projects briefs (EPBs) and approve them themselves.

Government commitment to environmental protection has fluctuated over the last decade, as evidenced by the funds allocated towards the sector. Much of the funding in the sector comes from donor countries. As argued by [28, 29], to make policy effective, it must be designed carefully with clearly measurable goals. Funds must be available for its effective implementation, then that must be reinforced by government will, and the capacity to measure policy outcomes. Thus, going by the findings, it can be concluded that the implementation of Zambia's environmental regulatory regime cannot achieve desired outcomes with such a wide range of problems, from a centralized system, through to lack of funds and material resources, and from lack of technical personnel capacity to enforce and measure policy outcomes to unclear goals. If these can be dealt with, driving environmental sustainability agenda would probably be much easier to achieve since there is already enough and well designed regulations in place. It can also be concluded that there are no clear market-based instruments (MBIs) for firms. MBIs such as environmental tax breaks, tradable permit systems or targeted subsidies are a cost-effective way to protect and improve the environment as these provide incentives to both firms and consumers to opt for cleaner production and greener products respectively [30].

While it is concluded that the Environmental Protection Fund (EPF) is one of the most successive stories of the Copperbelt Environment Project (CEP), it can be further said that the EPF still has challenges in objectively costing environmental liabilities of mining firms as well as enforcing compliance. Mining firms are simply not complying enough. While issues of surety, trust and transparency have been raised by these firms, most of them (70%) have 2 to 3 years arrears and are thus not up-date with EPF contributions and bank guarantees [31]. The contributions are very little compared to the environmental degradation caused. The current quantum method needs to be improved so that it reflects actual liabilities of any given firm in the extractive industry, including small scale mining companies, as these can be among some of the major sources pollution. The pool funds contributed are far too insufficient to manage the environmental damages of the associated mining activities. Environmental clean-ups can be very costly as evidenced in the CEP implementation.

Finally, all measures that have been put in place by government to ensure that the environmental degradation caused by mining activities are adequately managed, are not working effectively. There is still weak enforcement. These findings are generally consistent with the findings of the 2014 Auditor General's Report on Mining in Zambia.

References

[1] P. Modak and A. K. Biswas, Conducting environmental impact assessment for developing countries, New York, United Nations University Press, 1999.

[2] Organization for Economic Corporation and Development, Environmental outlook to 2050 report, OEC, 2011.

[3] DfID, EC, UNDP and Word Bank, Linking poverty reduction and environmental management: policy challenges and opportunities. Washington DC. The World Bank, 2002.

[4] International Panel for Sustainable Resource Management, Marrakech process and green economy initiative: working towards a sustainable consumption and production for a green economy: UNEP, 2010.

[5] http://www.equator-principles.com: Accessed 26 May 2015

[6] International Council on Mining and Metals Report on Zambia, 2014

[7] C. Makondo, Environmental Impact Assessment: a comparative analysis of the UK and the Zambian regulations (MSc Dissertation) 2009, Institute of Estuarine and Coastal Studies, University of Hull, United Kingdom.

[8] E.M. De Santo, Missing marine protected area (MPA) targets: How the push for quantity over quality undermines sustainability and social justice, Journal of Environmental Management Vol.124. 2013, pp.137-146.

[9] The Business of Sustainability: Building industry cases for corporate responsibility, Edited by Ulrich Steger, 2004.

[10] J.Lindahl, Environmental impacts of mining in Zambia: Towards better environmental management and sustainable exploitation of mineral resources, 2014, Geological Survey of Sweden.

[11] E.F. Lambin and P. Meyfroidt. Global land use change, economic globalization and looming land scarcity: proceeding of the National Academy of Sciences (USA) 108: 3465-3472, 2011.

[12] World Business Council for Sustainable Development, Effective biodiversity and ecosystem policy and regulation: business input to the COP10 of Convention on Biological Diversity. WBCSD, Geneva, 2010.

[13] L. Xianbing, B, Liu , T. Shishime, Q. Yu, J. Bi, and T. Fujitsuka, An empirical study on the driving mechanism of proactive corporate environmental management in China, Journal of Environmental Management, vol, 91, 2010, pp.1707-1717.

[14] D.G. Fraser, J. Andrew Dougill, W.E. Mabee, M. Reed. P McAlpine, Bottom up and top down: Analysis of participatory processes for sustainability indicator identification as a pathway to community empowerment and sustainable environmental management, Journal of Environmental Management Vol. 78 2006, pp.114–127.

[15] A.A .Hezri, Sustainability indicator system and policy processes in Malaysia: a framework for utilization and learning, Journal of Environmental Management, Vol. 73, 2004, pp.357–371.

[16] K. Zhang, and Z. Wen, Review and challenges of policies of environmental protection and sustainable development in China, Journal of Environmental Management, Vol. 88, 2008, pp. 1249–1261.

[17] R.G. Eccles, I. Ioannou, I., and G. Serafeim, The impact of corporate culture of sustainability on corporate behavior and performance- working paper, 2012.

[18] E. G., Hertwich, W. S. Pease, and C. P. Koshland, "Evaluating the environmental impact of productsand production processes: A comparison of six methods," Science of the Total Environment, 196, 1997. Pp.13-29.

[19] G., Winter, Business and the environment: A handbook of industrial ecology, Hamburg: 1988, McGraw-Hill.

[20] D., O'Rourke, L. Connelly, and C. P. Koshland, "Industrial ecology - a critical review," International Journal of Environmental Pollution 6 (2-3), 1996. Pp.89-112.

[21] E., Matthews, C. Amann, M. Fischer-Kowalski, S. Bringezu, W. Hüttler, R. Kleijn, Y. Moriguchi, C. Ottke, E. Rodenburg, D. Rogich, H. Schandl, H. Schütz, E. van der Voet, and H. Weisz, The weight of nations: Material outflows from industrial economies, Washington, DC: 2000. World Resources Institute.

[22] S., Nakamura and Y. Kondo, "Input-output analysis of waste management," Journal of Industrial Ecology, 6 (1), 2002, pp. 39-64.

[23] Duchin, F., "The conversion of biological materials and wastes to useful products," Structural Change and Economic Dynamics, 1(2), 1990, pp.243-261.

[24] C. J., Cleveland, "Biophysical economics: From physiocracy to ecological economics and industrial ecology," in J. Gowdy and K. Mayumi, eds. Bio-economics and Sustainability: Essays in Honor of Nicholas Gerogescu-Roegen, Cheltenham, England: 1999. Edward Elgar Publishing, pp.125-154

[25] http://zambia.co.zm/zol_articles/2014/02/16/reframing-the-lower-zambezi-copper-mining-project-debate.accessed 6/2/2015

[26] S., Erkman, "Industrial ecology: a historical view," Journal of Cleaner Production, 5(1-2), 1997, pp. 1-10.

[27] Duchin, F., and G. Lange, The Future of the Environment: Ecological Economics and Technological Change, 1994. New York: Oxford University Press

[28] J. Mitchell, "The Use (and Misuse) of Surveys in Policy Analysis," in Handbook of Public Policy, 2007.

[29] A. Gilbert, Criteria for sustainability in the development of indicators for sustainable development, Chemosphere: Elsevier Science, No. 9, vol. 33, 1996, pp.1739–1748.

[30] http://ec.europa.eu/environment/enveco/studies.htm#2 accessed, 6/2/2015.

[31] GRZ, Management of Environmental Degradation Caused by Mining Activities in Zambia. Auditor General's Report, 2014, Government of the Republic of Zambia (GRZ).

12

A comprehensive micro-thermal analysis of thermal desalination plants for improving their efficiency

Adel K. El-Feky

Atomic Energy Authority, Nuclear Research Center, Reactor Department, Cairo, Egypt

Email address:
Elfiqi02@hotmail.com

Abstract: The second law efficiency is a main tool for evaluating the irreversibility through a system. Models have been applied for multistage flash (MSF) and mechanical vapor compression (MVC) distillation for evaluating the entropy generation at each component in the process. The minimum work of separation (which is the work consider to extract a unit of pure water from a saline feed water) is considered through the study, the entropy generated due to irreversibility plus that of the in-equilibrium of the discharge brine and its mixing with the seawater are important during the calculation of the minimum separation work, its value for MSF is 0.729 kWh/m^3. Exergy is a useful concept in the analysis of systems involving both thermal and mechanical energies. The thermal performance of distillers is assessed based on the first and second laws of thermodynamics. This study focused on the analysis of the energy and exergy of MSF and MVC units. The exergy losses due to irreversibility for the subsystems of the units are evaluated, the specific exergy losses of the MSF unit is at the range of 63 kJ/kg. The exergy destruction in heat recovery and heat rejection sections, brine heater and all the other systems are calculated, these values are 61, 17, 10, and 12% respectively. The study showed that, the second law efficiencies of the MSF unit is around 4 % and for the MVC is around 7%, so these law efficiencies clarify that there are many ways to improve the plant performance by reducing the highest exergy destruction through these systems.

Keywords: Multistage Flash Distillation, Exergy, Second Law Efficiency, Separation Work, MVC

1. Introduction

The need for drinking water has grown fastly over the last century. Nowadays, around 1.7 billion people have shortage for accessing to water, particularly in the developing countries, especially in Middle East, South Asia and Africa. The shortage in the drinking water contributes in 70-80% of diseases in around 90 developing countries.

In order to overcome this shortage, it is important to improve both the water usage and the ways of producing the potable water from non potable one. In the western countries, the water leaves the houses is known as grey water. This type of water is mild contamination and can be reused, this reused water serves in lowering the overall water demand, in this situation the water supply sources will be extended the lifetime of these sources[1].

The continual research and development of desalination processes have resulted in a variety of commercial desalination methods. These processes are classified into major desalination processes that include multi stage flash, multi effect, vapor compression and reverse osmosis desalination; while the minor processes include solar humidification, freezing and electrodialysis desalination. Nowadays, the multi stage flash and reverse osmosis desalination are the commercial systems that choice for desalination from brackish water and sea water.

From the practical analysis of the operating data of many desalination plants, it is concluded that the second law efficiency of the studied plants is less than 10 %. From these studies, nowadays as it is known that the modern power plants have a second law efficiency of around 50 %, it is clear that there is a wide area for improving these desalination plants. This improvement can be done for the new and existing plants. The new experiences from other systems can be used to enhance the second law efficiency of the desalination plants to approximately the double. The enhancement of the second law efficiency will result in reducing the energy used in the desalination by half.

The thermodynamics analysis of ideal desalination

processes with actual plant operational data is of great importance to the desalination processes. The best tool to do this analysis is the second law. Through this study, there is two main parts for it, the first one is the separation process which is the main point in desalination and the second way is the analysis of the distillation process which are the vapor compression, multi effect and multi stage desalination system[2].

2. Literature Review

O. Hamed et al. [14] have been collected the available data for design and operation of most of SWCC distillers, and utilized it to simulate and analyze the thermal performance of these units. Through their study the impact of different design and operating parameters such as top brine temperature and number of stages on the thermal performance of MSF system is evaluated.

Kamal Mohammedi et al. [11] their study focus on the thermal performance of a small desalination unit, for evaluating the losses through the multistage flash unit, and the study focuses on the environmental impacts associated with the concentrated brine discharge to the Mediterranean sea.

Narmine et al. [7]investigated through their theoretical and experimental work with a MVC system of capacity of 5 m3/d that the potable water production rate increases by increasing the operating temperature, and the evaporator temperature has a good effect on its heat transfer coefficient.

Vesa [17] reviewed a MVC system design parameters, with a design capacity of 500 m3/d. The study analysis showed that the exergy efficiency of the heat exchanger is around 90 % which is too high compared to other systems. His study concluded that the pre-heaters were very effective although the temperature differences through them are low. This led to reduce the exergy destruction through that unit.

Alasfour et al. [4] their study was performed by using a steady state model for hybrid MSF-MVC system, through their analysis they concentrated on both the 1^{St} and 2nd laws of thermodynamics. Their study concluded that, increasing the temperature drop across the stage leads to increases the specific power consumption which decreases the rate of the productivity. But through the MVC system, increasing the brine temperature decreases the specific heat transfer area. Results indicate also, that increasing the temperature drop across stages leads to increase in exergy destruction.

Mabrouk et al. [18] investigated a hybrid MSF-MVC system; their study showed that the performance ratio (PR) is increased by 2.4 times than conventional MSF system. Also this study indicates that the minimum exergy destruction was reached at a suction pressure of 8 kPa.

3. Distillation Processes

The desalination methods are numerous, In this section, the distillation process will be discussed.

Distillation is a separation process in which a volatile component of a solution is evaporated and then condensed where is the salts remain. The pure water is evaporated by the transfer of heat to a body of saline water, then the pure water vapor is removed from contact with the saline water, then it is condensed by the removal of heat, where the pure water is formed.

3.1. Multi-Stage Brine Recirculation

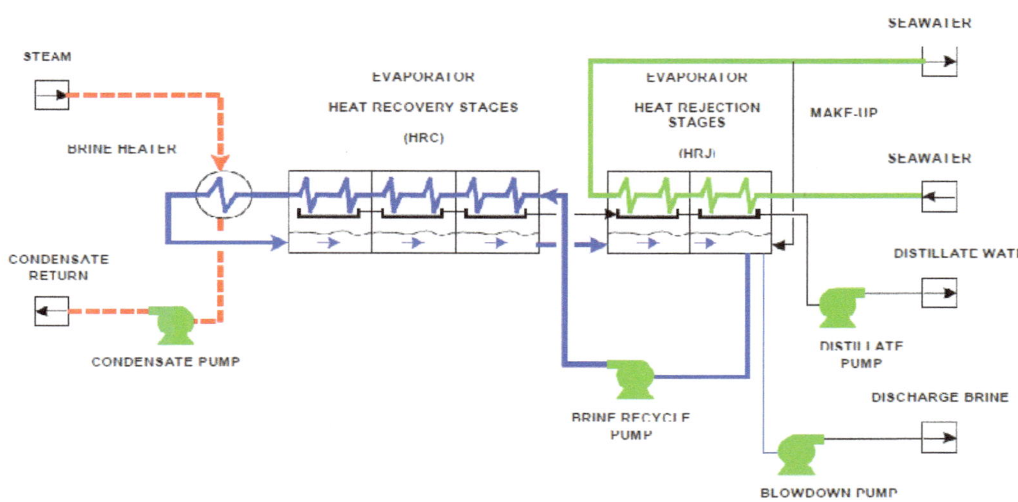

Figure (1). *Multi-stage Flash Desalination with Brine Circulation (MSF-BR)*

(MSF-BR), a schematic of the MSF-BR system is shown in figure (1). As showed in the diagram, the intake seawater stream is introduced into the condenser tubes of the heat rejection section [2,5]. The temperature of the intake seawater leaving the heat rejection should be close to the brine temperature of the last stage. The intake seawater divided into two streams, the first one is the cooling seawater stream which is generally discharged into the sea again; the second stream fed again to the bottom of last stage of the heat rejection section. The recirculated brine is taken from the pool of the last stage of the unit and injected to the condenser of the recovery section. This stream is heated by the latent heat of the vapor produced (flashing) to increase the feed stream, then this brine enter to the brine heater to increase the brine temperature to the maximum value (TBT).

Hot brine enters the flashing stages of the heat recovery section then to the heat rejection section, where a small amount of fresh water vapor is formed by brine flashing in each stage.

Most of the MSF system operates at temperatures below 100°C and under reduced pressures, also the temperatures of the brine to the sea and in the last stage is in the range of 30-40°C.

3.2. Mechanical Vapor Compression

(MVC) is one of the promising low temperature desalination systems. Such systems are characterized by number of advantages, such as: low corrosion rate, minimum scale rates, reliable and flexible operable systems and highly pure distillate [3].

In the mechanical vapor compression (MVC) desalination process, the mechanical compressor is used to recycle the latent heat from the evaporating side to the condensing side of the evaporating tube. The compressor is used in increasing the process pressure to a high value, where this compression causes the condensing of the vapor that supply the spraying droplets additional heat that increase the amount of generating vapor. The incoming saline water is preheated nearly to its boiling temperature corresponding to the evaporator vapor pressure; this is done by using a heat exchanger where it cools both the brine and distillate. Most of the energy consumed in this process is through the compressor work, as the compressor not only increases the pressure but it reduces the vapor pressure in the evaporator. This desalination method is used in remote areas, resorts and on ships; it contributes of around 2% of the world desalination capacity.

The capacity of SEE-MVC systems has increased over the years from the small production volume to present value of 5000 m3/d. the required source of power is the electricity only. The SEE-MVC consists of four main components: vapor compressor, preheater of intake saline water, evaporator condenser and pumps for brine and distillate [5,7,2]. Figure(2) below presents the flow process of the system

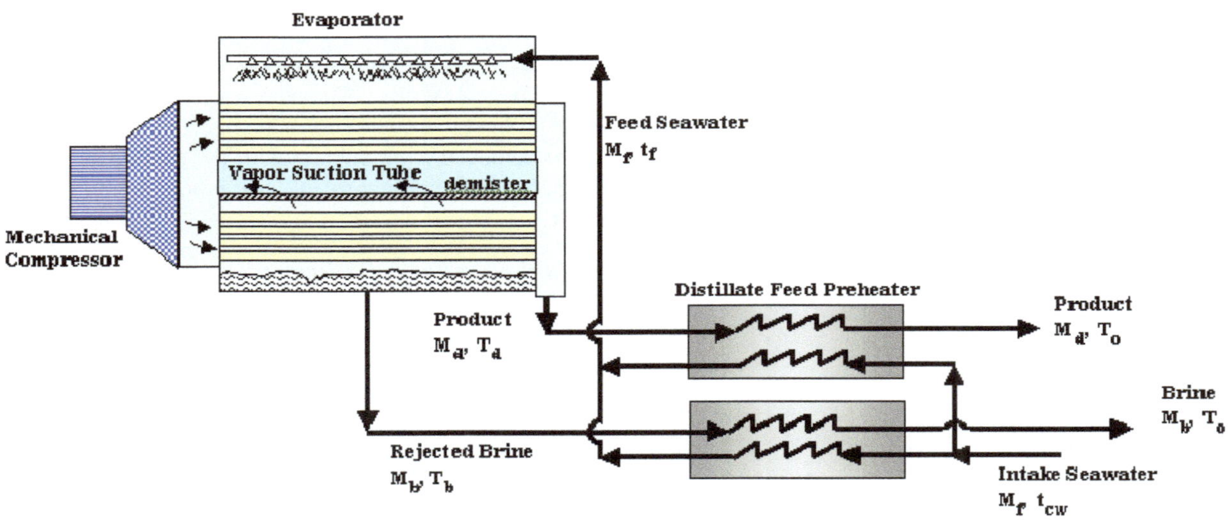

Figure (2). *Single Effect Evaporator with Mechanical Vapor Compression (SEE-MVC)*

A simple model is considered Aly et al. [7], and the design values and model output are listed in tables3, 4.

4. Thermodynamic Analysis

Through this section, three parameters will be discussed; the first items is to describe the main parameters affecting on the desalination performance, the second item will discuss the separation processes of a mixtures, and the third one is to discuss the second law efficiency for desalination system[5]

a. The main performance parameters

There are many parameters used to describe the performance of the desalination process, the first parameter, the performance ratio (PR) is known as the ratio of the water product mass flow to the heating steam,

$$PR = \frac{M_d}{M_s} \tag{1}$$

The second parameter, is the gain output ratio (GOR); it is known as the ratio of the enthalpy required for evaporating the distillate to the heat input to the system,

$$GOR = \frac{Md}{Q/h_{fg}(T_0)} \tag{2}$$

So, for the thermal desalination process; it is clear that both the performance ratio and gain output ratio are very similar. So the relation between the both parameters is the ratio between the latent heat of vaporization at the distillate temperature to it at steam temperature,

$$GOR = PR \times \frac{h_{fg}(T_0)}{h_{fg}(T_s)} \tag{3}$$

The third parameter, is the specific electric consumption (SEC); it is defined as the ratio input work (separation) to the water product flow rate,

$$SEC = \frac{W_{sep}}{M_d} \tag{4}$$

There is a difference between both the gain output ratio and the specific electrical ratio in its measuring units, so it cannot be comparable without a conversion factor.

The forth parameter, is the recovery ratio (R_r); it is known as the ratio between the water product flow rate to the feed sea water flow rate

$$R_r = \frac{M_d}{M_{sw}} \tag{5}$$

b. Separation Process of a Mixture

A desalination plant essentially separates incoming saline water into product water and brine. The product water contains a low concentration of dissolved salts and brine contains the remaining high concentration of dissolved salts. The plant requires a certain amount of energy in the form of heat or work [2, 16].

The specific Gibbs function g is defined as the combination property as, g = h − Ts, if it is assumed to represent the desalination system as a black box to represent the system allover;

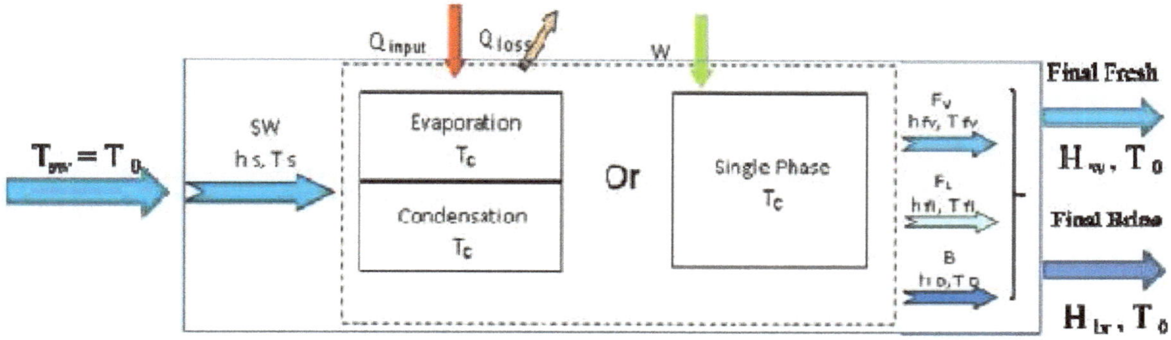

Figure (3). *Indicates the control volume which confines all inlets and outlets*

Both the separation work and heat required are calculated by the thermodynamics laws over the control volume selected,

From the control volume;

$$\hat{W}_{sep.} + Q + (m \times h)_{sw} = m_d \times h_d + m_b \times h_b \tag{6}$$

Where, m is the mass flow rate, h is the specific enthalpy, s_w is the seawater stream, m_d is the distillate product stream, and b is the brine stream

By dividing the above equation by the ambient temperature T_0,

$$\hat{S}_{gen.} + \frac{Q}{T_0} + (m \times S)_{sw} = m_d \times S_d + m_b \times S_b \tag{7}$$

Where, S is the specific entropy

From the specific Gibbs function g, and the above two equations

$$\hat{W}_{sep.} - \hat{S}_{gen.} \times T_0 + (m \times h)_{sw} - (m \times S)_{sw} = (m \times h)_d - (m \times S)_d \times T_0 + (m \times h)_b - (m \times S)_b \times T_0 \tag{8}$$

$$\hat{W}_{sep.} = (m \times g)_d + (m \times g)_b - (m \times g)_{sw} + \hat{S}_{gen.} \times T_0 \tag{9}$$

If the operation process is in the limit of reversible operation, i.e. the entropy term is zero

($S_{gen.} \times T0 = 0$, at the limit of reversible)

$$\therefore \hat{W}_{sep.}\,(least) = m_d \times g_d + m_b \times g_b - m_{sw} \times g_{sw} \tag{10}$$

If the heat is used to produce desalinated water for a desalination system instead of the work, so the heat of separation is the predominate parameter.

For Carnot engine,

$$Q_{sep.} = \hat{W}_{sep.} + Q \tag{11}$$

$$\frac{Q_{sep.}}{m_d} = \frac{\hat{W}_{sep.}}{m_d} + \frac{Q}{m_d} \tag{12}$$

$$\therefore \frac{\hat{W}_{sep.}}{m_d} = \frac{Q_{sep.}}{m_d}\left[1 - \frac{T_0}{T_H}\right] \tag{13}$$

For the reversible operation heat engine;

$$\therefore \frac{Q_{sep.}}{m_d} = \frac{\hat{W}_{sep.}}{m_d}\left[\frac{1}{(1-\frac{T_0}{T_H})}\right] \tag{14}$$

$$= \frac{\hat{W}_{sep.} + \hat{S}_{gen.} \times T_0}{m_d\left[1 - \frac{T_0}{T_H}\right]}$$

From the mass balance, salinity balance and above equations,

$$m_{sw} = m_d + m_b \tag{15}$$

$$m_{sw} \times X_{sw} = m_d \times X_p + m_b \times X_b$$

$$X_b = \frac{m_{sw} \times X_{sw} - m_d \times X_d}{m_b} = \frac{X_{sw}}{1-R_r} - \frac{R_r \times X_d}{1-R_r} \tag{16}$$

as, $\frac{\hat{W}_{sep.}}{m_d} = g_d + \frac{m_d}{m_d} x g_b - \frac{m_{sw}}{m_d} x\, g_{sw}$

$$\therefore \frac{\hat{W}_{sep.}}{m_d} = g_d + \frac{m_{sw} - m_d}{m_d} x\, g_b - \frac{m_{sw}}{m_d} x\, g_{sw} \tag{17}$$

It is important to observe from the above equation, that the minimum work of separation is affected by the salinity of both the feed seawater and the product streams and not affected by the way used for desalination.

$$\frac{\hat{W}_{sep}}{m_d} = g_d + [\frac{1}{R_r} - 1] \text{ x } g_b - \frac{1}{R_r}\, x\, g_{sw} \qquad (18)$$

The minimum separation work can be calculated from equation (18), for different water salinity and recovery ratios at a certain temperature. It is obvious from this equation that, at a certain degree of salinity increasing the recovery ratio from 0 percent to 100 percent the separation work is increased and it decrease with decreasing the salinity at the same recovery. Figure (4) below illustrates the minimum separation work at different salinities of the incoming feed water with pure product and different recovery ratio. These curves indicate also, the lowest and highest minimum separation works which corresponding to 0 and 100 percent recovery.

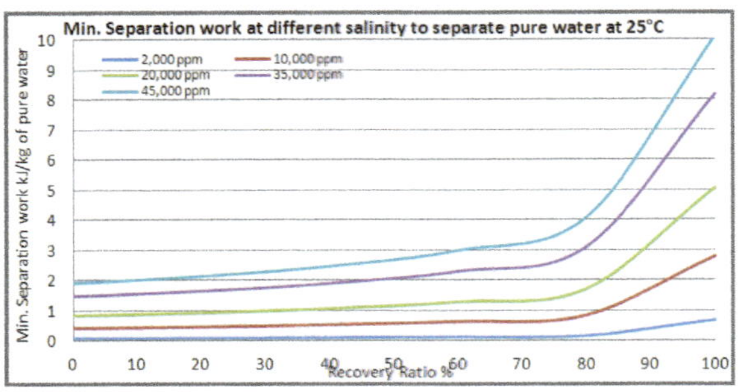

Figure (4). Indicates the minimum separation at different salinities with pure water product

It is concluded that, the minimum separation work increases strongly at higher recovery ratios greater than 80 percent so it is required to avoid the design stages of any desalination units to work under a recovery ratio of 80 percent. It is clear that the work required is dependant completely on the feed and product salinity (increase the feed salinity increase the required work and increase the product salinity reduced the required work).

5. Exergy Analysis of Desalination Plant

5.1. Msf Plant Description

The MSF unit consists of 22 stages, the flow diagram of this unit is shown in figure (5), it works at a salinity of 46,500 ppm and feed seawater temperature at 35°C and rate of 2397 kg/s, where a value of it is returned back to the pool of the last stage at a temperature of 43.3°C and at a rate of 808 kg/s, after mixing with the brine in its salinity goes to 64,828 ppm [2,11,14].

The circulated pump delivered a rate of flow of 3621 kg/s to the recovery section, so its temperature increases to 85°C, then it enters to the brine heater to raise its temperature to 90.8°C.

The brine heater uses steam at a rate of 37.02 kg/s at a temperature of 98.9°C.

The condensates vapor through all stages at a rate of 272 kg/s. The salinity of the brine discharge to the sea at a rate of 536 kg/s and salinity of 70,093 ppm.

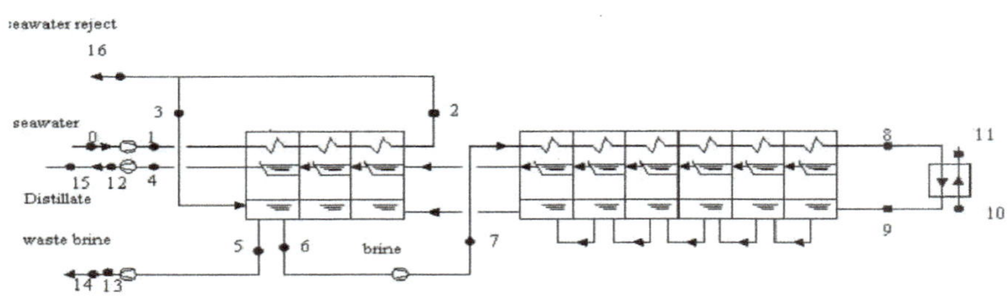

Figure (5). Illustrates the Multi-Stage Flash unit flow sheet process (MSF-BR)

5.2. Msf Plant Exergy Analysis

The Multi-Stage Flash (MSF) unit under this study is analyzed:

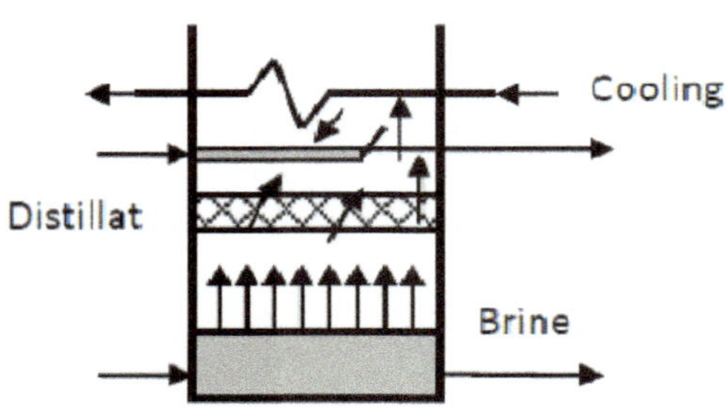

Figure (6). Illustrates the details of one stage of the MSF process

5.3. MSF Unit Overall Exergy Analysis [13, 15]

Total specific exergy stream consists of,

$$E_{total} = E_{th} + E_{ch} + E_{po} + E_{k.e}$$

Both the E_{ch}, E_{po}, $E_{k.e}$ can be ignored, so the total exergy can be expressed as,

$$E_{total} = E_{th} = (h - h_0) - T_0(s - s_0) \qquad (19)$$

$$\text{Exergetic efficiency } (\eta_{II}) = \frac{Wmin.}{Ein.} \qquad (20)$$

$$\text{Exergetic Destruction} = E_{in.} - E_{out.} \qquad (21)$$

For stage efficiency

$$\eta_{II}(\text{stage}) = \frac{Eusefull}{Ein.stage}$$

$$= \frac{(Ein.-Eout.)distillate + (Eout.-Ein.)cooling}{(Ein.-Eout.)brine} \quad (22)$$

For standard condition, at T_0= 308 K, P_0= 101.3 KPa, and salinity = 64,500 ppm (which is referred to the dead state). At these conditions, the specific heat of salt can be taken to be Cp_s = 0.8368 kJ/kg.K, also the enthalpy and entropy of the salt at the ambient conditions are hs_0 = 29.288 kJ/kg and ss_0 = 0.1009175 kJ/kg.K.

The enthalpy and entropy of salt at certain temperature T can be computed from the following equations [11].

$$h_s = hs_0 + Cp_s * (T-T_0) \quad (23)$$

$$s_s = ss_0 + Cp_s \ln(T/T_0) \quad (24)$$

The molar mass of NaCl = 58.5 kg/kmol and the molar mass of water 18 kg/kmol. For seawater of salinity of 64,500 ppm; the mass fraction of salt (mf_s) = 0.0645 and the mass fraction of water = 0.9355.

The sum of the Mole fraction (X) of salt and water at the feed seawater equal 1.

i.e.) $X_s + X_w = 1$

$$X_s = \frac{Mw}{Ms\left(\frac{1}{msf}-1\right)+Mw} \quad (25)$$

$$\therefore X_s = \frac{18}{58.5\left(\frac{1}{0.0645}-1\right)+18} = 0.02077 \simeq 2.077\ \%$$

$$\therefore X_w = 1 - 0.02077 = 0.9792 \simeq 97.92\ \%$$

So, the molar salinity of the feed seawater is around 2.077 %. The enthalpy and entropy of salt and water mixture (as assumed that the seawater is an ideal solution) can be calculated from these equations:

$$h = mf_s * h_s + mf_w * h_w \quad (26)$$

$$s = mf_s * s_s + mf_w * s_w \quad (27)$$

The exergy rate for a flow for any fluid stream can be calculated using (Cengel and Boles, 1998)

$$\text{Exergy rate } (E) = m[\, h - h_0 - T_0(s - s_0)\,] \quad (28)$$

By using the above equations, the exergy at each point of the system can be evaluated. This is illustrates in tables (1, 2).

Table (1). *Indicates the specific enthalpy and entropy 0f both Salt and Water*

State	T (K)	P (Bar)	hs (kJ/kg)	Ss (kJ/kg.K)	hw (kJ/kg)	Sw (kJ/kg.K)
0	308	1.013	29.288	0.1009175	146.727	0.50516
1	308	1.68	29.288	0.1009175	146.727	0.50516
2	316.3	1.15	36.233	0.1231691	181.407	0.61623
3	316.3	1.15	36.233	0.1231691	181.407	0.61623
4	314.4	0.0796	34.644	0.1181274	173.469	0.59107
5	316.3	0.0796	36.233	0.1231691	181.407	0.61623
6	316.3	0.0796	36.233	0.1231691	181.407	0.61623
7	316.3	6.35	36.233	0.1231691	181.407	0.61623
8	358	6.35	71.128	0.2268	355.854	1.13441
9	363.8	6.35	75.981	0.240248	380.207	1.20194
10	371.9	0.974	Brine H.	Brine H.	414.279	1.29463
11	371.9	0.974	Brine H.	Brine H.	414.279	1.29463
12	314.5	5.78	34.727	0.1183935	173.887	0.59239
13	316.3	2.92	36.233	0.1231691	181.407	0.61623
14	308	1.013	29.288	0.1009175	146.727	0.50516
15	308	1.013	29.288	0.1009175	146.727	0.50516
16	308	1.013	29.288	0.1009175	146.727	0.50516

Table (2). *Indicates the enthalpy and entropy of saline water*

state	T (K)	X (g/litre)	m (kg/s)	h (kJ/kg)	S(kJ/kg.K)	Sp. Exergy(kJ/kg)
0	308	46.5	2397	139.152	0.4791	0
1	308	46.5	2397	139.2141	0.4791	0.0621
2	316.3	46.5	2397	172.043	0.5844	0.4787
3	316.3	46.5	808	172.034	0.5844	0.4787
4	314.4	0	272	173.469	0.59107	10.521
5	316.3	70.093	536	171.231	0.58166	-3.475
6	316.3	64.828	3621	171.995	0.58426	-2.668
7	316.3	64.828	3621	171.955	0.58426	-1.985
8	358	64.828	3621	337.395	1.07557	12.035
9	363.8	64.828	3621	360.4838	1.13959	15.526
10	371.9		37.02	414.279	1.29463	389.002
11	371.9		37.02	414.279	1.29463	0
12	314.5	0	272	173.887	0.59239	11.046
13	316.3	70.093	536	171.2313	0.58166	-3.322
14	308	70.093	536	138.4953	0.47683	-3.912
15	308	0	272	146.727	0.50517	10.332
16	308	46.5	1589	139.152	0.4791	0

The intake stream salinity taken as the dead state "0", so the exergy at that point is zero and the exergies of the brine streams are negative "-ve" as it is clear in the above table due to its salinities are above the salinity of the dead state.

It is obvious that when the salinity is above the salinity above the dead state the fluid stream need an input work to reduce the stream salinity, and vice versa for the pure water, it can produce work.

The minimum exergy is the difference between the exergies of the output streams and the input stream,

- The min. separation work:

The Exergy (min.) rate = $E_{14} + E_{15} - E_0$ = 713.472 kW

= 2.62 kJ/kg

= 0.729 kWh/m3

@ T = 35°C (the min. separation work curves above)

- The exergy changes through the pumps:

For the recirculated Pump,

$$\Delta E_{rec.} = E_7 - E_6 = 2473.143 \text{ kW}$$

For distillate Pump,

$$\Delta E_{dist.} = E_{12} - E_4 = 142.8 \text{ kW}$$

For brine Pump,

$$\Delta E_{brine} = E_{13} - E_5 = 82.008 \text{ kW}$$

For intake Pump,

$$\Delta E_{intake} = E_1 - E_0 = 148.853 \text{ kW}$$

∴ E pumps = 2846.804 kW

The pump system efficiency is assumed to be 80 %, the total pumps exergy equal to:

$$\text{Total pumps E} = E_{pumps} / \eta_p$$

∴ Total pumps E = 2846.804 / 0.8 = 3558.505 kW

- The brine heater, the exergy supply to the brine by heating steam,

$$\text{Exergy rate due to steam } (E_{steam}) = m_s * (h_{fg} - T_0 * s_{fg}) \quad (29)$$

$$E_{steam} = 14340.0672 \text{ kW}$$

∴ E total in. = Epumps + Esteam = 17898.5722 kW

The Second law Efficiency of the MSF unit under study equal:

$$\eta_{II} = E_{min.} / E_{total\ in.} = 713.472 / 17898.5722 = 3.986\ \% \sim 4\ \%$$

it is clear, that the second law efficiency is very low for the MSF, as the useful value is only equal to 713.5 kW through very high consumed energy of 17,899 kW, i.e the total destruction energy of the unit is equal to 17,185.5 kW

For calculating the exergy destruction through the components of the MSF unit:

1. Energy destruction for the cooling brine,

$$E_{Des.\ Cooling\ brine} = E_2 - (E_3 + E_{16}) = 783.6494 \text{ kW} \quad (4.56\%)$$

2. $E_{Des.\ distillate} = E_{12} - E_{15}$ = 194.208 kW (1.13%)
3. $E_{Des.\ brine} = E_{13} - E_{14}$ = 316.24 kW (1.84%)
4. Exergy destruction for system pumps:

$$E_{Des.\ Pumps} = 3558.505 - 2846.804 = 711.701 \quad (4.14\%)$$

5. The exergy destruction through both the recovery and rejection sections (Edes.(rec.+rej.):

$$E_{Des.(rec.+rej.)} = E_{in.} - E_{out.} = 13480.1407 \text{ kW} \quad (78.44\ \%)$$

6. The exergy destruction through the brine heater

∴ $E_{Des.\ BH.}$ = 1759.943 kW (10.24 %)

It is clear that, the highest destruction in the MSF unit is through the hear recovery and heat rejection section by a value of 78%, so this indicates that the process through the flashing in both recovery and rejection is very low efficient. This followed by the exergy destruction through the brine heater section by a value of 16%, followed by the pumps and cooling brine by a value of 4.27% and 4.98% respectively as indicated in figure (7).

For all distillers, the major exergy destruction has been occurred in the recovery section where it can reach to around 60 percent.

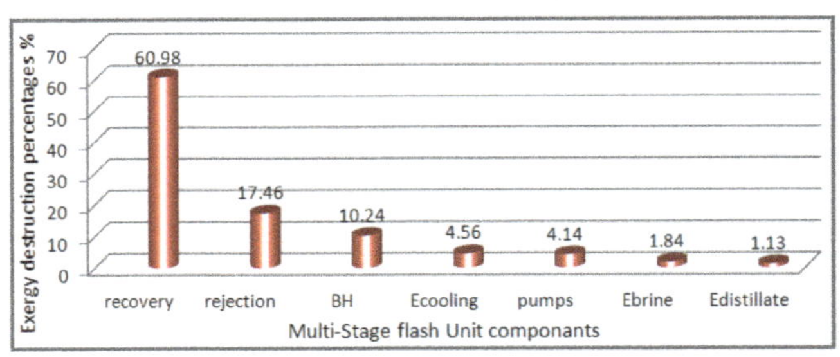

Fig (7). Exergy destruction through MSF unit Componants

So the specific exergy losses through the brine heater, the rejection section, and the losses through the outlet streams is lower than the recovery section. This is because of the greater the heat transfer area of the recovery section than the sum of the brine heater and rejection section area (greater by several times).

The unit under study is working at a top brine temperature of 90°C and at a performance ratio of 7.5 at these conditions, the specific exergy losses at the range of 63 kJ/kg. The next figure represents a comparison of the exergy losses between this study and other two plants, the differences between these units are due to many parameters may be salinity, pumps efficiency, plant capacity, top brine temperature and accurate calculation.

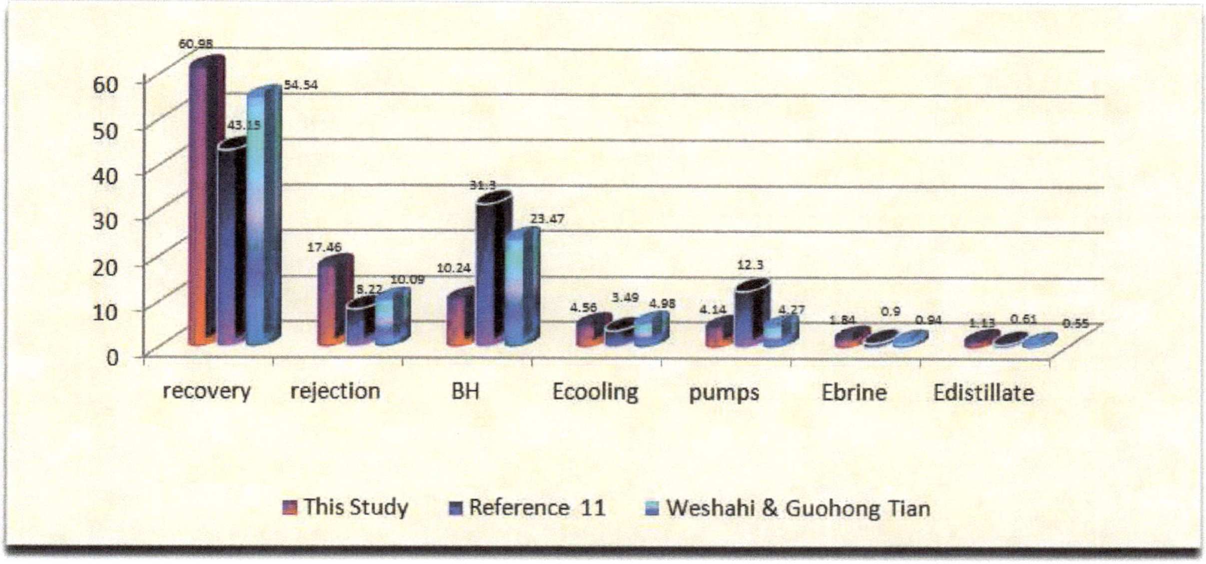

Figure (8). *indicates a comparison of the exergy losses between this study and other two plants*

5.4. Single Effect Evaporator Mechanical Vapor Compression (SEE-MVC)

A single effect evaporator (SEE-MVC) model is considered. A model developed by Narmine et al. [7] is chosen for this study, the design data and model results are shown in tables 3, 4.

Table (3). *Design data for MVC unit*

Parameters	value
Seawater inlet temperature	20°C
Seawater inlet salinity	35,000 ppm
Discharge brine salinity	10 ppm
Top brine temperature	69.5°C
ΔT evaporator-condenser	3.8°C
Recovery ratio (R	47%
Isentropic compressor efficiency	85%
Compressor inlet pressure	30 kPa

Some of the equations used for calculating the exergy generated and the exergy destruction through the unit [4,15,17,18]:

- The exergy balance for the evaporator can be expressed as,

$$(E_f + E_s)_{in.} = E_d + E_b + E_{des.evap.} \quad (30)$$

- For the specific exergy,

$$Espec. = \sum Ethermal + Echemical + Epressure$$

The chemical and pressure exergy can be neglected, so the specific exergy can be taken to be thermal only.

$$\therefore E_{spec.} = (h - h_0) - T_0(s - s_0) \quad (31)$$

The destructive exergy through the evaporator can be evaluated by:

$$E_{des.\ evap.} = E_{in.} - E_{out}$$

$$= m_s * (e_{Compo.o} - e_{cond.}) + m_f * e_f - m_b * e_b - m_d * e_v \quad (32)$$

- For the compressor,

$$W_{comp.} = m_d * (h_s - h_v) / \eta \quad (33)$$

$$E_{Des.\ comp.} = P_{shaft} - m_d (e_{comp.o} - e_v) \quad (34)$$

The exergy losses for the MVC system:

The rate of the work lost in a process is equal to the rate of dissipated exergy. So the rate of the exergy destruction D which is related to the created entropy production " S^{cr} " known by Maxwell-Gouy-Stodola (MGS) equation[3],

$$\frac{dD}{dt} = -T0 * \frac{dE}{dt} = -T_0 * m * \Delta X \quad (35)$$

where, the entropy creation S^{cr} and exergy destruction are losses and ΔX is the driving force related to the flow rate m.

$$\therefore X = \frac{1}{m}\frac{dS}{dt} = S^{cr} \quad (36)$$

For the unit under study:

As seen in the above table the unit operates at a temperature of 70°C

For a boiling point elevation corresponding to a temperature of 70°C and salinity of 35,000 ppm;

The BPE= 0.75°C

$$T_v = 70 - 0.75 = 69.25°C$$

$$\Delta T = 3.5°C, \quad Tc = 69.25 + 2.5 = 71.75°C$$

$$T = \frac{Tv+Tc}{2} = (69.25 + 71.75)/2 = 70.5°C$$

$$h_{fg} = 2331.5 \text{ kJ/kg} = 41967 \text{ J/mol}$$

$$\Delta X = h_{fg} * \Delta T / T^2 = 41967 * 2.5 / (343.5^2) = 0.88 \text{ J/mol. K}$$

Destructive exergy = $T_0 * \Delta X$ =293 * 0.88 = 260.6 J/mol.

= 260.6/(18 * 3.6) = (5/24) x 4 kW= 0.83 Kw

The second law efficiency (η_{II}) = 6.4 %

The model output data can be summarized in the following table,

Table (4). *Model output data*

Parameters	value
Specific electricity consumption	14 kWh/m3
Compression ratio (Cr)	1.3
Product water temperature	26.5°C
Discharge brine temperature	25°C
Second law efficiency	6.4 %

6. Conclusion

Increasing of the demand of fresh water is the main driving force for developing of the seawater desalination technologies all over the world and especially in the Middle East area. So this study is concluded in the analysis and the enhancement for the cases studied:

6.1. Analysis

1. The overall exergy efficiency of this MSF unit is only around 4 % which offers potential for improvement.
2. The exergy efficiency is the highest in the first stage and going down in the direction of the last stage.
3. For this MSF unit, the study indicates that around 78 % of exergy destruction happened through the distillers (stages). Most of it occurs through the heat recovery section, this is clear in figure (7), and the specific exergy losses at the range of 63 kJ/kg.
4. Periodically, the Acid cleaning showed be performed, this leads to enhance the plant performance and at the same time this reduced the exergy destruction,
5. The overall exergy efficiency of the MVC unit is around 7 % which offers potential for improvement.

6.2. Enhancement

1. Recovering the waste heat from the hot distillate water can enhance the exergy efficiency of the MSF units.
2. Using low grade steam as a heating source is efficient for thermal desalination plants.
3. The hot distillate extract can be used as a source of powering other processes such as MED.
4. Using high efficient pumps of (89%) can reduce the electrical consumed power as the destructive energy through pumping system reach 4 %.
5. Applied of the co-generation systems, by coupling the power plants with desalination plants, through this the waste heat of the power plants as a heat source for desalination units, this will enhance the performance of the desalination plants.

Nomenclature

Symbols

BH	brine heater
BPE	Boiling point elevation, °C
Cp	Specific heat capacity, kJ/kg. K
CR	Compression ratio
E	rate of exergy kJ/s
G	specific Gibbs function
GOR	gain output ratio
GR	Gain ratio
H	Enthalpy, kJ/kg
h_f	Saturated enthalpy, kJ/kg
hfg	Latent heat, kJ/kg
M	mass flowrate kg/s
Mf	mass fraction
MSF	Multi stage flash system
n	Number of stage
P	Pressure, kPa / bar
Ppm	Part per million
PR	performance ratio
Q	heat kJ
Rr	recovery ratio
S	entropy generation rate kJ/k
SEC	specific electric consumption
SEE-MVC	single effect evaporator
T	temperatur K
TBT, T_1	Top brine temperature, °C
Tf	Feed seawater temperature, °C
To	Environment state temperature, K
Tv	Saturated vapor temperature, °C
W	work kW
X	Salt concentration, ppm
ΔT	Temperature difference per stage, oC

Subscripts

0, amb.	dead state, ambient condition
B	Brine
comp.	compressor
cond. / c	condenser
cr	Creation
D	Distillate
Des.	destructive
evap.	evaporator
F	Feed
gen.	generation
H	heat reservoir k
in.	inlet, incoming
out.	Outlet
s	steam /salt
sep.	separation
Sw	Seawater
V	Vapor
W	Water

References

[1] S. Kazadi, Y. Hong, C. Chau, A. Chaudhary, A. Chaudhary, J. Park, J. Liu, M. Kim, D. Kim, S. Kim Desalination Powered By Entropy

[2] Yunus Cerci, Yunus Cengel, Byard Wood, Nafiz Kahraman, and E. Sinan Karakas Desalination and Water Purification Research and Development Program, 2003

[3] K. S. Spiegler, Y. M. El-Sayed the energetic of desalination processes, Desalination 134 (2001) 109 - 128

[4] Fuad N. Alasfour, Hassan K. Abdulrahim, the effect of stage temperature drop on MVC thermal performance, Desalination 265 (2011) 213-221

[5] Water desalination technologies in the ESCAWA member countries, United Nations, New York, 2001

[6] Huseyin Yapici, Nesrin Kayatas, Nafiz Kahraman and Gamze Basturk, Numerical study on local entropy generation in compressible flow through a suddenly expanding pipe, Entropy 2005, 7, 38-67

[7] Narmine H. Aly, Adel K. El-Fiqi, Mechanical vapor compression desalination systems-a case study, Desalination 158 (2003) 143-150

[8] Devin L Shaffer, Laura H Arias Chavez, Moshe Ben-Sasson, Santiago Romero-Vargas Castrillon, Ngai Yin Yip, and Menachem Elimelech, Desalination and reuse of high salinity shale gas produced water: Drivers, Technologies, and Future Directions, Environmental Science & Technology, 2013

[9] Goran Wall and Mei Gong, on exergetics, Economics, and Desalination, http://www.exergy.se/

[10] K Ghana Sundari, Veerabhadra Ponnaganti and T Michael N Kumar, Exergy analysis of a low temperature thermal desalination system, Int. J. Mech. Engineering and Robotics Research, vol. 2, No. 1, January 2013

[11] Kamal Mohammedi, Anissa Talamali, Youcef Smaili, Imane Saadoun, Aomar Ait-Aider, Environmental impact of seawater desalination plants: case study in Algeria, American Journal of Environmental Protection, 2013; 2 (6):141-148

[12] Mostafa H. Sharqawy, John H. Lienhard V and Syed M. Zubair, Thermophysical properties of seawater: A review of existing correlations and data, Desalination and water treatment, 16 (2010) 354-380

[13] Anwar Binamer, Second law and sensitivity analysis of large ME-TVC desalination units, Desalination and water treatment, (2013) 1-12

[14] Osman A. Hamed, Mohammad AK Al-Sofi, Monazir Imam, Ghulam M. Mustafa, Khalid Ba-Mardouf and Hamad Al-Washmi, thermal performance of multistage flash distillation plants in Saudi Arabia, presented in International desalination association world congress on desalination and water reuse, " the value of water in the 21st century," San Diego, USA, 29 August -3 sept. 1999

[15] O.A. Hamed, A. M. Zamamiri, S. Aly and N. Lior, thermal performance and exergy analysis of a thermal vapor compression desalination, Energy Conversion Mgmt Vol. 37, No. 4, pp. 379-387, 1996

[16] Karan H. Mistry and John H. Lienhard, Generalized least energy of separation for desalination and other chemical separation process, Entropy 2013, 15, 2046-2080

[17] J. M. Veza, Mechanical vapor compression desalination plants-A case study, Desalination, 101 (1995) 1-10

[18] Nafey A. S., Fath, H.E.S., Analysis of a new design of multi-stage flash-mechanical vapor compression desalination processes, desalination 204 (2007) 482-500

13

Isolating Bioactive Compound from Marine Prosobranch *Purpura persica* from Tuticorin Coast

Santhi V.[1], V. Sivakumar[2], S. Jayalakshmi[3], R. D. Thilaga[4], M. Mukilarasi[5]

[1]Research Centre and P.G Department of Zoology, J. A. College, Periyakulam, Tamil Nadu, India

[2]P.G and Research Department of Zoology, V. O. Chidambaram College, Thoothukudi, Tamil Nadu, India

[3]Department of Marine Sciences, Annamalai University, Parangipettai, Tamil Nadu, India

[4]P.G and Research Department of Zoology, St. Mary's College, Thoothukudi, Tamil Nadu, India

[5]Departmemt of Computer Science, Sakthi Engineering College, Chennai, Tamil Nadu, India

Email address:

velu.santhi62@gmail.com (Santhi V.), drvsivak@gmail.com (V. Sivakumar), jayacas@gmail.com (S. Jayalakshmi), rdthilaga12@gmail.com (R. D. Thilaga), mukil.shanthi@gmail.com (M. Mukilarasi)

Abstract: Antibiotics have revolutionized life saving medicine by providing cure for many number of life threatening diseases in human history, unexpectedly, many pathogenic microorganisms have developed resistant towards current antibiotic and this trend has become more and more serious. Hence the present study has been aimed to find out new marine derived antibiotic from Prosobranch mollusc *Purpura persica.* The whole body crude extract of methanol was partially purified by normal phase silica gel 160-120 mesh (Glaxo, Bombay) column chromatography with low polar to high polar solvent Hexane: Chloroform (F1); Chloroform (F2); Benzene (F3); Benzene: Methanol (F4), and Methanol F5). The antimicrobial activity of crude and eluted fractions were assayed against ten bacterial pathogens viz *Aeromonas hydrophila, Bacillus cereus, Escherichia coli, Pseudomonas aerogenosa, Salmnella typhi, Shigella flexneri, Vibrio cholera 0139, Vibrio cholera classical, Vibrio cholerae 01790 and Vibrio cholerae EITOR* and nine fungal pathogens viz. *Aspergillus flavus, Aspergillus terreus, Aspergillus niger, Aspergillus fumigatus, Fusarium moniliforme, Trichoderma sp. Penicillium citrinum, Penicillium oxallicum and Rhizopus sp.* respectively using the agar disc diffusion method. To find out the most probable antibiotic compound HPLC and GC-MS studies were carried out. Among the tested bacterial pathogens *S. typhi, P. aerogenosa, S. flexneri* and *B. cereus* and fungal pathogens *A. fumigatus, A.* terreus, *F. moniliforme* and *Trichoderma sp.* showed inhibition in growth by crude, F2, F3 and F5 fractions of *P. persica* respectively. The GC-MS and HPLC analysis revealed the presence chloridate cholest-5-en-3-01(3a) - carbano chloridate, a chloride compound 9, 12-octadecadienoyn chloride (z, z), and a cholest-5-en-3-ol(3)-carbonochloridate, eugenol, dibutyl phthalate, 1,2-Benzenedicarboxylic acid, diisooctyl ester and Phthalic acid, bis (7-methyloctyl) ester, and a steroid cholest-5-ene, 3-bromo-(3a), 2-piperidinone a monoterpene azulene, a fluro compound acetic acid, tri fluro- tetradecyl which were responsible for inhibiting the growth of microbes tested and the present test organism *P. persica* have great potential for developing useful drugs.

Keywords: Antibacterial Activity, Solvents, Inhibitory Zone, GC-MS Analysis, Test Pathogens

1. Introduction

The knowledge acquired in the past two decades and the discovery of new groups of antimicrobial compounds makes natural antibiotics the basic element of a novel generation of drugs for the treatment of bacterial and fungal infections [1, 2]. Different therapeutic applications of these compounds, from topical administration to systematic treatment of infections, have been developed by several biotechnological companies (http://www.inimaxpharma.com; http://biotech .deep13.com//alpha/alpha.html)

It is widely accepted among clinician medical workers,

microbiologist and pharmacologist that antibiotic resistant will, in the near future, leave health care professional without effective therapies for bacterial and fungal infections. The critical events are the emergence of *Staphylococcus aureus* with decreased sensitivity to methicilin [3], worldwide resistance to penicillin in *Staphylococcus pneumonia* and multiple resistances to *Mycobacterium tuberculosis*. The emergence among enterococci of resistance to another useful and widely effective antibiotic, vancomycin [4], might accelerate the spread of vancomycin- resistant genes via plasmids; throughout other species eventually limiting the efficacy of this drug. Consequently the priority for the next decades should be focused in the development of alternate drugs and as the recovery of natural molecules should be as natural as possible with a wide range of action over several pathogens easy to produce and not prone to induce resistant.

A very different kind of substances have been obtained from marine organism among other reasons because they are living in a very exigent, competitive and aggressive surroundings very different in many aspects from the terrestrial environment, a situation that demands the production of quite specific and potent active molecules. The knowledge of the physiological and bio chemical features of marine organism might contribute to the identification of natural products of bio medical importance. Over the last 30 years many structurally novel antimicrobial metabolites have been isolated from marine organisms.

Marine invertebrates offer good source of potential antimicrobial drugs [5, 6]. Studies on antimicrobial mechanisms and compounds of marine invertebrates may provide valuable information for new antibiotic discoveries and give new insights into bioactive compounds in molluscs. Among the invertebrates, the molluscs are very good source for biomedical important products [7]. Many classes of molluscs with bioactive compounds like antitumour, antileukemic, antibacterial, cytotoxic, anti inflammatory and antiviral properties have been reported [8,9]. These reports suggest that molluscs are the rich source for discovering novel lead compounds for the possible development of new types of antibiotics for pharmaceutical use. Keeping the importance of gastropods in terms of bioactive compounds with antibacterial properties, the present study has been undertaken to determine the antimicrobial activity of extracts from *P. persica* against various human pathogenic microorganisms.

2. Materials and Methods

2.1. Collection and Preparation of Samples

The mollusc *P. persica* was collected from intertidal rocky shore of harbour area of Gulf of Mannar, nearby Theraspuram Tuticorin, situated in the South east coast of India, during April 2015 to December 2015. The collected samples were rinsed with sterile sea water and carefully removed from their shells. The flesh was cut into small pieces and air-dried. The air-dried flesh was immersed in 100% A. R. Grade methanol for 10 days at room temperature. The extract from the solvents was filtered by using Whatman no. 1 filter paper and evaporated to dryness in rotary evaporator and the dried extract was stored at 0°C for further use.

2.2. Microbial Strains Used

Antimicrobial activity of tissue extracts were determined against 10 different bacterial pathogens, viz., *Aeromonas hydrophila, Bacillus cereus, Escherichia coli, Pseudomonas aerogenosa, Salomnella typhi, Shigella flexneri, Vibrio cholera* 0139, *Vibrio cholera classical, Vibrio cholerae 01790 and Vibrio cholerae* EITOR. These clinical strains were obtained from Basic Biomedical sciences, Bharathidasan University, Trichy.

2.3. Antimicrobial Susceptibility Assay

In vitro anti bacterial activity was assayed by the disc diffusion method of Bauer *et al.,* [10], Antibacterial activity was expressed in diameter zone of inhibition which was measured with the outer side of the disc to inner side of the inhibition zone. Each active extract was tested thrice for confirmation of activity.

Invitro antifungal activity was determined using the techniques of Kelman *et al.,* [11]. The fungi pathogens *Aspergillus flavus, Aspergillus terreus, Aspergillus niger, Aspergillus fumigatus, Fusarium moniliforme, Trichoderma sp. Penicillium citrinum, Penicillium oxallicum and Rhizopus sp.* were obtained from TNAU, Coimbatore. Crude extract was assayed for anti fungal activity following the above mentioned disc diffusion method.

After initial screening the extracts showing broad spectrum were fractionated using normal phase silica gel 160-120 mesh (Glaxo, Bombay) column chromatography with low polar to high polar solvent Hexane: Chloroform (F1); Chloroform (F2); Benzene (F3); Benzene: Methanol (F4), and Methanol F5). The fractions thus obtained were once again evaporated, concentrated, and assayed for antibacterial and antifungal activity. The extracts showing broad spectrum activity was examined for MIC by testing at different concentrations viz; 1mg/ml, 10mg/ml, and 100mg/ml.

The most potent fractions of test mollusc were subjected to HPLC and GC/MS to characterize the possible compounds responsible for antimicrobial activity.

GC-MS analysis: GC-MS analysis was carried out on a GC Clarus 500 Perkin Elmer system comprising a AOC 20i auto sampler and gas chromatography interfaced to a mass spectrometer (GC-MS).

2.4. Identification of Compounds

Interpretation on mass spectrum GC-MS was conducted using the data base of National Institute Standard and Technology (NIST) from Dr. Duke's phytochemical and Ethnobotanical databases.

2.5. Confirmation of Some of the Characterized Compounds by HPLC

In order to prepare the correct solvent system for the most potent fraction of test animal (F3 fraction of *P. persica*) the silica gel column purified extracts of test animals were repurified by alumina column and TLC for HPLC analysis. The isolated compounds were analyzed by HPLC (Shimadzu, Advanced VP binary series) using a C18 Hhypersil (ODS) column. Standards were run to confirm the presence of the same in the isolated compounds of the test animals.

3. Results and Discussion

3.1. Antibacterial Activity of Extracts from Purpura persica

The crude extract inhibitory zone of *P. persica* range varied from 2mm (*P. aerogenosa*) to 7mm (*S. typhi*) (Figure 9). The crude extract was partially purified through silica gel column chromatography. Of the five fractions F2, F3and F5 of *P. persica* exhibited wide spectral antibacterial activity in most of the bacterial strains tested (Figure 10, 11 & 12). F2 fraction was highly active against *S. typhi* (19mm) and it also showed inhibitory activity against *P. aerogenosa* (13mm and *S. flexneri* (12mm). F3 fraction the maximum inhibitory zone was obtained in *V. cholerae* classical and *B. cereus* (11mm). Maximum spectrum antibacterial activity was attained at 100mg concentration against *S. typhi* (13mm) and minimum against *V. cholerae* 0179 (3mm) in F2 fraction.

Maximum inhibitory zone was exhibited against *B. cereus* (9mm) and minimum against *E. coli* (4mm) by F3 fraction. Several marine molluscan extracts possessed broad spectrum antibacterial activities affecting the growth of bacteria, fungi and yeasts [12]. The crude whole body methanol extracts of *P. persica* showed wide spectrum antibacterial activity against seven pathogens out of ten pathogens tested. Rajaganapathy [13] also reported that the methanol extract from the whole body of *Hemifuses pugilinus* exhibited activity against *Bacillus subtilis, Escherichia coli* and *Klebsiella pneumonia.* Similarly Santhanaramasamy and Murugan [14] experimentally analyzed the methanolic extract of *C. virgineus* and *C. ramosus* and they also observed the broad spectrum antibacterial activity of body tissue extract. Thilaga [15] screened the antibacterial activity *of* a marine mollusc *Babylonia spirata* against *bacterial* pathogens.

Similar result was reported by Chellam *et al.*, [16] in chloroform extract of *Pterai chinensis* which inhibited eight fish pathogens and the acetone extract in the same animal showed broad spectral activity against all the fish pathogens tested. Abraham *et al* [17] studied the antibacterial activity in alcoholic extracts of holothurians species inhibited *S. typhi.* The present study with test animal P. persica also corroborates the earlier findings of suppressing the activity of S. typhi. Even then also difference in antibacterial activity found in the molluscan extracts may depend on the solvents used for extracts and the compounds extracted [18]. Gnanamhal *et al.,* [19] found that 0.07 mg of *Trochus radiatus* inhibited *Proteus mirabilis* and 0.15 mg of extract inhibited *Serratia marcescens*. Some of the peptides obtained from oysters inhibited *E. coli* at 330 mg/ml and *Vibrio alginolyticus* at 162 mg/ml [20].

3.2. Antifungal Activity of Extracts from Purpura persica

Crude methanol extract of *P. persica's* inhibitory range varied in between 2mm (*A. flavus*) and (*A. fumigatus*) 6mm (Figure 1) (Figure 13). Among the silica gel fractions, maximum activity was observed in fraction F2 (Figure 14). Maximum inhibition zone was obtained against *A. fumigatus* (24mm) (Figure 2) and minimum against *A. niger* (5mm) in fraction F2. It also showed higher degree of inhibition zone against *A.* terreus (23mm), (Figure 3), F. moniliforme (22mm) (Figure 4), *Trichoderma sp.* (20mm) (Figure 5), *P. citrinum* (19mm) (Figure 6). *A. flavus* (18mm) and *P. oxallicum* (17mm) (Figure 7) Fractions F1 and F5 (Figure 15) also shown inhibition zone against *A. terreus* (21mm) and *Trichoderma* sp. (16 mm) respectively. In MIC maximum antifungal activity was exhibited against *F. moniliforme* (12 mm), (Figure 8) *A. terreus* (12 mm) and minimum against *Rhizopus* sp. (2 mm) at 1 mg level and maximum against *A. fumigatus* (!9mm) and minimum against *A. niger* (1mm) at 100mg level in F2 fraction (Figure 16). F1 fraction exhibited a maximum antifungal spectrum against *Aspergillus terreus* (10mm) even at 1 mg.

As a result potential activity was found in *P. persica's extract* against *A. fumigatus, A.* terreus, *F. moniliforme* and *Trichoderma sp.* Prem Anand and Patterson Edward [21] reported moderate antifungal activity from the extract of various bivalve molluscs. Highest inhibitory activity was observed against *Aspergillus niger* with methanol extract of *Microcosmus curvus* by Karthikeyan *et al.*, [22]. Similar result was obtained by Chandran *et al.*, [23, 24, 25] in *Perna viridis methanol* extract against *Aspergillus flavus* and *Mucor* sp. Comparable work was that of Mohamed Hussain & Ananthan [26, 27] who observed that the chloroform extract of *Didemnum psammathodes and Didemnum candidum* inhibited *Penicillium* sp. whereas no antifungal activity was noticed against *A. flavus*, *A. fumigatus* and *A. niger*. Several marine natural products showed significant antifungal activity. Callipeltins J and K, MIC at 1 μm [28]; the triterpene glycoside holothurin B, MIC at 1.56 μg/ml [29]; the macrolids neopeltolide MIC at 0.62 μg/ml [30]); AP, a polypeptide type AMP isolated from the Chilean scallop *Argopecten purpuratus*, showed antifungal activity against *F. oxisporum* and *Saprolegnia parasitica* [31].

3.3. Characterization of Most Potent Fractions of Test Mollusc

As the F2 & F3 of *P. persica* showed more antimicrobial activity these fractions were subjected to HPLC and GC-MS to characterize the complete carbon skeleton responsible for probable antimicrobial derivative from the test molluscs.

3.4. GC/MS and HPLC

When F2 fraction was subjected to GC/MS analysis a steroid compound with chloridate cholest-5-en-3-01(3a) - carbano chloridate with maximum percentage of 72.31% and a chloride compound 9, 12-octadecadienoyn chloride (z, z), were identified as antimicrobial compounds. In F3 fraction (Figure 17, 18, 19 & 20) of *P. persica* only six antimicrobial compounds were identified such as a phenolic compound eugenol, three Plasticizer compounds, dibutyl phthalate, 1,2-Benzenedicarboxylic acid, diisooctyl ester and Phthalic acid, bis (7-methyloctyl) ester, and a steroid cholest-5-ene, 3-bromo-(3a)-. Among the anti microbial compounds present, 79.95% was the 1, 2-Benzenedicarboxylic acid, diisooctyl ester. An alkaloid compound 2-piperidinone was found to be in maximum (12.89%) when F5 fraction (Figure 21, 22, 23, 24 & 25) of *P. persica* was subjected to GC/MS analysis. Apart from that this fraction also had a monoterpene azulene, a fluro compound acetic acid, tri fluro- tetradecyl ester (4.88%) and an alcoholic compound 1-Octanol, 2, 7-dimethyl (5.57%).

Various characterized compounds were confirmed in accordance with the merging of retention time shown by the standard retention time in HPLC. The retention factor of 12.018min, 7.603min, 14.20min and 19.275 of the isolated compounds from F3 fraction of *P. Persica* on HPLC were completely identical to that of standard azulene (Figure 26) diethyl phthalate (Figure 27), oleic acid (Figure 28) and eugenol (Figure 29). Marine molluscan extract are usually complex mixtures of bioactive molecules mainly proteins, peptides and sterols.

The HPLC and GC/MS spectra from *P. persica's extract* provided a complete carbon skeleton of chloridate cholest-5-en-3-01(3a) - carbano chloridate, a chloride compound 9, 12-octadecadienoyn chloride (z, z), and a cholest-5-en-3-ol(3)-carbonochloridate, a phenolic compound eugenol, three Plasticizer compounds, dibutyl phthalate, 1,2-Benzenedicarboxylic acid, diisooctyl ester and Phthalic acid, bis (7-methyloctyl) ester, and a steroid cholest-5-ene, 3-bromo-(3a) alkaloid compound 2-piperidinone a monoterpene azulene, a fluro compound acetic acid, tri fluro-tetradecyl ester and an alcoholic compound 1-Octanol, 2, 7-dimethyl which might be responsible for the inhibition of *S. typhi, P. aerogenosa, S. flexneri, B. subtilis, A. fumigatus, A. terreus* and *F. moniliforme* in the present study.

Similar finding was reported by Way On Mudianta *et al.*, [32] in an Indonesian sponge Halichondria sp. has provided 3- alkyl piperidine alkaloids tetradehydrohaliclonacyclamine A, the mono-N- oxide and a C-2 epimer. Brominated indoles 6- bromo 2-methylthioindolin-3-one extracted from Australian Muricid *Dicathais orbita* has been identified as anticancer drug indole derivatives of 6, 6′ -dibromoindigo have been antimicrobial activity [33]. An alkaloid Batzelladine L and M isolated by Hua *et al.*, [34] inhibits *S. aureus*.

Antifungal activity of crude and various fractions of *Purpura persica* against different pathogens

Fractionated extract (F2) of *Purpura persica* against pathogens

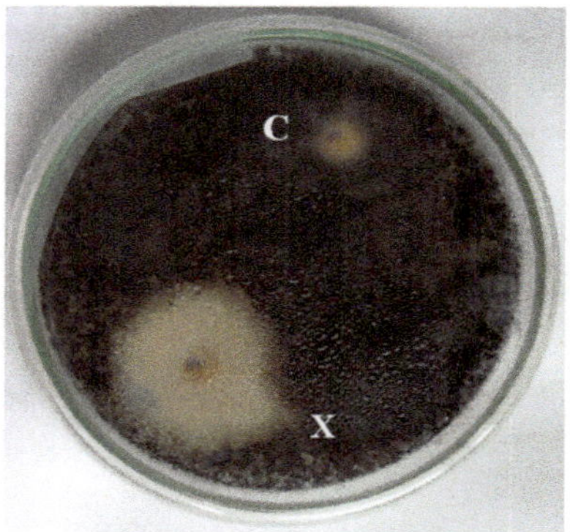

Figure 1. *Aspergillus fumigatus.*

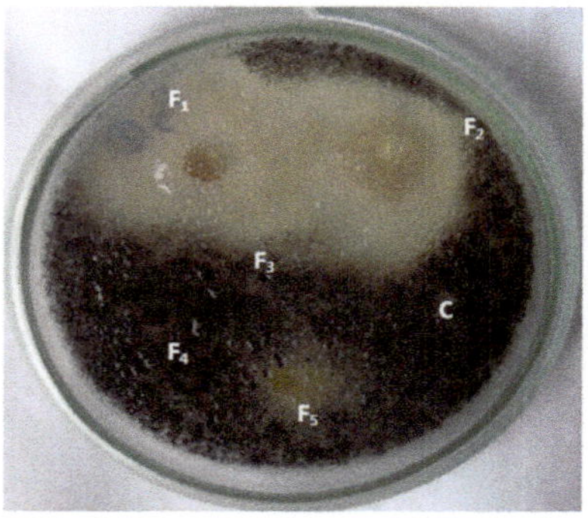

Figure 2. *Aspergillus fumigatus.*

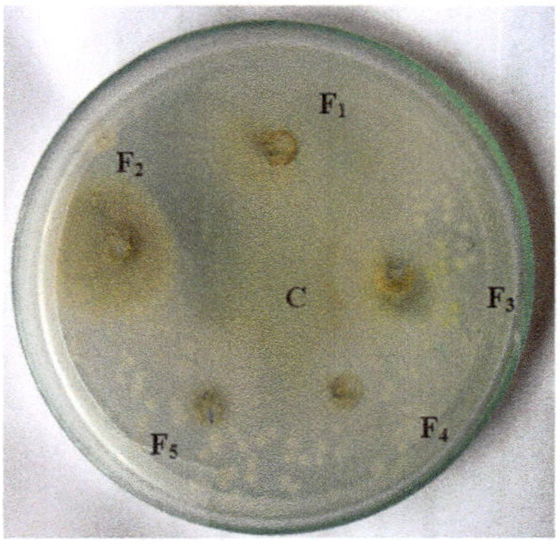

Figure 3. *Aspergillus terreus.*

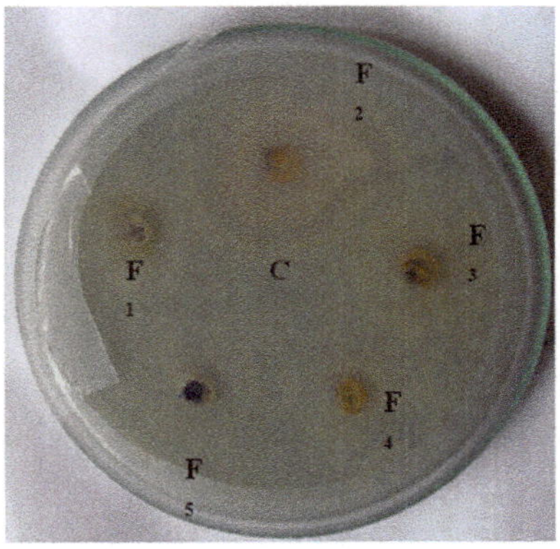

Figure 4. *Fusarium moniliforme.*

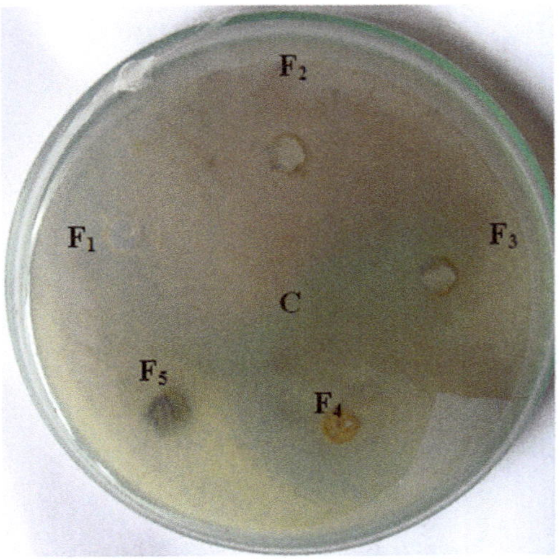

Figure 5. Trichoderma sp.

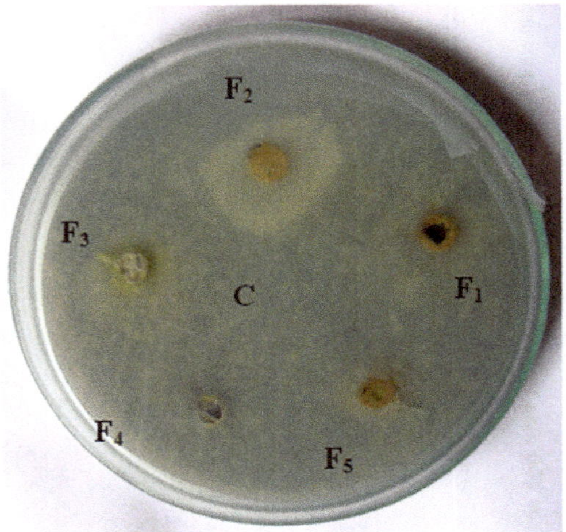

Figure 7. Penicillium oxalicum.

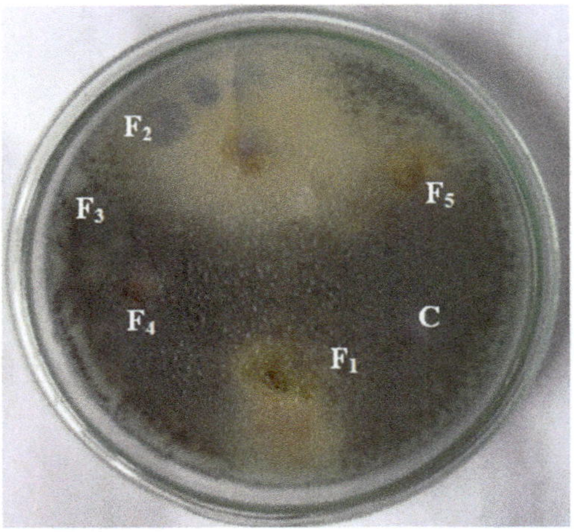

Figure 6. Pencillium citrinum.

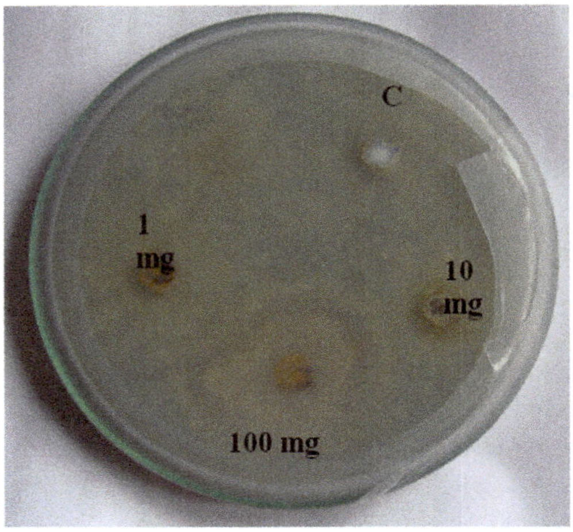

Figure 8. Fusarium moniliforme.

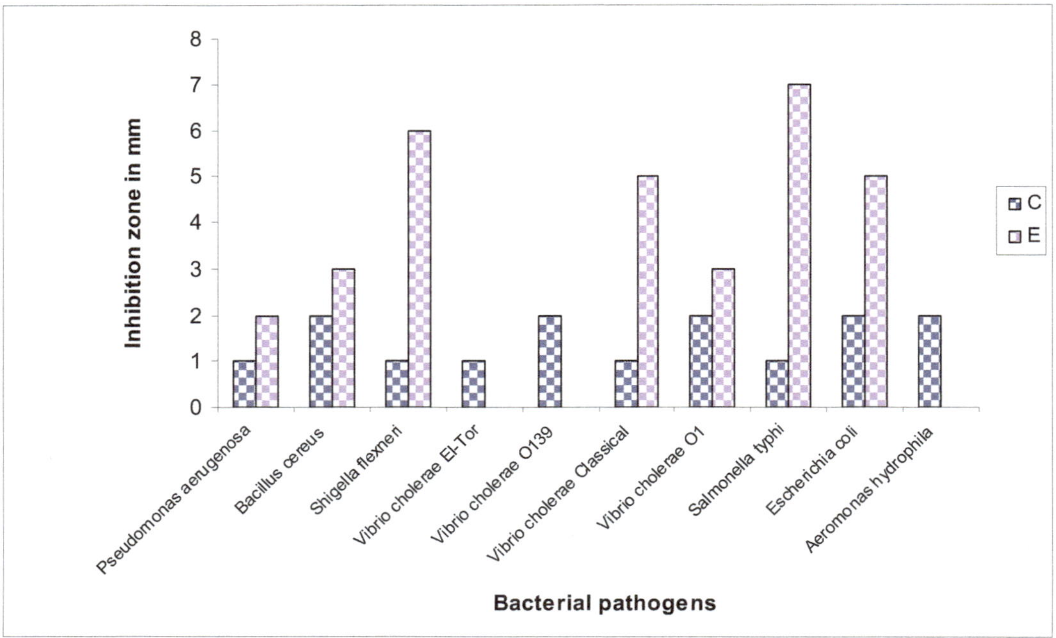

Figure 9. Antibacterial activity of crude extract of Purpura persica against different pathogens.

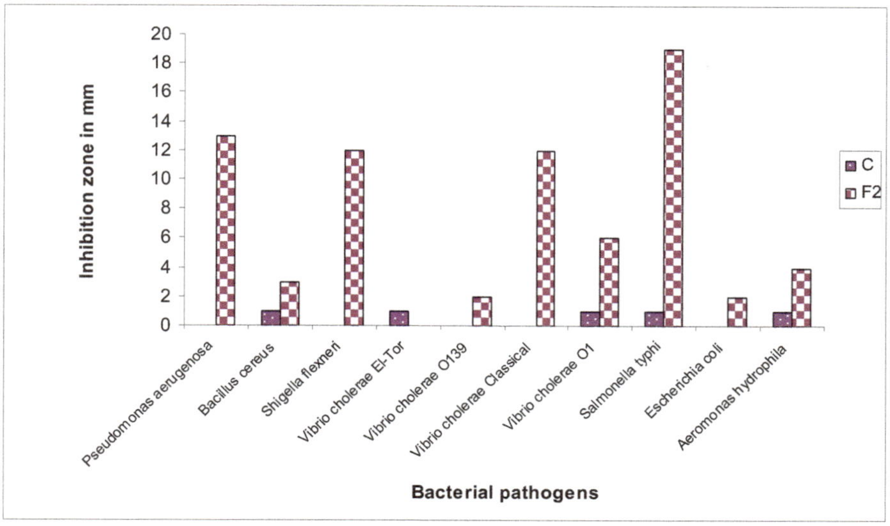

Figure 10. *Antibacterial activity of F2fraction of Purpura persica against different pathogens.*

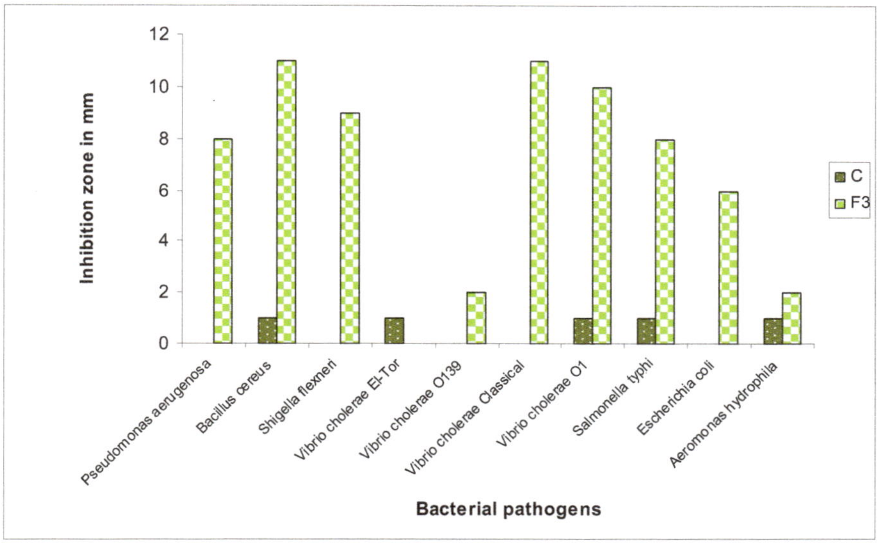

Figure 11. *Antibacterial activity of F3fraction of Purpura persica against different pathogens.*

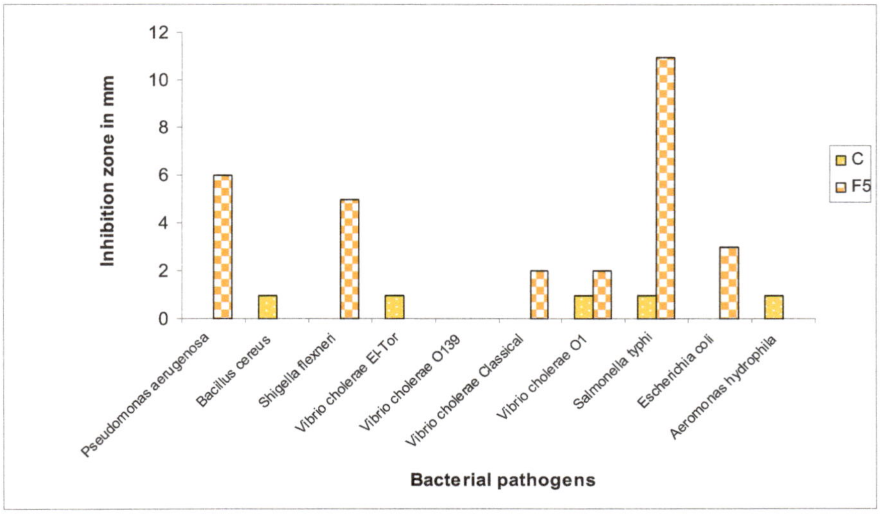

Figure 12. *Antibacterial activity of F5 fraction of Purpura persica against different pathogens.*

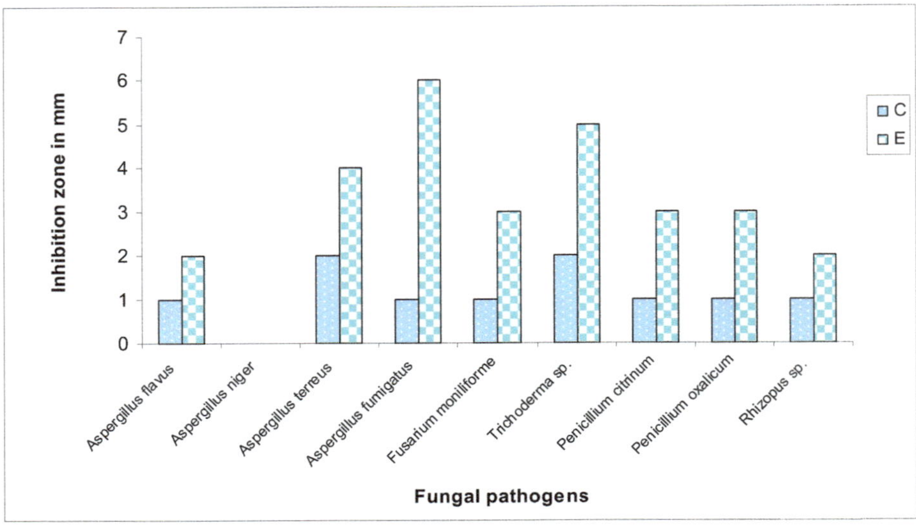

Figure 13. *Antifungal activities of crude extract of Purpura persica against different pathogens.*

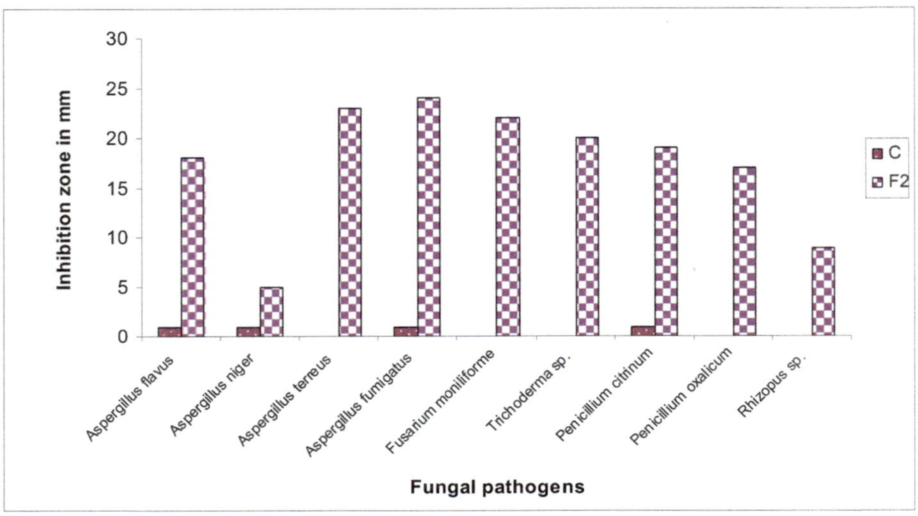

Figure 14. *Antifungal activities of F2 fraction of Purpura persica against different pathogens.*

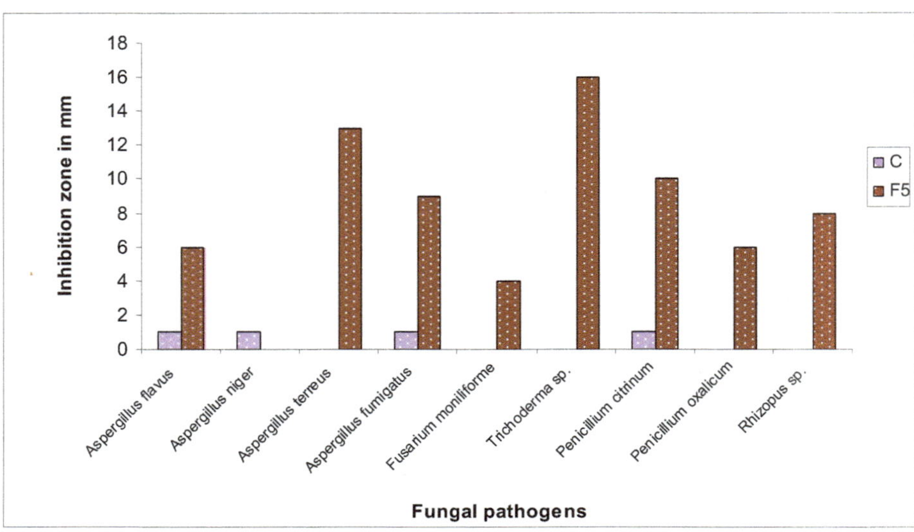

Figure 15. *Antifungal activities of F5 fraction of Purpura persica against different pathogens.*

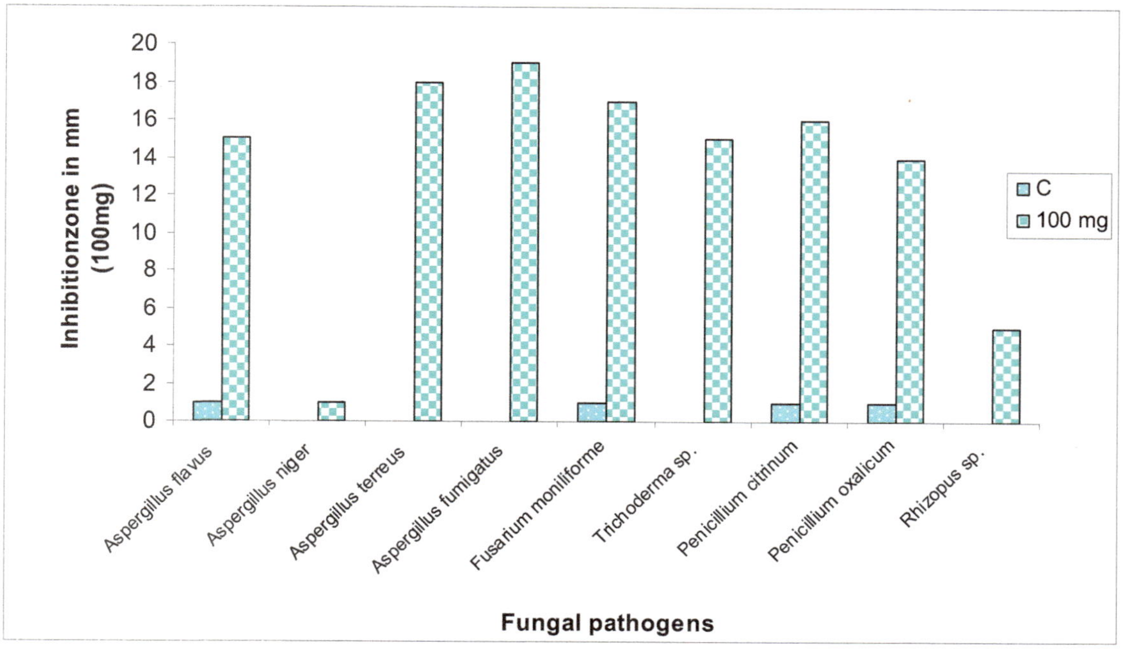

Figure 16. *Antifungal activities of minimum inhibitory concentration of Purpura persica against different pathogens.*

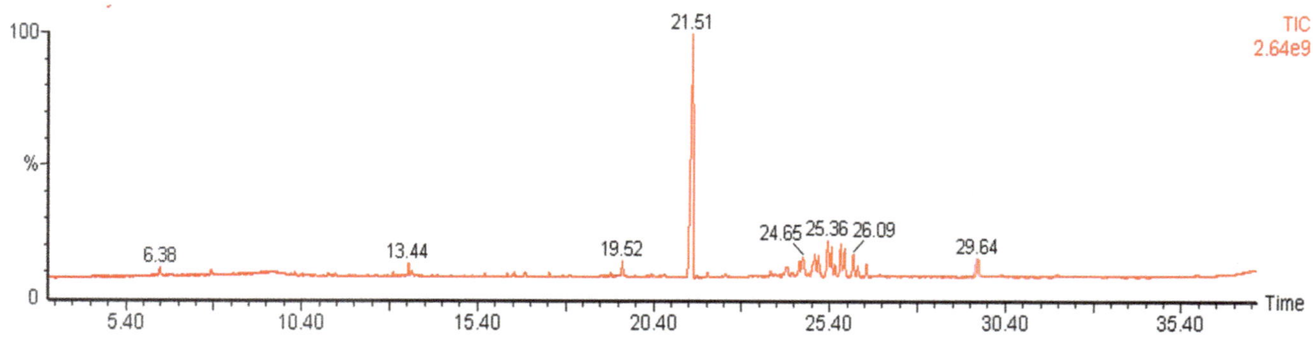

Figure 17. *Chromatogram of column extract (F3) of P. persica by GC-MS.*

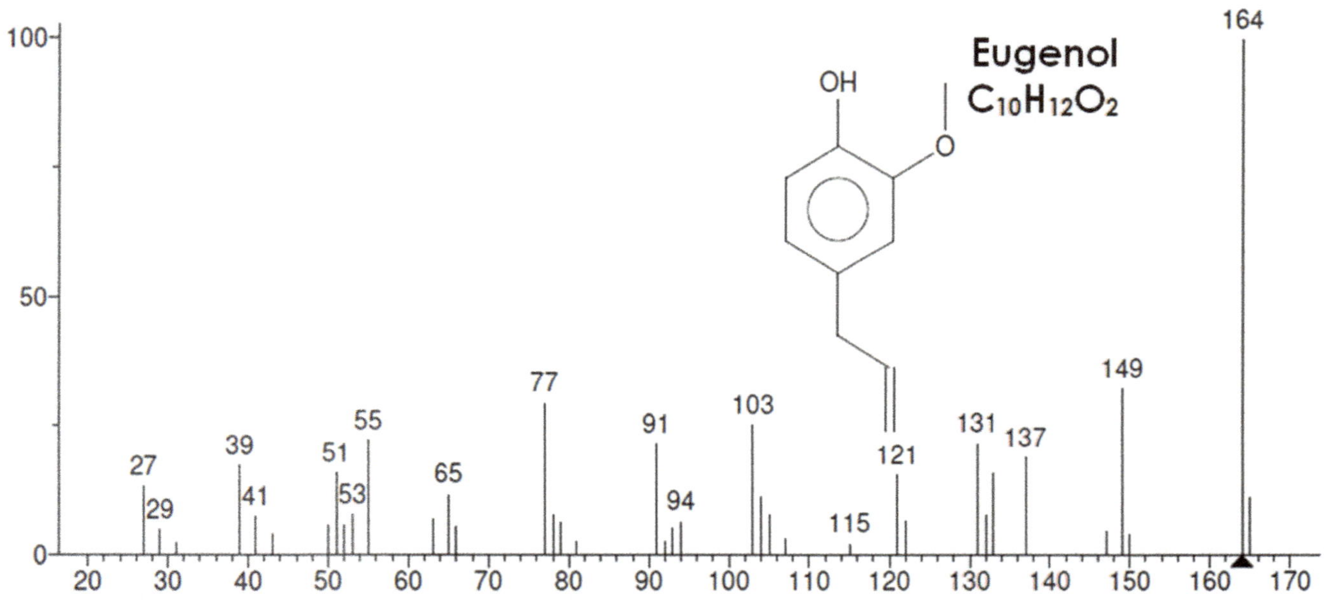

Figure 18. *Chromatogram Component in F3 fraction of P. persica – Eugenol.*

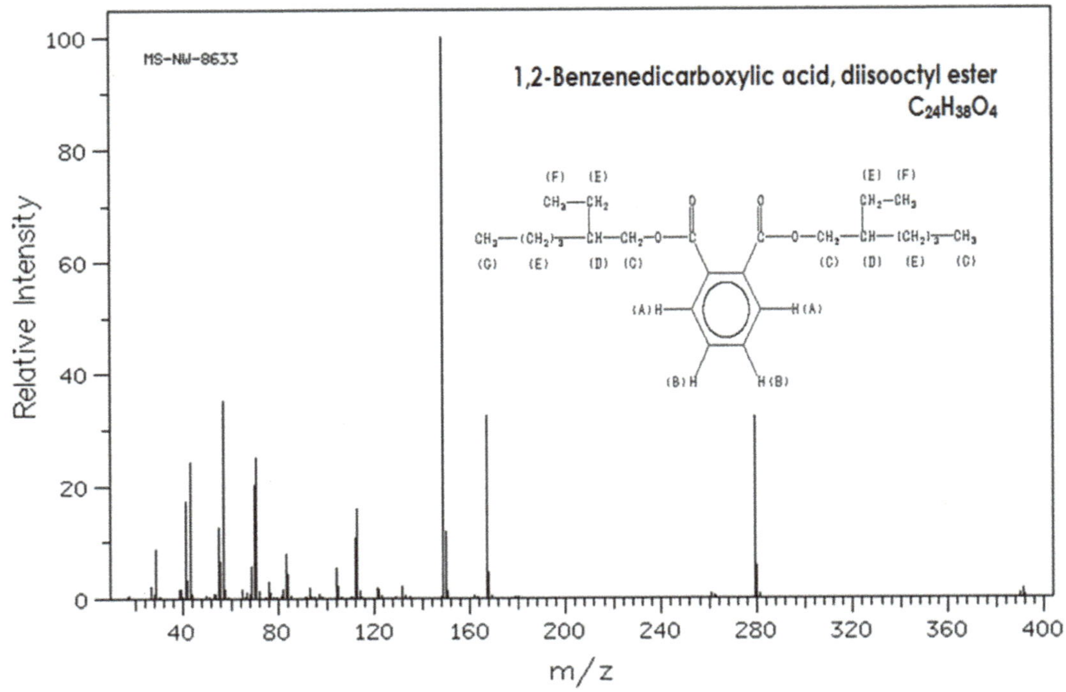

Figure 19. *Chromatogram Component in F3 fraction of P. persica - 1,2-Benzenedicarboxlic acid, diisooctyl ester.*

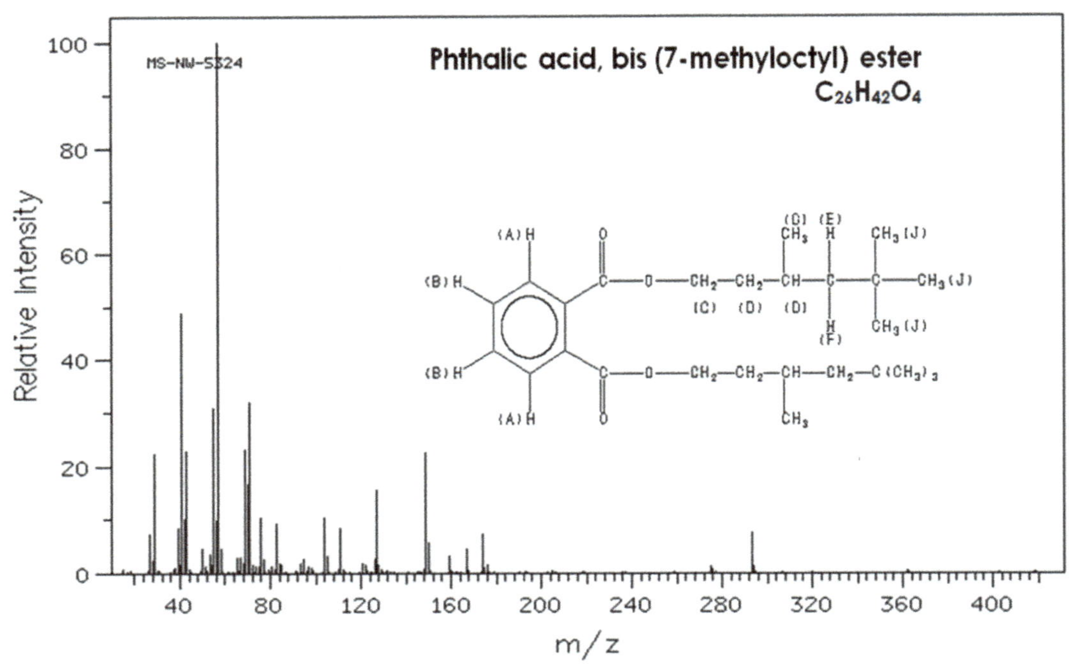

Figure 20. *Chromatogram Component in F3 fraction of P. persica Phthalic acid, bis (7- methyloctyl) ester.*

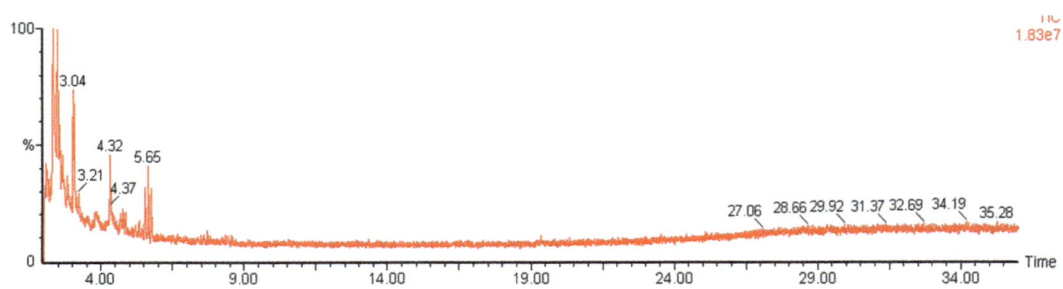

Figure 21. *Chromatogram of column extract (F5) of P. persica by GC-MS.*

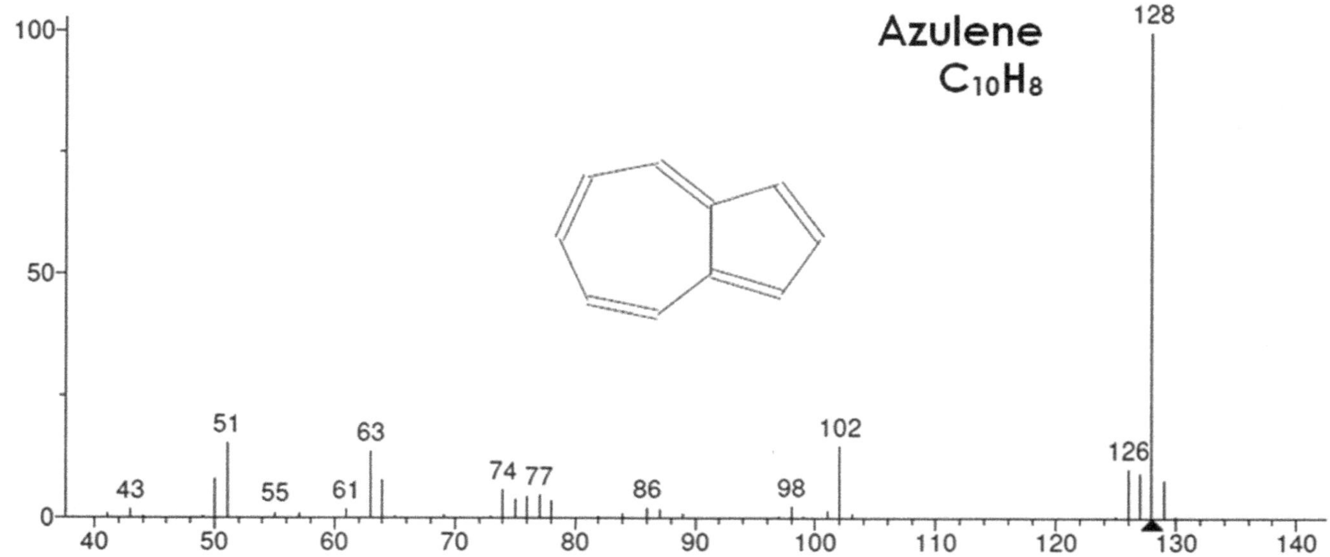

Figure 22. Chromatogram Components in F5 fraction of P. persica - Azulene.

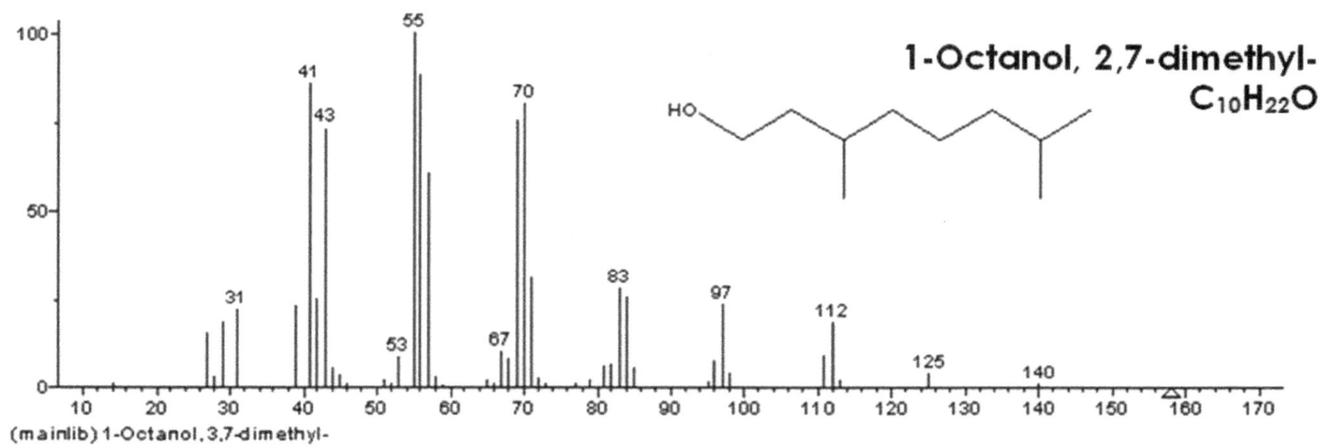

Figure 23. Chromatogram Components in F5 fraction of P. persica -1 Octanol, 2,7-dimethyl.

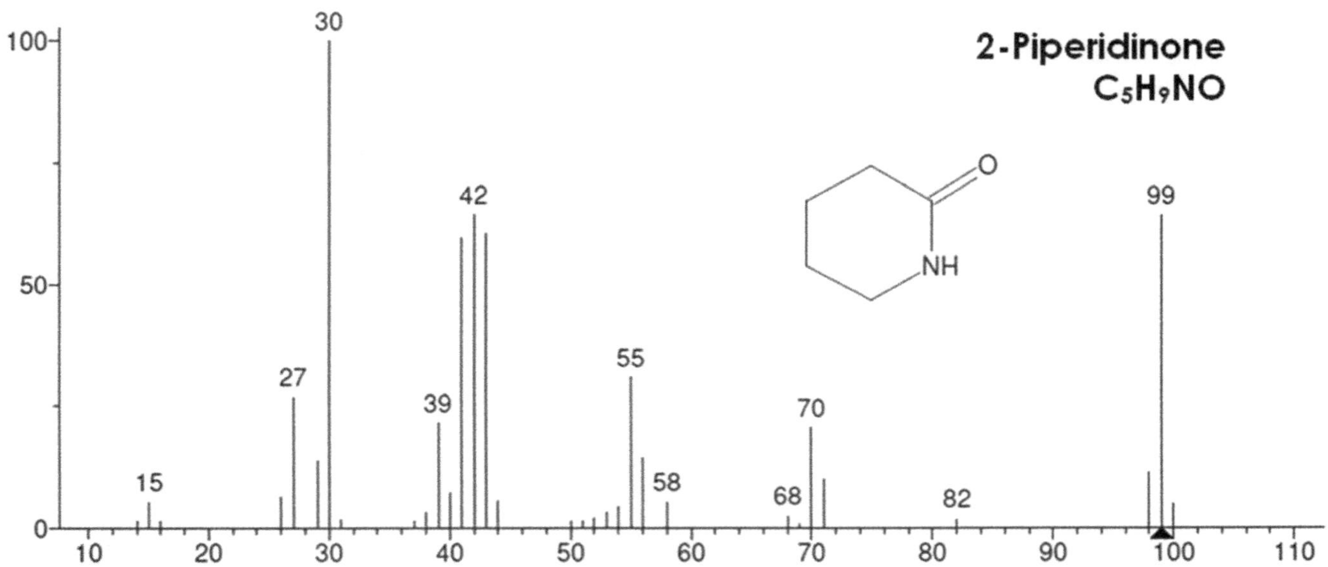

Figure 24. Chromatogram Components in F5 fraction of P. persica – Piperidinone.

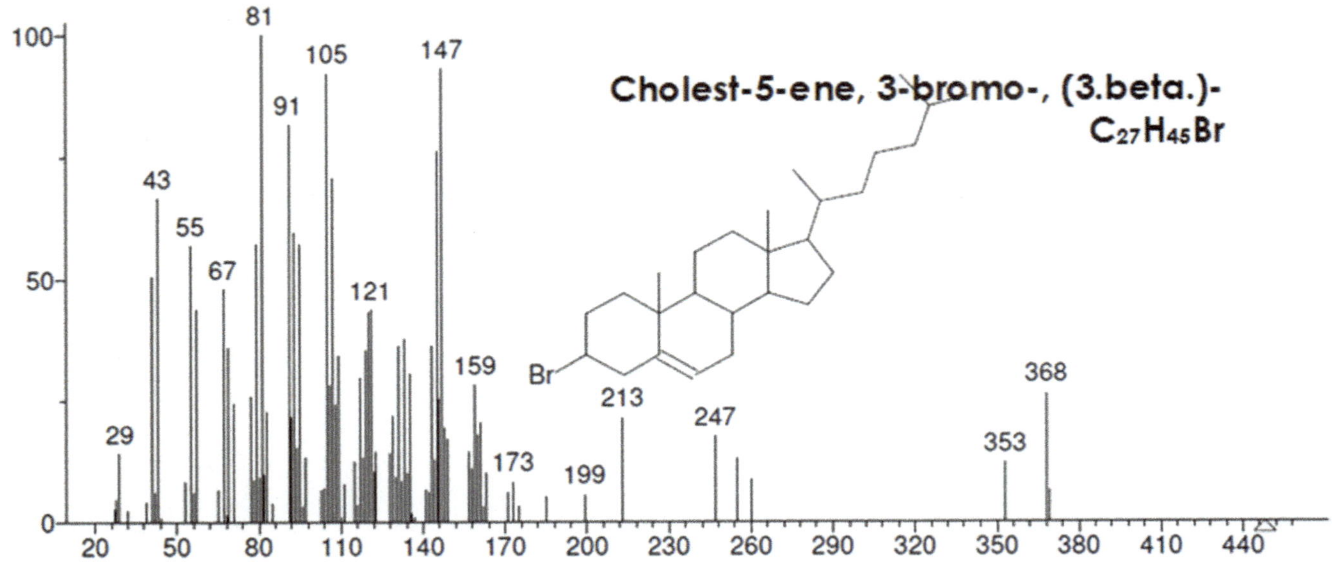

Figure 25. *Chromatogram Component in F5 fraction of P. persica* - Cholest-5-ene,3-bromo-,(3.beta.)-.

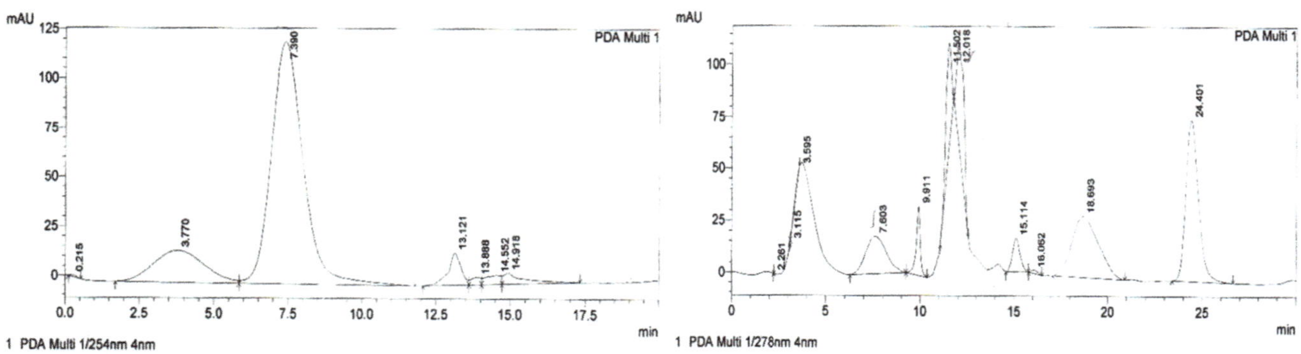

Figure 26. *HPLC chromatogram of isolated compound (TLC) of (F5 S1 F3) P. persica compared to the standard azulene.*

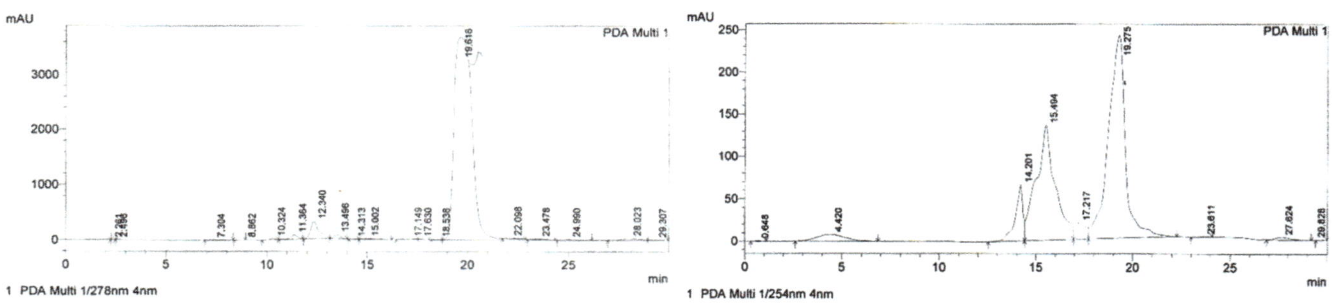

Figure 27. *HPLC chromatogram of isolated compound (TLC) of (F3 S1 F4) P. persica compared to the standard eugenol.*

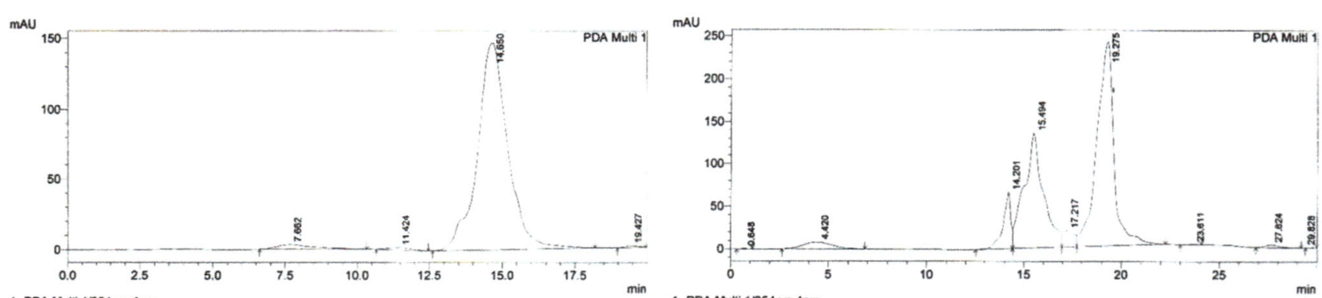

Figure 28. *HPLC chromatogram of isolated compound (TLC) of (F5 S1 F5) P. persica compared to the standard oleic acid.*

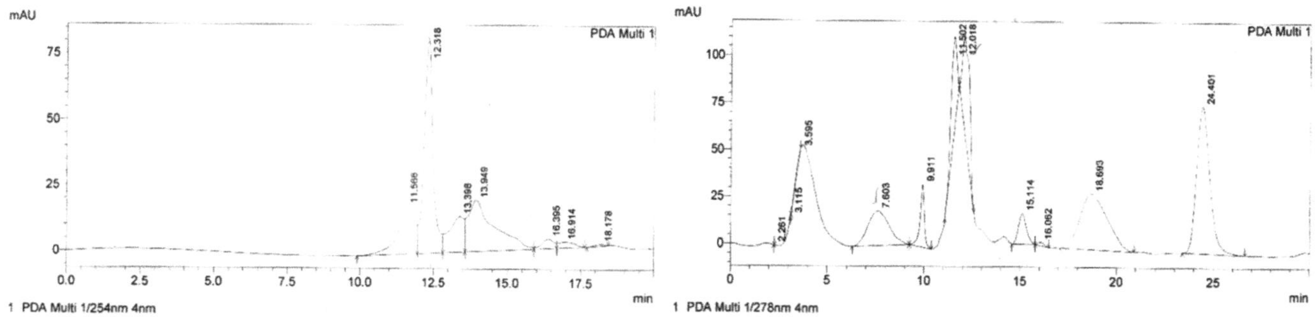

Figure 29. HPLC chromatogram of isolated compound (TLC) of (F5 S1 F3) P. persica compared to the standard diethyl phthalate.

4. Conclusion

Since *P. persica* is surviving in intertidal rocky shore with coral reef area, either to protect from their enemies or through food the animal might synthesis these chemicals which might be responsible for the inhibition of both bacteria and fungi. These antimicrobial compounds can be relatively synthesized, chemically modified, analyzed, and manipulated. However, these compounds are also primarily translational products of genes with potent biological activity and can be manipulated by techniques of modern molecular genetics confers the antimicrobial compounds *P. persica* an important role in expanding bridge between bioactive drug and molecular genetics.

Acknowledgements

The authors are grateful to the Jayaraj Annapackiam College, Periyakulam, the Principal V. O. C. College, CECRI, Karaikudi, IICPT, Thanjavur and CAS in Marine Biology, Annamalai University for providing the research facilities, support and platform to complete this project. Thanks are due to Prof. V. Sivakumar for his unremitting encouragement and valuable suggestions and acknowledge UGC, Hyderabad for their financial assistance.

References

[1] J. De Lucca, "Antifungal peptides potential candidates for the treatment of fungal infections". Expert opinion on Investigational drugs, 2000, 9, 2, pp. 273-299.

[2] R. E. W. Hancock, "Cationic antimicrobial peptides towards clinical applications. Expert opinion on Investigational drugs". 2000, 9, pp. 1723-1729.

[3] P. Proksch and W. E. G. Muller, "Frontiers in Marine Biotechnology". Horizon Bioscience", 2006 Norfolk, UK.

[4] R. B. Novak, E. Henriques, S. Charpentier, Normark, E. and Tuomanen, "Emergence of vancomycin tolerance in *Streptococcus pneumoniae.* Nature" 1999.399, 6736, pp. 590-591.

[5] S. S. Jayaraj, R. Thiagarajan, M. Arumugam and P. Mullainadhan, "Isolation, purification and characterization of (beta)-1, 3-gulcan binding protein from the plasma of marine mussel *Perna virdis"*. Fish shell fish Immunology 2008, 24: pp. 715-725.

[6] A. S. Shenoy, "Octopus a delicacy in Japan". Sea Food Exp. J., 1998, 20: 21-25.

[7] J. L. Balcazar, I. D. Blas, I. Ruiz ZarZuela, D. Cunningham, D. Ventrell and J. L. Muzquiz, "The role of probiotics in aquaculture", Vet. microbiol., 2006, 114: 173-186.

[8] J. W. Blunt, B. R. Copp, M. H. G. Munro, P. T. Nothcote, M. R. Prinsep, "Natural products from marine organisms and their associated microbes", Nat. Prod. Rep., 2006, 23, pp. 26-78.

[9] M. Santhana Ramasamy, and A. Murugan, Potential antimicrobial activity of marine molluscs from tuticorin, Southeast coast of India against 40 biofilm bacteria, J. Shell Fish Res., 2005, 24, 1 pp. 243-252.

[10] A. W. Bauer, W. M. M. Kirby, J. C. Sherris and M. Turck., "Antibiotic susceptibility testing by a standardized single disc methods". Am. J. Clin. Patol., 1996. 45 pp. 493-496.

[11] D. Kelman, Y. Kashma, E. Roserberg, M. Ilan, I. Ifrach and Y. Loya., "Antimicrobial activity of reef sponge *Amphimedon viridis* from the Red sea": evidence for selective toxicity. Aquat. Microb. Ecol., 2001, 24 pp. 9-16.

[12] J. Rajaganapathi, S. P. Thyagarajan and J. K. P. Edward, "Study on Cephalopod ink for anti retroviral activity". J. Exp. Biol., 2000, 38, pp. 519-520.

[13] T. P. Anand, and J. K. Patterson Edward, "Screening for antibacterial activity in the opercula of gastropods". Phuket Mar. Biol. Centre Spl. Pub., 2001, 25 pp. 215-217.

[14] M. Santhana Ramasamy and A. Murugan, "Potential antimicrobial activity of marine molluscs from tuticorin, Southeast coast of India against 40 biofilm bacteria", J. Shell Fish Res., 2005, 24, 1, pp. 243-252.

[15] R. D Thilaga., "Studies on some ecological aspects of *Babylonia spirata* (Linn.) among the Tuticorin Coast, Ph.D. Thesis". Manonmaniam Sundaranar University, Thirunelveli. 2005.

[16] C. Chellaram, K. Mary Elizabeth Gnanambal and J. K. Patterson Edward., "Antimicrobial activity of winged oyster *Pteria chinensis* (Pterioda pteriodae)", ndian Journal of marine sciences, 2004, 33, 4, pp. 369-372.

[17] T. J. Abraham, J. Natarajan, and S. A. Shanmugam., "Antimicrobial substance of potential biomedical importance from holothurian species". Indian J. Mar. Sci., 2002, 31, 2 pp. 161-164.

[18] N. Annamalai, R. Anburaj, S. Jayalakshmi and R. Thavasi, "Antibacterial Activities of Green Mussel (*Perna viridis*) and Edible Oyster (*Crassostrea madrasensis*)", Research Journal of microbiology, 2007, 2 (12): pp. 978-982.

[19] K. M. E. Gnanambal C. Chellaram and J. Petterson, "Antibacterial activity of whole body extracts of *Trochus radiatus* (Mollusca: Gastropoda)" Proceedings of the National seminar on Reef Ecosystem Remediations (NERER'05), SDMRI Research Publication no. 9, 2005 pp. 182-186.

[20] Y. Gueguen, A. Herpin, A. Aumelas, J. Garnier, J. Fievet, J. Fievet, Escoubas, "Characterization of a defensin from the oyster *Crassostrea gigas*" Recombinant production, folding, solution structure anti microbial activities, and gene expression. J. Biol chem. 2006, 28: pp. 313-323.

[21] J. K. Seo, J. M. Crawford, K. L. Stone, E. J. Noga,"Purification of a novel arthropod defensin from the American oyster, *Crassostrea virginica*" Biochem. Biophys. Res, Commun. 2005, pp. 338: 1998-2004.

[22] M. Karthikeyan, M. G. Ananthan and T. Balasubramanian, "Antimicrobial activity of crude extracts of some Ascidian (Urochordata Ascidiacea)", from Palk Strait, (South East Coast of India), World Journal of fish and Marine Sciences, 2009, 1 (4): pp. 262-267.

[23] B. Chandran, G. Rameshkumar and S. Ravichandran, "Antimicrobial activity from the gill extraction of *Perna virdis"* (Linnaeus, 1758). Global Journal of Biotechnology and Biochemistry, 2009, 4 (2): pp. 88-92.

[24] R. D Thilaga., S. Vimala and P. Subavathi, Isolation and characterization of bioactive compounds and antibacterial activity of marine gastropod *Phalium glaucum* (L) International Journal of pure and applied zoology, 201, 2, (3); pp. 218-223.

[25] N. Periyasamy, M. Srinivasan, S. Balakrishnan Antimicrobial activities of the tissue extracts of *Babylonia spirata* Linnaeus, 1758 (Mollusca: Gastropoda) from Thazhanguda, southeast coast of India. Asian Pac J Trop Biomed, 2012, 2 (1): pp. 36-40.

[26] M. Janaki, V. Santhi. Anita Kannagi, "Bioactive potential of Fusinous nicobaricus from Gulf of Mannar," International Journal of pharmaceutical research and bio-science, 2015, 4 (5): pp. 262-270.

[27] S. Mohammed Hussain and G. Ananthan, "Anti microbial activity of the crude extracts of compound Ascidian, *Didemnum candidum* and *Didemnum psammathodes"* (Tunicata: Didemnidae from Mandapam) (South east coast of India). Current Research Journal of Biological sciences, 2009, 1 (3): pp. 168-171.

[28] M. V. D Auria, V. Sepe, D orsi R. Bellotta, F. Debitus and C. A. Zampella, Isolation and structural eludication of callipeltins J-M: "Antifungal peptides from the marine sponge *Latrunculia* sp.", Tetrahedron 2007, 631: pp. 131-140.

[29] A. Kumar, K. Chaturvedi, P. K. Shuka, V. Lakshmi, "Antifungal activity in tritertpeneglycosides from the sea cucumber *Actinopyga lecanora"*. Bioorg. Med. Chem. Lett., 2007, 17: pp. 4387-4391.

[30] R. B. Kunze, B. Bohlendorf, H. Reichenbach, G. Hofle, "Pedein A and B: production, isolation, structure, elucidation and biological properties of new anti fungal cyclopeptides from *Chondromyces pediculatus* (Myco bacteria)" J. Antibiot. (Tokyo) 2008, 61: pp. 18-26.

[31] G. Arenas, F. Guzman, C. Cardenas, L. Mercado S. H. Marshall, "A novel antifungal peptide designed from the primary structure of a natural antimicrobial peptide purified from *Argopecten purpuratus* haemocytes" Peptides, 30: 2009, 1405-1411.

[32] Wayan Mudianta, Peter, L. Katavic, K. Lybette and Lambert, "Structure and absolute configuration of Z-alkylpiperidine alkaloids from an Indonesian sponge of the genus *Halichondria",* Tetrahedron, 2010, 66: pp. 2752-2760.

[33] K. Benkendorff, A. R. Davis, and J. Bremner, "Chemical defense in the egg masses of benthic invertebrates: An assessment of antibacterial activity in 39 molluscs and 4 polychaetes", J. Invertebr. Pathol., 2001, 78: pp. 109-118.

[34] H. M. J. Hua, Peng, R. E. Dc Dunbar, A. G. Schinazi, E. De Castro Andrews, Cuevas, L. F. Garcia-Fernandez, M. Kelley, M. T. Hamann, "Batzelladine alkaloids from the Caribbean sponge *Monanchora unguifera* and the significant activities against HIV-1 and AIDS opportunistic infections pathogens", Tetrahedron 2007, 653: pp. 11179-11188.

The Soil Heavy Metals Element Release and Migration Characteristics of Hengshihe River Both Sides for Guangdong Dabaoshan

Su Wen-ji[1], Fu Shan-ming[2]

[1]Academy of Civil and Transportation, South China University of Technology, Guangdong, China

[2]School of Environmental Science and Engineering, Guangzhou University, Guangdong, Guangzhou, China

Email address:

tbyswj@163.com (Su Wen-ji), fushanming@qq.com (Fu Shan-ming)

Abstract: This paper through soil environment investigation and sampling analysis of Hengshihe River both sides of Shao guan Dabaoshan mining, along the route of the acidic mining waste water flows, and selected 6 research division of the Liangqiao, Shui louxia, Tangxin, Yanghe-lianxin, Shangba and Xiaba village, to analyze the pollution source and discusses the pollution element migration patterns and characteristics of Dabaoshan mine. The results show that the Dabaoshan mine original high background values of heavy metals pollution and the secondary pollution caused by mining activities of superimposition and accumulation is the important source of pollution. The migration ability of heavy metal elements in soil is Cd > Cu > Pb > zinc. The Cd elements have to be considered as an optimization control of soil restoration governance elements in Dabaoshan pollution area.

Keywords: Soil, Pollution Source, Release, Migration, Background Value

1. Introduction

Soil is an important part of the ecological environment. This is a research focus in the soil ecological and environmental risk issue for soil heavy metal pollution caused by mining activities, and these have caused pollution is not ended with the production of enterprises and closed. So the work of restore soil management is a long way to go, the cumulative effect of soil heavy metals, especially the research for the soil heavy metals element release and migration characteristics is great significance.

Guangdong Dabaoshan mining of mineral resources leads to the deterioration of the environment and ecological imbalance in the region, cause potential harm to residents of health. Since the serious consequences of heavy metal pollution of Dabaoshan was reported many times by media, its soil environmental problems were paid close attention to the government and academia. The environmental pollution problems caused by Dabaoshan polymetallic mining in the past has not been attended because of mismanagement, local protection and people environmental protection consciousness, the low expenditure on environmental protection. In the late 1990s, people gradually realized that Dabaoshan mining can't post-natal environmental protection, environmental protection investment must synchronize with mining, ore dressing, in the aspect of soil environmental protection and pollution control has made positive progress, both in the scientific research, capital investment, management, supervision and technical support. They have made great achievements. Since the 1960s, the soil environmental protection work on the study of Dabaoshan mining area is roughly experienced the following three stages (Figure 1).

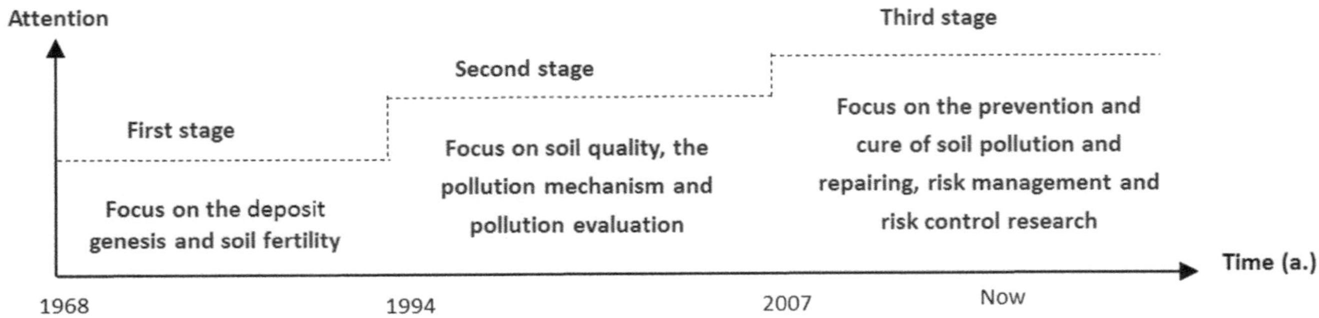

Figure 1. *The research phase of soil environmental for Dabaoshan.*

The predecessors' research work of Dabaoshan mining area, more focused on the deposit genesis [1-4], distribution characteristics and pollution evaluation [5-11], ecological risk and health risks evaluation [12-15], and so on. The researches show that soil heavy metal pollution elements of Dabaoshan mining area and the surrounding are mainly Cd, Pb, Cu and Zn. At present, the focus in Shangba reclamation and control of heavy metal pollution of soil improvement and vegetation remnant [16-18], more based on the State Soil Environment Quality Standard (GB15618-1995) and the Guangdong Shao guan red soil background values studies [19-23]. But the evaluation parameters and index only used by Guangdong Shao guan red soil background values is inadequate because the Dabaoshan mining area of heavy metal element enrichment characteristics. Few studies of predecessors for Dabaoshan mining area about heavy metals element background values and man-made pollution law of superposition in the relationship among heavy metal distribution characteristics, accumulation, characteristics analysis and evaluation [24-28].

Therefore, this paper from the Dabaoshan mine area background values of heavy metals, not affected by the mining activity on the control of the value and with mining activities led to the superimposition and accumulation degree of correlation of pollution, comprehensive study within the scope of the mining area and the surrounding influence elements of main heavy metal pollution in soil Cd, Pb, Cu and zinc, its release and distribution regular pattern, which will provide a scientific basis for ecological risk assessment and early warning.

2. The General Situation

The scope of this study area is to east longitude 113°40 '~ 113°50', north latitude 22°24'~24°35'. Geological environment impact area is about 65 km^2. The overall terrain for lower from north to south, the north is an elevation of 800~1200m mountain; the south is low mountain hills and alluvial plains. The ore of Dabaoshan mining is mainly composed of chalcopyrite, galena, sphalerite, pyrite, molybdenite, black (white) tungsten, bismuthinite and pyrrhotite, etc. This is a super-large polymetallic opencast mine in Guangdong province [29-30].

The ecological environment of mining stope is destructed serious, the soil layer is shallow, strong acid, and water storage is very weak. So most of the rain and slope infiltration water directly erode the surface soil, slope, ore-bearing bed, and in the form of surface runoff flow down quickly. The wastewater acid is very high; contain Cd, Cu, Pb, zinc and other heavy metals, and constantly moving from the mine to the downstream. At present, the mining waste discharge in two large tailing sand reservoirs of - Tie Long sediment tailing sand reservoir and FanDong (also known as CaoDuiken) tailings. Tie long tailing sand reservoir is located in the southwest of Dabaoshan mine, storage capacity is about 10 million m^3. Sediment tailing sand reservoir is to prevent the soil with the flash floods mine inflow Hengshihe River from the top of the mountain water, but the tailing sand reservoir full of silt already, the acid water of tailing sand reservoir overflow dam crest flow downstream perennial, and gather in Hengshihe River. FanDong tailing sand reservoir is located in the northeast of the stope, and it was built in the 1970s, used to collect washing and recycling use the waste water of ores dressing. The original capacity is 1.05 million cubic, later heightening the dam and its capacity becomes 1.25 million cubic. By decades of use, the capacity is shrunk to 2 million cubic. Since 2007, 2 tailings were capacity enlarging reform, and building up a new wastewater processing factory and the corresponding wastewater pipeline, and the scale is $Q=1.5\times10^4 m^3/d$.

The composition of soil parent materials in Dabaoshan mining area and the surrounding area has a great effect on soil chemical composition. The soil layer of the slope and Hengshi River upstream flood plain terrace at LiangQiao, Shui Louxia village is thinner. It is mainly comprised of coarse gravel soil, gravel and coarse sand. The soil layer of ShangBa and XiaBa village is thicker. It is an alluvial plain by Hengshi River downstream flow and Chen Gong river flows (not polluted by Dabaoshan mining) intersection formed with small particles to the clay and silt of river sediments deposited in there.

3. Sample Collection and Analysis

3.1. Sample Collection

To Dabaoshan mining as a starting point, along the both sides of Hengshi River, we collected the rock and soil samples at tailing sand reservoirs, Liang Qiao, Shui Louxia, Tang Xin, Yang He, Lian Xin, Shang Ba and Xia Ba village. (Figure 2) [32]. According to the actual level of soil profile, we collected topsoil, subsoil and weathered rocks for the residual soil, and ensure that the characteristics of the soil samples can represent each profile section; For both sides of alluvial soil and agricultural soil of Hengshi River, 104 surface soil samples (plough layer depth is 0-20cm) and 115 vertical section soil samples (sampling depth is 0-150 cm) were collected by 200-400m spacing for the parallel Hengshi River and 50-100m spacing for the vertical rivers flow. At the same time, 9 soil samples as contrast values for evaluate the effect of the mining activity on heavy metal pollution and they were collected by Chen Gong River which almost not influenced by mining activities, the soil forming environment and the tillage methods are similar or identical to Hengshi River both sides. Drilling with Luoyang shovel, soil samples in the sample bags, serial number and record the sampling point coordinates at the same time.

3.2. Sample Preparation

The samples were spread into a thin layer on clean paper loosely and evenly, and placed indoor of cool and ventilated place dry naturally, rejects all impurities such as plant roots, porphyrization repeatedly. All the soil sampled were passed the sieve of diameter of 1 mm, and then take some samples which are passed of 1mm sieve, further grinding, the let them pass the sieve of 200 mesh, sieve samples mixed after all, grinding into the sealing bag sealed samples, label and record for heavy metal content analysis.

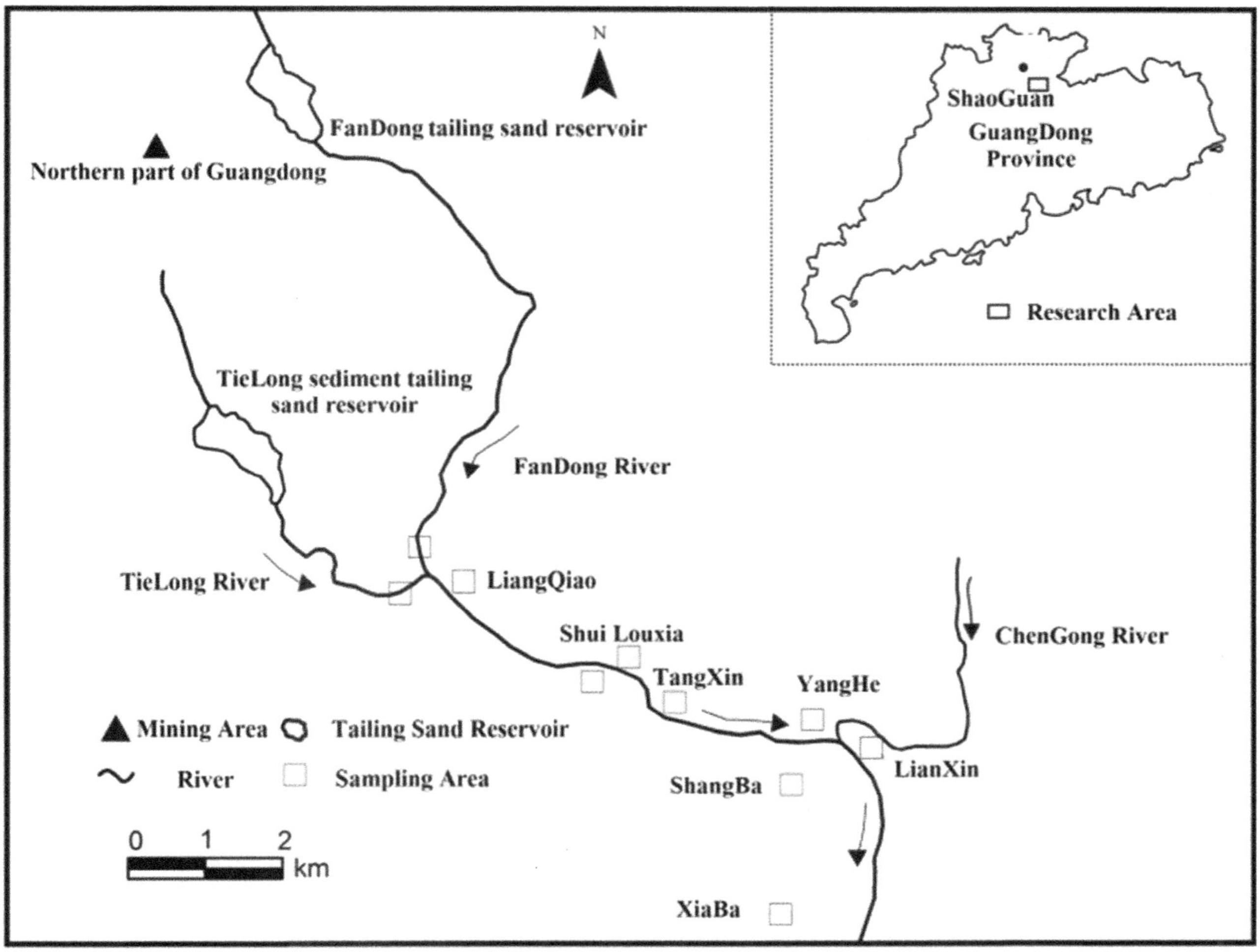

Figure 2. *Research and sampling area.*

3.3. Sample Analysis Test

The sample of 0.500g is weighed and set in polytetrafluoroethylene crucible, wetted by high purity water, soaked with 10 mL aqua regia overnight, then heated steam to nearly dry at electric heating plate. Take off the electric heating plate for a moment then adding 5 mL flu hydric acid and continue to heat for drying. Take off the electric heating plate again for a moment then adding 5mL perchlorate and continue to heat for drying. Removed, added 5% dilute nitric acid rinse crucible wall, dissolving in the salt at electric heating plate, remove and cooling it, and transfer the solution into 50 mL flask and prepared the test solution and shake well. In this experimental process, the parallel samples and standard samples, blank samples were joined, in order to ensure the accuracy of the data. All the content of heavy metal element of Pb, Cd, Cu and zinc for samples were determined with heavy metal concentration determination instrument ICP – MAS [33].

4. Analysis and Discussion

4.1. Analyze the Heavy Metal Release Source on Both Sides of Hengshi River

Soil parent material and material composition for soil chemical composition of parent rock is very obvious. Most mineral compositions of soil materials come from the primary mineral particles and all kinds of secondary minerals inherited of Pedogenic parent material. Therefore, the content of heavy metal elements in soil and its distribution characteristics not only depends on the influence of sewage irrigation on the soil heavy metal elements, but also with the soil of various mineral compositions in soil materials in the inheritance. Only in the research of soil mineral composition and source and analyzing the change of the soil heavy metal element content, we can correctly evaluated on the soil heavy metal.

4.1.1. Soil Heavy Metal Primary Release Source

1. Comes from the Pedogenic parent material——heavy metal in ore body of surrounding rock

The average content of heavy metal elements which is in the mining area and the periphery of various kinds of rock of Pb, Cu and zinc, Mn and Ni etc. were determined and analyzed by predecessors. the analysis results show that the content of heavy metal elements, such as Cu, Pb, zinc in different surrounding rock of the Dabaoshan mining area are number times to hundreds times higher than the average content in the earth's crust (Clarke value) (Tab. 1). It shows the area of surrounding rock is the sources of the original of Cu, Pb and zinc. So the Pb, Cu and zinc content in the soil which these rocks as pedogenic parent material is higher.

Table 1. *Geometric average content and the concentration Clarke value of heavy metal element from surrounding rock of Dabaoshan mining area [1].*

Rock Name	Cu		Pb		Zn	
	$\bar{x}$	Δc	$\bar{x}$	Δc	$\bar{x}$	Δc
arenaceous shale	253	4.4	2416	120	2366	29.6
manganic rook	5760	101	4260	213	4363	54.5
siltstone	657	11.5	728	104	542	36
clay rock	5750	23	4500	56	966	5.8
tuff	342	-	1157	-	1835	-
manganese sand shale	1675	37	4930	246	1510	15.8
Peaty limestone	27.5	6.8	109	12	124.4	6.2
quartz sandstone	45.6	1	20.6	2.9	58.1	3.8
secondary dacite-porphyry	24.1	1.2	25.3	1.3	23.1	-2.5
Granodiorite porphyry	10	-2	20	1	8	-7.5

Remarks: 1. $\bar{x}$ is the geometric average content of heavy metal elements (unit: mg/kg);
2. Δc is concentration Clarke, that is greater than Clarke value multiples, Greater than the multiples for the plus sign"+", Less than a multiple for the minus sign "-";
3. Clarke number: according to Vinogradov (1962), TuLiqian and FeiDeBo (1961);

2. Comes from the Pedogenic Parent Material--Heavy Metal in Ore

Material composition effects on soil chemical composition of ore are very obvious. Dabaoshan ore type is more, both from magmatic rocks and from sedimentary rocks. The heavy metal element combinations of difference ore type is very big different (Table 2). Mining area of Cu, Pb and zinc had formed separate deposits, Cu, Pb and zinc content in the ore deposit is higher than the non-mine soil, the average abundance value is much higher than the earth's crust. As a result, the mineral weathering and its exploitation will lead to the release of heavy metals, leaching, denudation, erosion and carry, gradually deposited along the mountain slope, Hengshi river on both sides of the terraces and flood land, caused enrichment of heavy metals by the mineralized bodies into the surrounding soil environment migration and diffusion, become an important source of soil heavy metal elements at Dabaoshan mine

4.1.2. Soil Heavy Metal Secondary Source Release

Table 2. *Geometric average of heavy metal element of Dabaoshan ore.*

Ore Name	Wallrocks Ore	Cu	Pb	Zn
Molybdenum ores	secondary dacite-porphyry	25.8	25.2	16.2
	granodiorite porphyry	21	10.0	18.9
	Limestone of Qiziqiao Formation	17.9	20.3	12.3
Brass ore	secondary dacite-porphyry	1847	243	58.4
	Limestone of Qiziqiao Formation	4591	414	38.5
Brass Lead-Zinc Ore	Limestone of Qiziqiao Formation	9815	6857	1090

1. From beneficiation activities—the heavy metals of mine tailing

The particle size of mine tailings from beneficiation activities as much as fine sand and silt contains a lot of pyrite, pyrrhotite, sphalerite, chalcopyrite, galena, a set of metal sulfides. Along with wastewater migrated to the downstream rivers and soil environment, become a direct source of pollution. Sample analysis shows that the heavy metal content of Pb, Cd and Cu in two tailings is difference very much. They are mainly caused by difference tailing ore source (Table 3).

Table 3. *The average content of heavy metal at Dabaoshan tailings (Unit: mg/kg).*

Sampling point	Pb	Cd	Cu	Zn
Fandong tailing	368.6	4.02	1058.74	1273.94
Tie long tailing	995.4	17.30	398.78	1318.3
Sampling point	Cr	Mn	Ni	
Fandong tailing	15.07	297.38	9.00	
Tie long tailing	41.07	362.21	14.20	

2. From the Wastewater of Tailings

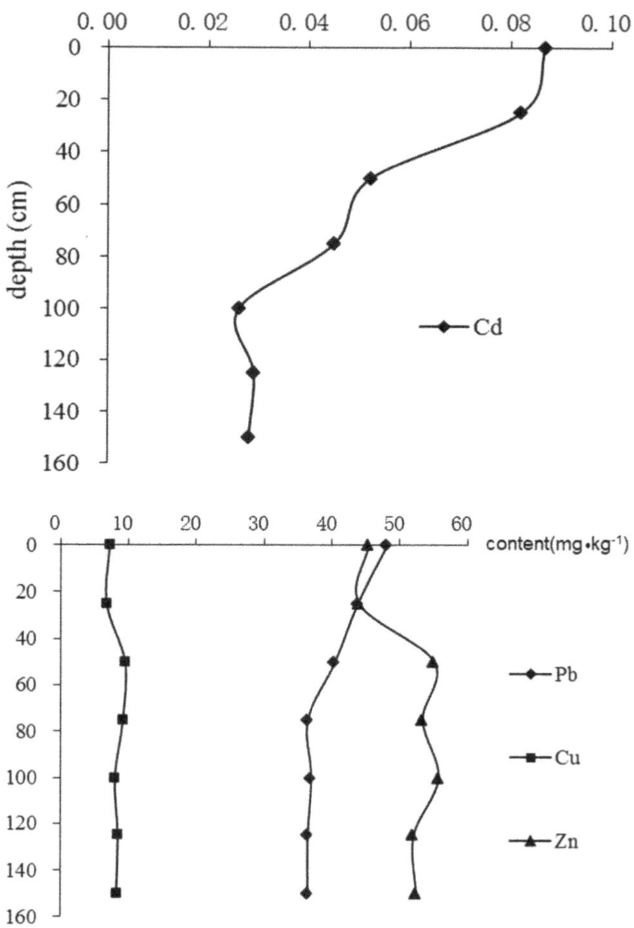

Figure 3. *The soil heavy metal content distribution on the vertical profile of Chen Gong River.*

A large number of acid mine water which rich heavy metals element such as Pb, Cd, Cu, zinc, Cr and Mn is released in the process of mining. The waste water overflow dam, import Hengshi River, and become an important source of pollution of the downstream water and the soil. Because FanDong tailing mainly pile up copper mine backfilling, while Tie Long tailing mainly pile up iron mine backfilling materials, thus cause the two heavy metal content in tailings of mine wastewater showed different distribution characteristics (Table 4).

Table 4. *The geometric average of heavy metal content of wastewater at Dabaoshan tailings (unit: mg/L).*

Sampling point	pH	Pb	Cd	Cu
Fandong tailing	2.93	0.45	0.14	6.60
Tielong tailing	3.14	0.43	0.48	3.53
Sampling point	Zn	Cr	Mn	
Fandong tailing	9.83	0.05	38.66	
Tielong tailing	28.4	0.04	31.37	

4.2. The Migration and Distribution Characteristic of the Soil Heavy Metal on Both Sides of Hengshi River

4.2.1. The Background Value, as the Background Value and Contrast Value

Because the environment pollution was caused by human activities, it is difficult to find the background value which can reflect the original material composition, properties and structure characteristics of the soil environment, and to determine the extent of the pollution caused by human activities. So although main soil types of Dabaoshan area is red soil, the background values of heavy metals at Shao guan area of Guangdong province can be used for reference, but in order to further research the distribution and migration characteristics of heavy metals in soil environment of Dabaoshan mine area. This paper, the Chen Gong river village as a contrast area which is not affected by mining activities but its pedogenic environment, atmospheric environment, farming method, pesticide and chemical fertilizer are same or similar to the study area. To choose the heavy metal element content average value of contrast area that the content at this the vertical section depth tend to stable invariant as the resembling background value, to choose the average content of soil heavy metal within the scope of the depth of plough layer of contrast area as contrast value to judge the degree of heavy metal pollution (Figure 3).

The Figure 3 shows that on the vertical section, the content of Pb and Cd in soil plough layer (0-30cm) is highest, with the increase of depth, heavy metal content showed a trend of fluctuation is reduced; And the content of Cu and zinc in the soil plough layer (0-30cm) is lower, with the increase of depth, the trend of the heavy metal content was increased fluctuate, the content reach the highest at 50cm depth, below 60cm depth, the content tend to stable invariant be stable gradually. Under section 100cm depth, Pb, Cd, Cu and zinc element content changes almost stable invariant, it shows that the soil heavy metal content is influenced by the outside is not obvious below the depth, it reflects the nature of soil heavy metal content. this phenomenon shows the condition of the soil by ChenGong river basin, the vertical migration ability of Pb and Cd element are relatively strong, the depth reach to 100cm; And the vertical migration ability of Cu and zinc element are relatively weak, only reach to 60cm. the soil heavy metal content of Pb, Cd, Cu and zinc elements is no longer clear change below 100cm. therefore, the heavy metal content within the scope of this depth is no longer affected by human activities. This reflects the natural state of soil content, so the heavy metal content in this depth would be a judge value to Hengshi river basin as the resembling background value of soil is affected by human activities (Table 5).

Table 5. *The evaluating value of soil heavy metals in the study area (unit: mg/kg).*

Evaluation Index	Pb	Cd	Cu	Zn
red soil background values of Shao guan of Guangdong [1]	34.38	0.03	14.38	48.75
As background values (100cm depth of soil by Chengong river)	37.03	0.03	8.50	63.8
plough layer contrast value (0-30cm depth of soil by Chengong river)	48.2	0.09	11.25	45.4
the state soil environment quality secondary standard (pH < 6.5)	250	0.3	50	200

Table 5 shows: the average content of Cd and Pb element of plough layer in the study area were higher than red soil background values of Shao guan of Guangdong, and as the resembling background value close to the red soil background values of Shao guan of Guangdong. So the contrast area plough layer affected by human activities.

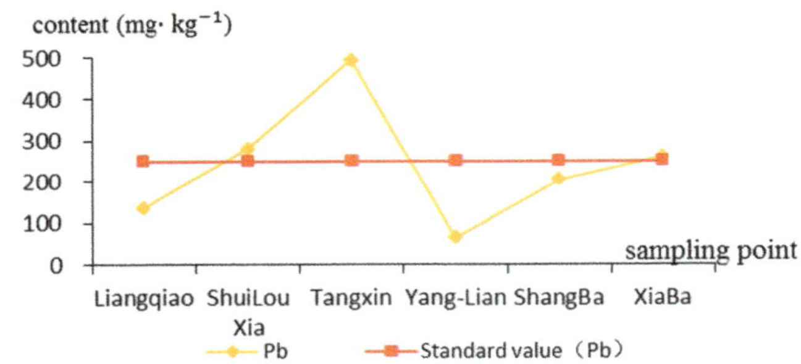

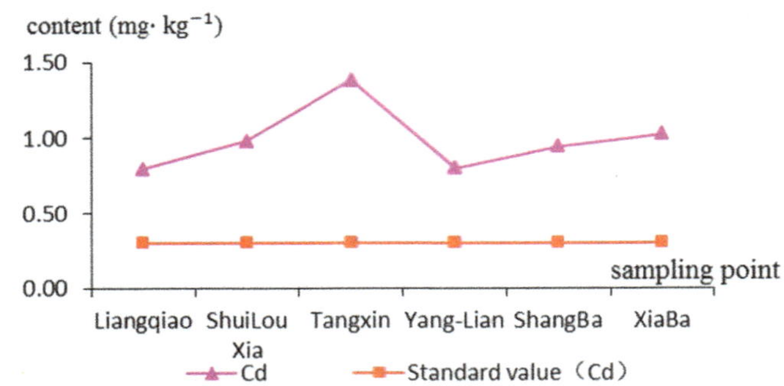

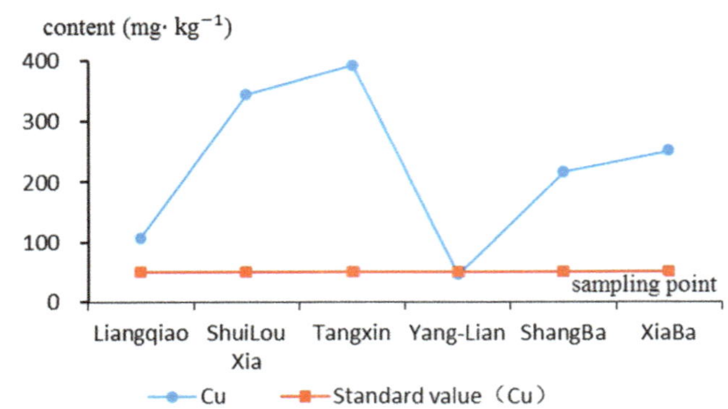

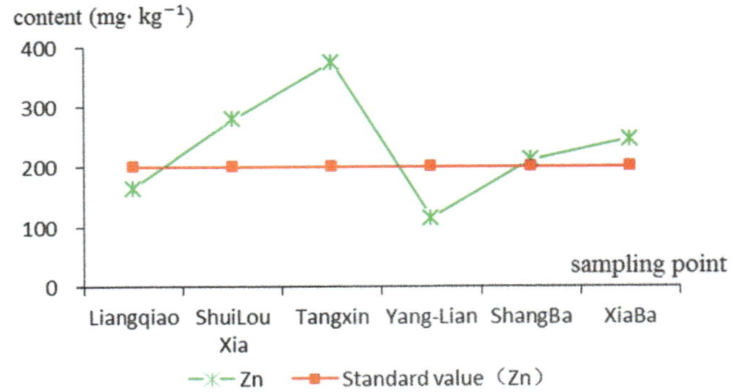

Figure 4. *The migration and distribution characteristics of the surface soil heavy metal element along the rivers.*

The average content of Pb, Cd, Cu and zinc element in the plough layer of the study area were significantly higher than the contrast value of Chen Gong River's village which is not affected by mining activities. This description that there are many higher content heavy metal from waste residue, acidic waste water discharge of Dabaoshan mining activities, along the Hengshi river irrigation farmland caused by pollution of heavy metal content.

4.2.2. The Migration and Distribution Characteristics of Soil Heavy Metal

Heavy metal element migration and distribution is influenced by many factors, such as PH, clay content, organic matter content and soil coexist elements, etc. there are 3 space patterns of migration for horizontal, vertical migration of rivers and vertical ground migration. From the theory analysis, with the horizontal longitudinal flow along the river, the characteristics of migration and distribution of soil heavy metal content is reduced gradually from the river upstream to the downstream areas; In vertical on the level of the river section, the more far away from the river on both sides of the soil, the content of heavy metal migration and distribution is characterized by reduced gradually. Content in the migration and distribution of the vertical section should be characterized by: heavy metal content is the highest soil, with the increase of the depth of the heavy metal content is a trend of decrease, gradually close to the background value or the resembling background value. But the soil of the Dabaoshan mining are affected by sewage irrigation, topography and soil heavy metal activity, it shows different characteristics of migration and distribution.

1. The surface migration and distribution characteristics of soil heavy metals

(1) Along the river migration: Tab. 6 and Fig. 4 can be seen Pb, Cd, Cu and zinc element average excess standard multiples are: 1.2 times, 2.8 times, 4.2 times and 1.2 times in the Hengshi River sewage influence area, surface soil in the study area have been the effects of Pb, Cd, Cu and zinc element, give priority to with Cd and Cu pollution, secondary is Pb and zinc pollution. The most polluted is TangXin village.

Table 6. The contrastive analysis to the surface soil heavy metal content along the Hengshi River.

element	index			
	Pb	Cd	Cu	Zinc
maximum	900.06	3.88	616.44	588.56
minimum	18.61	0.24	14.66	69.85
average	288.05	0.84	210.38	241.22
average excess multiples	1.2	2.8	4.2	1.2

(2) Vertical longitudinal flow cross the river migration: at Shangba village of about 14.9km away from the Tie Long tailings, the samples from vertical longitudinal flow cross the river were analyzed. The section line of Fig. 5 of about 14.9km away from Tie Long tailing, the left bank is mainly affected by Chen Gong river, the right bank is mainly affected by Hengshi river. Therefore, the left bank of the heavy metal elements Pb, Cu and zinc do not exceed the standard, the content of Cd also show only slightly overweight 1.4 times; the right bank section show that the heavy metal element content close to Hengshi river banks, the heavy metal content is higher. The right bank of the surface soil heavy metals content Pb does not exceed the standard, and near the banks of the river in the surface soil Cu and zinc content is more than standard values, a multiple of 3.1 times and 1.1 times respectively, the right bank of surface soil heavy metals Cd content than standard value, maximum is a multiple of 1.8 times standard valued. Thus, the surface soil heavy metal pollution status of this section line is given priority to with Cd pollution, Cu pollution is secondary.

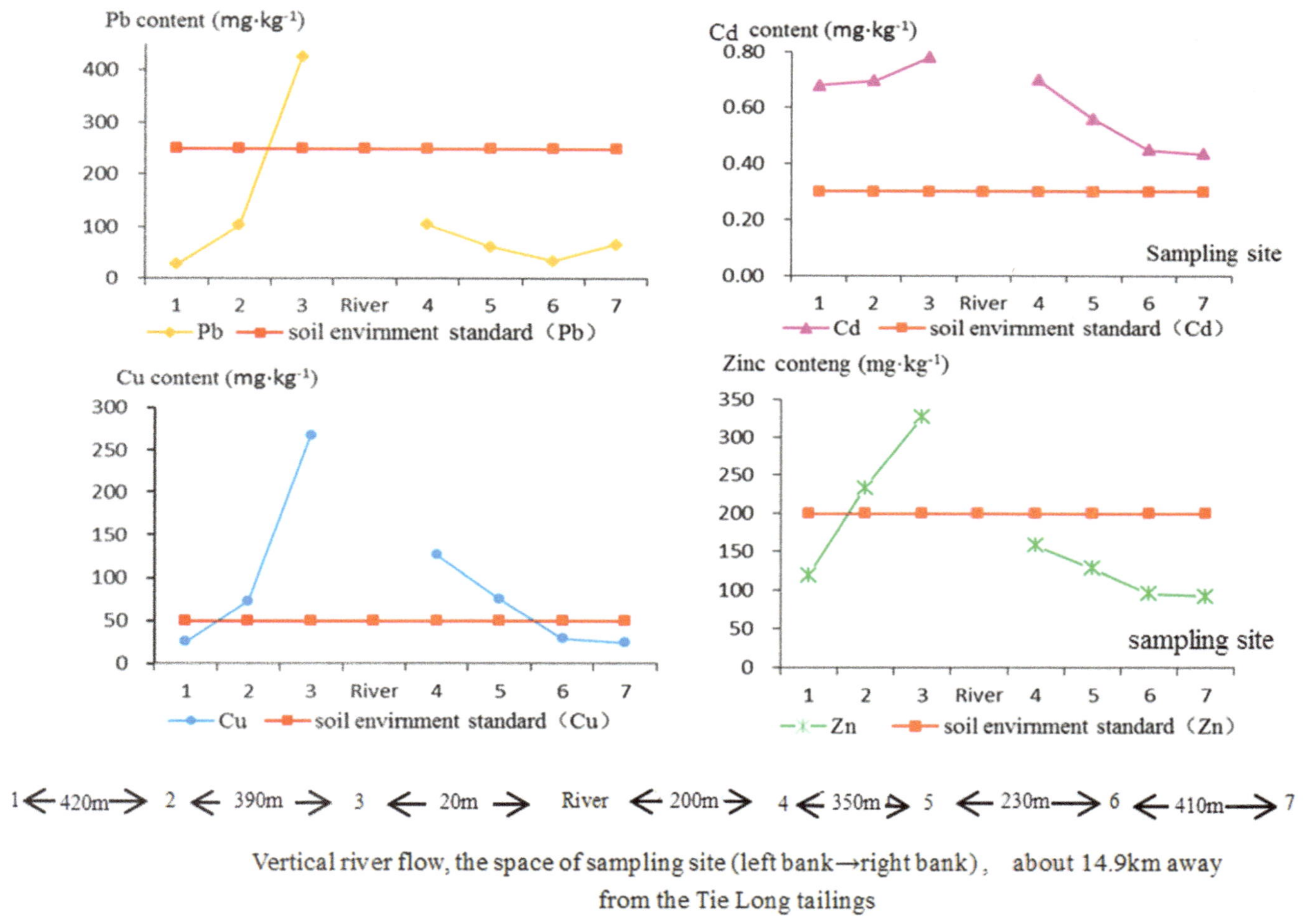

Figure 5. The migration and distribution characteristics of the surface soil heavy metal element for vertical rivers.

2. The migration and distribution characteristics for profile of soil heavy metal

Figure 6 is different distribution characteristics of heavy metal element for vertical profile at Shangba village of 7 sampling point. Because the study area is located in the alluvial plain, most of the sampling points are preferable to deep vertical soil samples of 150 cm.

As can be seen from the Figure 6:

(1) The Pb content of surface soil isn't larger than standard value, with the increase of depth; the content of Pb basic stable, its content is lower than the standard values. Illustrate the Pb element partition in the research the basic characteristics of no pollution;

(2) the Cd content of surface soil is larger than standard value, with the increase of depth, the content of Cd element reduce gradually, till 125cm, its content tend to be standard value, but 150cm depth, there are still a part of the Cd content of soil is not stable, this explain Cd element of

vertical migration distance is far, move faster, easier migration characteristics;

(3) the Cu content of most of the surface soil is larger than standard value, with the increase of depth, the content of Cu element gradually reduce, but till 150cm depth, there are still a part of soil Cu content is not stable, this explain Cu element of vertical migration distance is far, move quickly, easy to the characteristics of the migration;

(4) the zinc content of most of the surface soil is larger than standard value, with the increase of depth, the content of zinc element gradually reduce, till 125cm, its content tend to be standard value; till 150cm depth, there are still a part of soil Zinc content is not stable, this explain Cu element of vertical migration distance is far, the features of moving faster.

From the above analysis: in soil vertical section, the release of the heavy metal content and spatial distribution characteristics follow the above theory, but in longitudinal and transverse section along the river flow, is not follow the rules of the spatial distribution, long-term wastewater irrigation to carry a large amount of slime and suspended material into the soil, physical accumulation process of heavy metals are not single, but by dissolution, precipitation, coagulation, complexation, and adsorption all sorts of chemical reaction, such as formation of different chemical forms, thus the pollution caused by excessive. The degree of pollution is affected and controlled by many factors.

Mobility analysis: the heavy metals element of Dabaoshan sewage irrigation area, from upstream to downstream show different potential soil heavy metal migration sequence, especially in Shangba village sewage irrigation area, the migration ability of heavy metals are present a Cd > Cu > Pb > zinc. Although reveals that the total amount in Shangba Cd is not large, but its potential ability to migrate, the strongest polluting is the highest; Cu and Pb also has strong potential ability to migrate, zinc potential migration ability are relatively weak. Thus further illustrate in Shangba village of Dabaoshan, Cd element need to priority control of heavy metal elements at Dabaoshan sewage irrigation area.

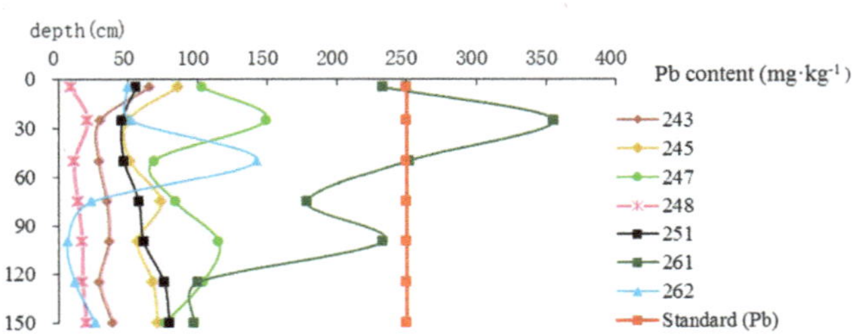

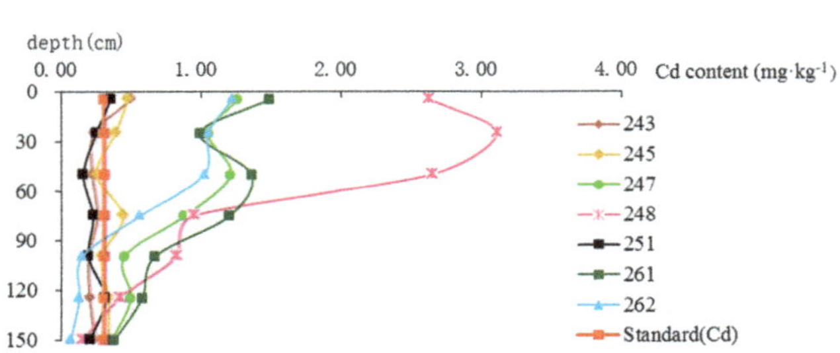

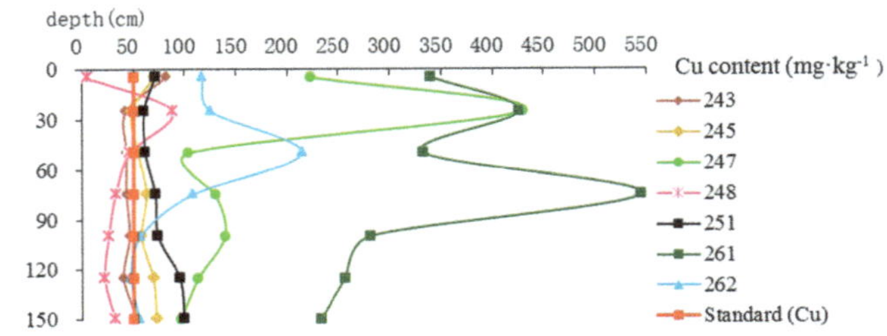

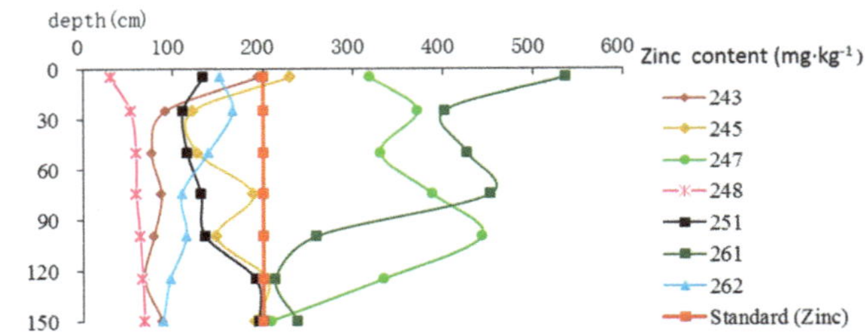

Figure 6. *Migration and distribution characteristics of soil heavy metal element for vertical section.*

5. Conclusion

1. The heavy metals pollution source of Dabaoshan mine is high background values and the superimposition and accumulation of pollution caused by mining activities.

2. The migration ability of heavy metals in Dabaoshan soil is Cd > Cu > Pb > zinc.

3. The Cd element has to be considered as an optimization control of soil restoration governance elements for Dabaoshan area.

Acknowledgements

Aid financially project: public welfare project of Ministry of Land and Resources (No: 201111020-7).

References

[1] Qiu Shiqiang. A preliminary study on the genesis of the Dabaoshan strati form polymetallic deposit [J]. Geological Review, 1981, 27 (4): pp. 333~340.

[2] Luo Nianhua. The Geological and Geochemical Features and the Origin of Dabaoshan Polymetallic Deposit in Guangdong Province. [J]. Journal of Guilin college of geology, 1985, 5 (2): pp 184~195.

[3] Liu Xiaoshan, Zhou Shunzhi. On the occurrence of middle Ordovician volcanic and analysis of ore-forming mechanism of siderite poly-metallic of deposit from Dabaoshan, qujiang county Guangdong province [J]. Journal of Nanjing University (Natural Sciences Edition), 1985, 21 (2): pp 349~362.

[4] Liu Houqun, Yang Shiyi, Zhang Xiulan. A Preliminary Study on the Genesis of the Dabaoshan Polymetallic Deposit in Northern Guangdong [J]. Journal of Geology, 1985, (1): 47~61. R. Nicole, "Title of paper with only first word capitalized," J. Name Stand. Abbrev.

[5] Zhuang Ming Zhen. Discussion on Metallogenic Condition and Ore Genesis of Polymetallic Deposit of Dabaoshan [J]. Geology and Prospecting, 1986, (5): pp 27~31.

[6] Huang Shujun, Zeng Yongchao, Jia Guoxiang. On the Genesis of Dabaoshan Polymetallic Deposit in Guangdong Province [J]. Geochemical, 1987, (1): pp 27~35

[7] Yao Dexian, Zen Lingchu. On Genesis of Dabaoshan Mineral Deposits [J]. Act Scientia rum Naturalium Universitatis Sunyatseni, 1994, 33 (3): pp 91~100.

[8] Wu Yonggui, LIN Chu-xia, TONG Xiao-li. Environmental impacts of acid mine drainage from the Dabaoshan Mine: I. Downstream aquatic ecosystem [J]. Ecology and Environment 2005, 14 (2): pp 165~168.

[9] LIN Chu-xia, LU Wen-zhou, WU Yong-gui. Environmental impacts of acid mine drainage from the Dabaoshan Mine: II. Agricultural ecosystem [J]. Ecology and Environment, 2005, 14 (2): pp 169~172

[10] CAI Mei-fang, DANG Zhi, WEN Zhen. Risk assessment of heavy metals contamination of soils around mining area [J]. Ecology and Environment 2004, 13 (1): pp 6~8.

[11] Zhou Jianming, Danzhi, Si Tuyue. Distribution and Characteristics of Heavy Metals Contaminations in Soils from Dabaoshan Mine Area [J]. Journal of Agro-Environment Science, 2004, 23 (6): 1172~1176.

[12] Yang Zhen, Hu Mingan. Environment Investigation of Heavy Metal Pollution from M inning in Dabaoshan [J]. Environmental monitoring and manage technology, 2006, 18 (6): 21~24.

[13] CHEN Qing-min, ZHANG Xiao-jun, HU Ming'an. Assessment of Aquatic Pollution of Heavy Metals in a Copper- iron Area [J]. Environmental Science and Technology, 2006, 29 (6): 64~67.

[14] Xu Chao, Xia Beicheng, Qing Jianqiao. Analysis and Evaluation on Heavy Metal Contamination in Paddy Soils in the Lower Stream of Dabaoshan Area, Guangdong Province [J]. Journal of Agro-Environment Science, 2007, 26 (Supplement): 549~553.

[15] Xu Chao, Xia Beicheng. Ecological risk assessment of heavy metals in paddy soils contaminated by acid mine drainage with toxicity characteristic leaching procedure [J]. Ecology and Environment 2008, 17 (6): 2264~2266.

[16] Xu Chao, Xia Beicheng, He Shimei. Characteristics of Cd, Zn, Pb, Cu Content of Paddy Soils in the L owerStream of Dabaoshan Area, Guangdong [J]. Act Scientia rum Naturalium Universitatis Sunyatseni, 2008, 47 (3): 122~127.

[17] Xu Chao, Xia Beicheng, Wu Haining. Speciation and Bioavailability of Heavy Metals in Paddy Soil Irrigated by Acid Mine Drainage [J]. Environmental Science, 2009, 28 (11): 2293~2296.

[18] ZOU Xiao-jin, QIU Rong-liang, HUANG Sui-hong. Immobilization and re-vegetation of heavy metal polluted soils in Dabaoshan, Guangdong Province by amendments [J]. China Environmental Science, 2008, 28 (9): 775~780.

[19] Qiu Yinghua, Wu LingFang, Liao Linjuan. Analysis of Vegetation Actuality and Restoration in Dabaoshan Mining Field, Guangdong Province. [J]. Forestry Science and Technology of Guangdong Province, 2010, 26 (5): 22~27.

[20] Rudnick and Gao. Composition of the Continental Crust [C]. Elsevier Pergamon, Holland and Turekian. 2004, (3), 1-64.

[21] Zhen Jiajia, Jiang Xiao, Zhang Xiaojun. Pollution Assessment of Heavy Metals in Soil around Dabaoshan Polymetallic Ore Deposit [J]. Environmental Science & Technology, 2008, 31 (11): 137-145.

[22] Zhou Jianming, Dang Zhi, Cai MeiFang. Speciation Distribution and Transfer of Heavy Metals in Contaminated Stream Waters around Dabaoshan Mine [J]. Research of Environmental Sciences, 2005, 18 (3): 5-10.

[23] Xu Lianfeng, Liu Tenghui. The Zonal Differentiation of Soil environmental Background Values and Critical contents in Guangdong [J]. J. South China g r. Univ., 1996, 17 (4): 58~62.

[24] Duan Xinchun, Wang Wenjin, Dang Zhi. Distribution of Heavy metals in water around the Daobaoshan mine [J]. Earth and Environment, 2007, 35 (3): 255~260.

[25] Zhou Xiaojin, Qiu Rongliang, Zhou Xiaoyong. Heavy metal contamination and health risk assessment in Dabaoshan, China [J]. Acta Scientiae Circumstantiae, 2008, 27 (4): 855~862.

[26] Li Yongtao, Thierry Becquer, C•cile Quantin. Effects of heavy metals on microbial biomass and activity in subtropical paddy soil contaminated by acid mine drainage [J]. Acta Ecologica sinica, 2004, 24 (11): 2430~2436.

[27] Abundance in Earth's Crust. WebElements.com. 2007-04-14.

[28] CHEN Sanxiong, XIE Li, CHEN Jiadong. Evaluation on soil heavy metal pollution at Dabaoshan mine in Guangdong province [J]. Journal of Nanjing Forestry University (Natural Science Edition), 2012, 36 (3): 59~63.

[29] Xu Youning, Zhang Jianghua, Ke Hailing. Cd contamination of farmland soil in a gold mining area and its environmental effects [J]. Geology in China, 2013, 40 (2) 636~643.

[30] Su Wenji, Xu Youning, Fu Shanming. The distribution regularity and accumulation risk of heavy metals in water and soil along the Hengshi River in the Dabaoshan mining area, Guangdong Province. Geological Bulletin of China, 2014, 33 (8): 1231-1238.

[31] SU Wenji, FAN Sheng. Analysis and evaluation of heavy metal pollution along the river of Dabaoshan mine area [J]. Hebei Journal of Industrial Science and Technology, 2014, Vol. 31 (3): 199-203.

[32] QIU Jinquan, FU Shanming, SU Wenji. Characteristics and Significance of Copper Eco-geochemistry in Soil of Hengshi River Basin in the Dabaoshan Mine, Guangdong Province [J]. Ecology and Environmental Sciences, 2015, 24 (10): 1695-1704.

[33] QIU Jinquan, FU Shanming, Xiao Fang, SU Wenji. Prediction and Ecological Risk Warning Assessment of Heavy Metals in Soil of Hengshi River Basin in the Sulfide Mine, Northern Guangdong Province, China [J]. Earth and Environment. 2015, Vol. 43 (3): 338-344.

Earthworm: A Potential and Sustainable Source for Soil Fertility

M. Kanchilakshmi, Arockiam Thaddeus

Research Centre of Zoology, Jayaraj Annapackiam College for Women (Autonomous), Periyakulam, Tamil Nadu, South India

Email address:
kanchilakshmik@yahoo.com (M. Kanchilakshmi), arockiamt@yahoo.co.in (A. Thaddeus)

Abstract: An altitude based study was undertaken in the present investigation to find out the status of ecological indices of earthworms in turn to assess the fertility of the soil. Evidences are ample to prove that the physico chemical factors of the environment /soil are responsible for greater ecological indices of earthworms. In the present investigation also it was proved that the earthworm abundance is responsible for the soil fertility for which supportive studies were made on 1. the population dynamics of earthworms, 2. Size of body pores on the skin of earthworms and the secretion of coelomic fluid and the 3. Physico chemical factors of the soil in the area surveyed. It was also found that the earthworm abundance was found to be greater in Meghamalai hills of Theni district compared to that of Sirumalai hills of Dindigul district but when the plains and hills of both the district were compared the earthworm population was found to be more in the plains. Atomic force microscopic studies were also made to measure the size of the body pores of the skin of earthworms of plains and hills. It was found that the number of body pores is more and their size is also larger in the earthworms of plain lands. As a result ceolomic fluid secretion was more. The coelomic fluid contains more than forty proteins and exhibits several biological activities and it supplements the soil with nutrients. Moreover, physical parameters of the soil samples of hills and plains such as rainfall and temperature and chemical parameters such as Ec, p^H, Nitrogen, Phosphorous, Potassium, Calcium, Magnesium, Iron, Zinc, Copper, Boron, were found to be conducive for the growth and multiplication of earthworms in the soil of the plain lands than the hills. Hence, it can be concluded that the earthworm population has got its influence on soil fertility; the more the number of earthworms, more availability of coelomic fluid, which indirectly influences the fertility of the soil, enhances the economy of the nation and the world at large. In total the global worming suppresses global warming.

Keywords: Earthworm Abundance, Hills, Plain Lands, Physico Chemical Factors, Coelomic Fluid, Soil Fertility

1. Introduction

Earthworm density and biomass are fluctuated seasonally with both physical and organic factors of soil. Earthworms are an important integral of the invertebrate community in most soils, both in terms of their contribution to overall below ground biomass on soil biogeochemical cycles [1, 2, and 3]. The distribution of earthworms in soil depends on components like soil moisture, availability of organic matter and pH of the soil. They appear in disparate terrain mainly those which are unilluminated and moist. Earthworms are generally absent or rare in soil with a very coarse texture, in soil with high clay content, or soil with pH < 4 [4]. Earthworm density and biomass alternated periodically with both physical and organic factor of soil have been known to influence the abundance and distribution of earthworms [5]. The coelomic fluid cells have been analyzed and were contrasted with the population abundance of the worms. The earthworm coelomic fluid cells (coelomocytes) have been analyzed by measuring its cell size micrometrically. Earthworms are obvious target organisms for environmental studies. In recent years, earthworms have been used extensively in ecotoxicology to evaluate the consequent of spread contaminants present in soils [6]. In this aspect, coelomocytes and immunological end points such as phagocytosis have been demonstrated as sensitive biomarkers for trace metal monitoring [7]. To this termination, numerous techniques have been evolved to procure the coelomocytes from the earthworms. The protein component of earthworm cuticle is related with an immensely much huge amount of polysaccharide material than is found in the vertebrate collagens. Earthworm's cuticle also belongs to a growing number of collagens [8].

The work on the population density (number/m^2) of earthworms based on the distribution in different localities and different ecosystem and their role in soil fertility tested on plant growth was plenty in the past; but work on coelomic fluid and their role in the replenishment of nutrients to the soil and their sustainability are meager. Moreover, the study on earthworm population and the nature of body pores on the skin of the earthworm based on altitude is scarce. Hence, the present study aims at comparing the population dynamics of earthworm at two different altitudes (plains and hills). Supportive studies were made such as population density in different altitudes, physico chemical nature of the soil of the area of study, size of the body pores and skin of the earthworms, and nature of the body pores based on the altitude by comparing a plain land, and a hill of 1500 meters height from the sea level. The earthworm population of two hill stations of two districts (Meghamalai hills of Theni district and Sirumalai hills of Dindigul district) was also surveyed and compared. Meghamalai hill latitude – 9°39'and 10°30' N and 78°30' E, Sirumalai hill latitude 10°05' and 10°09' N longitude 77°30' and 78°20 E.

2. Materials and Methods

2.1. Study Area

An extensive survey of earthworm was carried out in area located in the nearby Sirumalai hill station at Dindigul Taluk of Dindigul District and Meghamalai hill station at Aundipatti Taluk of Theni District. The earthworm survey was carried out during September 2013 to Aug 2014.

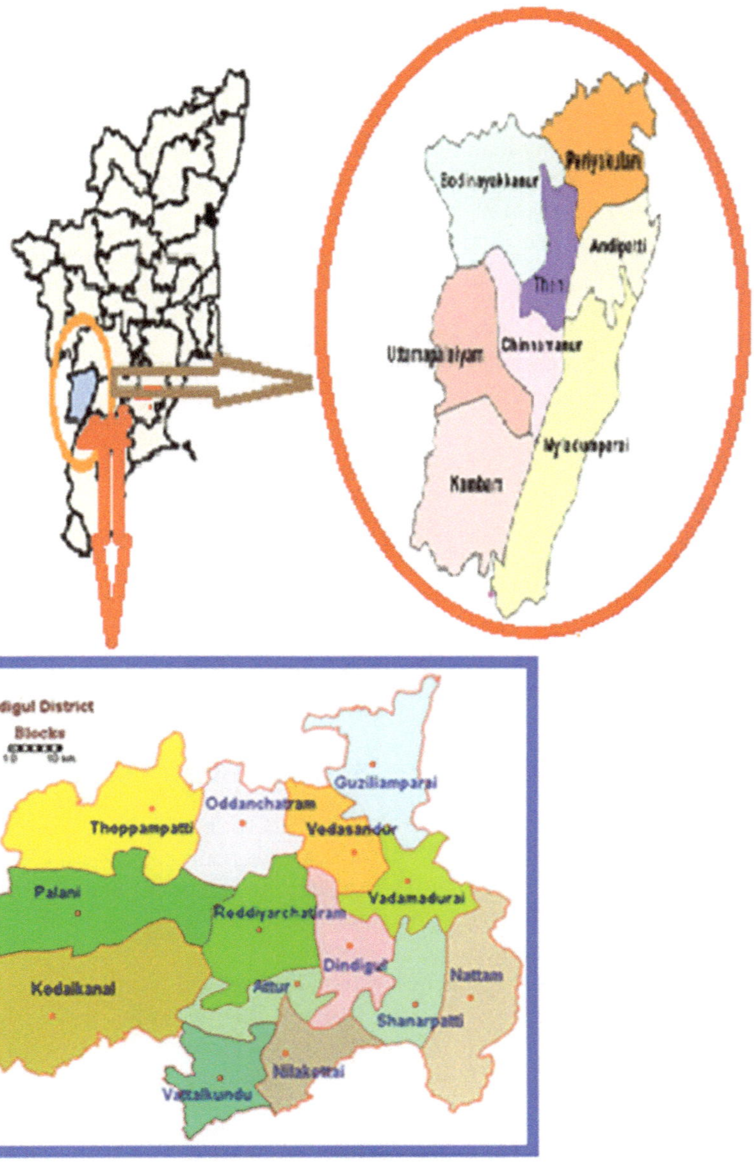

Figure 1. *Map showing the study area.*

2.2. Field Sampling

Four different sampling sites were chosen for the survey of earthworms. The sampling sites include two different ecosystems namely coffee plantation in Sirumalai hills, grapes plantation in Kamalapuram (Dindigul District) and Tea plantation in Meghamalai hills, grapes plantation in Odaipatti (Theni District). The sites are surrounded by stone hills and experience a distinct dry season from March to July. The average annual temperature was 21°C - 30°C.

2.3. Meghamalai

Generally called High Wavy Mountains, is a cool and misty mountain range situated in the Western Ghats in Theni district, Tamil Nadu, in South India. The place is situated at an elevation of 1,500 m above sea level and it is rich in flora and fauna. Mass of the rainfall is received during the North East monsoon in the months of October, November and December.

2.4. Sirumalai

It is a region situated 25 km (16 mi) from Dindigul is located at 10.35°N 77.95°E longitude and 40 km (25 mi) from Madurai, Tamil Nadu, India., There are many high hills in the area. Sirumalai is a dense forest region with a moderate climate throughout the year. The altitude of the hill is 1600 meters above sea level, and it contains diversified flora and fauna and a variety of animals. The survey and collection was done for one year starting from September 2013 to August 2014. The annual temperature is 22-32°C. Dindigul receives rainfall with an average of 812 mm (32.0 in) annually. The South west monsoon, with an onset in June and lasting up to August, brings scanty rainfall.

2.5. Collection of Earthworms

Earthworms were collected from different habitats by digging the soil and hand sorting method. The grasses and leaf litter were cleared from the surface of the sampling site per sampling occasion was marked. The soil was dug and organic matter was analyzed deliberately and was transported to the laboratory along with their native soil in wet cloth bags. The number of each earthworm species indicated per square meter was classified as abundant. The ecological categories of earthworm species were also recorded. The physicochemical properties of soil from each sampling site were also analyzed to establish the possible species habitat relationship.

2.6. Preservation

Collected worms were sacrificed using 70% ethyl alcohol Earthworms were identified to the species level with the help of monographs of many researchers. [9, 10, 11]

2.7. Harvesting of Coelomicfluid

Cold shock method

Earthworms were collected from different habitats by digging the soil and hand sorting method. The grasses and leaf litter were cleared from the surface of the sampling site per sampling occasion was marked. In the present study earthworms (*Eudrilus euginea)* were subjected to cold shock method for the collection of coelomic fluid. Earthworms were using of cold-shock method of ice packing and the fluid was collected in a clean dry test tube. The earthworm coelomocytes were analyzed with the help of Trinocular Microscope to observe the cells and are identified and photographed.

2.8. AFM Study

This AFM work was conducted in the Department of physics, UGC-DAE consortium, Devi Ahilya Vishwa Vidyalaya University, Indore, Madhya pradesh. North India. The skin pieces of earthworms of the size of approximately 0.5 cm × 0.5 cm were cut from dorsal skin and fixed over a magnetic holder by using double side adhesive tape. Nano scope E of AFM from Digital instruments, USA for imaging in contact mode, Triangular cantilevers of Si3N4, ˜100 micron, wide leg, with a force constant of 0.58N/M was used in the present investigation. Scan made on areas of 5 micron ×1 micron were selected on the earthworm skin surface for analysis. The body pores and roughness parameters of the sample (Earthworm skin) were determined by AFM.

2.9. Calculation of Ecological Indices

During random survey some habitats were selected and ecological indices were calculated. Diversity index (H'=∑ pi ln pi) where pi= relative abundance of the species which is estimated by pi=n_l/N, species dominance (C=∑$(pi)^2$), species richness (d=S-1 /ln N where S is the total number of species) and evenness (e=H' /ln S) were calculated for some selected ecosystem during random survey by standard methods [12, 13].

2.10. Soil Analysis

Soil samples were taken from each site for analysis. Temperature at 10 cm depth was measured by soil thermometer and pH, Electrical conductivity, carbon, Total nitrogen, and potassium, phosphorous were analyzed.

3. Result and Discussion

The occurrence and density of earthworms, analysis of soil sample, temperature, moisture, EC, organic carbon, organic matter, total nitrogen, phosphate, potassium and magnesium are presented in along with rain fall, temperature and humidity (Table 1). The maximum rainfall (mm) and Humidity (%) was recorded in theni district compared with dindigul district (Figure 2). The maximum earthworm abundance in meghamalai hills, and plain lands of Theni district. showing species composition of earthworms in various altitude and ecosystem. From Shannon abundance, diversity index and evenness index, it is apparent that the number of species of earthworm was high in meghamalai hill and lower in Sirumalai hills (figure 3, 4 and figure 5). The results of the present study also revealed that the evenness was recorded as high in meghamalai hills and was lower in Sirumalai hills (figure 6 and figure 7). Earthworms are very sensitive to soil p^H [14]. Acidic soils are characterized by higher opportunity of metal ions [15] that cause unfriendly conditions for earthworms. In the present study the clay soil with an electrical conductivity of 0.20, p^H 6.5 - 7.50 along with high moisture holding capacity appears to be favouarable for high earthworm abundance and diversity. This may be associated with the ability of clay soil to retain more moisture for a prolonged time due to its fine particles. The finding is in accordance with [16] various studies which have indicated a decrease in earthworm population in response to dry seasons [3, 17, 18]. However, in Sirumalai, low species diversity index was attributed to the peaty soil with dry soil which has low water holding capacity. Moisture is one of the key factors affecting the survival of all earthworm species [3, 19] Earthworm diversity was low in the plain land; only two to three earthworm species were found. The finding was in consonance with authors [20] who reported that urban and agricultural lands are generally characterized with low earthworm diversity. Human activities may disrupt and limit the development of the earthworm community in agricultural land. A significant decline in number of individual earthworms in summer can be attributed to changes in soil temperature and moisture [21]. The result of present study revealed that the diversity of earthworm species in plain lands of two districts was found to be high in Theni and Dindigul districts during NorthWest monsoon and lower in summer (figure 8 and figure 9). Similar results were shown by [22] who reported higher diversity indices and evenness in cultivated soil whereas [19] reported the same from vegetable soil. The results revealed that the evenness value was recorded high during northwest monsoon and lower in summer (figure 10 and figure 11).

Table 1. *Physico-chemical parameters of soil samples of Meghamalai hills (Theni district) and Sirumalai hills (Dindigul). Period of survey: September 2013 to Aug 2014.*

Parameters Habitats	**Sirumalai (Hill)**	**Meghamalai (Hill)**	**Dindigul (Plain)**	**Theni (Plain)**
	Coffee	**Tea**	**Vineyard**	**Vineyard**
Type of soil	Coastral with peaty soil	Clay soil	Clay soil	Red soil
Colour	Black	Brown	Brown	Red
Ec	0.28	0.20	0.40	0.28
pH	6.69	7.50	7.8	6.9
Temperature	17°-21°C	15-17°C	30-32	29-30
Nitrogen	7.3	8.3	74	65
Potassium	3.3	4.3	4	23
Phosphorous	1.95	3.0	50	182
Iron	11.8	12.1	5.2	4.7
Manganese	9.2	11.3	2.7	2.2
Zinc	1.62	-	0.81	0.81
Copper	3.5	2.56	0.79	0.71
Boron	0	0	0	0

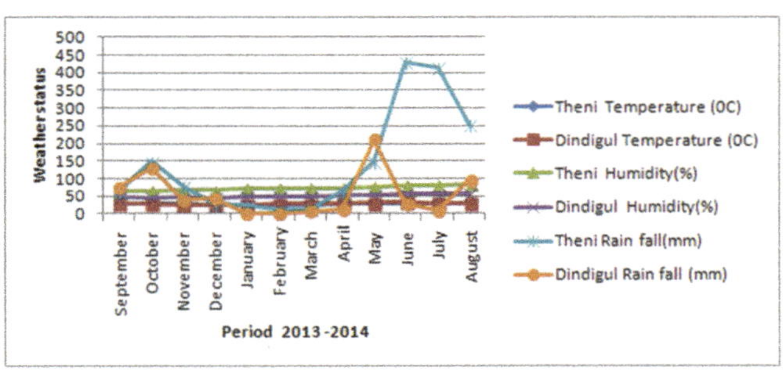

Figure 2. *Weather Condition in Theni and Dindigul District.*

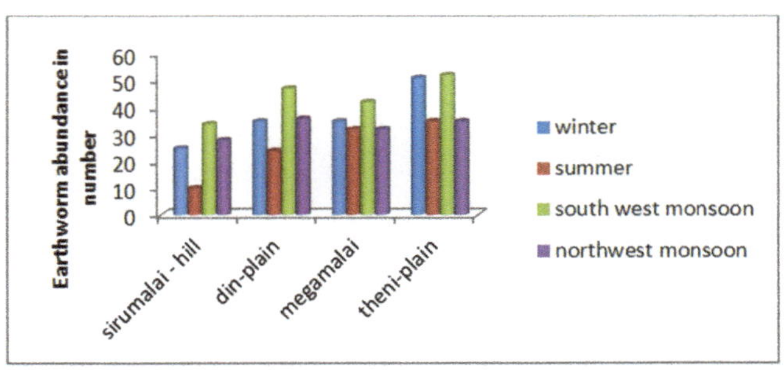

Figure 3. *Abundance of earthworm population in hills and plains (Theni and Dindigul Districts).*

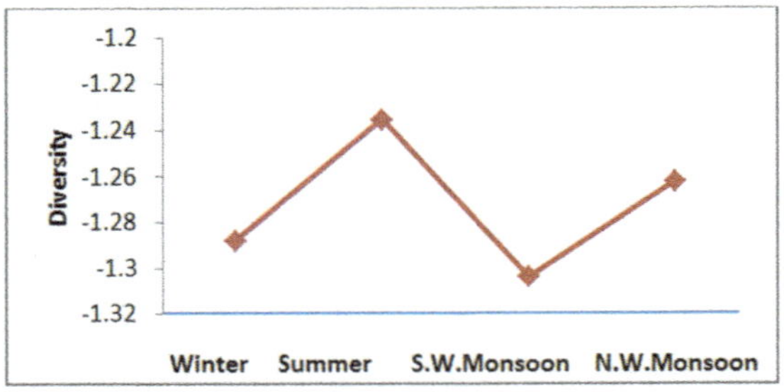

Figure 4. *Diversity of earthworm species in Meghamalai hills (Theni district).*

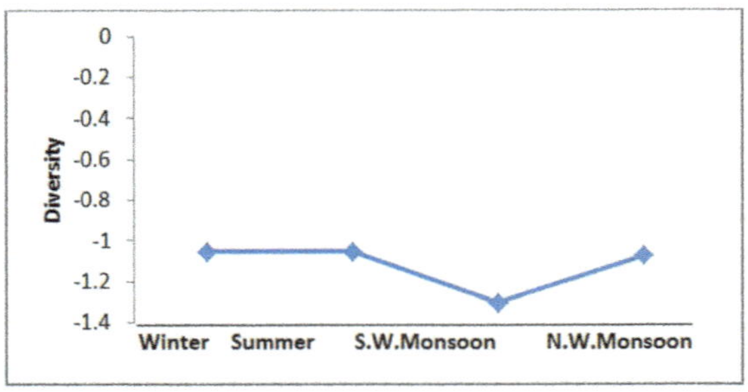

Figure 5. *Diversity of earthworm species in the Sirumalai hill of Dindigul district.*

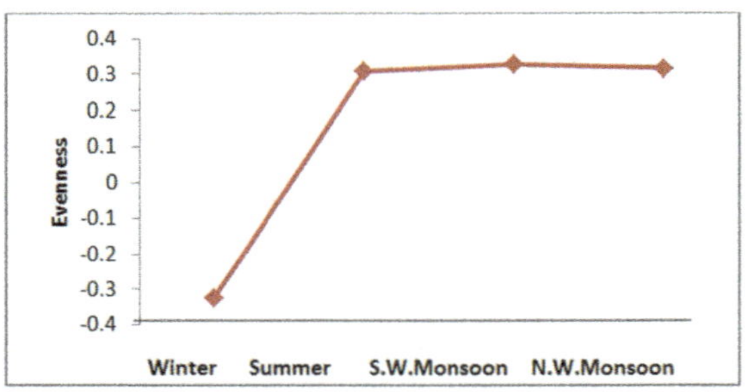

Figure 6. *Evenness of earthworm species of Meghamalai in Theni District.*

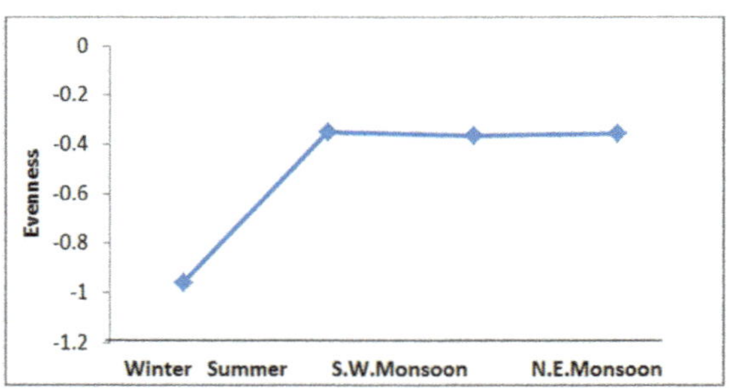

Figure 7. *Evenness of earthworm species from sirumalai in dindigul district.*

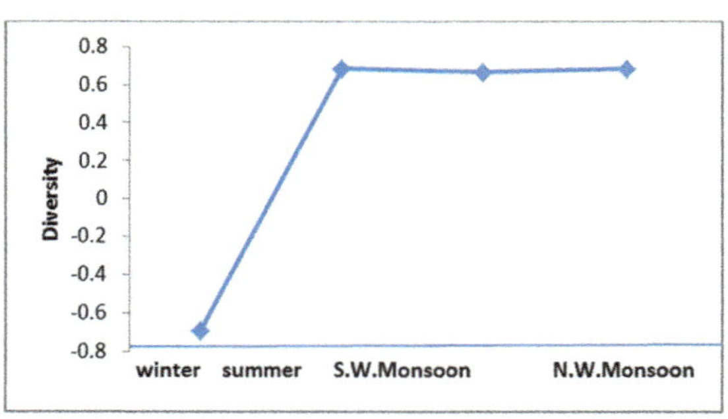

Figure 8. *Diversity of earthworm species of plain in Theni District.*

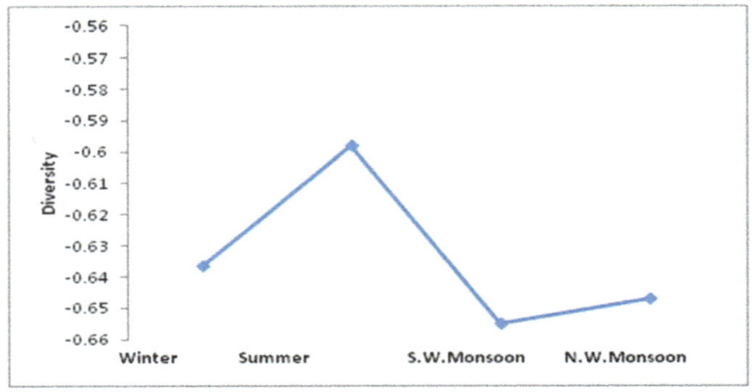

Figure 9. *Diversity of earthworm species from plain in dindigul district.*

Ample of evidences have been reported that the abundance of earthworm population in terms of ecological indices of earthworms are mainly due to ecological factors. But in the present investigation earthworm abundance or the ecological indices of the earthworms are responsible for the soil fertility. It was supported by the recent workers that [23, 24] soil structure, gas dynamics, water flow, and C & N departure and stabilization may be altered by the presence and community structure of earthworms. Earthworms are very sensitive to touch, light and dryness. Water-logging in the soil can origin them to come to the surface. Worms can tolerate a temperature range between 5°C to 29°C. A temperature of 20°C to 25°C and moisture of 50–60 percent is optimum for earthworm function [25]. Earthworm diversity is influenced largely by soil nutrients and rainfall patterns [26]. It was also supported by the contributors [1] that earthworm ecological groups have variable effects on the soil physical, chemical and biological properties.

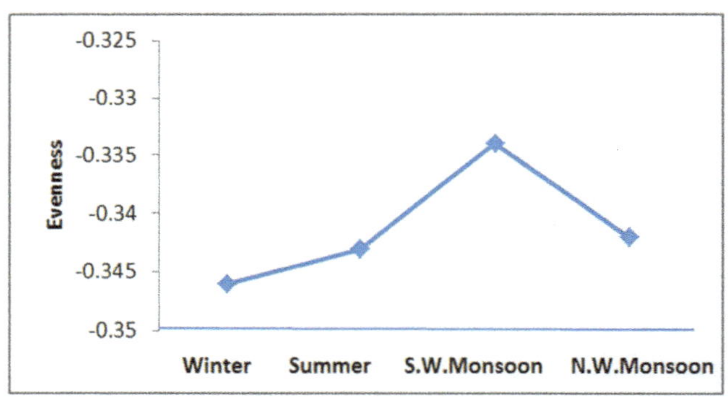

Figure 10. *Evenness of earthworm species in plain (Theni District).*

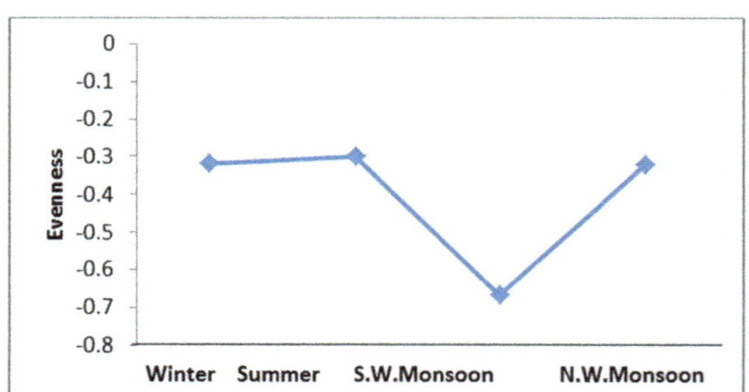

Figure 11. *Evenness of earthworm species in plains (Dindigul District).*

In the present investigation, Coelomocytes from the body fluid of the earthworm *Eudrillus euginea* were studied using Trinocular microscopy and were photographed. Cells were coinciding with the immune cells. Different size of cells of amobeocytes were found to be 6 μ m and 10 μ m and granulocytes of 5 μ m and 8 μ m respectively in the earthworms collected from hills and plain lands of Dindigul and Theni District Figure 12. AFM studies were also made on the earthworm skin and the size of the body pores were measured. The pore size was found to be 1.000 μ m, 1.500 μ m and 5.66 μ m in the earthworms collected from hills and plain lands of Dindigul and Theni District Figure 13. This is in concordance with the authors [27] who reported that all these coelomic cells are involved in various aspects of cellular and humoral immunity, the former by phagocytosis. Chlorocytes/ eleocytes resemble the invertebrate live cells in certain function. For example, they are involved in the metabolism and storage of glycogen and lipids. They also transport nutrient via the circulation into the coelomic fluids as well as various cells and organ [28]. Cold shock method of fluid collection was found to be the safest method of coelomic fluid. Cold shock method produced 1.5 ml of fluid collection in 30 minutes. [29, 30]. The phenomenon has been exploited as noninvasive methods of sampling coelomic fluids and coelomocytes using electric current [31].

The coelomic fluid cells have been analyzed and were compared with the body pores and globular proteins in the skin of the earthworms. Coelomocytes are considered as the immune cells of coelomate animals (Annelids, Mollusca, Arthopoda). These cells are of the types of leukocytes that have long been considered to constitute the major innate immune defense system of these animals. It has been found that coelomic fluid of the earthworms contains more than 40 proteins and exhibits several biological accomplishments as follows: cytolytic, proteolytic, antimicrobial, hemolytic, heam agglutinating, tumorolytic, mitogenic activities [32, 33, 34, and 35]. It was also reported [36] that the coelomic fluid gains the production of *Rhidium* by 20-25% due to an increase in soil organic matter. furthermore, the inclusion of coelomic fluid to soil used in growth of 1) *Seasamum indicum, Vinga radiate, Vinga mungo* increased plant growth rates [37, 38] 2) rice, bean, corn, tobacco, oats and peanuts lead to increase their manufacture and length of roots[39, 40]. This is in support to the present investigation that the more the number of earthworms the more will be the excretion of coelomic fluid which in turn improves soil fertility which was assessed by physico-chemical analysis of the soil.

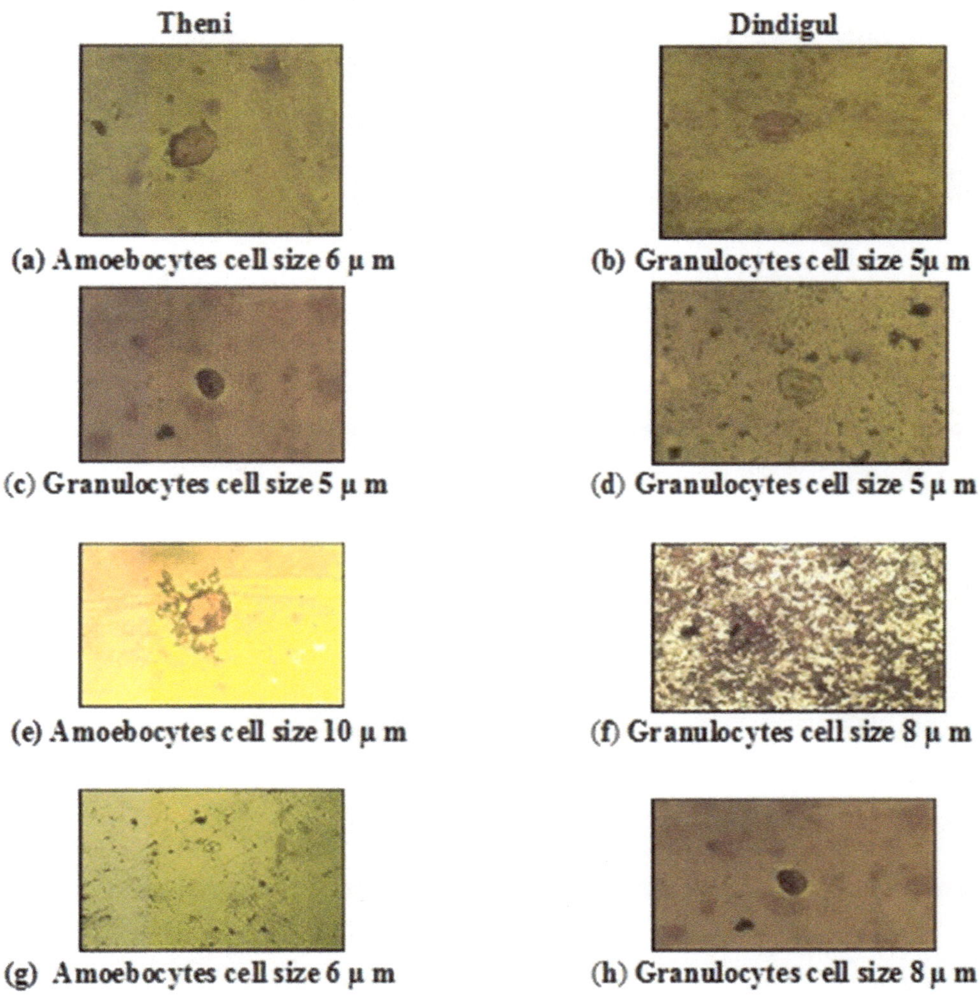

Figure 12. *Types of coelomocytes in coelomic fluid of earthworms.*

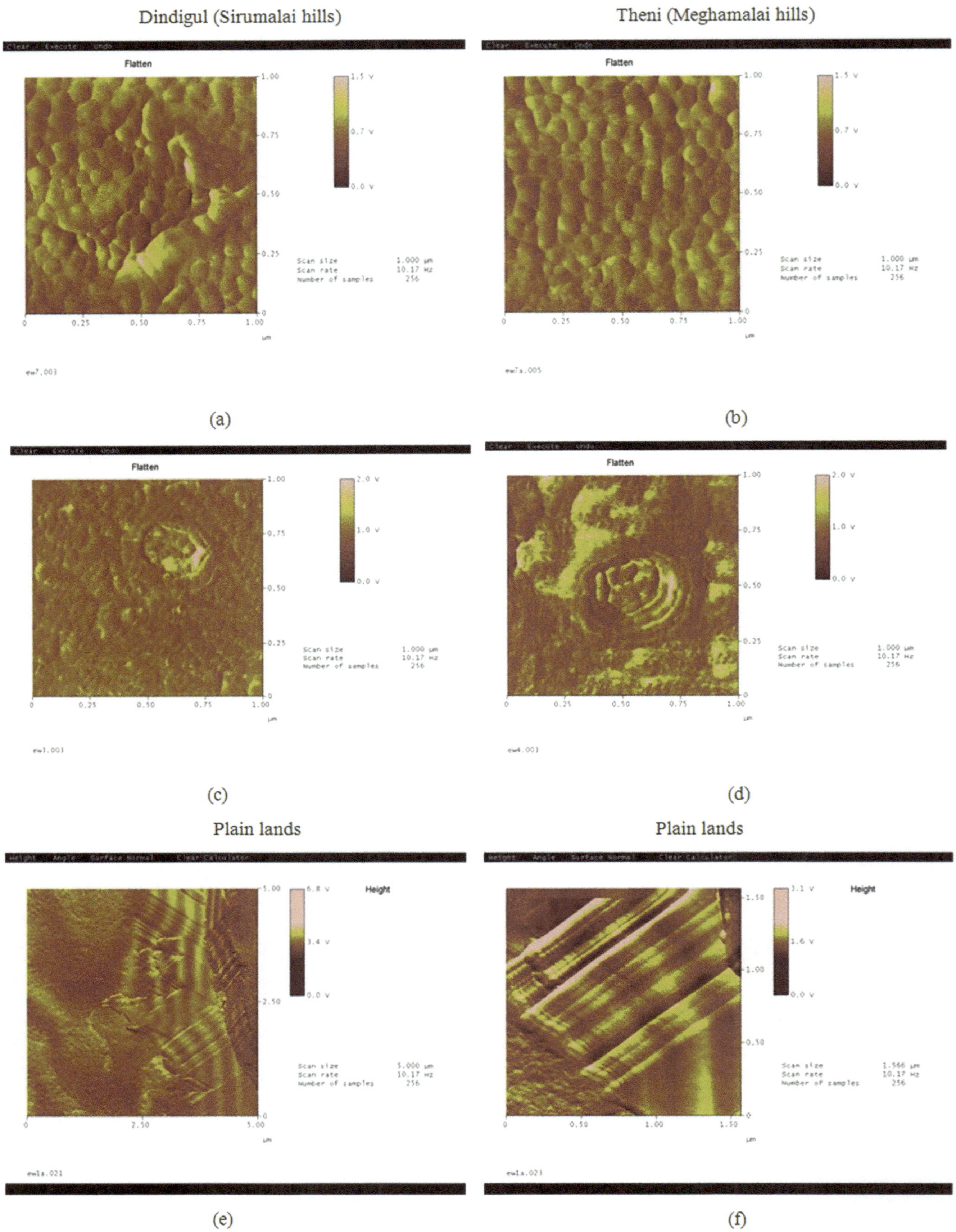

Figure. 13. *AFM studies on body pores of earthworms of two districts (Plain lands & hills).*

The high accessibility of feed and moisture content kept in this man made environment appear to be the positive constituents. [41] Fragoso et al., (1999) reported that the structural composition of earthworm communities varies depending on the type of agro ecosystem.

4. Conclusion

The altitude plays a major role in the population density of earthworms along with the climatic factors. It doesn't mean that the population density increases not only with the increase in altitude but also the physico chemical factors of the soil in the ecosystem plays their role. At the same time, it was proved in the present investigation that there is sustainability in the soil fertility in the region where there is earthworm abundance, and was supported by the biophysical studies on coelomiic fluid and the body pores of earthworms with the help of Trinocular microscopy and Atomic Force Microscopy. In nutshell greater the earthworm abundance in a particular area, greater will be the soil fertility of that area. Which indirectly enhances the economy of the nation and the world at large. In total the global worming suppresses global warming.

Acknowledgement

I thank my institution and my guide for providing all facilities to carry out this work. Thanks are due to University Grants Commission, New Delhi for funding and UGC – DAE Consortium, Indore, Madhya Pradesh for the lab facilities to do AFM studies.

References

[1] K. E. Lee, "Earthworms: Their Ecology and Relationships with Soil and Land Use". 1985. Vol. 5. Academic Press, New York.

[2] S. W. James, "Soil nitrogen, phosphorus and organic matter processing by earthworms in tall grass prairie". Ecology, 1991, 72, pp. 2102-2109.

[3] P. J. Bohlen, W. M. Edwards and C. A. Edwards. "Earthworm community structure and diversity in experimental agricultural watersheds in Northeastern Ohio". 1995, *Plant and Soil* 170, pp. 233-239.

[4] K. Gunathilagraj P. T. Ramesh, "Degradation of coir wastes and tapoica peels by earthworms". In Training Program in Vermiculture. New Delhi: Indian Council of Agricultural Research 1996 (ICAR).

[5] W. White, "An earthworm is born, starting". Publishing company, New York. 1975.

[6] M. A. Giggleman, L. C. Fitzpatrick, A. J. Goven, B. J. Venables, Effects of pentachlorophenol on survival of earthworm (*Lumbricus terrestris*) and phagocytosis by their immuno active coelomocytes. Environ. Toxicol. Chem. 1998. 17, 2341–2394.

[7] S. Senthil Kumar, M. V. Sriram Chandrasekaran and K. Haripriya., "Effect of vermicompost and fertilizer on the growth and yield of rose". *J. Interacademicia*, 2004. 8, pp. 207-210.

[8] P. Ville, P. Roch, E. L. Cooper, P. Masson, J. Narbonne, "PCBs increase molecular-related activities (lysozyme, antibacterial, hemolysis, proteases) but inhibit macrophage-related functions (phagocytosis, wound healing) in the earthworms". J. Invert. Pathol. 1995. 65, pp. 217–224.

[9] G. E. Gates, Burmese earthworm, "An introduction to the systematic and biology of magadrile oligochaetas with reference to southeast Asia", Transaction of the American philoshophical society, 1972 62, pp. 1-326.

[10] J. M. Julka, "The fauna of India and the adjacent countries. Megascoleccidae: octochaetidae; (Earthworm) *Haplotaxida, Lumbricina*": XIV, Zoological survey of india, culcutta., 1988. 400.

[11] J. Stephenson, Oligochaeta, the fauna of British India including Ceylon and Burma, Taylor and Francis Ltd. London, 1923XXIV +518pp.

[12] C. E. Shannon and W. Weaver "The mathematical theory of communication", university Illinois press, Urbana, 1949. pp. 117.

[13] E. H. Simpson, Measurement of diversity, Nature, London 1949. pp. 163-681.

[14] C. A. Edwards, Report on the second stage in the development of a standardized laboratory method for assessing the toxicity of chemical substances to earthworms. Commission of the European Communities, Brussels, Luxemburg, 1984.

[15] N. C. Brady, "The nature and property of soil". Collier McMillan Press, London1984.

[16] G. H. Baker, P. J. Cartera, J. P. Curry, O. Cultreric & A. Beck. "Clay content of soil and its influence on the abundance of *Aporrectodea trapezoides* Duges (*Lumbricidae*)". *Applied Soil Ecology* 1989, pp. 333-337.

[17] G. X. Gonzalez, Zou & S. Borges. "Earthworm abundance and species composition in abandoned 298 earthworm diversity and population density in kaki bukit tropical croplands: comparisons of tree plantation and secondary forests". Pedobiologia 1996, 40, pp. 385-391.

[18] I. S. Najar, and A. B. Khan. "Earthworm communities of Kashmir Valley, India". Tropical Ecology, 2011. 52, pp. 151-162.

[19] C. A. Edwards, and P. J. Bohlen. "Biology and ecology of earthworms". 3rd ed. Chapman & Hall, London, UK, 1996.

[20] K. M. Smetak, J. L. Johnson-Maynard & J. E. Lloyd. "Earthworm population density and diversity in different aged urban systems". Applied Soil Ecology 2007, 37, pp. 161-168.

[21] J. K. Whalen, R. W. Parmelee & C. A. Edwards. "Population dynamics of earthworm communities in corn agroecosystems receiving organic or inorganic fertilizer amendments". *Biology and Fertility of Soils* 1998. 27, pp. 400-407.

[22] G. Tripathi, P. Bhardwaj. "Comparative studies on biomass production, life cycles and composting efficiency of *Eisenia fetida* (Savigny) and *Lampito mauritii* (Kinberg)", *Biores. Technol.*, 2004. 92, pp. 275–283.

[23] A. J. Vanden Bygaart, C. A. Fox, D. Follow, and R. Protz. "Estimating earthworm influenced –soil structure by morphometric image analysis". Soil Science Society of America Journal 2000 64, pp. 982-988.

[24] R. V. Pouyat, M. M. Carreiro. "Contrasting controls on decomposition of oak leaf litter along an urban-rural land use gradient. Oecologia". 2003, pp135: 288–298.

[25] P. Hand. "Earthworm biotechnology. In: Greenshields R(Ed) resources and application of biotechnology": The New Wave Macmillan Press, Newyark, US, 1988, pp 214-257.

[26] C. Fragoso, & P. Lavelle. "Are earthworms important in the decomposition of tropical litter?" 1995 pp. 103-112. In: M. V. Reddy (ed.) Soil Organisms and Litter Decomposition in the Tropics. Oxford Publishers, New Delhi.

[27] R. P. Dales, Y. Kalaç. "Phagocytic defence by the earthworm. *Eisenia foetida* against certain pathogenic bacteria", Comp Biochem Physiol. 1992.101A, 3, pp. 487-490.

[28] E. B. Affar, M. Dufor, G. G. Poirier, D. Nadeau. "Isolation, purification and partial characterization of chloragocytes from the earthworm species *Lumbricus terrestris*". Mol Cell Biochem, 1998185: 123-133.

[29] Hideshi Kobayashi, Naoshi Ohta, Masato Umeda. "Biology of Lysenin, a Protein in the coelomic fluid of earthworm *Eisenia foetida*". International Rev. of Cytology. 2004, 236, pp. 45-98.

[30] Weidong Pan, Sianghui Liu, Feng Ge, Tao Zheng. "Reconfirmation of antimicrobial activity in the coelomic fluid of the earthworm *Eisenia foetida Andrei* by colorimetric assay". J. Biosci. 2003, 28 6, pp. 723-731.

[31] S. S. Hamed, E. Kauschke, E. L. Cooper. "Cytochemical properties of earthworm coelomocytes enriched by Percoll. In: A New Model for Analyzing Antimicrobial Peptides with Biomedical Applications". Beschin A, Bilej M, Cooper EL [Eds], IOS Press, Ohmsha, 2002. pp 29-37.

[32] A. Çotuk. and P. R. Dales. "The effect of the coelomic fluid of the earthworm *Eisenia foetida* sav. on certain bacteria and the role of the coelomocytes in internal defense". 1984., Comparative Biochemical Physiology, 78A 2: pp. 271-275.

[33] S. Lange, F. Nubler, E. Kauschke, G. Lutsch, E. L. Cooper, and A. Herrmann. "Interaction of earthworm hemolysin with lipid membranes requires sphingolipids". The Journal of Biological Chemistry, 1997. 272, 33: pp. 20884-20892.

[34] S. Lange E. Kauschke W. Mohrıng and E. L. Cooper. "Biochemical characteristics of *Eisenia* pore, a pore-forming protein in the coelomic fluid of earthworms". European Journal of Biochemistry, 1999. 262: pp. 547-556.

[35] E. L. Cooper and P. Roch. "Earthworm Immunity: a mode of immune competence". Pideobiologia. 2003, 47, pp. 676-688.

[36] M. A. Gopal, Gupta and G. V. Thomas. "Opportunity to Sustain Coconut Ecosystem Services through Recycling of the Palm Leaf Litter as Vermicompost: Indian Scenario" (A Technology/ Research Note). 2010. 26, 2: pp. 42-55.

[37] S. S. Hatti R. L. Londonkar, S. B. Patil, A. K. Gangawane and C. S. Patil. *J. Crop Sci.*, 2010a 1,1 pp. 6-10.

[38] S. S. Hatti, R. L. Londonkar, S. B. Patil, A. K. Gangawane and C. S. Patil. *J. Crop Sci.*, 2010b 1,1 pp. 1-5.

[39] M. J. Prabhu. "The Hindu Newspaper", 28th December, In: Science and Technology Section. (Cited from Zambare et al, 2008), 2006.

[40] R. M. Atiyeh, S. Suler, C. A. Edwards, G. Bachman. "Effects of vermicomposts and composts on plant growth in horticultural container media and soil". Pedobiologia, 2000 44, pp. 579-590.

[41] C. P. Fragoso, E. Lavelle, B. K. Blanchart, J. J. Senapati, Jimene, Martinez, T. Decaens & J. Tondoh. "Earthworm communities of tropical agroecosystems: origin, structure and influence of management practices". In: Earthworm Management in Tropical Agroecosystem. CAB International. Wallingford, U. K. 1999. pp. 27-55.

Water is Life: Situation Analysis of Access to Household Water Supply in the Wa Municipality, Ghana

Enoch Akwasi Kosoe, Issaka Kanton Osumanu

Department of Environment and Resource Studies, University for Development Studies, Wa, Ghana

Email address:
ekosoe@uds.edu.gh (E. A. Kosoe), kosumanu@uds.edu.gh (I. K. Osumanu)

Abstract: Far too many people in urban Ghana live without access to safe drinking water and this is a primary determinant of continuing poverty. Using questionnaire survey and key informant interviews, this study explored the situation of households' access to potable water supply in the Wa Municipality, a low-income urban area in north-west Ghana. This study highlights several important issues on access to water supply focusing on availability, accessibility and cost. A clustered sample of residential areas of the municipality was employed and the study was able to examine spatial disparities in access to water supply. The results indicate that access to potable water supply in the municipality is generally high as only 13 percent of households depend on open wells as their main source. Also, the private sector (individuals and non-governmental organizations) are major players in public water supply in the municipality. In spite of the progress made in access to potable water supply, distance and cost of water remain serious challenges confronting households. The study recognizes that private individuals create additional water facilities to augment public supplies, but the positive impacts on health and livelihoods could be greater if access to finance could be facilitated. Most importantly, water supply improvement strategies should incorporate an integrated vision, which sees adequate and quality water both as a goal in itself and as a contributor to economic and social development through gains in public health.

Keywords: Accessibility, Health, Households, Urbanization and Water Supply

1. Introduction

The importance of adequate water quantity for human health has been recognised for many years and there has been an extensive debate about the relative importance of water quantity, water quality, sanitation and hygiene in protecting and improving health (1 - 3). In spite of this recognition, international guidelines or norms for minimum water quantities that domestic water supplies should provide for households remain largely lagging. For instance, whilst the Millennium Declaration Goals included a target to halve the proportion of people who are unable to reach or to afford safe drinking water by 2015 (4) it did not specify in what quantity such water should be supplied (5).

Municipal water supply involves a great expense of time and energy, especially for those who have no water sources in or near their homes (6). This is because in many urban areas in developing countries, public water providers have failed to provide households with adequate water supply (7, 8). This failure has been attributed to low-quality and inadequate coverage, poor cost recovery, high labour cost and low labour productivity, and large state subsidy (9). But it has also been argued that public water providers give poor services because of political involvement in the sector (10). Again, (11) stressed that population growth, rapid urbanization, inadequate levels of economic development and the absence of appropriate management and technical capacities meant that an increasing number of urban population do not have access to water.

In Ghana, WHO/UNICEF has indicated that the country has achieved and exceeded its 2015 national MDG target of 78% for water coverage by 6% (12). But this does not reflect in low-income urban areas (13, 14). It is estimated that about 3.5 million people, many of whom are in low-income urban communities, still do not have improved sources of drinking water (14). Also, urban households do not have water flowing regularly through their taps from the municipal water supplies and have to supplement their limited supply with water from other similarly unreliable sources (2015). This has led to public water rationing, forcing urban residents to resort to private suppliers (16, 17). The water sector in Ghana has been constrained by interlinked political, administrative, economic, social and legislative factors (18). These factors, however, hinder a working water supply system in the country (19). Also, the high cost of treatment of water, high maintenance cost of water plants and equipment, water losses during distribution, increasing population and non-payment of water bills by consumers account for the country's public water sector inability to supply adequate water to their clients (20). Compounding the problem is the inability of the public water system to meet the increasing water demands with respect to deficiency in water quality and quantity, as well as unaffordability and inadequate investment in water infrastructure (16, 17). This has led to the involvement of the private sector to improve water supply services in the country but this move, in itself, is not a panacea to the problem of inadequate household water supply (18, 19). The issue of privatization has been controversial because consumers do not understand why companies should make profits out of a basic necessity (21). The real issue is better management and efficient delivery of clean drinking water. The political will for reform is essential, but the responsibilities of government agencies for supplying water have not been defined clearly. Partnerships between the government, government agencies and citizens meant to create awareness of water issues, ranging from fighting corruption, conservation and reaching consensus on appropriate water rates, to creating laws to protect resources and mechanisms to resolve water disputes have equally failed to achieve good results (22).

In this paper we present the findings of a water survey which was designed specifically to assess the situation of household access to water supply in the Wa Municipality of North West Ghana. The resulting data, together with information drawn from relevant literature, are used to explore some of the issues and concerns that have been raised with respect to access to water supply in low-income urban areas of developing countries. Here we focus on three main issues. First, we consider the spatial distribution of potable water sources within the municipality. Second, we highlight what constitutes safe access to water and its relevance to households. Third, we examine water availability and cost in relation to the various potable water sources.

2. The Urban Water Supply Challenge in Context

Water is key to urban life. While access to water supply improves health, it also has implications for a number of other development goals (Figure 1). According to the 2003 World Water Development Report (6), problems of poverty are inextricably linked with those of water - its availability, proximity, quantity and quality - and as such improving access to water could make a major contribution towards achieving the eradication of extreme poverty and hunger. Moreover, improved access to water supply directly affects sanitation and hygiene and indirectly reduces mortality rate amongst the under fives. Given that an estimated 88 percent of diarrhoeal diseases - the second leading cause of death in the under fives - can be attributed to unsafe drinking water, inadequate sanitation and poor hygiene (23), improving access to safe water (and sanitation) will be central to mortality reduction.

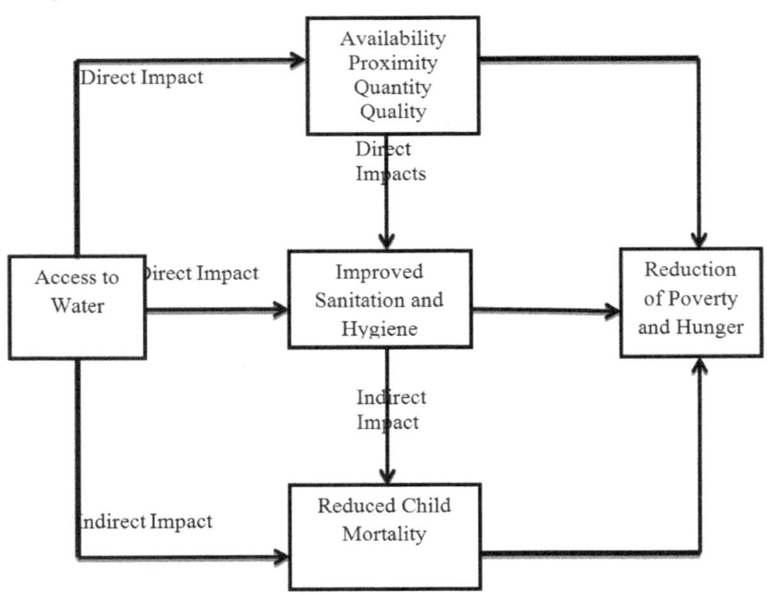

Figure 1. *Impacts of access to water supply.*

Although access to safe drinking water can have a strong positive effect on human health and well-being, the development and management of water resources as a whole also has significant health implications, usually a combination of negative and positive (23). The balance between these effects, and protection of the health of vulnerable groups, is in the hands of those who plan and implement interventions, which can be strongly influenced by the health sector and health arguments. It has been noted that, in most low and middle-income countries, there is virtually no reliable information on who has safe drinking water; there are merely estimates of the percentage of the population - total, rural and urban - who have access to an improved water source (24). Going by the currently accepted definition, individuals are deemed to have access to an improved water source if they have a household connection or access to a public standpipe, borehole, protected well, protected spring or rainwater collection facility, capable of providing 20 litres per capita per day, at a distance of not more than 1,000m. An unimproved source is taken to be any unprotected well or spring, vendor-provided water, tanker truck provision of water and bottled water (25). The latter is considered to be unimproved owing to concerns over the quantity of supply and not its quality.

One of the problems is that the accepted definition places the emphasis on the quantity of water and the distance to source, but makes no provision for the important aspects of quality, regularity of supply, cost, or the number of people using the source. As such, having a source of water close by does not necessarily mean that an individual has access. WHO/UNICEF acknowledge that the definition of an improved water source is problematic and that the assumption that certain technologies are better for health than others may not be true in all cases, noting that "in some locations an unprotected household well may provide a better supply of water, both in terms of quantity and quality of water, than a household connection which may be subject to intermittence and poor water quality" (26). Another problem is that in urban areas, rapid population growth and migrations motivated by cultural, economic, political, and environmental factors strain existing water infrastructures and create enormous problems in planning and constructing new infrastructure. Residents of many of the low-income cities enjoy only intermittent access to piped water, often of dubious quality and only from public taps at substantial distances from their homes. Others depend on water vendors for small volumes of costly water of unsure quality. Where providers cannot guarantee water quality at the point of supply, or where it cannot be guaranteed at the point of use, because of contamination during collection, transport, and storage, consumers face significant health risks.

Given the failure to reduce the numbers of people without access to basic water supply during the 1990s, and the financial implications of even the apparently modest international development target of halving the proportion of people not served with improved drinking water by 2015, it is evident that "business as usual" cannot provide a satisfactory response. Approaches that rely solely on time and resource-intensive centralized solutions will leave hundreds of millions of urban residents in developing countries without access to safe water far into the foreseeable future. There should be an outcry for immediate, concerted efforts to confront the reality that water supply coverage rates in the developing world barely keep pace with population growth.

3. The Study Area and Methodology

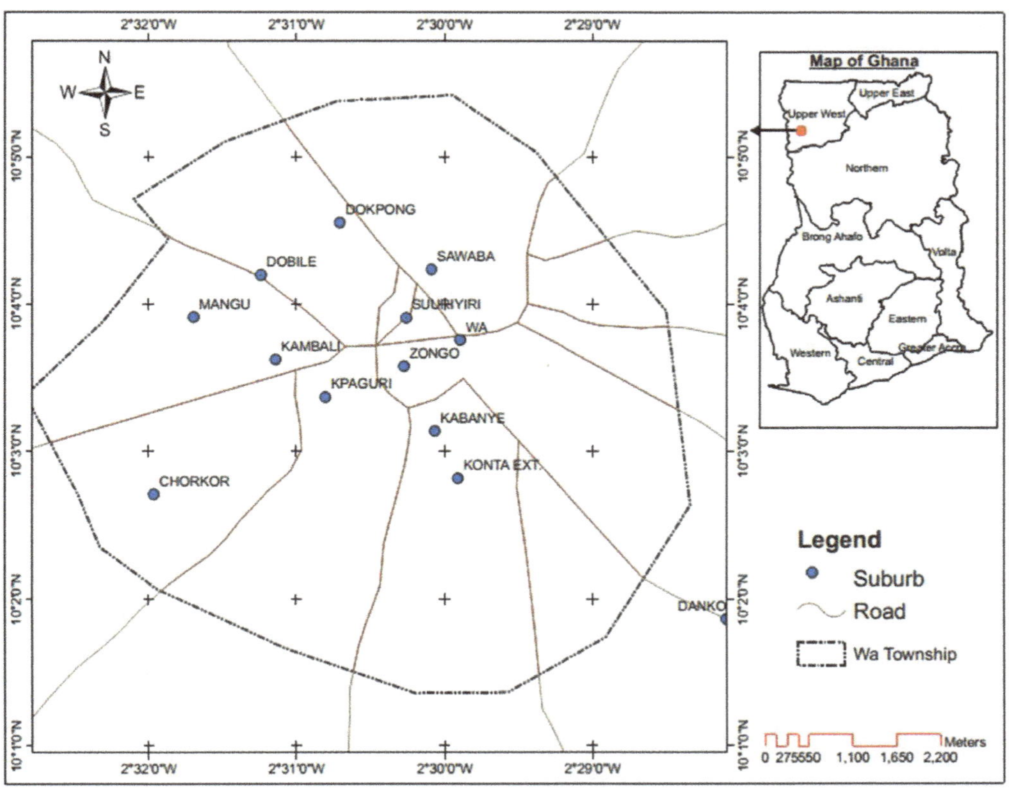

Figure 2. *The study context.*

The locational context of this study is Wa, which is the regional capital of the Upper West Region of Ghana. Wa lies between latitude $1^0$40'N to $2^0$45'N and on longitude $9^0$32'W (Figure 2), thus covering an area of approximately 1,180 square kilometres, which is about 32 percent and 2.56 percent of the region and nation respectively. Wa's population was estimated to be 135,638 (Female 65,887 / Male 69,751) by the year 2010 with a growth rate of 2.7 percent per annum (27). The spatial distribution of the population displays a typical character of a young municipality – a heavy concentration of population in Wa town surrounded by smaller towns and rural settlements. Using the 2010 Population and Housing Census (27) figures, Wa's population is 50 times higher than the next populous settlements (Busa, Sagu, Charia, Kperisi and Boli) each with populations below 3,000 people. The significance of this type of distribution is that Wa town provides the highest level services (first level services and functions) in health, education, finance, administration of justice and security, commerce and transportation, amongst others, to its hinterland and patent services for resource mobilization, peace building and community needs identification.

Apart from the use of documented sources, the study also generated first-hand information from the field. The study relied on qualitative and quantitative approaches taking into consideration the data demands of the study. Household questionnaires were used as its primary data collection instrument. This approach is deemed appropriate when the object of the research is to explore attitudes or reactions of a group or community in response to some commonly experienced aspects of their environment. Through such interactive discourse, participants are able to offer insights on the perspective of the enterprise, revealing clues to the social contexts that shape their opinions. In all, One hundred and fifty-four respondents were interviewed using the household questionnaires.

A two-stage sampling methodology was adopted in the selection of households for the interviews. The first stage was the clustering of residential areas according to town planning classifications into three core zones. Zone One covered the unplanned residential areas, which covers mainly the traditional settlements before urbanization and their immediate surroundings. Zone Two is made of the planned suburbs which have most government bungalows and private residential apartments occupied mainly by high and middle class households. Zone Three comprises the Newly Developing Areas, part of which is made of the surrounding villages which have been absorbed into the urban agglomeration as a result of urban sprawl. Since households in Zone One are more extensive than in Zones Two and Three, 40 percent of the total sample was apportioned to Zone One and 30 percent each to Zones Two and Three. Within the selected residential areas, blocks were created based on the number of houses and interviewers selected households to interview by systematically walking through the blocks and interviewing one household in every tenth (for Zone One) and twentieth (for Zones Two and Three) house. The questionnaires were administered to household heads or their representatives. In a house where there were multiple households, only one household was interviewed.

In addition to the household surveys, in-depth interviews were organized with Ghana Water Company Limited (GWCL), Community Water and Sanitation Agency (CWSA), Private Water Providers and the Wa Municipal Assembly. Purposive sampling was used to select staff of these institutions and agencies. This is because these are officers knowledgeable in issues related to water supply and management. The interviews covered themes on the number of water provision facilities and their distribution within the municipality to enable the spatial mapping of the facilities as well as their detailed knowledge about the enterprise.

Hand-held global positioning system (GPS) receivers were used to pick geographic coordinates of various water points used by GWCL and other water facilities in the municipality. The coordinates were taken in the Degree Minute Second (DMS) format. These were converted to Decimal Degree (DD) in Microsoft excel spreadsheet using the formula: $=D+\frac{M}{60}+\frac{S}{3600}$, where D = Degree, M = Minutes and S = Seconds. The coordinates were imported to ArcMap software and converted to a Shapefile. The projection parameters were changed from the default World Geographic System 1984 (WGS 84) to projected Universal Transverse Mercator Zone 30 North. It was then overlaid with existing country Shapefiles and layout maps were produced showing how the various water points are spatially distributed within the municipality.

4. Results and Discussions

4.1. Spatial Distribution of Potable Water

The main water sources used by households in the Wa Municipality are boreholes both non-mechanized and mechanized (sometimes supplied through public taps), open wells and piped water from the Ghana Water Company Limited (GWCL). Of the households interviewed, 42 percent depends on public taps supplied by mechanized boreholes for their water needs whiles 12 percent gets their water supply from GWCL's Services. The distribution across residential zones depicts a high concentration of piped water in Zone Two and boreholes in Zone Three (Figure 3). Respondents assigned a number of reasons for their main water source ranging from being the only water source available in the community to being the cheapest. Besides the open wells, all the water sources are potable and this presents a much better picture compared to other towns in Ghana. This confirms the conclusion of the 2008 Ghana Demographic Health Survey (28) that the Upper West Region has the highest potable water coverage in Ghana. This finding also indicates a possible good health status for households in Wa.

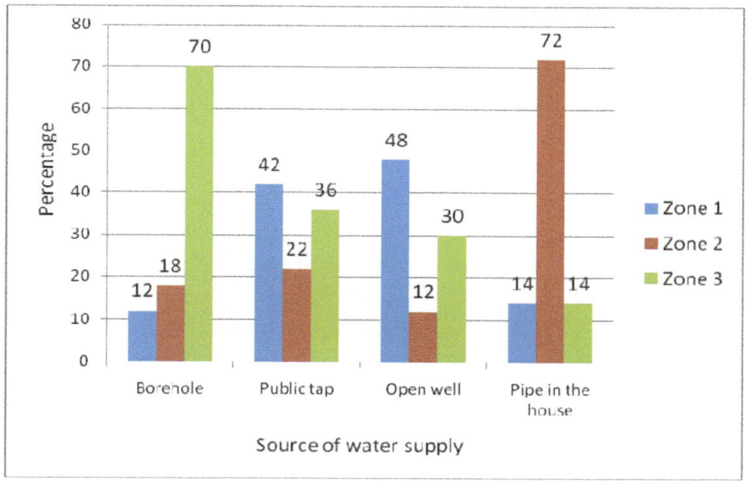

Figure 3. *Source of water supply for households by residential zones (%).*

The Ghana Water Company Limited has 16 boreholes (Figure 4) from which they serve the whole of the Wa Municipality with potable water. These 16 boreholes are able to generate and distribute 1,000 cm^3 to its customers (including households, institutions and business establishments). From Figure 4, it can be observed that five of the boreholes are located at the outskirts of Wa Township. These five boreholes have a total generation capacity of 700 cm^3 whiles the other 11 have a total generation capacity of 850 cm^3. The water company has two water storage facilities in the municipality - a concrete water tank and a steel water tank. The two have a combined storage capacity of 2,091 cm^3 made up of 1,818 cm^3 and 273 cm^3 respectively.

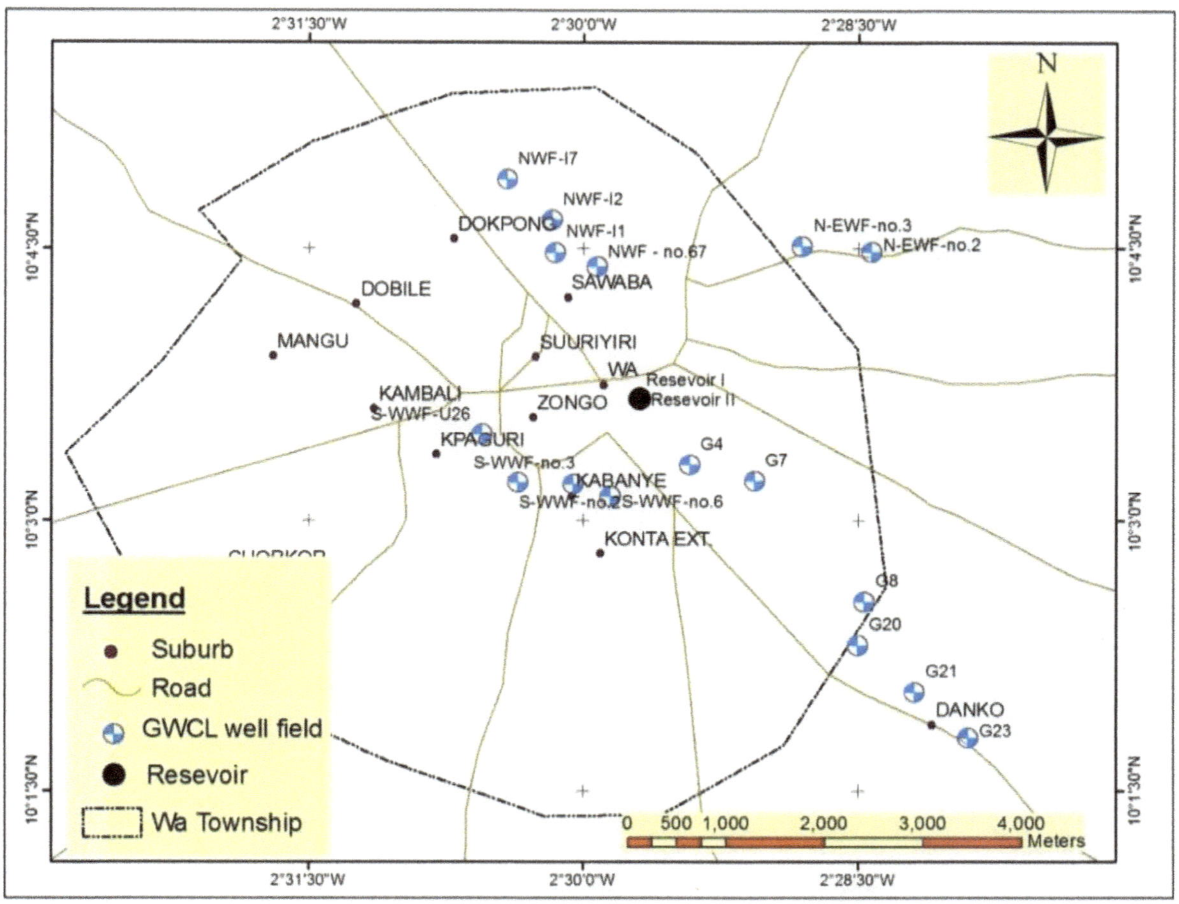

Figure 4. *GWCL water points for potable water supply in Wa.*

Though, in terms of numbers, the number of boreholes seems to be adequate, total yield from these boreholes is not able to support continuous supply of water for the municipality. As such the water company is forced to operate a rationing programme, leading to the supply of water once in every four days for different zones. The company has divided Wa into four zones - North, South, East and West – for purposes of the water rationing and each zone gets water supply one day in a four-day cycle. The North and South consist of areas comprising settlements in Zone Two, the West is made up of suburbs in Zone One while the East covers parts of Zones One and Three.

Analysis of the household data indicates that only 12 percent of the households interviewed depend on GWCL's water supply service. The low percentage of people patronizing GWCL's service could be a direct fall out of the company's inability to produce enough water to supply the municipality. This is evidenced by the number of boreholes constructed across the entire municipality by the Wa Municipal Assembly (Figure 5), Non-Governmental Organizations (NGOs) and private individuals to provide potable water to residents. The field survey indicated that the Wa Municipal Assembly has constructed a total of 40 boreholes in various locations as shown in Figure 5. Twenty-eight boreholes from the total 40, verified to have been provided by the Wa Municipal Assembly, are non-mechanized whereas 12 are mechanized (see Figure 5). Refreshingly, further enquiry revealed that the assembly has plans to mechanize all the boreholes it has constructed across the municipality to improve potable water service provision. These boreholes are managed by Water Committees (WCs) in the various suburbs. Also, some individuals have constructed open wells in their houses to supply them with water, but most of these wells dry up during the dry season leading to shortage of water during those periods.

Private individuals, unlike the Municipal Assembly, focus mainly on constructing mechanized boreholes rather than non-mechanized boreholes, with as many as 35 privately owned mechanized boreholes mapped in the study as compared to 13 non-mechanised privately owned boreholes (Figure 6). Interestingly, Mangu, a community located in the northern part of the Wa Township (Zone Two), alone accommodates four of the total number of boreholes. Konta Extension, Kpaguri and Dokpong (also in Zone Two) on the other hand have 3, 2 and 1 respectively of such boreholes. This in a way is an indication of the uneven spatial distribution of potable water supply in the Wa Municipality which could have a telling effect on the residents of these areas in the long term as the population continues to grow.

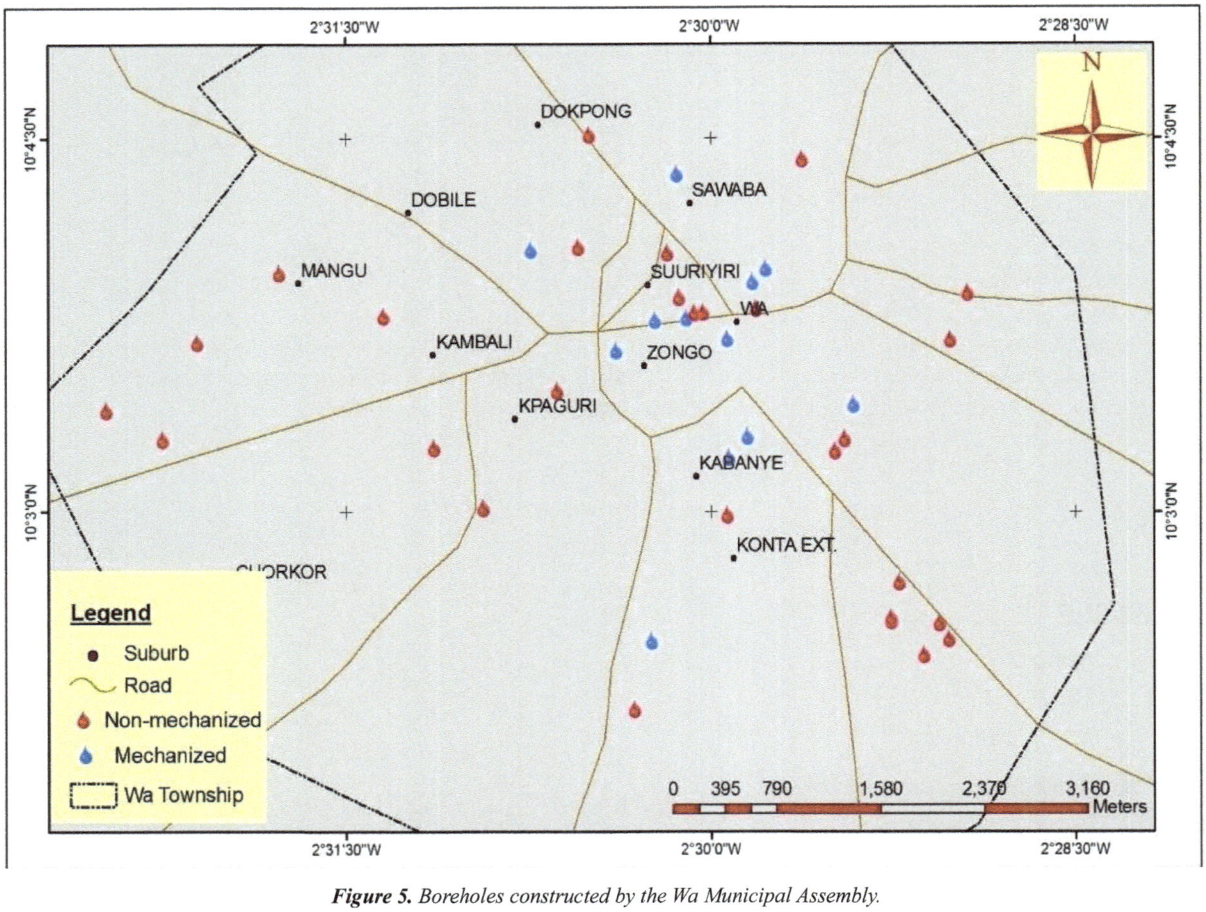

Figure 5. *Boreholes constructed by the Wa Municipal Assembly.*

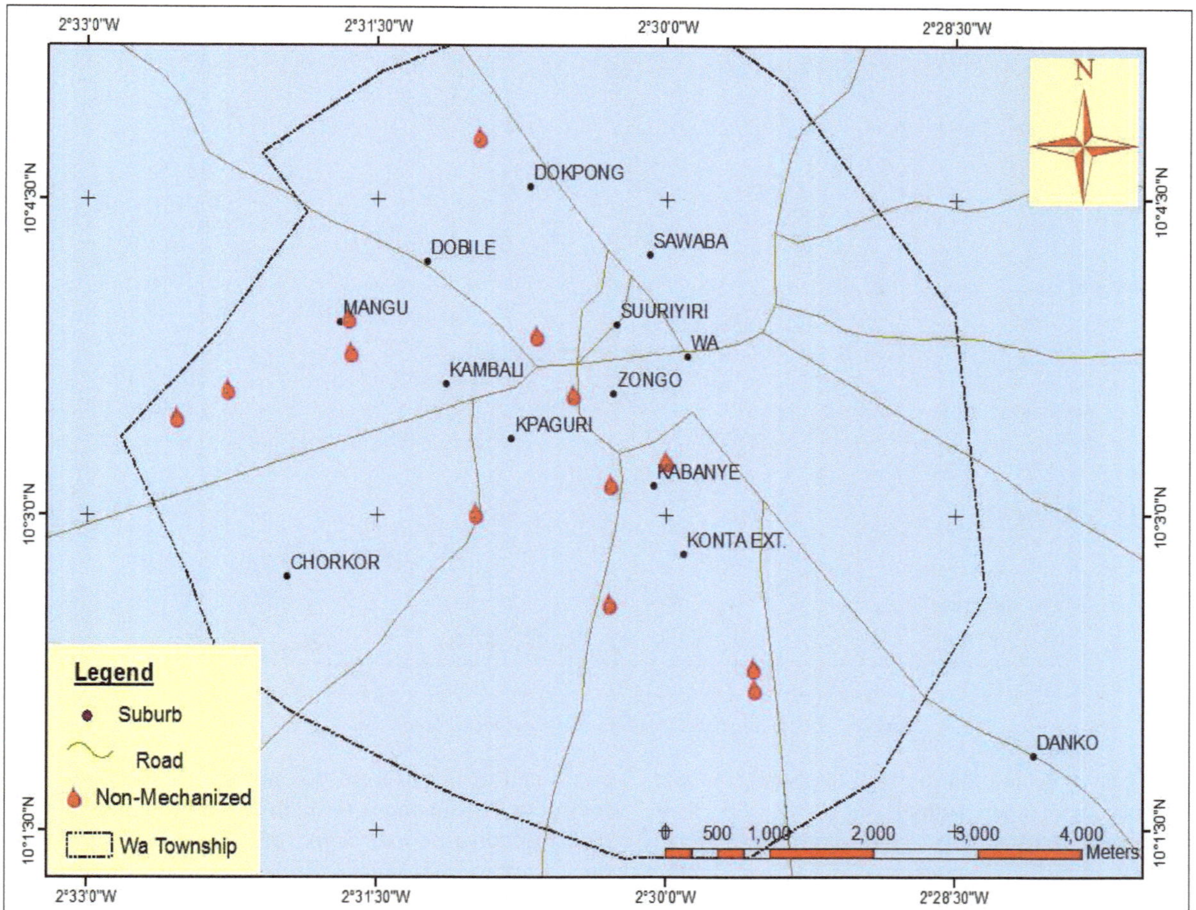

Figure 6. *Distribution of non-mechanized boreholes by private individuals.*

The household survey revealed that 33.1 percent of households, mainly in Zone Three, depend on non-mechanized boreholes for potable water supply. According to these households, the main reason for their dependence on this source is that non-mechanised boreholes are the only available water sources in their area. However, most of these non-mechanized boreholes have low yields thereby affecting the quantity of water used by those households which affects the acceptable minimum use of water of 20 litres per day per person prescribed by the UN. On the other hand, a seemingly positive outcome from the study was that 41.6 percent of the households interviewed depend on mechanized boreholes for their water needs. Reasons assigned by these respondents included water source being closer to their homes, water source being the only available source within their vicinity, and the source being relatively easier to obtain as compared to the non-mechanized boreholes in terms of effort.

Accounting for the spatial distribution of privately mechanized boreholes, it was realized that Kpaguri alone has as much as 31.4 percent of the privately owned mechanized boreholes recorded in the study. In sum, the entire southern part of Wa comprising Kpaguri, Konta Extension, Zongo, Kabanye, Kambali and Chorkor have as much as thirty (30) mechanized boreholes making up a whopping 85.7% of all privately owned mechanized boreholes in Wa. This is shown in Figure 7. Most of the respondents within the southern part of Wa maintained that their communities were initially not connected to GWCL's services and even in recent years the connectivity is very low. On the other hand, the northern part of Wa comprising Dokpong, Sawaba, Dobile, Suuriyiri and Mangu collectively accounts for only five mechanized boreholes (Figure 7), with most households within these communities being connected to GWCL's services whiles others buy water from vendor-provided services. This is consistent with the views expressed by officials of the Wa Municipal Assembly that in sitting their boreholes, they lay much emphasis on spatial considerations and that accounts for most of their mechanized boreholes (8) being located in the northern part of Wa (Suuriyiri, Zongo, Sawaba and Dobile).

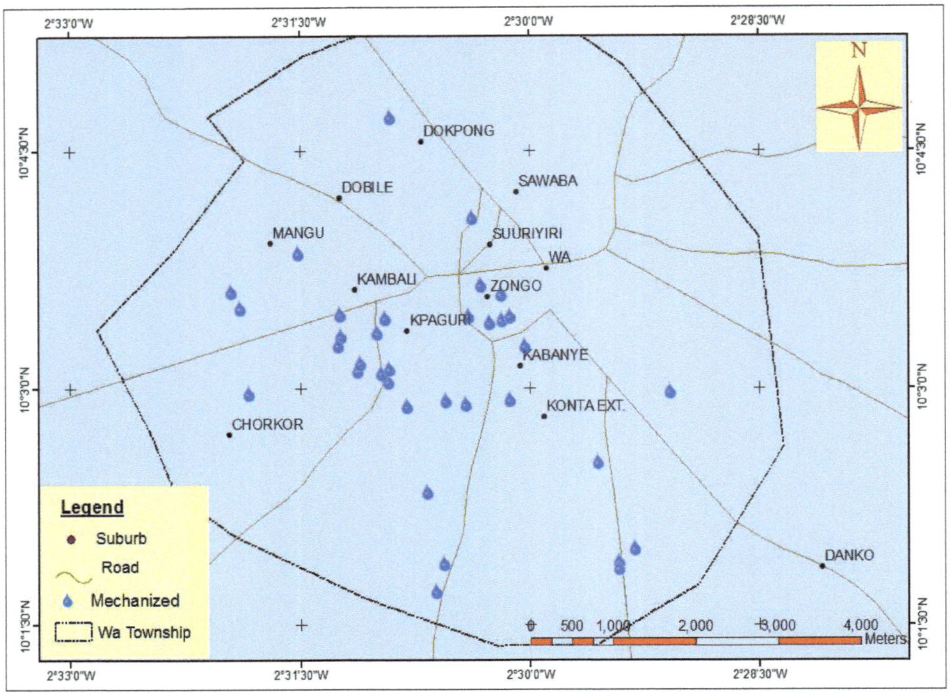

Figure 7. Distribution of privately owned Mechanized Boreholes.

Some respondents revealed that beyond the boreholes and the pipe borne water, other sources of water for their households are over-head storage tanks and open wells. Though the study identified a number of over-head storage tanks constructed to store water for onward distribution to standpipes, toilets, sinks, among others, these storage tanks in strict technical terms are not water sources but rather attachments to other potable water sources. Majority of these storage tanks are plastic tanks erected above the ground and connected to mechanised boreholes by PVC pipes. Water is transported from these through the PVC pipes to the stand pipes for distribution. Some of these over-head tanks also obtain water from tanker services. The distribution of these tanks is also skewed towards the southern part of Wa (Figure 8). A plausible explanation to this skewedness is the fact that most of these storage tanks are constructed as attachments to mechanized boreholes.

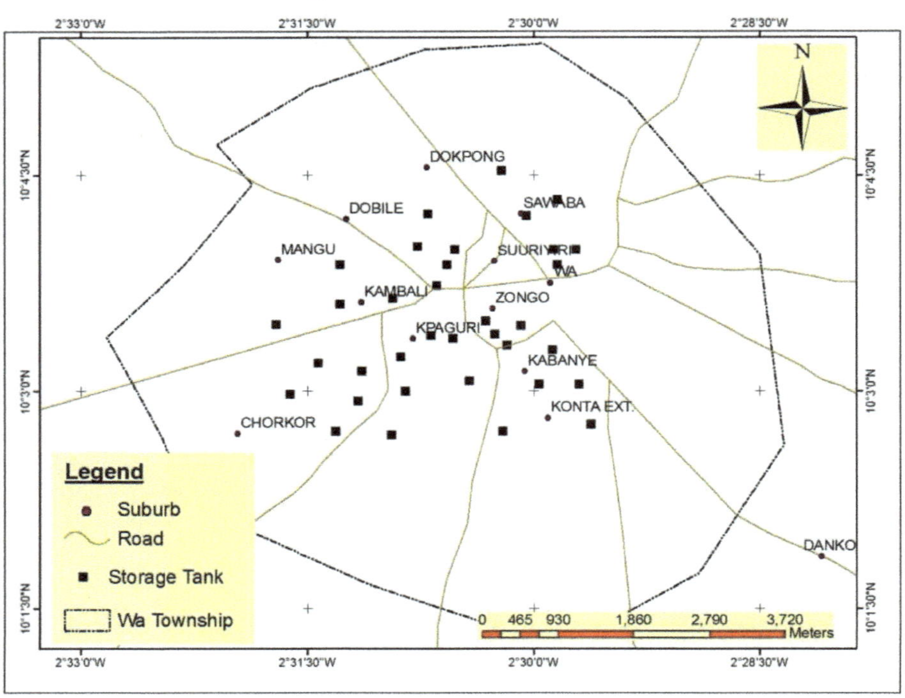

Figure 8. Distribution of water tanks.

There are 95 open wells constructed across the length and breadth of the Wa Municipality to provide water. Most of these wells are constructed by private individuals and, therefore, are found in individual homes where houses close by these homes also patronize them. Others are also community owned and, therefore, serve the water needs of the entire community. The distribution of these open wells, as depicted in Figure 9, is not even across the municipality. Indeed suburbs such as Zongo, Kambali, and Sawaba (all in Zone One) have more wells than other areas. Specifically, there are over 40 wells in Zongo, Kambali and Sawaba, as compared to 11 in the other areas. The wells seem to be concentrated around certain areas which may be due to the fact that those areas do not have access to GWCL's services. Another reason is the fact that people in those areas with fewer wells available do not like well water as most respondents saw it as unsafe for drinking.

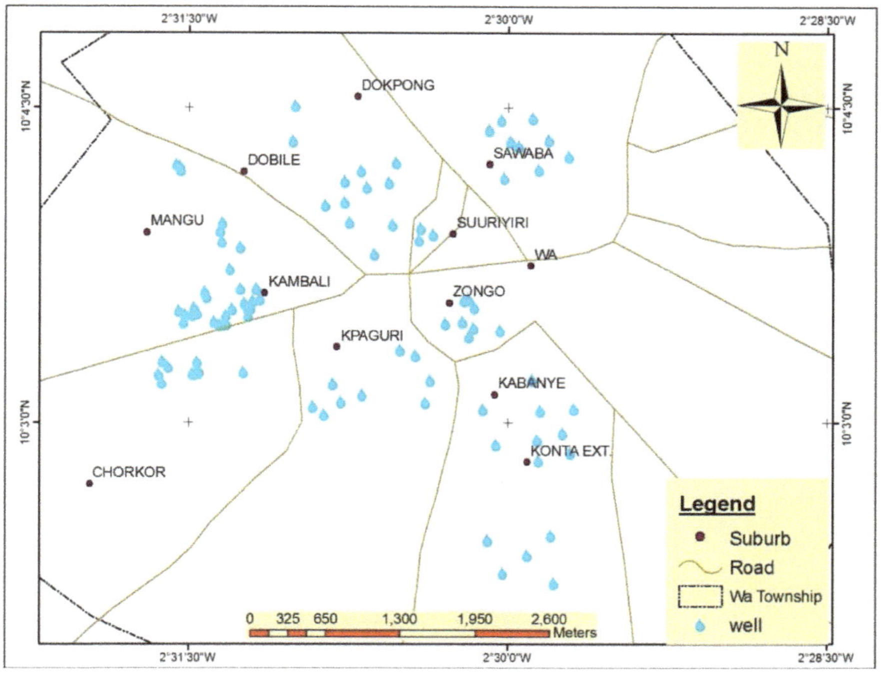

Figure 9. *Distribution of wells.*

4.2. Potable Water Availability and Accessibility

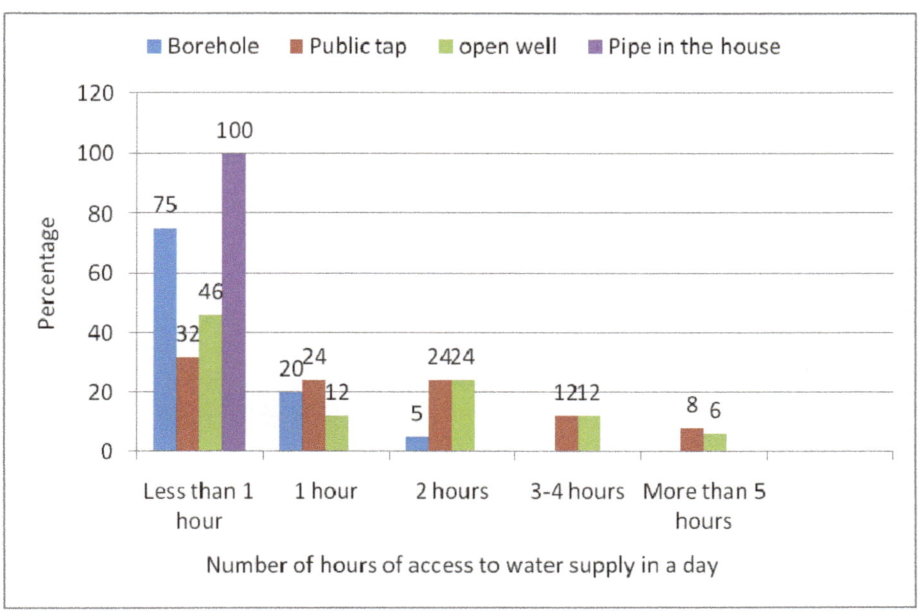

Figure 10. *Water availability source of water supply (%).*

Unlimited access to potable water supply is a pre-requisite to good health and increased productivity thereby reducing poverty in the long run (World Bank, 2006). Accessibility is a concept that has many dimensions and perspectives. However, this study considered accessibility from three perspectives: water availability, cost per unit fetched/used and distance taken to access the water facility. The findings reveal that households are, to a large extent, well placed to increase their productivity as 95 percent of them maintained that water from their principal source was available throughout the year, especially for households with piped water in their homes and borehole users (Figure 10), while the other 5 percent have intermittent water supply, especially during the dry season. This measure of availability took into consideration the number of hours users have access to the water facility in a day and the frequency of breakdown of the water facility.

Though the availability of water on hourly basis seems acceptable, the water availability is also influenced by the number of times the water facility breaks down because it interrupts supply. With as much as 74.7 percent of respondents depending on both mechanized and non-mechanized boreholes as their main source of water supply, the level of availability of water may not be totally optimum all year round considering that these boreholes do break down sometimes in the course of the year. Only 39.1 percent of the respondents indicated that the boreholes from which they obtain water had never broken down within the past one year (Figure 11), whiles the others reported that they have had their main source of potable water breaking down at least once during the past year. The frequency of break-down of boreholes was found out to be a function of whether it is mechanised or not. Though, virtually all boreholes were identified, as a matter of reality, to have the potential to become faulty, non-mechanized boreholes broke down more frequently than the mechanized ones. One major explanation to this occurrence that run through the responses was that the non-mechanised boreholes broke down often due to pressure on them.

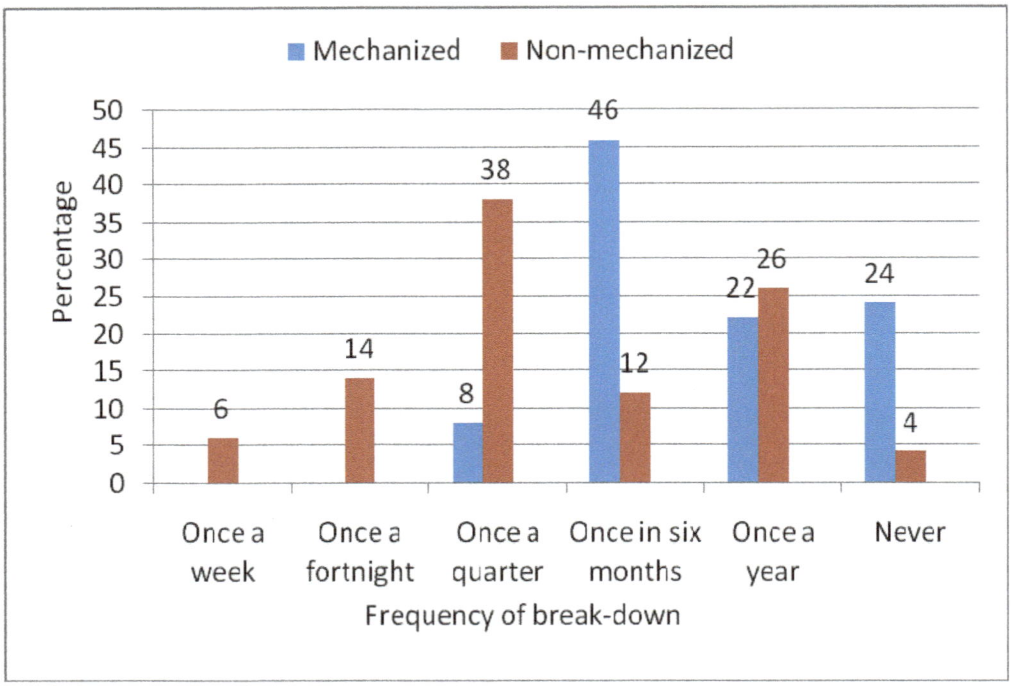

Figure 11. *Frequency of boreholes break-down by type (%).*

Cost is also a factor of accessibility and influences water use patterns of households. Where the price of water is considered to be cheap, the usage rate is higher and where the cost is seen as expensive, the relative rate of patronage is low (Cairncross, 2000). Respondents in the study mentioned different rates as the cost of water they incur on monthly basis. Contingent on their average monthly income and their household sizes, respondents judged the cost of water to be either cheap or costly, with only 15.7 percent of them indicating that they did not pay for the water they use because their water facilities are provided by NGO's and other development partners, as such it is to a large extent free. According to them, they only contribute for the maintenance and repairs in the event of break-downs. Similarly, users of open wells do not pay for water. However, 22.7 percent of respondents (borehole users) paid between GH¢2.00 and GH¢30.00 per month as shown in Figure 12. Unlike users of boreholes, the monthly amounts paid by subscribers of GWCL's services varied drastically, ranging from GH¢5.00 to above GH¢50.00 per month. The highest amounts paid for water were reported by households who obtain water from tanker supplies. Tanker water suppliers charge GH¢25.00 for 12 drums of water, which an average household uses for a two-week period.

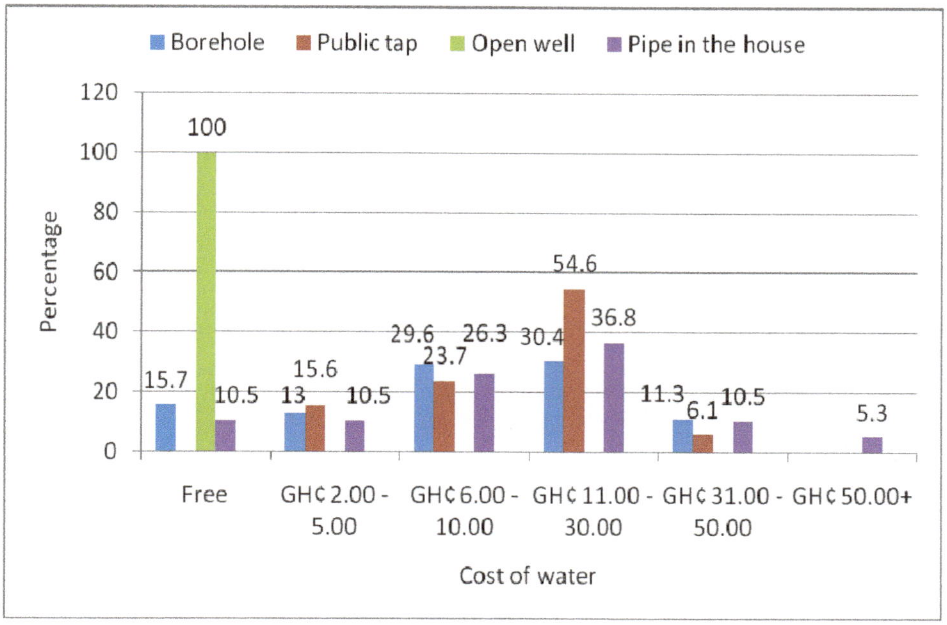

Figure 12. *Cost of water per month by source of water supply (%).*

Distance to a water facility and access to the water have an inverse relationship. Thus people's water usage rate is influenced by the distance they cover to access the water (Cairncross, 2000). If people consider the distance to be far, the water they use daily is affected negatively. Apart from those with wells and home connections from GWCL, the average distance covered by households to access water was 125 meters. About 23 percent of households covered 100 meters or less to access water, whiles 24.8 percent covered 300 meters or more to do same (Figure 13).

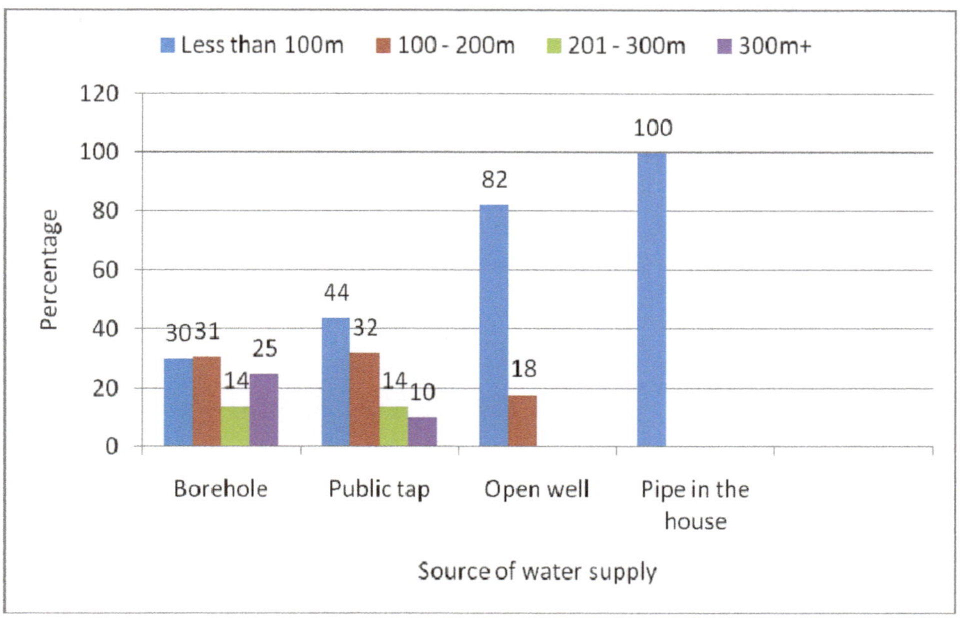

Figure 13. *Distance covered to access water by source of water (%).*

Respondents enumerated a number of challenges they face in accessing potable water. These are high cost of water, irregular flow of water, long queues, long distance, longer time taken to repair broken down water facilities and high chemical inputs (for water from GWCL). Figure 14 shows the challenges enumerated by respondents and their respective percentages.

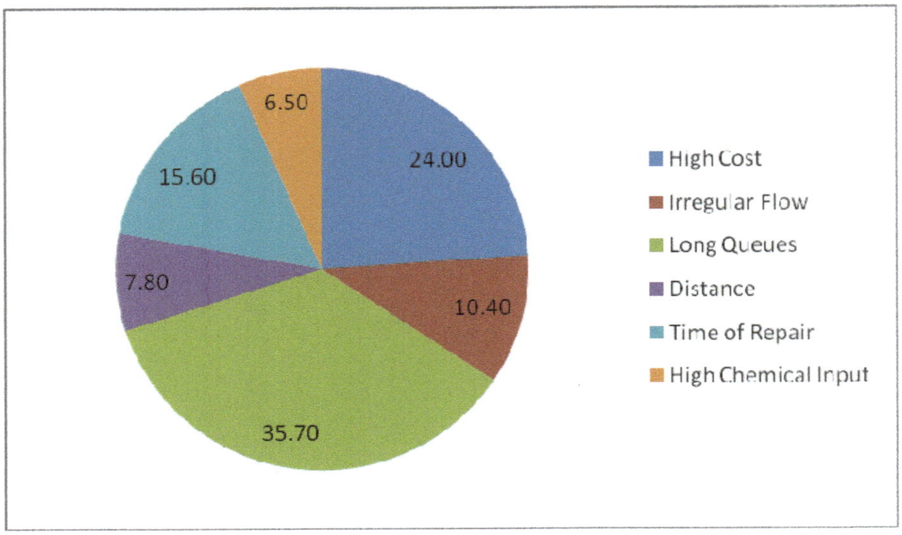

Figure 14. *Challenges of accessing potable water (%).*

GWCL is faced with some challenges which prevent it from delivering proper services to its customers and the public as a whole in Wa. Some of the challenges include operating limited number of boreholes (a situation which does not support continuous supply of water), lack of surface water projects and low investment in the water sector. The municipal assembly has some plans in place in addressing the challenges confronting potable water provision which include sourcing funding for the extension of surface water from the Black Volta near the border with Burkina Faso, construction of limited water systems (mechanizing) on existing boreholes and new boreholes, and enforcing by-laws governing drilling of boreholes per the guidelines of the Water Resource Commission (WRC).

5. Conclusion

The progress of effective water supply delivery processes in Ghana, as it is in many developing countries, has been stalled by inadequate and inefficient implementation of pro-poor policies as far too many people in urban areas of the country live without access to safe drinking water. This is a primary determinant of continuing poverty. But the situation in the Wa Municipality seems different as only 13 percent of households in the municipality do not have access to potable water supply. Superficially, this would appear to mean that coverage of water supply in the municipality is adequate, but it is more important to have a look at the situation given the significance of universal access to potable water supply. Especially, the study highlights that issues of distance and cost regarding access to potable water remains a challenge in the municipality. Appreciation of the linkages between water and health is not new. The case is clear and the effectiveness of actions well known. However, in view of the financial constraints and shortages of trained health professionals in a poor area like the Wa Municipality, the issue of access to water supply needs to be given the priority and attention it deserves. This study has highlighted several important issues on access to water supply in low-income urban areas which are different and require different strategies to tackle. Progressive expansion of improved water supplies is important but often fails to address the immediate needs of the most disadvantaged, and cost is always a concern in projects aimed at developing strategies to address these challenges. The findings further reveal that private individuals create additional water facilities to augment public supplies, but the positive impacts on health and livelihoods could be greater if access to finance could be facilitated. Most importantly, water supply improvement strategies should incorporate an integrated vision, which sees adequate and quality water both as a goal in itself and as a contributor to economic and social development through gains in public health. Policy makers must be able to recognize and build on these linkages so that water supply and other policy goals can be made mutually supportive.

References

[1] Cairncross, A. M. (1990). Health impacts in developing countries: new evidence and new prospects. *J. Inst. Water Envt. Mgt.* 4(6):571-592.

[2] Esrey, S. A., Feachem, R. G, Hughes, J. M. (1985). Interventions for the control of diarrhoeal diseases among young children: improving water supplies and excreta disposal facilities. *Bul. Wld. Hth. Org.* 63(4):757-772.

[3] Esrey, S. A., Potash, J. B., Roberts L., Shiff, C. (1991). Effects of improved water supply and sanitation on ascariasis, diarrhoea, dracunculiasis, hookworm infection, schistosomiasis, and trachoma. *Bul. Wld. Hth. Org.* 69(5):609-621.

[4] Sorenson, B. S., Morssink, C., Campos, A. P. (2011). Safe access to safe water in low income countries: water fetching in current times. *Soc. Sci. Med.* 72:1522-1526.

[5] Howard, G., Bartram, J. (2003). *Domestic Water Quantity, Service Level and Health.* New York: World Health Organization.

[6] United Nations (2003). *The 1st UN World Water Development Report: Water for People, Water for Life*. Retrieved on 16 August 2014. Available: http://www.unesco.org/water/wwap/wwdr/index.shtml

[7] Nickson, A. (2002). The role of the non-state sector in urban water supply. In Paper for the *Making Services Work for Poor People*. World Development Report (WDR) 2003/04 Workshop, 4–5 November in Oxford.

[8] Mwanza, D. (2004). *African Public Utilities not Performing Efficiently*. Paper presented at the 12th Union for African Water Suppliers Congress, 16–19 February, 2004 in Accra, Ghana.

[9] Panayotou, T. (1997). The role of the private sector in sustainable infrastructure development. In: Gomez-Echeverri, L. (ed.), Bridges to Sustainability: Business and Government Working Together for a Better Environment. *Yale School of Forestry and Environmental Studies Bulletin Series 101*. Yale University, New Haven, pp 54-72.

[10] Schwartz, K. H., Schouken, M. (2007). Water as a political good: revisiting the relationship between politics and service provision. *Water Pol.* 9(2):119-129.

[11] Biswas, K. A. (2006). Water management for major urban centres. *Water Res. Devt*. 22(2):183-197.

[12] WHO/UNICEF (2013). *Joint Monitoring Programme for Water Supply and Sanitation: Progress on Drinking Water and Sanitation – 2013 Update*. Geneva: WHO Press.

[13] Monney, I., Buamah, R., Odai, S. N., Awuah, E., Nyenje, P. M. (2013). Evaluating access to potable water and basic sanitation in Ghana's largest urban slum community: Old Fadama, Accra. *J. Envt. Earth Sci.* 3(11):20-36.

[14] Osumanu, I. K. (2013). Health changes of urban poverty and water supply in Northern Ghana. In: Wagner JR (ed.), *The Social Life of Water*. New York & Oxford: Berghahn Books, pp. 158-179.

[15] Owusu-Sekyere, E., Aasoglenang, T. A., Bonye, Z. S. (2014). Household water supply vulnerability in low income communities in Ghana: experiences from Aboabo in the Kumasi Metropolitan Area. *Int. J. Envt. Prot. Pol.* 2(1):9-18.

[16] Addo, K. K., Mensah, G. I., Bekoe, M., Bonsu, C., Akyeh, K. (2009). Bacteriological quality of schate water produced and solid in Teshie-Nungua suburbs of Accra, Ghana. *Afr. J. Food Agr. Nutri. Devt.* 9(4):1019-1030.

[17] Stoler, J., Weeks, R. J., Fink, G. (2012). Sachet drinking water in Ghana's Accra-Tema Metropolitan Area: past, present and future. *J. Water, San. Hyg. Devt.* 2(4):223-240.

[18] Osumanu, I. K. (2008). Private sector participation in urban water and sanitation provision in Ghana: Experiences from the Tamale Metropolitan Area (TMA). *Envt. Mgt.* 42:102-110.

[19] Suleiman, L., Cars, G. (2010). Water supply governance in Accra: authentic or symbolic. *Water Pol.* 12:272-289.

[20] Ahimah, J. K., Ofosu, S. A. (2012). Evaluation of the quality of sachet water vended in the New Juaben Municipality of Ghana. *Int. J. Water Res. Envt. Eng.* 4(5):134-138.

[21] Singh, A. I. (2006). Governing water wisely. *UN Chron.* 4:24-25.

[22] Osumanu, I. K. (2010). Community involvement in urban water and sanitation provision: the missing link in partnerships for improved service delivery in Ghana". *J. Afr. Std. Devt.* 2(8):208-215.

[23] Bartram, J., Lewis, K., Lenton, R., Wright, A. (2005). Focusing on improved water and sanitation for health. *Lancet,* 365:810–812.

[24] Satterthwaite, D. (2004). The millennium development goals and poverty reduction". In: Satterthwaite, D. (ed.), *The Millennium Development Goals and Local Processes: Hitting the Target or Missing the Point?* London: International Institute for Environment and Development, pp. 7-46.

[25] O'Hara, S., Hannan, T., Genina, M. (2008). Assessing access to safe water and monitoring progress on MDG7 target 10 (access to safe water and basic sanitation): Lessons from Kazakhstan. *Water Pol.* 10:1-24.

[26] WHO/UNICEF (2005). *Water for Life: Making it Happen*. Retrieved on 18 July 2014. Available: http://www.wssinfo.org/pdf/JMPJ)5_text.pdf

[27] Ghana Statistical Service (2012). *Ghana: Population and Housing Census 2010*. Accra: Ghana Statistical Service.

[28] Ghana Statistical Service (2012). *Ghana: Demographic and Health Survey 2008*. Accra: Ghana Statistical Service.

Indicators for Sustainable Development Strategies and Components of Tourism Regions in Egypt A Study of Alexandria and the Northwest Coast

Waleed Hussein Ali

Fayoum University, Faculty of Engineering, Department of Architecture, Fayoum, Egypt

Email address:
Waleed.a.hussien@gmail.com

Abstract: Sustainable urban development for tourism has been a priority for Arab and European institutions since the mid 1990's. In recent years, the tourism industry has grown in importance because of its positive effect on overall (economic, social, cultural, political and psychological) development[1]. Many economists, sociologists, anthropologists, psychologists, geography and administration experts are showing increased interest in the various services-related activities undertaken in different fields of tourism, and in tourism's relation to sustainable development; issues that have also become the subject of social and economic demand. The social demand is a reflection of the human desire to move and travel, while the economic aspect concerns the employment opportunities available for young people of both sexes in the various fields of tourism, which can gradually contribute to the resolution of the unemployment problem. Moreover, tourism plays a significant part in accelerating the extension of urbanization by planning cities and tourist resorts, constructing and developing roads and utility networks, investing available economic resources and by optimizing the use of environmental or local resources to manufacture popular products. Since it helps to increase foreign currency inflows from abroad, tourism is also an important source of national income for tourist-attracting countries.

Keywords: Tourism Development, Tourism Region, North Coast Region, Development Indicators, Sustainable Development

1. Research Objective

This study aims to explain the relationship between crucial sustainable development and the development of tourism regions in Egypt. It reviews sustainable development strategy indicators for the Northwest Coast and Alexandria regions and attempts to identify their positive and negative features.

2. Research Methodology

An inductive approach is used to study the available data and information on planning and tourism regions in Egypt. Following analysis of this data, indicators for the sustainable development of the Northwest Coast and Alexandria regions (selected because of their great importance to the development process in Egypt) are identified.

3. Targeted Regional Development[2]

The integrated development of economic, social, cultural, urban and environmental sectors in a specific region within the framework of a comprehensive national plan that defines development strategy, identifies general development indicators, and defines the size of investments in the development process.

Tourism Development: is a component of regional development (it is one aspect of the economic development of a region).

As a driving force behind a successful process of

1 Mohamed Youssry Ibrahim De3bbes – Sena3et al seya7a bayn al nazareya wa al tatbeeq – First Edition, 2003.

2 Mohsen Abdel Sa7eb Al Mozafar (previous reference)

sustainable tourism development, regional development leads to environmental, social and cultural sustainability that reduces disparities or gaps in existing community systems and allows for continued development while safeguarding the right of future generations to community resources.

Tourism Development: Objectives and Relation to Regional Development

Table 1. *Tourism Development: Objectives and Relation to Regional Development*

Tourism Development Objectives	Regional Development Objectives
Securing the continuous and balanced growth of State revenues from tourism Contributing to the development and conservation of the environment	Maximizing possibilities for exploiting untapped resources
Strengthening links between production in the tourism industry and production in other sectors	Inventory and evaluation of various resources and economic opportunities with a view to promoting economic integration and applying the concept of sustainable development
Raising the employment rate by creating new employment opportunities in the tourism industry and related sectors	
Upgrading and developing community services in all fields (education, health, social, recreational and cultural) through the establishment of colleges and institutes specialized in hotel management, tourism guidance and the management of public enterprises	Realizing social justice and focusing on the human factor by promoting education and scientific research (which are both the goals of development and one of its primary resources)
Underscoring the positive social and cultural impact of tourism	Creating regional equilibrium by eliminating economic, social and cultural disparities among different regions of the State
Ensuring that tourism sector plans contribute to local and urban development by developing new tourist attraction areas in the isolated or marginalized areas of different localities and linking these to developed regions (integration of uninhabited spaces into the developed regions)	implementing the spatial dimension and population settlement objectives of the national and regional development strategy
Working on formulating tourism development strategies that are consistent with national interests	

Source: Heba Mostapha Hassan Mohamed, *Taqyeem dor khettat al tanmeya al seya7eya fi 3amaleyat al tanmeya al eqleemeya fi misr*, Unpublished MA thesis, Faculty of Urban and Regional Planning, Cairo University, 2012

Sustainable Tourism: focuses on the point of encounter between the needs of tourists and those of the host region, thus protecting and enhancing future development opportunities. The management of all resources is geared towards fulfilling economic, social and spiritual needs, while simultaneously preserving the existing culture, environment, necessary biodiversity conditions, as well as all life systems and necessities.

3.1. The Concept of Eco-Tourism and Sustainability3

Eco-tourism is an educational, learning and cultural process that aims at acquainting tourists with the distinctive features and attributes of an environment and facilitating their integration within that environment. Sustainable tourism refers to the optimal exploitation of tourist sites by ensuring that the number of visitors corresponds to the nature of the site and those tourists are aware of the location's significance and of the suitable behavior to adopt, thereby preventing damage to both the site and the tourists. Sustainable tourism satisfies the needs of tourists while protecting tourist sites and creating additional employment opportunities for the local community. It strives to manage all available resources (economic, social, natural and esthetic) in accordance with the location's culture and heritage, and recognizes the need to maintain ecological balance and proper biodiversity conditions. The World Tourism Organization (WTO) focused on the concept of sustainable tourism in its 1980 Manila Declaration, at its 1982 meeting in Acapulco, Mexico, at its 1985 General Assembly meeting in Sofia, and in Cairo in 1995.

3.2. The Principles of Sustainable Tourism4

We may conclude that the principles of sustainable tourism are the following: Tourism planning, development and management must constitute a component of the overall sustainable development strategies for any given region or country. To provide optimal benefits, tourism planning and management should be undertaken in an integrated, unified manner that ensures the involvement of different government agencies, private foundations and citizens (groups and individuals).

These agencies, foundations, groups and individuals should adhere to certain ethical and other principles, such as respect for the host region's culture, environment and economy and for its traditional lifestyle and manners (including its political configuration). Tourism planning and management should focus on sustainability and aim to protect the host region by making optimal economic use of

3 Dr, Fahd Abdel Karim Ali Torkistani, *Al seya7a al bee'eya, mafhoomha wa 3anaserha*, 2008

4 *Tasyeer aleyat ta7qeeq al seya7a al mostadama ma3 altatbeeq 3ala alwa7at al dakhla wa al kharga be misr,* First International Symposium (Urbanism and Sustainable Tourism),Institute of Urban Management Techniques, University of M'sila, Algiers, 7 / 8 December, 2011

its natural and human environment.

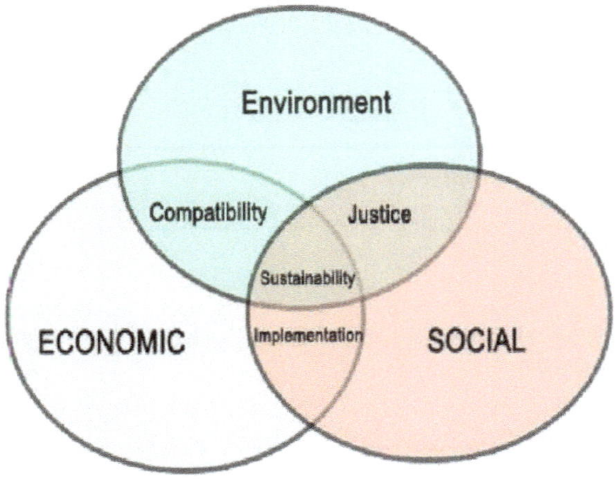

Figure 1. *Dr, Fahd Abdel Karim Ali Torkistani, Al seya7a al bee'eya, mafhoomha wa 3anaserha, 2008*

Special care should be taken to ensure that profits from tourism are equitably distributed among tourism promoters, members of the host community and the host region.

Studies and information on the nature of tourism and its impact on local inhabitants and on the cultural environment should be available prior to, and during, the development process. In particular, this information should be accessible to members of the local community, enabling them to participate in, and influence, overall development trends.

Prior to embarking on a tourism development process, or on any other project, it is crucial to complete an integrated analysis of environmental, social and economic plans that takes into account local environment and community requirements.5

An inspection, control and rectification program should be in place during all phases of tourism development and management, enabling local inhabitants and others to benefit from available opportunities and to adapt to future changes that will affect their lives.

3.4. The Concept of Eco-Tourism and Sustainability

Refers to an educational, learning and cultural process that concerns a given environment's components; it is a means for acquainting tourists with the host environment, and how best to adjust to it. Sustainable tourism refers to the optimal exploitation of tourist sites by ensuring that the number of tourists entering a site corresponds to its nature and that visitors are aware of the location's significance and of how to behave in a responsible and knowledgeable manner, thereby preventing damage to either party. 6

3.5. Traditional Tourism Development and Sustainable Tourism Development

A comparison between traditional and sustainable tourism development is presented in the table below.

Table 2. *Traditional Tourism Development and Sustainable Tourism Development*

	Traditional Tourism Development	Sustainable Tourism Development
Concept	Rapid development	Development in stages
Limitations	No limitations	Has certain limitations and a defined absorptive capacity
Time Span	Short-term	Long-term
Type	Quantitative tourism	Qualitative tourism
Management of Development Processes	External	Through local inhabitants
Development Strategies	Unplanned development	Planning precedes development
Planning	Partial planning of unconnected, isolated sectors	Comprehensive, integrated planning
Construction Methods	Focus is on building vacation units	Construction allows for environmental conditions & land use plans
Architectural Style	Traditional urban buildings	Local architectural styles
Adopted Programs	Project planning programs	Planning programs are based on the sustainability concept
Tourist Attributes	Tourist groups, large numbers	Individuals and small tourist groups
Duration of Stay	Short stays	Long stays
Demeanor	Noise and annoying sounds	Sensible and calm behavior
Repeat Visits	Single visit to the location	Probability of repeat visits to the location
Class of Tourist	Different educational levels	Highly educated and cultivated

Source: Atta Allah Farouq (2003), p. 63

Sustainable tourism meets the needs of tourists while also protecting and preserving tourism regions and creating employment opportunities for local communities. It strives to manage all available resources (economic, social, natural and esthetic) in keeping with the location's culture and heritage, and recognizes the need to maintain ecological balance and proper biodiversity conditions.[7]

3.6. Tourism Region Definition

According to Al Roubi (1987), a tourist region8 is an area (or a plot of land) that contains key attraction elements drawing tourists to visit it and stay in it for a specific period, and to take part in various tourist activities. It can be either a final destination, or a stopover point on the way to another destination. This explains the importance of interconnecting

5 *Tasyeer aleyat ta7qeeq al seya7a al mostadama ma3 altatbeeq 3ala alwa7at al dakhla wa al kharga be misr,* First International Symposium (Urbanism and Sustainable Tourism), Institute of Urban Management Techniques, University of M'sila, Algiers, 7 / 8 December,

6 *Tasyeer aleyat ta7qeeq al seya7a al mostadama ma3 altatbeeq 3ala alwa7at al dakhla wa al kharga be misr,* First International Symposium (Urbanism and Sustainable Tourism),Institute of Urban Management Techniques, University of M'sila, Algiers, 7 / 8 December, 2011

7 General Authority for Urban Planning, *Estrategiat al tanmeya le mo7afazat al gomhoreya, Iqleem al eskandereya,* Chapter 3, 2008

8 Nabil Al Roubi, *Al Takhteet al seya7i,* Cairo, pp. 159, 160

tourist resources and visit durations, according to the purpose of the trip and the tourists' preferences.

4. Planning for Tourism Development

Successful tourism development requires planning in accordance with a clearly defined national tourism development policy that is based on scientific approach. This is because tourism is a multifaceted activity linked to many other sectors of the development process. A scientific approach to tourism development planning realizes the following9:

1. Coordination and harmonization between the tourism sector and other sectors within the overall development system, due to the complex interconnection and mutually dependent relations among the different sectors (cultural, economic, environmental ...).
2. Creating a balance, within the tourism sector itself, among sometimes-conflicting needs that in most cases have to compete over limited resources
3. Adopting clearly defined, integrated approaches that benefit all the participants in tourism development during the various planning and implementation phases that expands the base of participants and ensures effective interaction in order to realize plan objectives
4. Laying the foundation for legislature that promotes and ensures successful tourism development

4.1. Tourism Planning Definition

Tourism planning is the process of drafting a blueprint of the estimated future tourism activity in a given country over a specific period. Formulating tourism plan objectives and achieving rapid and regular tourism development necessitates inventorying the country's tourism resources as part of the preparation and implementation of a coherent development plan that covers all branches of tourism activity and all tourism regions in a country.10

Figure 2. *Nabil Al Roubi, Al Takhteet al seya7i, Cairo, pp. 159, 160*

Tourism planning should not be perceived to be solely the province of official government bodies; it should be a joint work program among government agencies, the private sector and individuals. From the formulation of desired objectives up to the application and implementation of the tourism plan, tourism planning must be a joint operation, undertaken by all organizers of the tourist sector; government agencies responsible for supervising the sector, tourist service providers (institutions and businessmen), the consumers of these services (tourists), and the host community11.

Successful tourism planning depends on a number of factors, including:

- Ensuring that the tourism plan is an integral part of a comprehensive national social and economic development plan
- Creating an equilibrium among the various economic sectors
- Tourism sector development should be considered one of the strategic options for economic development
- Including the tourism industry among the production sectors in the State's economic structure
- The State must define the level of growth and the size of tourist flow required
- Clearly defining the roles of the private and public sectors in the development process
- Focusing on the relation between tourism planning and economic activity as a whole, and on its connection to environmental preservation
- A number of tourist development pioneers view the following four relationships as necessary factors for successful tourism planning:

The relationship between planning and economic activity:

In any region, successful tourism activity is closely linked to the scale of economic activity capable of engendering a constant increase in income and to the local economy's ability to absorb and use this income.12

The relationship between planning and the environment:

We believe that tourism and the environment are one and the same, because the tourist flow is directly associated with factors that attract tourists, such as climate, stunning natural landscapes, beaches etc, as well as with man-made tourist attractions (historical and archeological sites, modern tourist attractions, and others). This is why the preservation of environmental assets is so crucial.

The relationship between planning and tourist site visitors:

It is important to teach local residents to provide tourists with the information they need to enjoy their stay intellectually; the combination of psychological enjoyment and intellectual enrichment increases tourists' desire to re-

9 Ibrahim Bazazou, Takamol al eqleem al takhteeti wa al aqaleem al seya7eya, published article, menotourism.com

10 Fahd Abdel Karim Ali Torkistani, Al seya7a al bee'eya, mafhoomha wa 3anaserha, First Edition, 2008

11 Majalet game3at teshreen lelderassat wa albo7outh al 3elmeya – selselat al 3oloom al'eqtessadeya wa al qanooneya, Vol. 28, Issue 3 (2006)

12 General Authority for Tourism Development website - www.tda.gov.eg

Molakhass qera'at estratejeya al tanmeya le 'eqleem al eskandereya – mo7afazat matruh – al moqtara7aat wa al mortakezaat

visit the region.13

The relationship between planning and the influx of foreign currency:

Because the objective of planning is to increase the State's foreign currency resources, the success of tourism planning (especially in developing countries) is measured by its ability to increase foreign currency influx - whether through revenues from international tourism or through the investment of capital in tourism projects.

4.2. Tourism Planning Significance and Objectives

Tourism planning, a scientific approach to the organization and management of all tourist activity types and components, plays a primordial role in the development of tourist activity. It provides a common framework for decision-making in tourism resource management and identifies methods and trends to be adopted by the parties responsible, thus facilitating their work and preventing the waste of time and effort.

Tourism planning leads to coordinating and uniting the efforts expended by all parties responsible for the development of the tourism sector, reducing the duplication of decisions and actions and contributing to the successful achievement of general and specific objectives of tourism activity. Consequently, tourism planning is affected more by political, social and natural vacillations than it is by factors of production and other economic forces.

At the regional level, tourism planning is less specialized and detailed than it is at the local level, but much more so than at the national level - although the degree of specialization also depends on the size of the country and of the region. The national plan of a small country may be as detailed as the regional plan of a larger country (smaller countries may therefore not need separate regional and national plans).

4.3. Tourism and Urban Development Approaches and Methods

Tourism and urban planning and development approaches and methods differ from one country to another, and from one area to another, depending on the existing circumstances, the policies adopted and the components of individual tourist sites and tourism regions. Barbaza listed the following approaches to the development of tourism activity:

4.3.1. Intensive and Planned Topical Development

This approach was widely adopted in East European countries where local seaports controlled sea traffic on the Black Sea coast, in Varna and Burgas. To increase their foreign currency income during the post World War II era, the governments of these countries drafted programs aimed at the growth and promotion of tourism in various regions. Detailed market analyses and evaluations of the absorptive capacity of beaches were undertaken, and the necessary land plots were provided (a special system for managing development lands was established). To create partnerships between the private and public sectors, owner consortiums were set up, and the State provided adequate government funding.14

4.3.2. Comprehensive Development

The comprehensive development experiment was applied in East European countries. It involves the development of both the existing centers and new functional groups, and linking these to highways and utility networks, leading to the unification and growth of the region. The process takes place within the framework of a balanced development of the local population and a region's available economic resources, and covers all regional components (physical, economic, social …), while focusing on environmental preservation and sustainability.[15]

4.3.3. Integrated Development

In this approach, a single party (the State) is responsible for realizing an integrated, homogeneous development process conducted according to a clearly formulated development strategy, with well-defined policies. The reason is that the vast technical and financial resources necessary for this type of development are available to the State, but not to individual investors. Integrated development has been implemented in many countries and regions. It can take different forms, depending on development management methods, funding sources, or the population benefitting from development:

Balanced development: Based on the unified management of numerous projects

Management through a single agency that provides the required funding and employment opportunities, with local agencies assisting in certain areas (such as public utilities and other infrastructural works).

Rapid development: Through technical and financial cooperation between agencies and institutions.

Functional development: Through integrated urban planning, that turns the tourism region into a unified, integrated entity.

Motivational development: Independent parties (development companies) independently develop a given region based on previously laid and declared development plans. These companies play the part of polarized catalysts for development; they provide the momentum required for their development projects in keeping with their economic vision within the framework of a general regional and State plan, and according to conditions defined by the responsible agency, as follows:

- Independent (single-handed) development through a large company that supplies basic tourism facilities

---

13 General Authority for Tourism Development, *Estratejeya le tanmeyat al seya7a ?atta 3am* 2017

14 Fatena Al Wattar, Al malame7 altakhteeteya le ta7qeeq afaaq al tanmeya al mostadama, Al mobadarat wa al 'ebda3 al tanmawi fi al madeena al 3arabeya, Jordan, 2008

15 Mohamed Youssry Ibrahim De3bbes – *Sena3et al seya7a bayn al nazareya wa al tatbeeq* – First Edition, 2003

- Initial success in providing tourism facilities and activities engenders confidence among investors and development agencies and attracts more investors, development agencies and tourists, which in turn increases tourist demand for the development of complementary and secondary facilities. This creates room for active participation and motivates different actors to participate in the development process.
- The absolute growth phase (according to the free market theory), which depends on the status and capabilities of operating conditions

In the author's view, the integrated development approach is the most suitable one for the development of tourist regions in developing countries. Integrated development takes into account tourism promotion and growth programs that aim to increase foreign currency revenues, provides the necessary funding, creates employment opportunities and engages the participation of local agencies. Moreover, this approach entails integrated urban planning aimed at transforming tourist regions into unified, integrated entities.[16]

4.4. *Standards and Methods for Identifying Tourism Development Components*

Determining spatial frameworks for the implementation of present and future tourist development plans has become one of the prerequisites of tourism development. A tourism region is the spatial framework within which tourism sector plans are implemented; it can be used to measure the spatial dimension of tourism sector plans. It is extremely important, during the initial stages of formulating a country's tourism plan, to define the concept of 'tourism region'. This definition will allow the identification of a region's potential human and natural resources and thus help to determine the basic requirements for implementing the desired tourism development.[17]

4.5. *Factors Affecting the Demarcation of Tourism Regions*

Natural Components: Refers to the availability of natural and man-made tourist attraction elements. Natural tourist attraction elements include climate, topography, natural resources, mineral springs, mud baths, medicinal hot water springs, nature reserves etc. Man-made elements include historical sites and archeological monuments, cultural landmarks such as museums and art galleries, cultural and traditional events such as exhibitions, festivals etc. A region's natural components are used to measure the actual absorptive capacity of available resources in order to implement a balanced exploitation of resources that ensures their continuity (ongoing development), and to protect the distinctive characteristics of certain regions. The more varied the tourist attractions of a region, the more unique it becomes as a tourist destination and the greater its influence on the regional development process. [18]

Urban Components: Refers to the availability of necessary infrastructural elements (potable water networks, sewage systems, all types of communication networks), transportation facilities (airports, seaports, railways, road networks ...), various types of tourist accommodation (number and class of rooms), in addition to indispensable complementary tourist services, such as restaurants, travel agencies, clubs and theatres.

Economic Components: Refers to the economics of the tourism sector and its importance in each region. This is reflected in the compatibility between available elements of tourism supply and demand, including factors that attract tourism investment, because capital is a crucial component in the establishment of facilities, hotels, travel agencies and centers, as well as the equipment, tools and buildings needed for these facilities.

Social Components: Refers to one of the most important factors affecting the success of tourism development; the traditions and values of the population involved in tourism, including personnel working in the tourism sector (in official agencies such as tourism departments and ministries and other sectors, as well as in other facilities working in the industry). This component underlines the importance of learning, training, educating and comprehending the nature of tourism and its requirements.

Administrative and Political Factors: The social and political attributes of regions and internal administrative borders should both be taken into account during the demarcation of tourism regions, not only to facilitate management and administration but also because tourism development plans are implemented within these administrative borders.

Future Dimensions of Regional Development: may modify the thinking on demarcation (State's stance towards tourism development, possible discovery of new tourism resources or trends, new tourist centers etc)[19]

5. Proposals for the Demarcation of Egyptian Tourism Regions

Numerous attempts to divide Egypt into tourism regions have been made by specialists and others involved tourism. Below is a review of some of these demarcation proposals and the elements upon which they are based:

5.1. *Some Proposed Demarcations*[20]

Dr. Ahmed Al Essawy's Proposal (1987):

1. Marina Zone: The country's northern zone, it includes the Delta, Alexandria, Mersa Matruh and

16 General Authority for Urban Planning, *Al mokatat al estratigi al qawmi lel tanmeya al 3omraneya wa manateq al hay'a al 3ama lel takhteet al 3omrani – Al itar al 'eqleemi letanmeya mo7afazat eqleem al eskanderiya*

17 Central Agency for Public Mobilization and Statistics (CAPMAS), *General Census for Population, Housing and Establishments*, 2006, and 2008 Estimations

18 Ahmed Fawzi, Madkhal ila 3elm al seya7a, Dar Al fekr al gamee3ee, 2008, pp. 159 - 160

19 Majalet game3at teshreen lelderassat wa albo7outh al 3elmeya – selselat al 3oloom al'eqtessadeya wa al qanooneya, Vol. 28, Issue 3 (2006)

20 Magda Mohamed Gom3a, Goghrapheyat misr al seya7eya, 2000, p. 671

extends southwards to Siwa and the Qattara Depression

2. Memphis Zone: The central zone, it includes Greater Cairo, Fayoum, Beni Suef, Minya and Bahariya, as well as the Sinai and the Suez Canal
3. Located south of the Memphis Zone, it extends southward to the borders of the Nubian Zone and stretches laterally from the Red Sea coast in the east to the Libyan border in the west. Aswan is its principal capital city.21

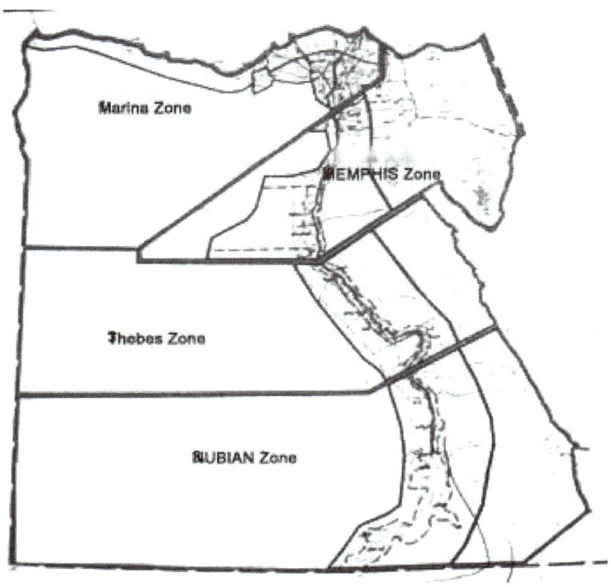

Figure 3. Dr. Ahmed Al Essawy's Tourism Region Demarcation Proposal

Source: Magda Mohamed Gom3a, *Goghrapheyat misr al seya7eya*, 2000, p. 671

It is to be noted that the proposed demarcation does not consider administrative borders, and that each region covers an extremely large area, making it difficult to build an integrated infrastructure. The proposal focuses on ensuring that each region has a waterfront (sea or Nile front). The demarcation disregards the element of natural homogeneity among regions and the similarities among individual regions' natural attributes (Kharga Oasis is split into two separate regions, for instance). The proposal does not take into account social aspects and the homogeneity of population characteristics, an essential component of tourism development.[22]

5.2. Egyptian Ministry of Tourism Proposal

1. Cairo Region: Cairo, Giza, Fayoum and Beni Suef governorates
2. Alexandria Region: Alexandria Governorate, extending from Abu Qir in the East to Agami in the West
3. Delta Region: Includes Egypt's northern governorates, from the shores of the Mediterranean and Lake Borolos in the north to the northern borders of south of Cairo
4. Upper Egypt Region: Minya, Assiut, Sohag, Qena and Aswan governorates to the Sudanese border in the south
5. Suez Canal Region: Port Said, Ismailia and Suez governorates
6. Northwest Coast Region: (Mediterranean Sea Region), includes Mersa Matruh
7. Oases Region: (Western Desert Region), includes Siwa, Bahareya, Farafra, Dakhla and Kharga Oases
8. Sinai Region: North Sinai and South Sinai governorates
9. Red Sea Region: Includes the Red Sea Governorate and extends from the south of Suez all the way to Egypt's southern border with the Sudan

The basis of this demarcation proposal is that (within a comprehensive tourism framework and according to the demand for different types of tourism and the use made of these), each region provides a unique type of tourism in line with its distinctive natural, demographic, economic and architectural characteristics.

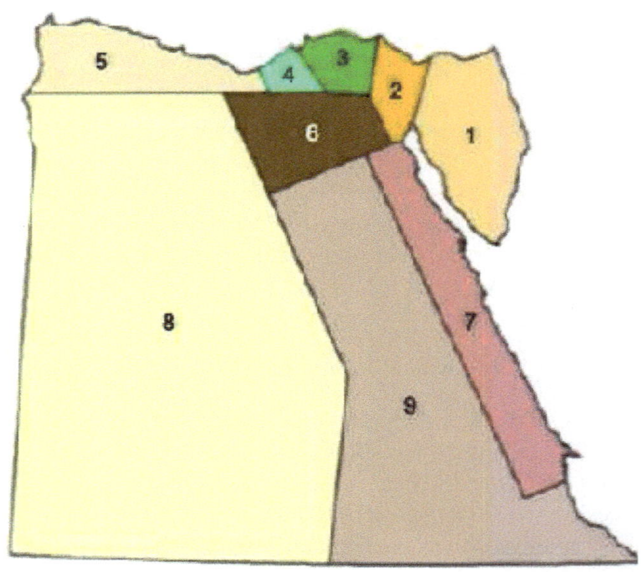

Figure 4. Egyptian Ministry of Tourism Region Demarcation Proposal

Source: Tourism Development Authority

It is to be noted that, in some of the regions, this proposal does not heed administrative borders. Also, no attention is paid, in some of the regions, to the diversification of tourism resources and natural resources. The proposal does not allow for linking the development of Red Sea Coast beach tourism to the development of cultural and archeological tourism in the Nile Valley. Moreover, the proposal overlooks the need of coastal regions, such as the Red Sea and Northwest Coast, for land buffer zones. Finally, the Oases Region does not include desert areas, and there is therefore no opportunity to exploit

21 Magda Mohamed Gom3a, Goghrapheyat misr al seya7eya, 2000

22 Mohamed Youssry Ibrahim De3bbes – *Sena3et al seya7a bayn al nazareya wa al tatbeeq* – First Edition, 2003

and develop safari tourism. [23]

5.3. The Japan International Cooperation Agency (JICA) Proposal[24]

The demarcation of tourism regions in this proposal is based on a number of premises:

- Ensuring that the demarcation of regions is in keeping with national objectives and tourism development and planning goals
- Homogeneity of natural attributes and of tourism demand and supply factors
- Egypt is divided into five major tourism regions, which are subdivided into 13 tourism sub-regions, as follows:

1. Cairo Region: Cairo, Giza, Suez, Central Delta and Fayoum
2. Nile Valley Region: Divided into two sub-regions (Central Nile Valley and the Upper Nile Valley)
3. Red Sea Region: South Sinai Governorate and the Red Sea Governorate
4. Mediterranean Sea Region: The north coast from Rafah in the East to Al Salloum in the West; it is divided into three sub-regions (North Sinai, Alexandria and Mersa Matruh)
5. Oases Region: Includes all the Western Desert oases, and is divided into two sub-regions (Siwa and the Western Desert)

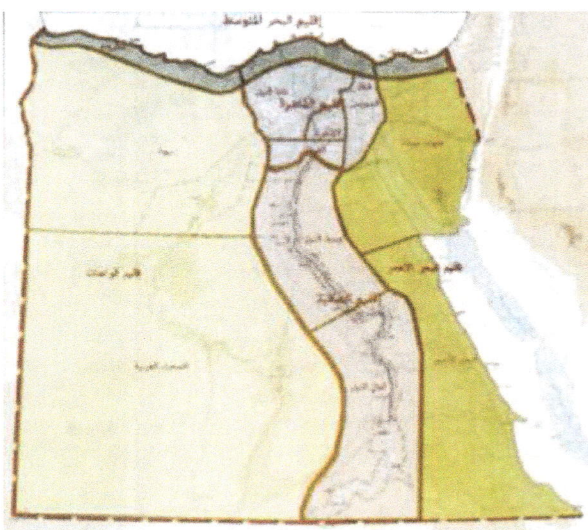

Figure 5. JICA Demarcation Proposal

Source: JICA study team - The study On Tourism Development Projects In Egypt

It is to be noted that this demarcation proposal only takes into account the homogeneity of natural attributes and tourism economics; it overlooks the factors of population traits and urban characteristics and disregards administrative borders in some of the regions.25

5.4. Indicators for Measuring Tourism Development Components

On the basis of the above review of factors affecting successful tourism development and some of the tourism region demarcation proposals that depend on the optimal exploitation of development components, we can identify indicators for measuring tourism development components by evaluating the considerations adopted by the demarcation process, as illustrated in the table below:

Indicators for Measuring Tourism Development Components

Table 3. Indicators for Measuring Tourism Development Components

Indicators for Measuring Tourism Development Components	Considerations Adopted in the Demarcation of Tourism Regions	Tourism Development Components
Diversification of natural components and natural attributes in the various regions Diversification of tourism activity in the various regions	- Availability and diversification of tourism resources within each region: historical / archeological / religious sites & interesting natural recreational sites (mountains and hills, nature reserves …), to attract tourists to each region and promote sustainability	Natural Components
Size and type of tourism demand Number of tourist nights Average stay duration Hotel capacity & class Room occupancy rate Degree of accessibility of tourist regions to regional road networks, airports and seaports Revenue generated by tourism Size of investment in tourism sector Size of tourism workforce	Availability, in each region, of necessary factors for tourist activity (tourism demand and supply & factors that attract tourism development investments) Availability of the human resources required to create the necessary employment opportunities at all levels of tourism activity, to ensure a strong, interdependent community with a solid social backing	Economic Components
Degree of local population participation in economic activities Proportion of local population in the workforce Number of graduate	Similitude among the local population concerning the importance of learning, training, education and recognizing the nature of tourism and its requirements	Social components

23 General Authority for Urban Planning, Ta7deeth estratejeyat al tanmeya al shamela lel sa7el alshemali algharbi wa zaheeroh al sa7rawi (2010 – 2032), April 2011

24 JICA study team - The study On Tourism Development Projects In Egypt

25 Egyptian Cabinet Information and Decision Support Center, Wasf misr belma3loumat, Seventh Issue, 2008

degree holders Proportion of inhabited area to total area Principal tourist cities in a region Hotel capacity and class Centralization of tourist services Population density in inhabited areas Degree of accessibility of tourist regions to regional road networks, airports and seaports Per capita share of paved roads (km / person) Per capita share of potable water produced	Availability of inhabited areas within a region, as a component of tourism development and as an on place urban base Existence of focal points for tourism development, in the form of principal established tourist cities; crucial bases to rely on in interacting with newly-established tourist centers Drawing on the trained workforce in densely populated cities (especially principal tourist resorts) to supply the needs of proposed tourist areas, leading to a re-distribution of the workforce	Urban Components

Source: Heba Mostapha Hassan Mohamed, *Taqyeem dor khettat al tanmeya al seya7eya fi 3amaleyat al tanmeya al eqleemeya fi misr*, Unpublished MA thesis, Faculty of Urban and Regional Planning, Cairo University, 2012

6. Planning Regions in Egypt[26]

From the regional perspective, Egypt is divided into seven planning regions, each with its own Higher Committee and Regional Planning Authority.

The seven regions are Cairo, Alexandria, the Suez Canal, the Delta, Northern Upper Egypt, Assiut and Southern Upper Egypt, distributed among governorates as follows:

1. Cairo Region: Cairo, Giza and Al Qalyubia governorates
2. Alexandria region: Alexandria, Al Beheira and Mersa Matruh governorates
3. Suez Canal Region: Northern Sinai, Southern Sinai, Port Said, Ismailia, Suez and Al Sharqiyah governorates, in addition to part of the Red Sea Governorate
4. Delta Region: Kafr Al Sheikh, Al Gharbiya, Al Menoufeyeh, Dammietta and Al Daqahliya governorates
5. Northern Upper Egypt Region: Beni Suef, Fayoum and Minya governorates as well as part of the Red Sea Governorate
6. Assiut Region: Assiut and New Valley governorates
7. Southern Upper Egypt Region: Sohag and Qena governorates, Luxor, Aswan, and part of the Red Sea Governorate

It is to be noted that, due to its linear shape, the Red Sea Governorate is divided into three planning regions. Nevertheless, it is included in the Southern Upper Egypt region in the present studies.

***Figure 6.** Planning Regions in Egypt*

Source: Khareetat al tanmeya wa al ta3meer le gomhoureyat misr al 3arabeya

There is a marked difference between the demarcation of planning and tourism regions. In the view of the Egyptian Ministry of Tourism, the main difference between the Alexandria planning region and the Alexandria tourism region is the Northwest Coast Region. Due to the unique nature and distinctive tourism constituents of the region, the Ministry of Tourism has underlined the relative national importance of the Northwest Coast Region. This study therefore focuses primarily on the Alexandria and Northwest Coast regions in order to realize a number of important objectives:

- Re-organization of urban conglomerations in the region
- Extending urbanization to the region's desert areas
- Organizing transportation facilities and linking new development plans to existing urban areas
- Using the many different tourism constituents available to provide new types of tourism and to diversify the tourism product
- Linking the Mersa Matruh tourism region to the New Valley and Delta regions through road networks and traffic axes.

6.1. Alexandria Region

The Alexandria Region is the second of Egypt's seven regions. In addition to Alexandria Governorate, the region includes Mersa Matruh and Al Beheira governorates (Figure **5**). The total area of the region is 53330 Feddans (about 2.22% of Egypt's total land area). According to the final 2006 census, the region's population is approximately 194.9 million (62.12% of Egypt's population). According to the final 2006 census, there is a total of 11coastal cities in the region, distributed as follows: two cities in Al Beheira Governorate, the city of Alexandria, the New Borg Al Arab city and seven cities in Mersa Matruh Governorate.

26 Estrategiat al tanmeya wa al ta3meer le gomhoureyat misr al 3arabeya

Figure 7. Alexandria Region

Source: Estrategiat al tanmeya le mo7afazat al gomhoreya, Iqleem al eskandereya, Chapter 3, General Authority for Urban Planning, 2008

Social Indicators:

1. With a population of 0.84 million and a population growth rate of 2.2% over the past ten years, Alexandria is the coastal city with the largest population in this region, and in Egypt generally. Mersa Matruh comes next, with a population of 8.109 thousand. Due to the rapid coastal development (tourist resorts) underway in the governorate that attracts migrant workers from neighboring governorates, the population growth rate in Mersa Matruh (7.7% over the past ten years) is one of the highest in the region.
2. Economic Indicators:
3. The Alexandria region is rich in industrial, agricultural, fish and animal resources. The region has a huge industrial base (existing and new industrial complexes). There are a variety of industries in the region; agro industries, heavy industries, and petrochemical and construction material industries. In addition to recreational beach tourism (the most important tourism resource in the region, particularly in Alexandria and Mersa Matruh governorates), the region is rich in archeological tourism resources spread across its different governorates (Pharaonic, Greek, Roman, Coptic and Islamic antiquities).
4. Among the governorates in the region, Alexandria is the governorate with the greatest diversity of economic sectors (industrial, services and commercial sectors), which explains why the population of Alexandria City is so much larger than that of other cities in the region. In Al Beheira Governorate, agriculture is the leading economic activity, while Mersa Matruh is best known for the tourism services it supplies (in addition to agriculture).27
5. The Region's Infrastructure:
6. The Alexandria Region (particularly in Alexandria and Al Beheira governorates) has an excellent land transportation network (road and railway) that connects the region to the nation's capital and to most capitals and major cities in neighboring governorates. The local transportation network also connects most villages and towns of the region's governorates.
7. The national and regional road network includes the international coastal highway running parallel to the country's northern coastline, from Al Salloum all the way to Al Arish. At present, this is the principal land route linking existing (or planned) tourist resorts on the Northwest Coast to other sections of the national road network.
8. The national railway network includes the Cairo – Tanta – Alexandria – Mersa Matruh – Al Salloum railway line. The railway network pattern in Alexandria and Al Beheira governorates is spread out, while in Mersa Matrouh Governorate it runs along a lateral line.
9. There are two seaports in the Governorate of Alexandria (the East Port and the West Port), in addition to the Mersa Matruh Commercial Seaport. Three international and national airports in the region contribute to promoting tourism development in the region: Al Nozha and Borg al Arab Airports in the Governorate of Alexandria and Al Alamein Airport in Mersa Matruh Governorate.

6.2. Regional and Urban Studies

The study of the regional and urban condition of coastal regions was conducted using:

1. Egyptian governorate development strategy reports issued by the General Authority for Urban Planning in 2008; a collection of comprehensive reports covering the development strategies for Egypt's governorates and the seven planning regions (Cairo, Alexandria, the Suez Canal, the Delta, Northern Upper Egypt, Assiut and Southern Upper Egypt). Each report includes a statement explaining the position and role of the region in which the governorate is located within the national spatial development system and plan. Each report also presents the framework for, and expected outputs of, the governorate's strategic development plan, according to its natural, human and urban attributes, based on official State-approved data and statistics.
2. Regional plans drafted and approved for each region:

First: Alexandria Region Development Strategies

A study of the general strategies formulated for the Alexandria Region shows that they are based on three primary principles:

- Enhancing the economic efficiency of existing activities by raising performance levels, increasing production, judicious investment and production decisions, upgrading existing infrastructure networks, and focusing on operations of replacement and renewal operations of existing assets.
- Encouraging population migration from the region's overcrowded governorates towards urban conglomerations in desert areas and new cities in the

27 Industrial Development Authority, Derassat al gadwa al awaleya le moqtara7 mamar al tanmeya bel sahara'a al gharbeya – derassa ta7leeleya le me7war aleskandereya, 2009

region

- Ensuring that new investments promote regional and national objectives by directing them toward production, services and urban sectors.
- Following the formulation of the development strategy, a number of particularly important projects were set up in the region, including:
- Tourism development of the North Coast, from Alexandria to Al Salloum (an area full of distinctive, long-term tourism development opportunities); Detailed studies have been carried out on developing of the area of the Northwest Coast sector that extends from the eastern borders of Mersa Matruh to Al Salloum by establishing service centers and integrating them into beach activities. These centers would be set up at specific points on public beaches and open areas as well as in areas with no access to public beaches.28
- Upgrading fishing harbors and expanding fish farms and enhancing their efficiency.

Second: Northwest Coast Regional Development Plan (2007)

Figure 8. *Alexandria Region Development Strategy*

Source: Estrategiat al tanmeya le mo7afazat al gomhoreya, Iqleem al eskandereya, Chapter 3, General Authority for Urban Planning, 2008

The regional development report defines the Northwest Coast Region as the area between 29°E and 25°30'W longitude and 30°N, 30° S latitude, a 50 km wide area that extends along the seashore for a distance of about 500 km. Total area of the region is approximately 20 thousand km2 (2% of the country's total area). The regional development report thus only includes Mersa Matruh Governorate in the Northwest Coast region; the governorates of Alexandria and Al Beheira (which are part of the region) are omitted.[29]

Its many beaches and the desert buffer zone that extends behind the seashore from a point 61 km west of Alexandria all the way to Al Salloum make the Northwest Coast rich in natural potential.

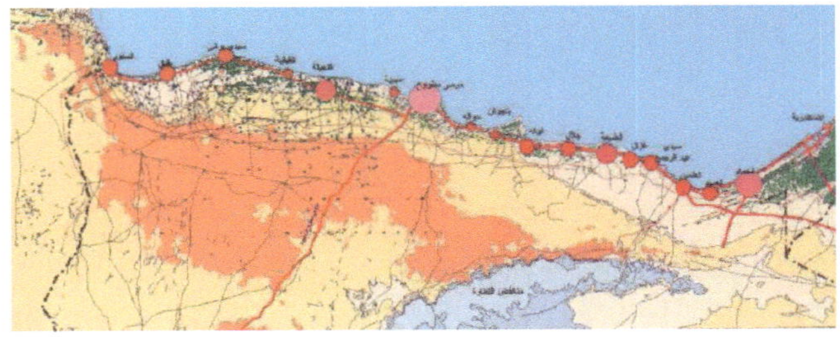

Figure 9. *Northwest Coast Regional Development Plan (2007)*

Source: Northwest Coast Regional Development Plan (2007), General Authority for Urban Planning, 1998

6.3. Regional Plan Objectives

The Northwest Coast Regional Development Plan aims at:

- Formulating a general strategy framework for realizing comprehensive development of all economic sectors and that is not limited to a single economic activity
- Judicious use of coastal urbanization to prevent random, unplanned development and its negative effect at regional and national levels (disparities between cities in Mersa Matruh Governorate and cities in other regions, and lack of equilibrium between the coastal strip - one of the region's components and the remaining components)
- The special attention accorded to the North Coast has encouraged large investment expenditures to build summer vacation units (concentrated mainly on the coastal strip) that are used only for limited periods of the year

The concept of equilibrium referred to in these objectives denotes the following:

- Developing economic and social sectors in accordance with existing demand, while stressing the presence of mutually interconnected small and large-size projects and not squandering available development resources and components
- Allocating a number of prime locations on the Northwest Coast for large projects that have the capacity to attract a high rate of economic activity and large populations
- Developing urban and other facilities necessary to support the coastal sector so as to enhance workforce and population stability.30

28 Egyptian Cabinet Information and Decision Support Center, *Wasf misr belma3loumat*, Seventh Issue, 2008

29 Ministry of State for Environmental Affairs, Egyptian Environmental Affairs Agency, *Al tawseef al bee'ee le mo7afazat matruh*, 2008

30 General Authority for Urban Planning, Al edara al 3amma lel bo7outh wa al

6.4. Regional Plan Concept

The regional plan concept was conceived to support the urban development strategy of dividing the Northwest Coast into sub-regions (planning sectors) according to the similarity of attributes, economic components and problems among the sub-regions, necessitating the adoption of appropriate urbanization policies for each region.

Accordingly, the Northwest Coast was divided into three planning sectors:

Al Hammam / Ra's Al Hekma – Ra's Al Hekma / Al Negeila – Al Negeila / Al Salloum.

We can therefore suggest three principal urban centers as administrative, economic and service sector leaders of the sub-divisions on the regional level. Naturally, the regional functions of these principal urban centers will vary according to the different resources and possibilities available to each planning sector.31

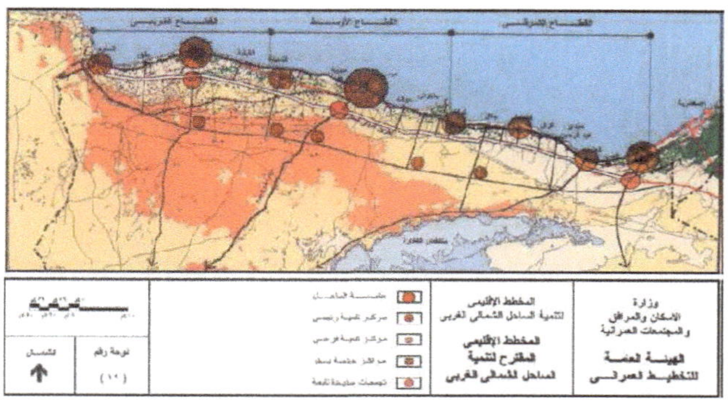

Figure 10. *proposed Northwest Coast Regional Development Plan*

Source: Northwest Coast Regional Development Plan (2007), General Authority for Urban Planning, 1998

Third: Development Methods Currently in Use in the Region

1. Infrastructure projects in the governorates are implemented according to the five-year plans proposed in the State budget. These cover all the kinds of infrastructure (roads, potable water, sewage, electricity and communications).
2. Law No. 119 / 2008 (the unified building code) states that the General Administrations for Planning and Urban Development in the governorates are to draft and approve detailed city plans within a two-year period from the issuance of the law. In cases where no general plans exist (or pending the approval of general strategy plans), the General Administration for Planning and Urban Development lays down temporary rules and conditions for detailed plans regulating the urbanization process. After submission of these rules and conditions to the local People's Assembly, the governor concerned issues a decree for their implementation pending the drafting and approval of general strategy plans.
3. Plans for the coastal cities of the region (such as Rasheed, Alexandria and New Bog Al Arab) were approved according to the above. Strategy and detailed plans for other cities have not yet been approved.[32]

7. Conclusion

This study aimed to explain the relationship between crucial sustainable development and the development of tourism regions in Egypt. It reviews sustainable development strategy indicators for the Northwest Coast and Alexandria regions.

It discussed the Tourism Development: Objectives and Relation to Regional Development with The Principles of Sustainable Tourism "social, economic, Environment".

And also discussed the Factors Affecting the Demarcation of Tourism Regions: Natural Components, Urban Components, Economic Components, Social Components, Administrative and Political Factors, Future Dimensions of Regional Development. The paper gave a standards and Methods for Identifying Tourism Development Components, and measure the Indicators for Measuring Tourism Development Components, and finally study the north coast and west Alexandria regions in Egypt in order to sustain the tourism development in both regions.

8. Results and Recommendations

This study deduces that the tourism development process is a process of change during which the exploitation of resources, the directing of investments (during a given period), the thrust of technological development and institutional change all form a coherent, integrated whole, thereby increasing possibilities for linking the present to the future to provide the basic needs of tourists.

Tourism regions play a crucial part in a system where all tourist sites are integrated. A tourism region does not necessarily only refer to a geographic area. Rather, it refers to the integrated tourist facilities and services within several tourist sites that, together, form a tourism region. Bringing together all the natural and urban components within a particular geographic area facilitates integrated, comprehensive tourism development of that area. It also facilitates the use of marketing cooperatives to promote efficient marketing of the tourism region, and makes it easier to formulate integrated marketing policies for the entire region. Tourism depends essentially on effective marketing mechanisms, which are not easily realized in the context of individual, isolated sites or through small-size tourism marketing programs.

---

derassat al bee'eya wa al tabee3eya – al derassat al bee'eya wa al tabee3eya le wa7at siwa (by Mahmoud Mohamed Nasr Allah, National Research Center),2009.

31 General Authority for Urban Planning, Ta7deeth estratejeyat al tanmeya al shamela lel sa7el alshemali algharbi wa zaheeroh al sa7rawi (2010 – 2032)

32 General Authority for Urban Planning, Estratejeyat al tanmeya lel sa7el alshemali algharbi (2022)

This paper concludes that the major indicators of sustainable tourism development are the following:

Natural components of a region; diversification of natural components is considered one of the most important indicators of regional development

Economic components of a region; refers to a number of indicators that contribute to sustainable development of tourism regions, such as the existing demand for a region, the size of investments directed to it, and the bed capacity (hotels) available in it.

Social components of a region; refers in particular to the degree of participation of the local population in economic activities and the size of the workforce employed in the tourism sector.

Urbanization components of a region; the degree to which tourism regions are connected to regional road networks, airports and seaports, hotel capacity and class, and the centralization of tourism services are among the most important indicators of successful sustainable development of tourism regions.

Development strategy for the Alexandria Region is based on three main principles:

- Formulating a general strategy framework through which to realize comprehensive development of all economic sectors and that is not limited to a single economic activity
- Judicious coastal urbanization to avoid random, unplanned development and its negative effect at regional and national levels (disparities between cities in Mersa Matruh Governorate and cities in other regions, and lack of equilibrium between the coastal strip - one of the region's components - and the remaining components)

Because the Northwest Coast Region's positive indicators for future development stand to contribute to the growth of Egypt's gross national product, this study recommends focusing attention on tourism development, and optimizing future development plans, in this region.

The study further recommends taking full advantage of integrated development projects and applying them to the development of the North Coast, while underlining the urgent need to formulate a flexible legal framework for investment to remove obstacles facing investors in the implementation of their projects.

References

[1] Ahmed Al Galad, Al tanmeya al seya7eya al motawassela, Alam al kotob, Cairo, 2000

[2] Ahmed Fawzi, Madkhal ila 3elm al seya7a, Dar Al fekr al gamee3ee, 2008

[3] Central Agency for Public Mobilization and Statistics (CAPMAS), General Census for Population, Housing and Establishments, 2006, and 2008 Estimations

[4] Egyptian Cabinet Information and Decision Support Center, Wasf misr belma3loumat, Seventh Issue, 2008

[5] Estrategiat al tanmeya le mo7afazat al gomhoreya, Iqleem al eskandereya, Chapter 3, General Authority for Urban Planning, 2008

[6] Fahd Abdel Karim Ali Torkistani, Al seya7a al bee'eya, mafhoomha wa 3anaserha, First Edition, 2008

[7] Fatena Al Wattar, Al malame7 altakhteeteya le ta7qeeq afaaq al tanmeya al mostadama, Al mobadarat wa al 'ebda3 al tanmawi fi al madeena al 3arabeya, Jordan, 2008

[8] General Authority for Tourism Development website - www.tda.gov.eg

[9] General Authority for Tourism Development, Estratejeya le tanmeyat al seya7a 7atta 3am 2017

[10] General Authority for Urban Planning, Al edara al 3amma lel bo7outh wa al derassat al bee'eya wa al tabee3eya – al derassat al bee'eya wa al tabee3eya le wa7at siwa (by Mahmoud Mohamed Nasr Allah, National Research Center),2009.

[11] General Authority for Urban Planning, Al mokatat al estratigi al qawmi lel tanmeya al 3omraneya wa manateq al hay'a al 3ama lel takhteet al 3omrani – Al itar al 'eqleemi letanmeya mo7afazat eqleem al eskanderiya

[12] General Authority for Urban Planning, Estrategiat al tanmeya le mo7afazat al gomhoreya, Iqleem al eskandereya, Chapter 3, 2008

[13] General Authority for Urban Planning, Estratejeyat al tanmeya lel sa7el alshemali algharbi (2022)

[14] General Authority for Urban Planning, Ta7deeth estratejeyat al tanmeya al shamela lel sa7el alshemali algharbi wa zaheeroh al sa7rawi (2010 – 2032)

[15] General Authority for Urban Planning, Ta7deeth estratejeyat al tanmeya al shamela lel sa7el alshemali algharbi wa zaheeroh al sa7rawi (2010 – 2032), April 2011

[16] Industrial Development Authority, Derassat al gadwa al awaleya le moqtara7 mamar al tanmeya bel sahara'a al gharbeya – derassa ta7leeleya le me7war aleskandereya, 2009

[17] JICA study team, The study On Tourism Development Projects In Egypt 1F. Lawson & Bovy: Tourism & Recreation Development, 1979

[18] Magda Mohamed Gom3a, Goghrapheyat misr al seya7eya, 2000

[19] Majalet game3at teshreen lelderassat wa albo7outh al 3elmeya – selselat al 3oloom al'eqtessadeya wa al qanooneya, Vol. 28, Issue 3 (2006)

[20] Ministry of State for Environmental Affairs, Egyptian Environmental Affairs Agency, Al tawseef al bee'ee le mo7afazat matruh, 2008

[21] Mohamed Youssry Ibrahim De3bbes – Sena3et al seya7a bayn al nazareya wa al tatbeeq – First Edition, 2003

[22] Molakhass qera'at estratejeya al tanmeya le 'eqleem al eskandereya – mo7afazat matruh – al moqtara7aat wa al mortakezaat

[23] Tasyeer aleyat ta7qeeq al seya7a al mostadama ma3 altatbeeq 3ala alwa7at al dakhla wa al kharga be misr, First International Symposium (Urbanism and Sustainable Tourism), Institute of Urban Management Techniques, University of M'sila, Algiers, 7 / 8 December, 2011

The Analysis of Sustainable Waterfront Development Strategy

Chien-Hua Chen

The Administrative Office of Keelung City, Taiwan, R.O.C

Email address:
kl037@mail.klcg.gov.tw, countryhandchen@yahoo.com.tw

Abstract: This paper discusses the phenomenon and historic background of the waterfront development. It emphasizes the need for system programming and sustainable management consideration pursuing the development of estuarine or coastal land. It also explains that the success or failure of a waterfront development should be evaluated through a set of multi-aspect indexes. According to different types of waterfronts, their spatial needs are also specific. In the paper, based on concepts of Alvin Toffler's Third Wave, the framework of three stages and six types of waterfront development is devised to discuss its evolution and functions. The development strength of waterfront is divided into three degrees originating the sustainable development concept. In development strategy aspect, this paper analyzes the original meaning and content of waterfront development. Finally, Keelung port city is then taken as an example of to illustrate the waterfront land use development while examining the problems of its current strategies.

Keyword: Waterfront Development, System Programming, Sustainable Management, Development Strategies

1. The Genesis and Historical Context of the Waterfront Development

The waterfront is the origin of human culture and economy. A lot of creatures depend on water for life. Mankind is no exception. Therefore, the earliest villages were close to water sources. This was not only because of our dependence on water but also due to traffic concerns. The rise and fall of many cities was related to transportation. In the very beginning, villages close to water turned into fishing villages. In the age of navigation, they became the posts of trade. In the industrial age, they became container ports. The old ports then became the new waterfront appearance. Such a process represents the diversity of waterfront use and its importance to human life and commercial activities [1].

With the advancements in Industrial Revolution Era and shipping communication, many industrial districts are established by the sea, for these industrial products can be efficiently transported. Communication network is established as well. In company with large-sized watercrafts and the advent of information revolution epoch, industrial structure has been altered. Following this, not a few of the new arisen high-tech industry is built in the suburb, instead of utilizing the previous harbor districts, therefore substantive idle land came into being in the waterfront of Post-Industrial Revolution Era. The unused land in the industrial districts and traffic occupation poses many economic and social problems to the urban development. Nevertheless, by reason of their favorable geographical positions in approaching downtown relatively, the land also bears opportunities to the urban development. The influence of the redevelopment of waterfront land caused by industry transformation is cosmopolitan. Hoyle (1999) [2] suggested that the redevelopment of waterfront land is an inevitable global phenomenon for a sustainable developed harbor city, from advanced countries to developing countries from cosmopolitan cities to small towns. And the success experience of Baltimore inner harbor renewal starting at 1960s even became an imitation object to other occident Post-Industrial Revolution countries. Consequently, this tide engulfed Australia, East Asia, Africa, and Latin America.

Hayuth (1998) [1] adopted a comprehensive view in deciphering the development and trend of harbor and city in the Post-Industrial Revolution Era. He finds out that the process of change and interfacial relation between harbor and city pivots the attitudinal transit of the public and the innovation of technology. The change of the two factors brings out the spatial, economic and ecological change to the waterfront (See Figure 1).

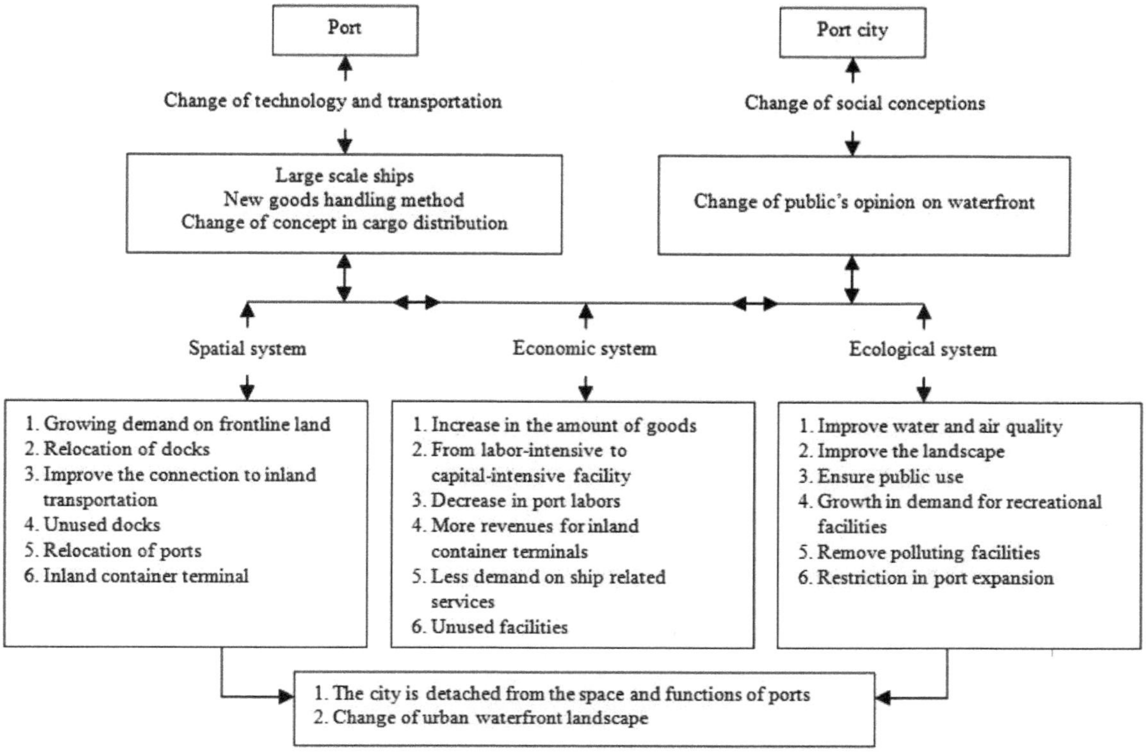

Figure 1. *Trend and development of the interface relationship between the port and city in the post-industrial revolution era.*

As waterfront land is enormously developed, another view has been proposed which focused on the protection of coast ecology. In the past, the development of estuarine and coast mostly put emphasis on engineering technical, but considerations of system planning and ecologically sustainable management were often ignored. As human beings are realizing the importance of environmental protection, it is absolutely necessary to consider the sustainable management of waterfront development in the future. Vallega (2001a)'s [3] further analysis found that global climate change and the Waterfront Recovery Agitation have attracted the international community's attention to the sustainable development of coast, and also catalyzed the international role of waterfront as a global link (See Table 1).

Table 1. *Summary of coastal control measures and waterfront development process.*

Phases	Coastal control measures	Waterfront development process
Takeoff phase	Takeoff: 1960s to mid1990s	Takeoff: late 1960s to mid 1990s
External environment	Coast erosion; the perspective of coastal function prolongation	Transformation of international labor market
Development conception	Global climatic variation	Waterfront revival theory (Baltimore tide)
The direct cause	Coast erosion; collision in using the coastal districts	The desertion of harbor and industrial districts
Aim	Environmental protection; selective economic development	Develop commerce and function of traveling and rest
The key issue	Social perception of global climatic variation; rational resource usage; consensus of decision makers	Interaction between city and harbor
Maturity phase	After mid1990s	After mid1990s
External environment	Global climatic variation and sea level rise; necessity of coastal ecosystem protection ; necessity of programming coastal function	Globalization; international tourism
Development conception	Sustainable development of coast; conformable coastal control; ecological system supervision; protection of cultural heritage; justice of spanned-generation	Post modern programming; transnational tourism; role of regional center; stress on cultural heritage
The direct cause	Sea level rise and its aftermath; ecological protection; increase of coastal functions and their collision	Globalization of information and media; internationalization of tour; link of global waterfront
Aim	The conception of the sustainable development of coast application	Promotion to the international role of waterfront; proper development of traveling and rest functions; arrangement of new functions
The key issue	Conformity of the environment and development; conformity of decision-making systems; conformity of law frame	Interaction of city and region; interaction of city and coast

Vallega (2001a) [3] believed that the waterfront century has passed two stages:

1 First stage (1960-1990) with the downfall of modernist, the updated version of waterfront has gained vital importance for saving local GDP and employment rate.
2 The second stage begins at the middle of 1990s, the postmodernism affected by the sustainable development concept rises; under the Chapter 17 of United Nations Agenda 21, the waterfront development and the management of waterfront conformability begins to interact with each other.

The tremendous developments of waterfront, in some perspective, reflect the changes in the global industrial structure as well [4-6]. The transformations of international labor market and the desertion of harbor industrial district in the Post-Industrial Revolution Era catalyze the theory of waterfront revival. The renewal tide of "Baltimore type" creates many waterfronts which focus on the functions of commerce development, travel and rest, and brings about the delicate changes in the interaction between city and harbor [7]. Moreover, the soaring of international tourism caused by the globalization of information and the new arisen industries alike draw the attention of many countries to the potential of touring industry. The post-modern layout not only lays stress on transnational tourist demand, but also takes into account the role of regional center of waterfront. And waterfront's emphasis on cultural heritage, driven by the push of internationalization of tourism, also becomes an important topic [8-9]. The link of global waterfront brings its international role to the light day by day, and proper development of traveling and rest function of waterfront becomes a consensus. The new functions lead the interaction between city and coast to be reinterpreted.

However, simultaneously, just with the vast development of waterfront land, many scholars who concern the ecological environment start to realize the crises of human overuse of resources. First of all, global climatic variation engenders the rise of sea level, and followed by the coast erosion and the collision of the use of coastal land. The consideration of environmental protection appears in the development of waterfront. These factors convince mankind that "selective economic development" is the most favorable alternative to the blessing of him. It becomes a consensus to all the nations that resource should be rationally used. Stress is placed at the coast ecological system protection and comprehensive programming of coastal functions. One by one, the notion of coastal sustainable development, conformability of coast supervision ecological system supervision, cultural heritage protection, and justice of spanned-generation appear [10-14]. The conception of sustainable development, which is paid respect in the development of waterfront, precipitates the conformity of environment and development, the conformity between decision-making systems and the conformity between law frames.

The above description roughly sketched out the development reasons and historical background of waterfront phenomenon; wherein, Hayuth's and Vallega's analysis emphasized the phenomenon of waterfront update in the post-industrial era. In fact, the impact and involved aspects of the development of waterfront are deeply studied on a continuous basis. As making extensive use of waterfront land for economic factors, human beings are also aware of the fragility of the environment, thus in this context the concept of coastal management have emerged. In the future, it is necessary to be more prudent when engaged in waterfront development; wherein, systematic and comprehensive coastal planning and management plays an important role in exerting the function and benefits of waterfront areas, and shall be necessarily taken into consideration for the future development strategy.

2. Multiple Assessment Standards for Successful Development of Waterfront

Although waterfront development has become a global trend, but because different countries have different development strategies, the achievement and evaluation of waterfront development were also greatly different. Although there were a lot of successful cases, cases of failure can also be found everywhere [15-18]. As for how to judge whether a waterfront development project is successful, there are different standards from different points of view, and views of judgment from multi-angles and stands have been mentioned in many documents with their own discourse points. In *Design and Development of Urban Waterfront* it was mentioned that to judge whether a waterfront development project is successful, we need to explore separately from such angles as the municipal government, the public and the development agencies [19].

The book implied that from the angle of municipal movement, there are three main standards to judge whether a project is successful:

1 contribute to economic growth.
2 help to increase employment.
3 help to improve the outlook of a city.

From the angle of the public, there are also three main standards judgment standards:

1 whether the general public can enjoy this construction project.
2 whether the construction project contributes to the city cityscape elegance.
3 whether it is economically advantageous.

In the viewpoint of development institutions, there two main standards:

1 there are enough development profits.
2 to get social recognition.

In addition, in *City, Capital and Water* it was mentioned that to judge the success of a waterfront development project needs from the angle of the government, the planner team, the developers, and the people [20]. Based on this view, the paper proposed four main categories of assessment criteria, and worked out the following table (See Table2):

Table 2. *Different Categories of Assessment Criteria.*

Category	Government	Planner team	Developers	People
Position and angle	The public interest	Satisfactory development	Development interests	Improved living standards
Assessment criteria	• Economic growth • Increase employment • Improve the outlook of city	• Planning vision • Vision of an ideal city • Sustainability development	• Development profits	• Comfortable environment • Living standard

The above four categories respectively represent the position and angle of public interest, satisfactory development, development interests, and improved standards of living, and have their different assessment criteria on the so-called success. An ideal waterfront development project is expected to achieve the desired target of each of the above-mentioned four categories. First, the government departments representing the public interest is expected to promote the public interest as the highest priority target , so it is hoped to achieve such effects as economic growth, increasing employment, improvement of city outlook, and etc. as the waterfront development project is promoted. Second, the role of planner team is also very important; a good planner can adhere to the planning ideal, analyze the characteristics and patterns of a city from a professional point of view, propose the future development vision for the city, and if necessary, act as a bridge for the communication between the government departments and other units. In addition, with the environmental and ecological changes, the concept of sustainable development should also be included in the overall development system; as for developers, they naturally take development interests as the most important consideration. Many waterfront development projects, especially large-scale waterfront development projects, need to introduce private capital, in that often non-public sectors have ways to independently bear all the funding and risks; as for whether a waterfront development project is worthy of investment, for developers, the most important assessment criteria is whether there is development profits; in other words, projects with no development profit will be naturally impossible to attract private sector's interest in investment; and usually when spurred by the development profits, the efficiency of the private sector, in general terms, will be improved [21-22]. Finally, the largest number of people represent improved living standards, because the most important judgment standard whether the planning is successful depends on the feelings of people living in that environment. However, we also find that in many development projects, people almost have no influence and even without participation mechanisms and channels; in fact, from many new planning concepts, it is gradually discovered the importance of attention to folk wisdom, but from a practical perspective, we see that the in many development projects, because the development results were not supported and recognized by the people, in the end the development would easily lead to a rapid decline. An ideal waterfront development project is hoped to create a win-win situation, but in the reality of development projects, we can easily find that many waterfront development projects cannot achieve the ideal, and an inappropriate waterfront development strategy is one of the important reasons why it is unable to reach the ideal.

3. The Classification of Waterfront from the Third-Wave Concept

Because different types of waterfront have different space requirements, it is essential to make classification of the waterfront before waterfront development. There are many different classification methods for waterfront; in *Design and Development of Urban Waterfront*, waterfront is approximately divided into the following classifications in views of location and function [19]:

1 Ecological protection zone.
2 New development.
3 Rebuilding of old industrial areas.
4 Connected to residential area.
5 Connected to downtown area.
6 For leisure and travel.

In 1985, according to development patterns, Breen & Rigby classified water into commercial type, cultural and educational environment type, historic type, entertainment type, housing type, and work type, of which work type refers to the commercial port and fishing port needing labor and machinery [23], but as for the current trends of waterfront development in different countries, particularly the waterfront mainly focusing on service industry as the main development pattern, its function and development tend to be diversified, so as long as there is no problem in compatibility, such development shall be basically rather easy to create a multiplier effect, thus to make waterfront development more successful.

According to waterfront activities, Vallega mentioned 12 kinds of major waterfront activities [24]; Toffler in his Third Wave series [25], clearly expounded his "third-wave" concept: The first wave refers to the agricultural era in 10,000 BC, the mankind got rid of nomadic lifestyle, and people settled down in fixed lives area, and began to develop towns and their own culture. The second wave was starting from the 18th century; taking the opportunity of "Industrial revolution", people left the farm and flocked to the cities to make a life in the factory. The third wave was another big change of the human mind -- "Information revolution", where information technology and social needs were the powerful driving force for its development, the whole world melted together, and people sought for cooperation without national boundaries.

Past studies divided waterfront into many types, but didn't further elaborated the functions or association of various waterfront types; this study will attempt to construct three stages and six types classification and evolution model for waterfront development, illustrate the functions and relevance of different waterfront types, and explain its functional development process.

After consulting the classification of waterfront activities proposed by Vallega(2001b) [24] and based on the "The 3rd Wave" concept raised by Toffler, A. in 1980, this paper has divided the development of ports into three development stages of added-value labor, added-value production and added-value service Combining the six primary functions of aquaculture and fishing, offshore fishing, distributed processing, logistics and trade, leisure and tourism and research experience to the classification of the functions of the waterfront, the paper has proposed the 12 classes of activities including the fishery, traditional fishing, operation on fishing grounds, goods distribution, processing intermediary, international logistics, trades stream, recreation, tourism, cultural preservation, marine research and ecosystem investigation. Embedding the value chain of waterfront evolution as well as the concept that "Industrial Revolution" and "Information Revolution" have stimulated the alterations in industrial structure, this paper has induced the three stages and six types evolution mode of the port development as is shown in Figure 2 and Table 3. This paper explains that developing functions and performances of different types of waterfront, and it has introduced the sustainable developmental concept hoping to develop the construction based on it, and it has shown how to combine urban development policy with waterfront development to finally reach the four sustainable development objectives of "basic needs of life, logistics production, ecosystem experience, sustainable life" [26-27].

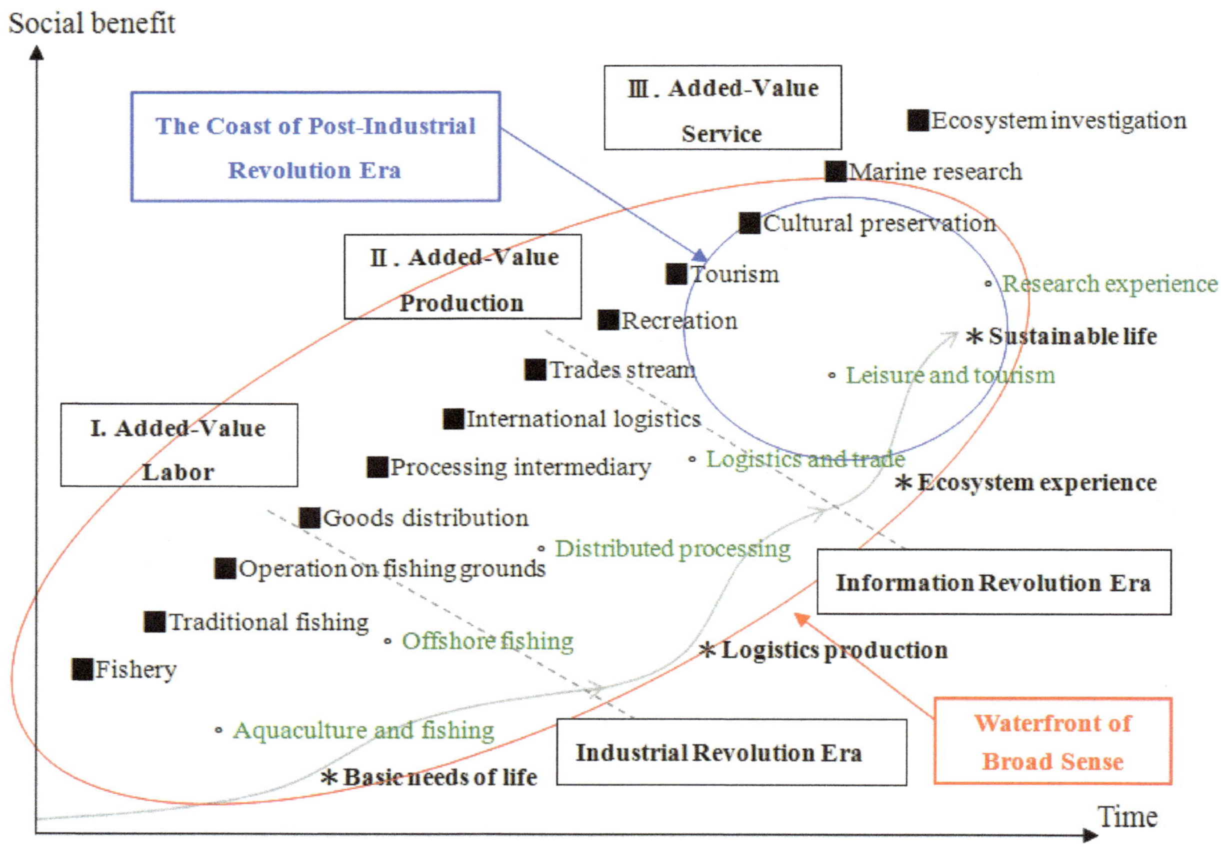

Figure 2. *Diagram of three stages and six types of port develop.*

Table 3. *Three stages and Six Types of Port City Development.*

Development stage	**I. Added-value labor (The first stage waterfront)**	**II. Added-value production (The second stage waterfront)**	**III. Added-value service (The third stage waterfront)**
Waterfront functions	Aquaculture and fishing; Offshore fishing	Distributed processing; Logistics and trade	Leisure and tourism; Research experience
Performance	Fishery Traditional fishing Operation on fishing grounds	Goods distribution Processing intermediary International logistics Trades stream	Recreation Tourism Cultural preservation Marine research Ecosystem investigation
Demand level	The demand to "basic needs of life" The demand to "sustainable life"	The demand to "logistics production"	The demand to "ecosystem experience"

The development and application of waterfront land is related to several important revolutions, first of all in agricultural time, the changes in the waterfront took place very slowly, and the industrial transformation period is also very long. The emphasis of the relation between humanity and waterfront is the "life" and the function styles are aquaculture and fishing, and offshore fishing. The primary performances are fishery, traditional fishing and operation on fishing grounds, the industry type belongs to "Added-value labor" and this paper has treated this waterfront development as the first stage waterfront. With the coming of "Industrial Revolution Era" industrial structure has changed greatly and the change period relatively accelerates, the emphasis of the humanity and waterfront relation is the "logistics production" its function styles are distributed processing, logistics and trade, the primary performances are goods distribution, processing intermediary, international logistics and trades stream, and the industry type belongs to "Added-value production". This paper has entitled this waterfront development as the second stage waterfront. With the coming of "Information Revolution Era" the changes in industrial structure become more intense; the change period becomes even faster, the emphasis of human and waterfront relation is "ecosystem" its function styles are leisure and tourism, research experience, its primary performances are recreation, tourism, cultural preservation, marine research and ecosystem investigation, and the industry type belongs to "Added-value service". This paper has treated this waterfront development as the third stage waterfront. In the future, the functions of the waterfront will be further diversified, and the development of waterfront will stop the enlargement without limitation with the rise of sustainable development, and the concept of protecting waterfront will be paid special attention, the development and management of the waterfront supplements each other, and the relation between city and different stages and types of waterfront will be established on "sustainable life" [26-27].

Actually many of the documentaries about waterfront concern about the third stage waterfront, the "Added-value service" type, or the "Post-Industrial Revolution" stage waterfront. But waterfront in a general sense includes the entire waterfronts that are mentioned above. The waterfront applications of different stages have their own time background and space demand according to the theory of waterfront development. And even the waterfront of the same city may often have different types. The space layout, infrastructure and the implantation of traffic of the former waterfront all aim at the special needs of that particular time. Therefore it should not be transformed hurriedly when the function of the inherent stage is still active. On one hand, this consideration is because the cost for compulsive transformation will be relatively higher, and on the other, the transformation of the urban industry and population structure cannot be finished in a short time. When the urban industry population still needs the current type of waterfront, the compulsive transformation will cause a negative impact on the urban industrial structure, even cause rebound. Therefore the transformation will only happen after the obvious shrinking and degrade of the waterfront function under current stage [28-29].

Besides, combining the concept of conservation and sustainable development, this study proposed that waterfront development can be divided into three categories according to the development strength, namely maintenance, renovation and renewal. Human kinds develop and make use of waterfront for different reasons, but as people realize the importance of environment and ecology, waterfront management concepts are paid more and more attention to, and waterfront is no longer endlessly developed. At the moment waterfront is unavoidably developed, we try to moderate the consumption of waterfront resources, based on the concept that maintenance is superior to renovation, and renovation is superior to renewal. First, undeveloped waterfront should be kept it original unless absolutely necessary and precious waterfront ecology should be especially maintained; second, for developed waterfront, we should respect its original development function, so we renew the waterfront only when the original staged function of the waterfront has obviously declined, and modest waterfront management measures are also absolutely necessary.

4. Origin, Significance and Content of Waterfront Development Strategy

The factors influencing the waterfront development are complicated. From the cases in the world view, the success of a waterfront development project needs very careful scheming of all aspects including development strategy, planning, design, operation, management, and etc., each of which are indispensable; usually serious mistakes of a single part involves may lead to failure of the entire project, so to discuss this issue, we must make an integral consideration. In the book *Design and Development of Urban Waterfront*, it is mentioned that success factors of waterfront development are as follows [19]:

1 Deliberate development strategy (Strategy).
2 Careful planning and innovative design (planning).
3 The main building is the key (Architectural Design).
4 Attach importance to management to maintain the attraction of waterfront area (operation and management).

Waterfront development includes political decision-making, urban planning, architectural design, development operation and management; the land development strategy mentioned in this paper focuses not only on the development of points, but also on the development of lines and surfaces.

The waterfront land development strategy and urban development strategies are closely related; more precisely, waterfront land development strategy is one part of urban development strategy [30]. At different stages and for different problems, we need different development strategies; for analysis of waterfront land development strategy, this paper will explain in two aspects: the origin and significance of strategy, and the contents of strategy:

(A) The origin and significance of strategy

Strategy originates to solve the problems we face; for the problems we face, the paper will consider from three surfaces: the past, the present and the future:

1 Why such a problem was created in the past?
2 How do we solve the problem now?
3 What is our vision in the future to solve these problems?

With respect to the thinking aspects of these three questions, the significance of development strategy can be divided into the following three points:

1 Make up for the deficiency of existing mechanisms: Many problems ignored in the past resulted in development deficiency, so since some problems have been created, we proposed the development strategies to make up for the deficiency of existing mechanisms.
2 Deal with new problems appearing during the development: For different time and space conditions, it is necessary to adjust the functional orientation, spatial structure, development direction and other major issues during the development, and propose the development strategies to cope with the new problems.
3 Depict conception for the future development prospects: Based on concern and understanding of the economic, environmental, and social changes, and in combination of the experience operation and theoretical research and analysis, propose development strategies to depict conception for the future development prospects of the solved problems.

Opposite to waterfront development strategy, another is to think about no waterfront development strategy, i.e. despite the free development of waterfront. In fact, despite the land with idle waterfront brought opportunities for urban development, we also learned from many development cases that no waterfront development strategy or improper development strategy may still cause the failure of the waterfront development.

As for the waterfront development strategy, in the sight of development experiences of countries in the world, generally government occupied an leading position; although a good strategy couldn't ensure the success of development projects (still need to the cooperation of other relevant factors), in the sight of development experiences of countries in the world, laissez-faire market operation or wrong strategies almost caused failure without exception, of development cases in London, England were most representative [31]. Taking Canary Wharf in London as an example; the government completely abandoned planning and control early by laissez-faire market operation, and this in some ways meant contempt of the status of planning teams and planners. As a result, certain specific locations with a profit-making nature indeed succeeded in a short term, but due to lack of development strategy for the whole region, specific locations with nonprofit-making nature cannot be taken care of and their developmental delay lead to failure of the whole region. In addition, taking the waterfront development in Copenhagen, Denmark as an example; the development was originally controlled by the government, and later due to the development efficiency, the government agency established a company, thus the development was in the charge of private sectors and enterprises; as a result, the efficiency was indeed considerably improved, but the development project was criticized for ignoring the interests of people living in the development region. The criticism also adversely affected the entire waterfront development project [32]. And then take the waterfront development in the Rocks and Woolloomooloo, Australia as examples; due to lack of support from trade union federation, the working class residents and the middle-class conservatives, the development projects in the two regions were frustrated. Many waterfront development strategies ignored people's rights, and put excessive emphasis on the efficiency; generally, the interests of the people living in the development region were not taken seriously, especially the interests of local low-income residents were often trampled and sacrificed; on this point, even Baltimore, the city often used as a model of success was of no exception. Through interview statistics, American scholar Krausse, G. H. (1995) [33] explored the feelings of residents in Newport, Rhode Island, the U.S. on tourism development and waterfront redevelopment, as well as changes of demographic and social structure. The development project made a large number of residents migrate outward. Such a phenomenon was greatly related to the fact that local residents' opinions were not included in the early development strategies. The four articles written by Brett, E. A. (2000), Ryan, C., Cooper, C. (2000), Bassett, K., Griffiths, R. and Smith, I. (2002), Brusbett, K. (2003)[34-37] all emphasized the importance of residents' participation during the development and planning process; many papers also noted the drawbacks of failure in including the public's opinions in the planning phase; and some literature about the waterfront development repeatedly mentioned the importance of including the public's opinions. In this case, the waterfront development in Canada developed a unique form because of its developed community organizations, and it was much precious that the rights of people living in the development region were considered [2, 38]. In the future this trend will be attached much importance on to make development strategies more perfect.

(B) Contents of strategy

The conditions of waterfront in each city are not the same, so the design of waterfront land development strategy shall be based on the conditions and problems of each city's waterfront in each; therefore, when designing the waterfront development strategies, the first things shall be to make SWOT analysis of the city's waterfront development conditions, in order to make a systematic understanding of the strengths, weaknesses, opportunities and threats to be engaged in waterfront development. Only after the above analysis, can we work out the content of waterfront development strategy, which is mainly divided into the five categories:

1 Waterfront development orientation: In a city, there may be many waterfront development units of different types and levels, and waterfronts of different type and level shall have different orientation, and the waterfront development orientation shall take into comprehensive consideration of the international community, the country's city division, and the city's development vision. Different types of waterfront also have different space requirements, so the waterfront development orientation is the primary content to design the waterfront development strategy, so as to make a clear direction for the waterfront development. In terms of the three stages and six types waterfront development and development strength concerned in this paper, it is necessary to find out the waterfront development orientation after a comprehensive analysis, and to distinguish which shall be orientated as the first, second and third stage of waterfront? Which need to be maintained? Which remain the original function and only renovation to improve the original function? And which need renewal and reconstruction?

2 Waterfront development layout: besides analyzing the waterfront according to the conditions and orientation, it is also necessary to analyze the waterfront development layout, so that the city's waterfront can be considered from a point level to a line and surface level. Considering whether the waterfront development involves functional or regional conflicts, and how to melt waterfront development layout into the urban development layout, the layout will naturally include in transportation planning, infrastructure, water quality and other relative constructions. In this general direction of layout, individual waterfronts develop gradually and the city will ultimately obtain development advantages.

3 Dimension and depth of waterfront development: The considered dimension and depth of waterfront development will affect the future development patterns and success or failure. An advanced dimension of waterfront development, developed in that it originally only considered the economic side, but now takes into account of environmental side, social side and sustainable development; and as for the angle from which it considers changed from the previous government to the stand of planner team, developers, and people. In the future, a successful waterfront development strategy will emphasize deep and diverse thinking.

4 Selection of the implementation mechanisms: Among the waterfront development experience in various countries, the implementation mechanisms of waterfront development, as for the management system, can be generally divided into the overall management type and decentralized management type, but most successful development experience indicated that it is easier to succeed if there is an overall management unit to control the overall situation of the development; As for the development systems, it is flexible depending on the local government's financial situation, basically necessary to classify the attribute of the development units, and then hand them over to different development agencies. According to the attribute, the development units of non-profit-making can be handed over to government agencies; the profitable development units can be handed over to the developers; whether choose a third sector organization of public-private partnerships or non-profit agencies as a development mechanism can be also adjusted according to the local political and economic systems; a large-scale waterfront development plan often takes decades to implement, so there is inevitable financial risks, so if different units are separately developed under the control of the co-ordination department, it can not only average the risks, but also can improve the development efficiency.

5 Decisions on the development order: The waterfront land development cannot be independent of urban development, so as for the external conditions of waterfront development, transport planning, infrastructure, water quality and other constructions are like umbilical cord and nutrients supplying waterfront land development. When these projects can be completed first, the waterfront land development will be easy to succeed; otherwise, the waterfront land development will be easy to mess up. Sometimes, the base condition of waterfront development is not all point-like, and sometimes it is linear; or several development units constitute a surface base type, and the development of different bases often influences each other, therefore, it is quite necessary to gradually ensure to the success rate of all development units. It is suggested that waterfront development units be divided into a number of development blocks. About development order of waterfront developed by different departments, it is generally recommended to first start from unit small and easy to succeed, in order to drive the development of the whole region through accumulating small victories into larger ones.

5. The Analysis of Waterfront Development Strategy-Taking Keelung City as an Example

This paper will take Keelung as an example to illustrate the above analysis on waterfront development strategy. Keelung is in the northeast of Taipei City, the capital of Taiwan, with the current population of 400,000. In 1970s, Keelung was one of the five metropolises in Taiwan, and Keelung Port locates at the city center of Keelung. The industrial development of Keelung used to highly count upon the container import-export industry of Keelung Port. In 1990s, Keelung Commercial Port has brought more than half of revenue taxes for Taiwan due to the continuous prosperity and the prevailing of shipping commerce, and it also brings along the local industrial boom indirectly. According to the master thesis "Study on the economic impact of Keelung Port container import-export industry on Keelung City" written by Tsai (1999) [39], the industry directly or indirectly related to container import-export industry of Keelung Port has reached 74.41% of the total in 1998, and the industrial population directly or indirectly related to it has reached 82.21% of the total population. While due to intense competition from the outer world as well as 95% of the urban lands are hills which have left limited development hinterland, Keelung City has many problems to overcome while matching towards the 21st century. While Keelung Port hangs together with the city center of Keelung City, Keelung used to be the most important ports in north Taiwan, the harbor depth and the development hinterland cannot meet the large-scale of the vessels due to the competition from other cities. After an overall consideration, Taiwan government has officially decided to build a Taipei Newport 60 km west of Keelung in 1997 (See Figure 3), this policy has directly influenced Keelung City. The economic structure of Taiwan has gradually changed to depend more on the Added-value service. A lot of deserted or low-use rate land floods along the city waterfront with the downfall of industries related to Keelung Port and the littoral industrial district. Furthermore the fishery of Keelung used to play a major role in Taiwan, but in recent years, the yield of fishery in fishery harbor reduces greatly with the recent year global climatic variation, ocean pollution, and overfishing. The traditional fishery harbor function cannot maintain the development of fishing village, the outflow of population and the desertion of harbor facilities become more and more serious [40]. The decay of activities in Keelung waterfront has directly influenced the local economic development and the civil quality of life; it has also generated much chaos in society. These factors have made Keelung Port face industry transformation pressure. The 29.6 long shoreline of Keelung City are the beach and rocky shore with abundant of ocean resources. Cooperating with the historic sites, this deflecting, rich coast and tidal zones have great potential for sight-seeing, the development of tourism has become an import option for Keelung industry transformation.

To improve the usage on the waterfront land, Keelung City has made many waterfront development plans on the more than 10 km long coastline. But the current Keelung City waterfront development plans have many conflicts among them themselves because they belong to different ruling units and the each does things in its own way. Waterfront development plans lack the true basic object which is good for the long-term development of the city. What's more, the scales and properties of these waterfront development plans are different, their location-chosen and development manners are not all reasonable, which have restricted the more reasonable development of Keelung City waterfront land. Actually the low usage rate of Keelung City waterfront land has impacted the urban competitive power, In the future, Keelung Port needs to establish a set of favorable urban development policy to meet the demand of city transformation development, and it must make great changes in the waterfront development mode.

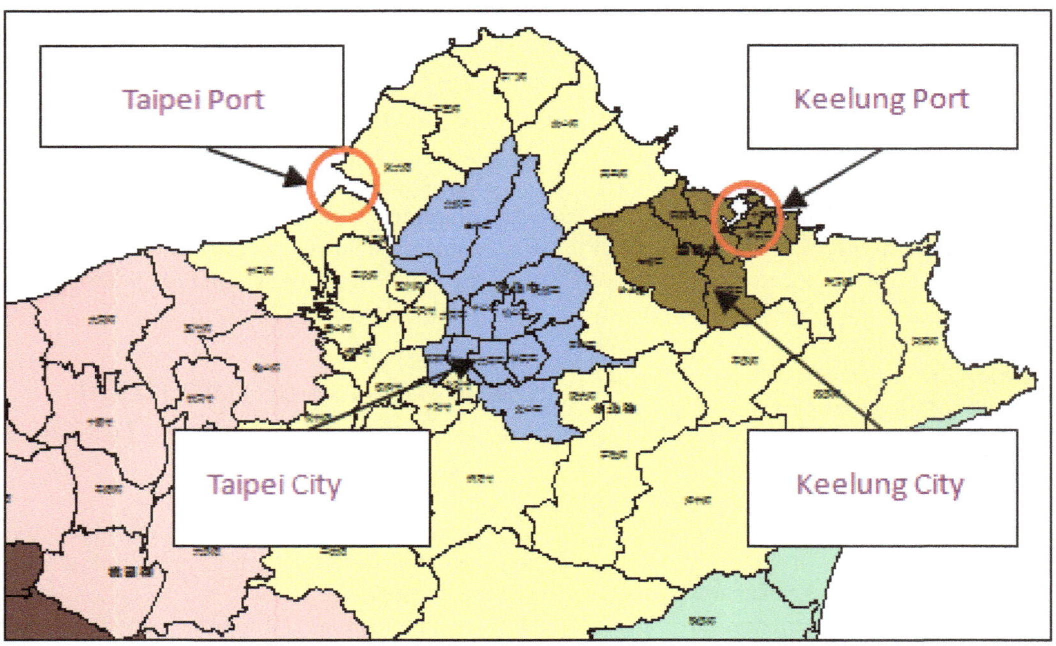

Figure 3. *The location diagram of the Taipei Port and Keelung Port.*

In terms of the number of existing vessels, Keelung city doesn't need so many fishing ports any longer . To make the fishing ports more efficiently used, according to research and analysis of this paper, among the fishing ports in Keelung, we only need to reserve Waimushan fishing port, Chengbing fishing port, and Batoutze fishing port as enough for existing vessels. Those the future function of which has obviously declined, such as Dawulun fishing port, Changtanli fishing port, and Wanghaihsiang fishing port, can be developed towards the third stage waterfront. Among them, the function of Chengbing fishing port is declining continuously; although it is maintained as the first stage waterfront at present, in the future a part of it will be released and developed as the third stage waterfront. As for the second stage of waterfront development, since the construction of Taipei Newport, it is expected that the container traffic will be difficult to rise in the future, so it is expected to be able to handle future container traffic only if leaving the West Coast container port which maintains a strong container function as the second stage waterfront; the East Coast container port the container function is reducing is currently maintained as the second stage waterfront, and in the future a part or all of it will be released as the third stage waterfront [41]. In addition, the Harbor-surrounding downtown trading area located in the inner port is fairly a large renewal development project, including waters and land two parts, with a total area of 97.08 hectares, of which the land area accounted for 71.31 hectares and is divided into seven units. Excluding financing for external infrastructure facilities, the amount invested in hardware construction of bases is estimated more than USD one billion [42]. Due to function recession of the inner port, in the future the inner port will be renewed as the third stage waterfront; Dawulun Waimushan coast, Hoping Island Geological Park, and Changtanli Tidal Flat Ecological Park will be maintained and conserved for there are precious coastal and embroidery rock zoology, biological traces and fossil which is rare rocky coast zoology, intertidal zone zoology and etc. valuable coastal ecosystems.

Actually, concerning the scattered locations of the present waterfront development in Keelung harbor city, there are many advantages in transformation to tourism. The text summarizes as follows:

1. Waterfront development locations are temperate distribution.
2. Close to the downtown.
3. Diversity of waterfront.
4. Waterfront possess many rare ecosystem.
5. A good number of traveling and rest sites around the waterfront.

After referring The Report on the Overall Review of the Enlargement and Alteration of Major Program of Keelung City (1998) [43], The Comprehensive Development Program of Keelung City (2000) [44] and a master's thesis written by Wen-Hsin Lin—Research on Fishing Port Multifunctional Development Strategies - A Case Study in Keelung (2006) [45], and other literature, this paper made SWOT analysis of the functions of all Keelung waterfronts and fishing ports, and made classifications like this: fishing ports suited to the first stage development include Waimushan fishing port, Chengbing fishing port, Batoutze fishing port; those more suited to sustain the second phase development include West Coast Container Port and Keelung East Coast Container Port; those more suited to sustain the third stage development include Dawulun fishing port, Harbor-surrounding downtown trading area, Bisha fishing port, Changtanli fishing port, and Wanghaihsiang fishing port.

Classified by the standards of maintenance, renovation and renewal, Dawulun Waimushan Coast, Hoping Island Geological Park and Changtanli Tidal Flat Ecological Park are suitable to be maintained; Dawulun Fishing Port, Waimushan Fishing Port, Keelung West Coast Container Port, Keelung East Coast Container Port, Bisha Fishing Port, Batoutze Fishing Port, Changtanli Fishing Port, and Wanghaihsiang Fishing Port are suitable to be renovated; and Harbor-surrounding downtwon trading area is suitable to be renewed.

According to the development blocks, the research, after analysis, broadly divided into Dawulun Waimushan fishing port block, harbor-surrounding downtown trading block, Hoping Island Chengbing fishing port block, Batoutze Bisha fishing port block, and Changtanli Wanghaihsiang fishing port block [46-47]. In terms of spatial patterns and internal and external conditions, considering the compartment of waterfront development patterns, harbor-surrounding downtown trading block can be used as "Harbor waterfront of local color", due to "Renewal Program of Central Business District of Keelung City" launched by Keelung City Government [42] and this region combining the city business center specialty. As for Hoping island Chengbing fishing port block, the Keelung City Government launched the "National Door Shanhaikwan Program Integrated Program Of Landscapes Around keelung City Bisha Fishing Port" [48], which divided Hoping Island into three development units: Hoping Island geological park, romantic town, and mountainous products street; combined with the future restructuring and development of Chengbing fishing port, it can be used as "Town waterfront of culture fascination". As for Batoutze Bisha fishing port block, according to the plan made by the Keelung National Museum of Marine Science and Technology [49], and combined with local Ocean University and etc. academic institutions, it can be used as "Integrated educational science and technology garden waterfront".

In terms of the development order of the blocks, according to the analysis of this research, The Bisha fishing port included in Batoutze Bisha fishing port block has transited to a tourism fishing port to be developed as the third stage, and there will be built an world-class maritime museum to its east, the waterfront land in this block is advantaged and should be the best place of top priority to develop. Following it, the Changtanli Wanghaihsiang fishing port block, located in the south-east of the maritime museum is also close, it has a great potential to be developed as the third stage waterfront, so it should be considered as the second priority to develop. With the only beach of the city and precious coastal and embroidery rock zoology, and not difficult to develop, The Dawulun Waimushan fishing port block can be regarded as the third priority to develop. And then followed was Hoping island Chengbing fishing port block; because of a larger area to develop and some difficulties to overcome, it can be considered as the fourth priority. Finally shall be the harbor-surrounding downtown trading block which is the most difficult to develop (See Figure 4 and Table 4).

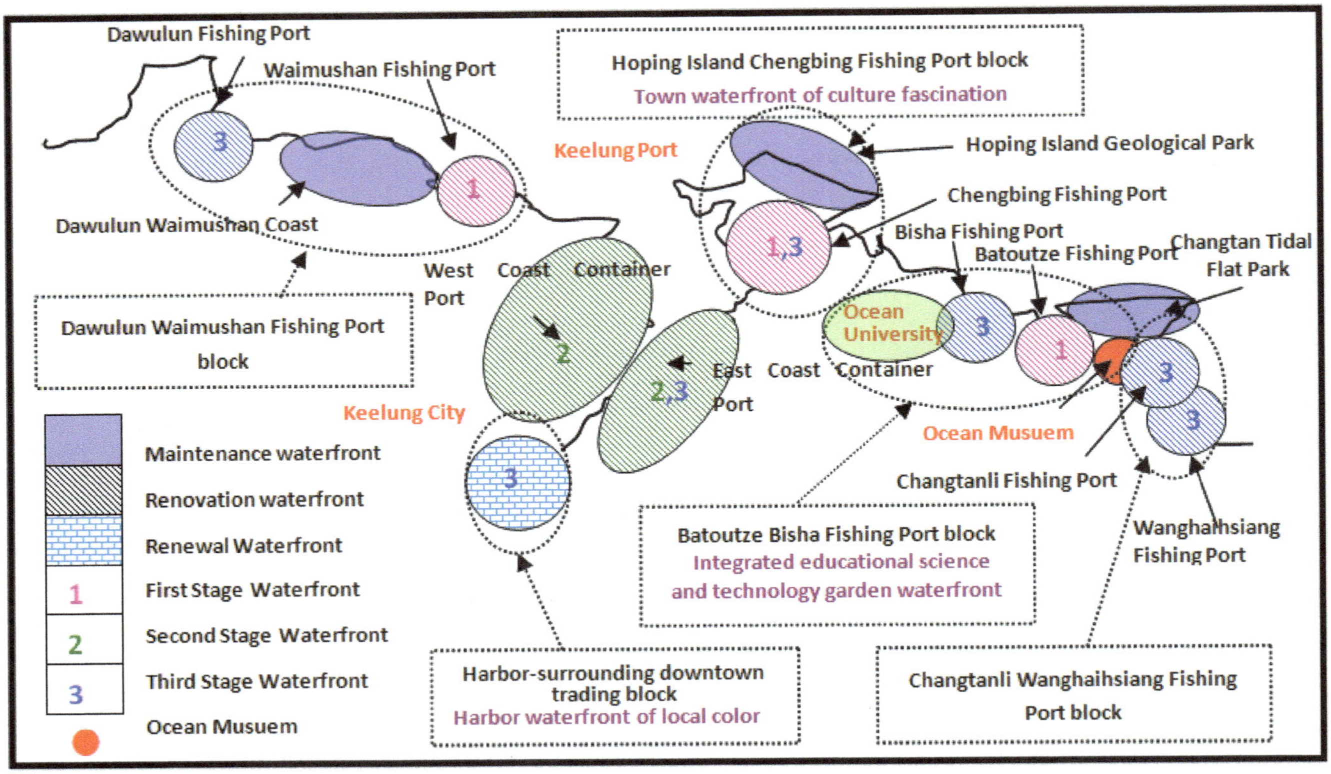

Figure 4. *Diagram of distribution of waterfront development in Keelung City.*

Table 4. *Table of phases of waterfront development of Keelung City.*

Development area and development block		Stage of development	Development intensity	Explanatory comment
Dawulun Waimushan Fishing Port block	Dawulun Fishing Port	3	Renovation	Fishery port function weakens and transformation into the third-stage waterfront
	Dawulun Waimushan Coast	-	Maintenance	Contains precious coastal and embroidery rock zoology and need preserving and maintaining
	Waimushan Fishing Port	1	Renovation	Remains an active fishery harbor function and need be kept as first-stage waterfront
Harbor-surrounding downtown trading block (Harbor waterfront of local color)	West Coast Container Port	2	Renovation	Remains an active container function and need be kept as second-stage waterfront
	Harbor-surrounding downtown trading area (all together seven renewal development units)	3	Renewal	Inner harbor function is weakening and will become the third-stage waterfront in a new way
	East Coast Container Port	2 ,3	Renovation	Container function weakening and kept as the second-stage waterfront. Parts of the district will be released as the third-stage waterfront
Hoping Island Chengbing Fishing Port block (Town waterfront of culture fascination)	Chengbing Fishing Port	1 ,3	Renovation	Fishery harbor function weakening and kept as the second-stage waterfront. Parts of the district will be released as the third-stage waterfront
	Hoping Island Geological Park	-	Maintenance	Containing biological traces and fossil which is rare rocky coast zoology and requiring maintenance and preservation
Batoutze Bisha Fishing Port block (Integrated educational science and technology garden waterfront)	Bisha Fishing Port	3	Renovation	Fishery harbor function weakens and transformation into the third-stage waterfront
	Batoutze Fishing Port	1	Renovation	Remains an active fishery harbor function and need be kept as first-stage waterfront
	Changtanli Tidal Flat Ecological Park	-	Maintenance	Containing rare intertidal zone zoology and requiring maintenance and preservation
Changtanli Wanghaihsiang Fishing Port block	Batoutze Fishing Port	3	Renovation	Fishery harbor function weakened and transformed into the third-stage waterfront
	Wanghaihsiang Fishing Port	3	Renovation	Fishery port function weakened and transformed into the third-stage waterfront

Despite of the excellent development conditions of Keelung waterfront, after reviewing with the waterfront land development strategy discoursed in this study, the waterfront development plan currently proposed by the Keelung City Government and etc. units still has many strategic errors:

1 There is no coordinated planning and management units for these waterfront development plans.
2 The waterfront development plan lacks overall consideration.
3 The development orientation of each waterfront development unit is not clear.
4 The considerations for waterfront development lack dimension and depth.
5 Transportation planning network and other infrastructure are not complete.
6 Public participation mechanisms and channels are not sound.
7 The operational management mechanisms should to be improved.

Taiwan's waterfront development has been behind advanced countries in the world for nearly three decades, but from another point of view, because of its slow start, it can absorb much experience of advanced countries in the development; however, the experience of successful waterfront development cannot be completely copied after all. During waterfront development, on the one hand we can refer to the development experience of similar cases around the world; on the other hand, it is essential to comprehensively consider the basic conditions of individual cases, the organizational structure of local government, the prevailing economic and fiscal conditions, and so other factors. The waterfront development of Keelung Harbor is also in its infancy, so we may learn the development experience of many foreign counties. The waterfront land development strategy analysis in this paper can be provided for Keelung government agencies and relevant units wishing to engage in waterfront development as reference.

6. Conclusion

Waterfront development is a worldwide phenomenon, and there are also a lot of trial error and learning processes. The waterfront transition problem faced to Keelung Harbor is an epitome of many other port cities facing the same problem; in the future the considered dimension and depth of waterfront development will continue to evolve, and we need to think from more diverse angles. When we know the preciousness of waterfront, waterfront development project should not only consider the factor of engineering technical. In particular, the consideration of systematic planning and sustainable management is absolutely necessary. Waterfront development involves very complex factors, so it is necessary to refer to cases of countries in the world, in order to minimize the possibility to make the same mistakes. For waterfront development, a good strategy is the first step; as different types of waterfront involve different concepts in space requirements, this paper attempts to use *Third Wave* developed by Alvin Toffler in 1980, to construct the three stages and six types classification and evolution relationship of waterfront development, and then incorporate sustainable development concept to classify the waterfront development strength into maintenance, renovation, and renewal, in order to illustrate the functional structure and development mode of different stages. In addition, Keelung port city was taken as an example to illustrate how to apply these ideas to the current Keelung waterfront land development strategy, thus to view the defects of Keelung waterfront land development strategies. Because waterfront development in different cities has its own unique conditions, it is not easy to set an absolute standard, but from the cases of countries in the world, there is still theoretical and spatial commonality among successful cases, worth learning. The successful experience of others was a result under certain political, economic, and geographic conditions, so when we learn the successful experience, we should learn its principles and ideas instead of copying its appearance and styles, so as to naturally improve the success rate of waterfront development. These concepts and principles are important implications for those countries wanting to develop waterfront plan.

References

[1] Hayuth, Y. (1998). Changes on the waterfront: a model-based approach. In: Hoyle, B.S., Pinder & MS Husain, D.A. (Eds.), Revitalising the Waterfront: International Dimensions of Dockland Redevelopment. Belhaven Press, Great Britain, pp.52-64.

[2] Hoyle, B. (1999). Scale and sustainability: the role of community croups in Canadian port-city waterfront Change. Journal of Transport Geography, 7:65-78.

[3] Vallega, A. (2001a). Urban waterfront facing integrated coastal management. Ocean & Coastal Management, 44:379-410.

[4] Al-But'hie, I.M., Saleh, M.A.E. (2002). Urban and industrial development planning as an approach for Saudi Arabia: the case study of Jubail and Yanbu. Habitat International, 26(1): 1-20.

[5] Alters, W.K.K. (2002). Local government and the decentralization of urban regeneration policies in the Netherlands. Urban Studies, 39(8):1439-1452.

[6] Chen, K.Y. (2007). System Planning for the Development of Keelung Port City Waterfront, Master's thesis, Graduate Institute of Harbor and River Engineering, National Taiwan Ocean University, Keelung.

[7] Hoyle, B. (2000b). Global and local change on the port-city waterfront. The Geographical Review, 90(3):395-417.

[8] Gladstone, D.L. (1998). Tourism urbanization in the United States. Urban Affairs Review, 34(1):3-27.

[9] Huang W., Corbett J.J., & Jin D. (2015). Regional economic and environmental analysis as a decision support for marine spatial planning in Xiamen. Marine Policy, 51:555-562.

[10] Gordon, D.L.A. (1997). Managing the changing political environment in urban waterfront redevelopment. Urban Studies, 34(1):61-83.

[11] Cuesta Camacho, D.E. (1998). Environmental injustices, political struggles: race, class and the environment. Duke University Press, Michigan United States.

[12] Gordon, D.L.A. (1999). Implementing urban waterfront redevelopment in an historic context: a case study of The Boston naval shipyard. Ocean & Coastal Management, 42(10-11):909-913.

[13] Flood S., Schechtman J. (2014). The rise of resilience: Evolution of a new concept in coastal planning in Ireland and the US, Ocean & Coastal Management, 102:19-31.

[14] Korbee D., Mol A.P.J. & Tatenhove J.P.M.V. (2015). Ecological considerations in constructing marine infrastructure: The Falmouth cruise terminal development, Jamaica. Marine Policy, 56:23-32.

[15] Baird, A. (1999). Analysis of private seaport development: the port of Felixstowe. Transport Policy, 6(2):109-122.

[16] Dovey, K., Sandercock, L. (2002). Hype and hope: Imagine Melbourne's docklands. Cities, 6(1):83-101.

[17] Keith, B., Ron, G. & Ian, S. (2002). Testing governance: partnerships, planning and conflict in waterfront regeneration. Urban Studies, 39(10):1757-1775.

[18] Keith, J. (2004). Waterfront redevelopment: a critical discourse analysis of the policy-making process within the Chatham maritime project. Urban Studies, 41(4):817-832.

[19] Chang, T.C., Feng, H., Peng, C.C. (2001). Design and Development of Urban Waterfront, Tong Ji University Press, Shanghai.

[20] Malone, P. (1999). City, capital and water. Tsang Hsin Press, Taipei.

[21] Batley, R. (1996). Public-private relationship and performance in service provision. Urban Studies, 33(4/5):723-751.

[22] Huang, W.C., Kao, S.K. (2014). Public-private partnerships during waterfront development process: The example of the world exposition, Ocean & Coastal Management, 92:28-39.

[23] Breen, A., Rigby, D. (1985). Caution: Working Waterfront - the Impact of Change on Marine Enterprise. The Waterfront Press, Washington.

[24] Vallega, A. (2001b). Ocean governance in post-modern society: a geographical perspective. Marine Policy, 25:399-414.

[25] Toffler, A. (1980). The third wave. Bantam Press, New York.

[26] Huang, W.C., Chen, C.H., Kao, S.K. & Chen, K.Y. (2011). The concept of diverse developments in port cities, Ocean & Coastal Management, 54:381-390.

[27] Chen, C.H. (2011). A Study on the Development Concept and Strategies of Waterfront, PhD thesis, Graduate Institute of Harbor and River Engineering, National Taiwan Ocean University, Keelung.

[28] Gleave, M.B. (1997). Port activities and the spatial structure of cities: the case of Freetown, Sierra Leone. Journal of Transport Geography, 5(4):257-275.

[29] Kojima H., Kubo T., Kinoshita A. (2013). Integrated coastal management as a tool for local governance of coastal resources: A case study of Munakata coastal zone. Ocean & Coastal Management, 81:66-76.

[30] Huang, K.Y., Chu, H.Y. (2003). Theory and Planning Method of Ecological City, Science Press, Beijing.

[31] Church, A. (1988). Demand-led planning, the inner-city crisis and the labour market: London Docklands evaluated In: Hoyle, B.S., Pinder & MS Husain, D.A. (Eds.), Revitalising the Waterfront: International Dimensions of Dockland Redevelopment. Belhaven Press, Great Britain, pp.199-221.

[32] Desfor, G. (2004). Flexible urban governance: the case of Copenhagen's recent waterfront development. European Planning Studies, 12(4):479-496.

[33] Krausse, G.H. (1995). Tourism and waterfront renewal: assessing residential perception in Newport, Rhode Island, USA. Ocean & Coastal Management, 26(3):179-203.

[34] Brett, E.A. (2000). Participation and accountability in development management. The Journal of Development Studies, 40(2):1-29.

[35] Ryan, C., Cooper, C. (2004). Residents' perceptions of tourism development: the case of Raglan, New Zealand. Tourism Review International, 8(1):1-15.

[36] Bassett, K., Griffiths, R. & Smith, I. (2002). Testing governance: partnerships, planning and conflict in waterfront regeneration. Urban Studies, 39(10): 1757-1775.

[37] Brusbett, K. (2003). , "People and government traveling together" community organization urban planning and the politics of post-war reconstruction in Toronto 1943—1953. Urban History Review, 27(2):44-59.

[38] Hoyle, B. (2000a). Confrontation, consultation, cooperation? Community groups and urban change in Canadian port-city waterfronts. Canadian Geographer, 44(3):228-243.

[39] Tsai J.L. (1999). Study on the economic impact of Keelung Port container import-export industry on Keelung City. Master's thesis, Graduate Institute of Traffic and Transpotation, National Chiao Tung University, Keelung.

[40] Keelung City Government. (2001a). The History of Keelung Fishery Development. Keelung City Government, Keelung.

[41] Chen, W.S. (2001). Research on Taipei Port Development Strategy Evaluation, Master's thesis, Graduate Institute of Harbor and River Engineering, National Taiwan Ocean University, Keelung.

[42] Keelung City Government. (2001b). Renewal Program of Central Business District of Keelung City. Keelung City Government, Keelung.

[43] Keelung City Government. (1998). The Report on the Overall Review of the Enlargement and Alteration of Major Program of Keelung City. Keelung City Government, Keelung.

[44] Keelung City Government. (2000). The Comprehensive Development Program of Keelung City. Keelung City Government, Keelung.

[45] Lin, W.H. (2006). Research on Fishing Port Multifunctional Development Strategies - A Case Study in Keelung, Master's thesis, Graduate Institute of Harbor and River Engineering, National Taiwan Ocean University, Keelung.

[46] Huang, W.C., Chen, C.H., Kao, S.K. (2005a). The resource integration and development strategy of Keelung waterfront. In: Keelung Waterfront Development Symposium. Graduate Institute of Harbor and River Engineering, National Ocean University, Keelung.

[47] Huang, W.C., Chen, C.H., Kao, S.K. (2005b). The fishing port transformation and comprehensive plan of Keelung port city waterfront. In: Fishery Industry Development and Fishing Port Development Forum. Graduate Institute of Building and Planning, National Taiwan University/Graduate Institute of Harbor and River Engineering, National Ocean University, Keelung.

[48] Keelung City Government. (2004). National Door Shanhaikwan Program Integrated Program of Landscapes around Keelung City Bisha Fishing Port. Keelung City Government, Keelung.

[49] Keelung City Government. (2005). The Traffic Improvement Planning Surrounding Keelung National Museum of Marine Science and Technology. Keelung City Government, Keelung.

19

Soil Fertility Status and Productivity Trends Along a Toposequence

Siraj Beshir[1, *], Mulugeta Lemeneh[2], Endalkachew Kissi[3]

[1]Madawalabu University, Department of Plant Science, Bale Robe, Ethiopia
[2]International Livestock Research Institute (ILRI), Addis Ababa, Ethiopia
[3]Jimma University, Department of Natural Resource Management, Jimma, Ethiopia

Email address:
siraj.beshir@mwu.edu.et (S. Beshir)

Abstract: The Impact of landscape positions and land use type on properties as well as land productivity are commonly observed in Ethiopia. The study was therefore conducted to quantify soil fertility and productivity trends under three land uses along the gradient (lower, middle and upper slope positions) of Nadda Assendabo watershed, Southwest Ethiopia. Thus field survey was covered a total of 54 composite soil samples using split plots (for soil physical properties) and split-split factorial arrangements with randomized complete block design from woodlot, grazing and crop land following topographic positions then different parameters were analyzed. The highest mean soil moisture content (56.68%) and porosity (71.09%) were observed for woodlot at the lower slope while the highest soil bulk density (1.74 g/cm^3) was observed for soil crop land at the upper slope where as the lowest (0.94 g/cm^3) was for woodlot at the lower slope. Sand, silt and clay fractions were significantly affected by interaction effects of slope and land uses ($P \leq 0.01$). Soil pH, electrical conductivity, organic carbon, cation exchange capacity, total nitrogen, available phosphorus, base saturation percentage and exchangeable basic cations for all land uses were increased down slope while they were affected significantly by slope steepness, land use and soil depth ($P \leq 0.01$). Therefore, electrical conductivity at middle and upper slope in grazing land and woodlot; total nitrogen, available phosphorus in crop land at lower slope was significantly different ($P \leq 0.05$) due to interaction effect. Furthermore, above ground biomass and grain yields of sorghum, maize and teff were also calculated on farm fields using split plots arrangements. Thus both yields were significantly affected by interaction effect of slope and crop type. Finally, based on the results woodlot, grazing and crop land were recommended for upper, middle and lower slopes respectively.

Keywords: Toposequence, Biomass and Grain Yield, Productivity Indicators

1. Introduction

Land is an essential resource particularly for agrarian society like Ethiopia. It is also a scarce resource with a carrying capacity that can be stretched only to a limited extent with the help of technology. As population grows rapidly, imbalance between supply for land resource and demand for it emerges. The consequences of imbalanced relationship are brought about either by natural or human factors such as deforestation, land degradation, soil erosion and conflict over land uses as a result of limited care of the resources and lack of awareness of the long-term effects (Buzuayehu *et al.,* 2002). Some of the conservation measures like terracing, bunds and counter cultivation could be used for sustaining land productivity (Desta, 2003). In addition, small holder farmers use woodlot as means for reducing soil loss and resolving boundary conflict that may be raised due to land fragmentation into patch (Chris *et al.,* 2004). The trees planted for this purpose could be used for construction, fuel wood and reducing pressure on utilizing crop residues (Bezuayehu *et al.,* 2002).

A decline in soil fertility is becoming one of the major challenges for establishing sustainable agriculture in sub Saharan African countries (Muchena, 2008) This is enhanced by changes in land use, alteration of the ecosystem and susceptibility of the land to external pressure which significantly affect soil physical, chemical, and biological properties. Due to these trends agricultural productivity per unit of land is declining through time and food production could not keep pace with population growth (Roy *et al.,* 2003).

Mostly, soil fertility parameters are influenced by rugged topography, steep slopes and land mosaic which results in exacerbating soil erosion rate through its morphological characteristics (Azene, 1997; Demel, 2001). On sloping lands, more than one-half of the soil particles that are dislodged by raindrops during rainfall are carried downhill (Brubaker *et al.,* 1993; Brady and Weil, 2002). Erosion increases dramatically because the increased angle facilitates water flow and soil movement. It is generally accepted that an increase in slope will increase soil erosion by water and landslide because they lead to an increase in overland flow volume and velocity. Runoff on lower slopes flows slowly and forms a water layer quickly deep enough to act as surface mulch (Wischmeier and Smith, 1978).

Beside, Ethiopia is a country where more than 85% of its labour force is engaged in agriculture (CSA, 2010). So working in the agriculture sector to improve the level of its productivity is a key issue to meet the requirements of the continued population growth. According to (FAO, 2010, as cited in Taffesse, 2011), the country population density per land area was 82.95 people per square kilometer and this resulted in the fragmentation of land into patch dynamics, which influences nutrient flow and further land productivity. In sight of this, Southwestern Ethiopia has a strong potential for increased agricultural productivity due to high intense of woodlot which its tree species has less allellopathic effect on fragmented landscape (JZARDS, 2011). However, environmental challenges, mainly deforestation and the consequential soil erosion and nutrient depletion, have profound effects, and threatened the livelihoods of the farmers.

Poor land management practices coupled with the rugged topography, erosive rainfall and more intense of gully and rill formation in Omo Nadda Woreda (Mainly Nadda Assendabo watershed) affected the agricultural practices in the area (SLMP, 2010). As a consequence, an increase in sediment load in down slope and reduction of agricultural productivity is becoming a challenge for small holder farmers. Some research have been done in the study area covered the area of land use/land cover dynamics and its impact (Amanuel, 2011); comparative analysis of soil nutrient balance at farm level (Abebayehu *et al.,* 2011) reported that land use change and absence of nutrient balance brought low soil productivity. However, no study has been done on the land use planning, soil fertility and land productivity following the gradients. Therefore, the objective of this study was designed to investigate impact of topographic position and land use on soil fertility and its productivity in Omo Nadda woreda specifically Nadda Assendabo watershed, Southwest Ethiopia.

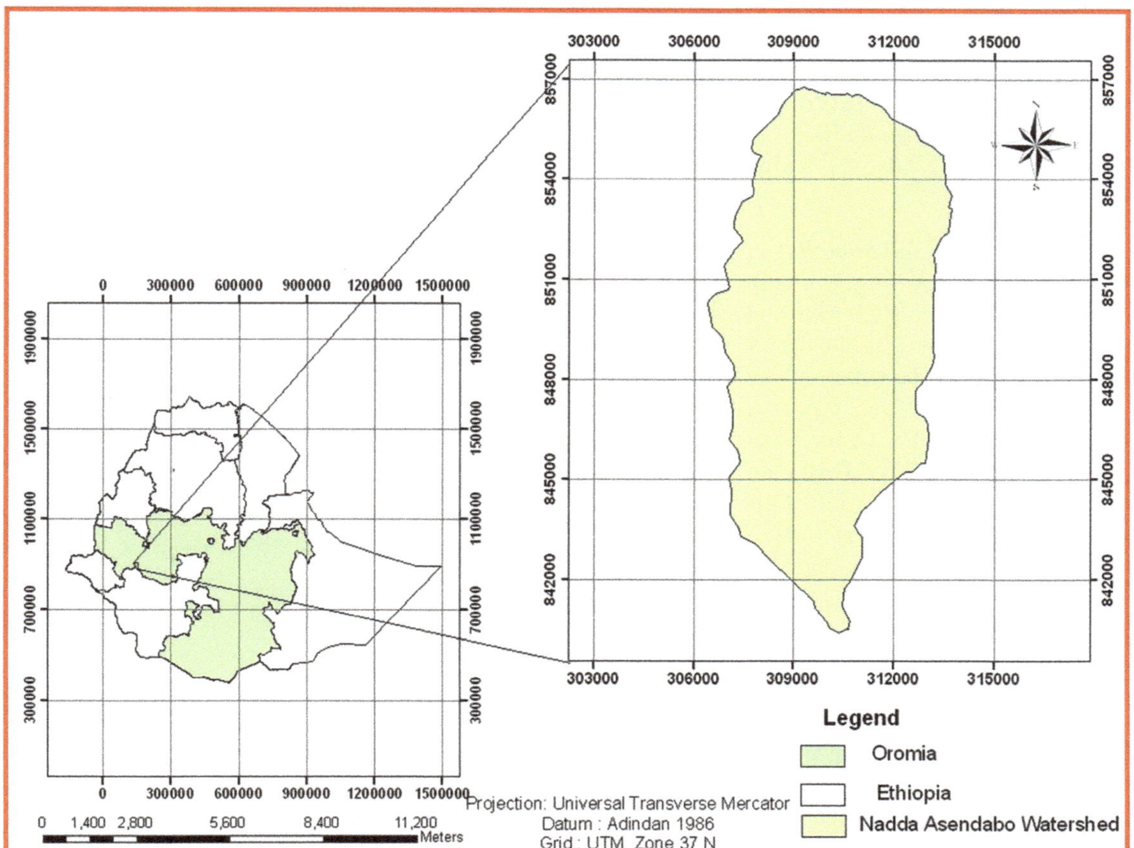

Figure 1. Location map of the study area.

2. Materials and Methods

2.1. Description of the Study Area

Nadda Asandabo watershed is located in Omo Nada Woreda, Jimma zone of Oromia Regional State. It is located close to Gilgel Gibe dam-I at about 260 km South West of Addis Ababa. Geographically it is located between 7^{0} 36' 00.87" - 7^{0} 41' 05.72" N latitude and 37^{0} 16' 55.88" - 37^{0} 14' 40.73" E longitude. It covers an area of 8,012 ha. This study was mainly focused on four kebels; Doyo yaya, Biso Gombo, Goro seden, Guddeta bula. The area is characterized by gentle, flat and undulating topography with the altitude ranging from 1650 – 2200 m a.s.l. According to Van Ranst *et al.,* (2010), the major reference soil groups in the Gilgel Gibe catchment are Nitisols, Acrisols, Ferralsols, Vertisols and Planosols. All these soils have a significant clay Bt level of dark reddish brown colour. They generally come from the alteration of the basaltic basement and are found on the hills and ridges. According to the surrounding elder's descriptions, 50 years ago the area was covered by indigenous trees such as *Poducarpus* and *Juniperus spps*. A later agricultural land expansion has resulted in destruction of forest trees and treats to wildlife (SLMP, 2009).

2.2. Methodology

This study was carried out immediately after cropping season specifically for soil sampling from the field in autumn and winter of 2013/2014. This was because of fall is the ideal time of a year to routine soil sampling of crop land. Additionally, cover crops were collected so that humidity standardized for surrounding environment. Furthermore, at this time soil and organic material were separable.

2.2.1. Experimental Design and Soil Sampling

Randomized complete block design with split-plot factorial conducted for soil moisture, bulk density and porosity whereas split-split plots for the left soil fertility parameters. The main plots, subplot and sub subplot were slope ranges, land uses and soil depth respectively. However, split plots arrangement was followed for surface soil moisture, bulk density and porosity. Prior to soil sampling, the landscape was divided into three gradients using GPS and clinometers namely; lower (0-15%), middle (15-25%) and upper slopes (≥25%) followed by selection of three land use types (i.e crop land, grazing land and woodlot) from each gradient. Each treatment was then replicated three times and the soil samples were collected from the depth of 0-20 and 20-40cm using auger. Five samples (in zigzag pattern) were collected to form composite soil samples (54), which were transported to the laboratory for further processing. Plant roots and shoots were handpicked and discarded. Then soil samples were air dried, mixed well before analyses.

2.2.2. Procedures for Soil Laboratory Analysis

(i). Soil Physical Properties

Soil moisture at the sampling time was determined by oven dried at 105°c for 24 hrs. Then it was expressed by weight as the ratio of the mass of water present to the dry weight of the soil sample as described in Gravimetric method and bulk density was determined by core method (Blake, 1965) after drying a defined volume of soil in an oven at 105 °C for 24 hours. Finally, soil pore space was calculated from measured bulk density values using the following equation:

Soil porosity (PS) $= 1 - \frac{Db}{Dp}$ whereas, Db=Bulk density; Dp=Particle density

For soil texture, initially the soil sample was pretreated with H_2O_2 (30%) to remove any organic material and sodium hexametaphosphate to disperse complex Ca^{++}, Al^{3+}, Fe^{3+}, and other cations that bind clay and silt particles into aggregates. The density of the soil suspension was determined with a hydrometer method (Day, 1965) to read in grams of solids per liter after the sand settles out and again after the silt settles. A correction was made for the density and temperature of the dispersing solution. Percentage of particle size classes were identified according to the USDA textural triangle.

(ii). Soil Chemical Properties

The pH was measured in H_2O, in ratio of soil to water (1:2.5), using pH meter whereas Ec measured by conductivity meter using suspension of 1:5 soil: water ratio. Organic carbon was determined by following Walkley and Black method (1934). Then, percent soil organic matter was calculated by multiplying the percent organic carbon by a factor of 1.724. Total Nitrogen (TN) was determined by the micro-Kjeldahl digestion, distillation and titration method (Bremmer and Mulvane, 1982). Available phosphorus (AP) was determined using the standard Bray II. Cation exchange capacity was determined at soil pH level of 7 after displacement by using 1N ammonium acetate method in which it was, thereafter, estimated titrimetrically by distillation of ammonium that was displaced by sodium (Chapman, 1965). Total exchangeable bases were determined after leaching the soils with ammonium acetate (Thomas, 1990). Amounts of Ca^{2+} and Mg^{2+} in the leachate were analyzed by EDTA and K^{+} and Na^{+} were analyzed using flame photometrically. Percent base saturation was calculated by dividing the sum of the base forming cations (Ca, Mg, Na and K) by the CEC of the soil and multiplying by 100.

2.2.3. Biomass and Grain Yields Assessment

Three crop lands (Maize (*Zea mays*), Teff (*Eragrostis tef*) and Sorghum (*Sorghum bicolor*)) were selected for this study since they are commonly practiced in the watershed. These lands and varieties were selected following discussion with farmers, elders and experts. A plot of 2 m x 2 m was temporary established using randomized complete block design with split plot factorial arrangement from the whole selected crop land considering its representativeness. Then for increasing precision of observation each crop land were replicated five times in anticipated ranges of elevation. Finally, the above ground biomass and yield was measured using spring balance in unit of kg after all grain yield was converted to kg/ha and biomass to t/ha for result presentation.

2.3. Statistical Analysis

Prior to statistical analysis treatments were arranged in factorial randomized complete block design format with slope steepness, land use and soil depth as factors. Three way analysis of variance (ANOVA) was performed to assess the significance of differences in soil parameters between slope steepness, land uses and soil depth, using the general linear model (GLM) procedure of the statistical analysis system (SAS Institute, 1996). A post hoc separation of means was done by LSD test after main effects were found significant at $P<0.05$.

3. Results and Discussion

3.1. Soil Physical and Chemical Properties Along the Toposequence

3.1.1. Surface Soil Bulk Density, Moisture Contents and Porosity

The maximum mean of soil moisture content at the sampling time was observed in woodlot, whereas the minimum in crop land. Therefore, interaction effect of slope and land use significantly affected ($P\leq0.01$) response of soil moisture and its mean comparison was significantly different ($P\leq0.05$) in crop land at the middle slope, crop and grazing land at the upper slope (Table 1). Woodlot had high leaf cover while others land uses had little leaf cover. This can result in the larger differences in infiltration rate, drainage and evapotranspiration between different land uses. In addition, this was due to the flow of water is low at the lower slope as there is no elevation differences and, thus, it gets time to infiltrate. It is agrees with the findings of Yang (2001). On other hand, the highest mean of soil porosity in woodlot might be come from high organic matter and finer soil particle in this land use type (Table 1 and 2).

Therefore, interaction effect of slope, land uses and soil depths significantly affected ($P\leq0.01$) bulk density and its mean comparison was significantly different($P\leq0.05$) for all cases except in crop land at the middle slope, grazing land and woodlot at the upper slope (Table 1). Furthermore, soil total porosity affected by these interaction effect in grazing land and woodlot at the lower slope and crop land at the upper slope ($P\leq0.01$) (Table 1).

Table 1. *Interaction effects of slope and land use on soil MC, BD and porosity.*

Slope (%)	Land use	MC	BD	Porosity
	CL	29.75c	1.15f	56.6c
0-15	GL	42.11b	1.06g	60b
	Wl	56.68a	0.94h	71.09a
	CL	22.35e	1.52c	40.18f
15-25	GL	28.35d	1.35d	48.89e
	Wl	39.96b	1.21e	55.27dc
	CL	17.37g	1.74a	29.95g
≥25	GL	20.72f	1.55bc	40.16f
	Wl	29.2cd	1.59b	46.31e
LSD (0.05)		1.21	0.05	4.8
CV (%)		8.48	7.28	6.77

GL=grazing land, CL=crop land, Wl=woodlot, MC=soil moisture content BD=bulk density; Means within a column followed by same letters in superscripts are not significantly different from each other at P=0.05.

Table 2. *Interaction effects of slope, land use and soil depth on soil particle.*

Slope (%)	LUT	Soil Depth	Sand	Silt	clay
	Cl	0-20	13.37m	45.3a	41.33f
		20-40	11.7o	39.3d	49c
0-15	Gl	0-20	15.16l	41.17c	43.67ed
		20-40	14.67m	26l	59.33a
	Wl	0-20	10.33p	39.3d	50.37b
		20-40	15.67k	24.33m	60a
	Cl	0-20	28e	41c	31j
		20-40	13n	42.33b	44.67ed
15-15	Gl	0-20	28.34d	36.33f	35.33h
		20-40	19j	37.33e	43.67ed
	Wl	0-20	28.67c	27k	44.33ed
		20-40	24.67i	30i	45.33d
	Cl	0-20	39.36a	31.96h	28.68k
		20-40	27f	35.33g	37.67g
≥25	Gl	0-20	26.67g	29.33j	33.75i
		20-40	27f	37.67e	35.33h
	Wl	0-20	35.67b	25.67l	38.66g
		20-40	26h	32.33h	41.67f
LSD (0.05)			0.24	0.41	1.02
CV (%)			11.32	9.45	3.28

Means within a column followed by same letters in superscripts are not significantly different

3.1.2. Soil Particle Distribution Along the Toposequence

Table 2 clearly shows that sand, silt and clay fractions differed significantly along three slopes, land use types and both soil depths ($P \leq 0.01$). Similarly, the sand and clay fraction strongly influenced by interaction effect of slope steepness, land use types and soil depth ($P \leq 0.01$) (Table 2). As a result, mean comparison of sand soil particle in woodlot at the lower slope (surface soil), crop land and grazing land (subsurface soil) at middle slope, crop land (surface), grazing land for both depths and woodlot (subsurface) at the upper slope was significantly different ($P \leq 0.05$). The values of silt/clay ratios were 0.63, 0.68 and 0.72 for crop land, grazing land and woodlot soils respectively (Table 2). The decreasing clay content can be used as an indicator for reducing degree of weathering. This finding agrees with the works of Van wambeke (1962) who used silt to clay ratio to estimate the degree of weathering of soil pedon, and postulate that the lower the ratio, the higher the degree of weathering. By increasing weathering, the silt fraction changed into clay fraction so that clay content increased, thus the silt/clay ratio was to be lower. Furthermore, results of this study were showed that the ratio of clay contents in soil particle distribution for all land uses increased from surface to sub-surface along the gradient. This result agree with the works of Ashanafi *et al.* (2010); Esteban *et al.* (2000) and Sheleme (2011) who reported that the texture of the subsurface horizons became finer with soil depth, due to migration of clay from surface to lower horizons.

Silt fraction affected significantly by interaction of land use and soil depth ($P \leq 0.05$), slope and land use and slope and soil depth ($P \leq 0.01$) (Table 2) but mean comparison in grazing land and woodlot, surface and subsurface soil depth was not significantly different. In addition, clay fraction in crop land (surface soil), grazing land and woodlot (subsurface soil) at the lower slope, crop and grazing land (surface soil) at the middle slope as well as all land uses at the upper slope was significantly different ($P \leq 0.05$). This result agreed with the findings of Van wambeke (1962).

In addition it was supported by the works of (Ogeh and Ukodo, 2012) which revealed that sand fraction generally dominated the soils along the toposequence. Within the horizons, the sand content decreased with depth in all the profiles, silt content increased with depth in all the profile.

3.1.3. Soil pH, EC, SOM, TN and AP, Exchangeable Bases, BS and CEC

Results of the present studies showed that soil pH increased down slope and with soil depth (Table 3). Statistically, pH was significantly affected by interaction of slope steepness, land use and soil depth in crop land, grazing land and woodlot (subsurface soil) grazing land (surface soil) at the upper slope (Table 3). Relatively lower pH values in the soil of the crop fields and grazing land, as compared to those under woodlot, might be due to depletion of basic cations by the harvested crop biomass, over grazing and leaching. Gebeyaw (2007) has also reported that a lower pH value in cultivated land was attributed to a high rate of organic matter oxidation. This is important to produce organic acids and provide H^+ to the soil solution, and thereby reduces soil pH values. This explanation was also partially supported by Tana (1996) and Butros *et al.* (2010) who reported that soil pH was increased with soil depth along the gradient due to carbonate content of the soil and high intensity of rain fall.

Table 3. *Interaction effects of slope, land use and soil depth on soil physico-chemical properties.*

Slope (%)	Land use	Soil depth (cm)	pH	Ec (dS-1m)	OM(%)	TN (%)	AP (ppm)	Ca	Mg	Na	K	CEC	BSP (%)
								cmol+Kg- of soil					
0-15	Cl	0-20	6.51f	0.042fgh	6.25cbd	0.39ed	4.56efg	12.61jik	5.04de	0.34fg	0.56eghf	25.33f	72.94fe
		20-40	6.6e	0.05ef	5.47cebd	0.32fheg	4.14g	16.07ef	5.67dc	0.44de	0.69ed	34.67c	66.57ijk
	Gl	0-20	6.78d	0.066d	8.98a	0.49cb	7.61a	16.93ecd	6.6c	0.52dc	0.75cd	33.44dc	74.74dce
		20-40	6.85c	0.079c	6.87b	0.4ced	6.28b	18.4bcd	8.96b	0.64ba	0.92cb	42.18b	68.67hgi
	Wl	0-20	6.91b	0.204b	9.63a	0.65a	5.29cd	19.9b	9.39b	0.61b	0.89cb	42.29b	73.16fe
		20-40	6.98a	0.355a	8.86a	0.5b	4.73defg	21.31a	12.23a	0.71a	1.75a	47.09a	76.81bc
15-25	Cl	0-20	5.53l	0.025jkl	4.62ef	0.23hi	2.71hij	13.19jih	3.1gf	0.24hi	0.43ih	17.84g	96.21a
		20-40	5.87j	0.028ijk	4.83efd	0.29fhig	2.37hij	14.23ihfg	4.13fe	0.3hg	0.51ghf	25.05f	77.44bc
	Gl	0-20	5.73k	0.037ghi	5.79cebd	0.33feg	4.92def	14.15jihfg	5.27dce	0.34fg	0.55eghf	25.93fe	78.79b
		20-40	5.93i	0.046fg	6.45cb	0.29fhig	5.89bc	15.54efg	5.82dc	0.48de	0.64egdf	30.28de	74.67dce
	Wl	0-20	6.2h	0.046fg	5.81cebd	0.37fe	4.17g	16.8ed	6.18dc	0.44de	0.68edf	33.3dc	71.68feg
		20-40	6.28g	0.061de	6.59b	0.47cbd	4.9def	19.04bc	8.7b	0.58bc	1.01b	39.39b	74.46dce
≥25	Cl	0-20	5s	0.014l	2.59g	0.13j	2.3hij	7.41k	2.49g	0.14j	0.18j	16.57h	61.68m
		20-40	5.13r	0.019kl	3.29gf	0.21ji	1.45jk	9.93jk	1.82g	0.18ji	0.26ij	18.45g	66.1o
	Gl	0-20	5.17qr	0.031hij	3.29gf	0.21ji	2.06jk	12.36jihg	2.94gf	0.23hi	0.29ij	21.47g	73.68ij
		20-40	5.23np	0.035ghij	4.54ef	0.27hig	3.26h	14.14jihfg	4.27fe	0.33fg	0.38ih	26.03fe	73.59lk
	Wl	0-20	5.2opq	0.037ghi	4.67ef	0.26hig	2.57hij	14.96ehfg	5.09de	0.28hg	0.39ih	29.67f	69.85n
		20-40	5.4m	0.043fg	4.94ced	0.29fhig	2.82hi	15.89ef	6.02dc	0.4fe	0.48gh	34.3dc	66.64ij
LSD (0.05)			0.05	0.011	1.56	0.09	0.68	2.22	1.36	0.08	0.18	4.37	3.04
CV (%)			3.1	8.1	13.82	10.95	13.55	6.02	13.75	8.83	14.77	4.86	7.56

Land use are the same as Table 3; BSP= Base Saturation Percentage; Means within a column followed by same letters in superscripts are not significantly different from each other at P=0.05

Interaction effect of slope and land use on Ec was significantly different in grazing land (surface) and woodlot (surface and subsurface) at the lower slope comparatively with others ($P \leq 0.05$) (Table 3). Following this scenario, Ec were higher in the woodlot with a mean of 0.355 $dS^{-1}m$, whereas the lowest 0.014 $dS^{-1}m$ in the crop land at the upper slope (Table 3). Generally, Ec increased from surface to subsurface for all land uses, following sequential landscape gradients (Table 3). This result was in agreement with the findings of Doerge *et al.* (1999) and Butros *et al.* (2010) indicating greater clay particle percentage, loss of Calcium (Ca), Magnesium (Mg), Sodium (Na) and Potassium (K) containing soluble salts after deforestation and cultivation and increased porosity of the soil under these land uses resulted in reduction of soil electrical conductivity. This result also supported by the works of (Alemayehu and Sheleme, 2013) which indicated crop with high biomass and large physiological activities assists soil erodibility.

SOM in crop land at the lower slope was significantly different ($P \leq 0.05$) from middle and upper slope due to interaction effects of slope with land use and soil depth (Table 3). The lower organic matter observed in crop and grazing land comparing to woodlot could be explained by the seasonal cover in the former following over cultivation, free grazing and absence of soil nutrient balance. Removal of the surface soil rich in organic matter by soil erosion which undoubtedly could have accelerated with the removal of the plant cover is also expected to contribute to the lower organic matter contents observed in the crop and grazing land. The maintenance high levels of organic matter under woodlot are apparently attributed to the presence of vegetation and the associated high biomass and litter fall in the soils. This result was in agreement with the works of (Pan and Bhardwaj, 2013) forest has the highest carbon pool where as the agricultural crop had the lowest carbon pools in comparison.

Soil TN in woodlot at the lower slope (surface and subsurface) was significantly different ($P \leq 0.05$) from other land use type at the three slopes due to interaction of these factors (Table 3). In line with soil organic matter contents, the highest mean soil total nitrogen (0.65%) was observed in the woodlot followed by the grazing land (0.5%), where as the lowest value (0.13%) of total nitrogen was recorded in the crop lands (Table 3). Reduced input of plant residues into the soils also has contributed to the depletion of organic matter thereby enhanced rate of nitrogen in crop land. As the area receiving high mean annual rain fall, leaching could be another reason for the decline in TN in cropped fields. Nitrate ions which are not adsorbed by the negatively charged colloids dominate in most soils and, thus, move downward with drainage water and are readily leached from the soil. Removal of crop residues and subsistence agricultural system affects nutrient balance more negatively (Fitsum *et al.*, 1999).

AP was significantly different ($P \leq 0.05$) in grazing land (surface) at lower slope comparatively with land use in all slope along a toposequence due to interaction effects of slope, land use and soil depth. Reduction of AP contents could be related to degree of its fixation which occurs at low pH levels where iron and aluminum activity actually increases (Brady and Weil, 2002). According to the reports of Maja (2011) for health vegetation, increase in elevation was associated with a decrease in soil $NH4^{+}$ and $PO4^{3-}$ concentrations because the soil fungal: bacterial ratio was lowest at the lower elevations.

Exchangeable cations, Ca, Mg, Na and K were significantly different ($P \leq 0.05$) in woodlot (subsurface) at the lower slope due to interaction effects of slope, land use and soil depths (Table 3). This finding agreed with (Maja, 2011) reports, less amounts of soil erosion by water and high resistance of cations from leaching at foot slope and area covered by perennial vegetations improved level of exchangeable cations in the soil.

Cation exchange capacity was greater in woodlot land use type and at the lower slope (Table 3). This results supported by (Brady and Weil, 2002) cation exchange capacity increased with the ability of plant root to anchor soil mass and not rugged landscape which is less vulnerable to landslide due to elevation differences. Thus, this study was revealed that interaction effects of slope, land uses and soil depths along a toposequence was indicated CEC was significantly different ($P \leq 0.05$) in woodlot (subsurface) at the lower and middle slope from other values observed (Table 3)

Observed base saturation percentage in crop land at middle (surface) and upper (subsurface) slope was significantly different ($P \leq 0.05$) from others due to interaction effects of slope steepness, land use and soil depth (Table 3). This result was agreed with findings of Hikmatullah *et al.* (2003).

3.2. Crop Biomass and Grain Yield Along the Toposequence

Productivity of above ground biomass for sorghum and maize at the lower slope and teff at the upper slope was significantly different ($P \leq 0.01$) from others due to the interaction effects of slope and crop type along the toposequences. Mean grain yields were ranged from 495-1070 kg (teff), 1890-5500 kg (maize), 2235-2360 kg (sorghum) along the landscape. This result was significantly different ($P \leq 0.05$) in grain yield obtained from maize at all slope and teff yields at upper slope due to interaction effects of slope and crop type. It was agreed with works of Grabherr, 2009, as cited in Fabien, 2012).

Table 4. *Interaction effects of slope and crop type on biomass and grain yields.*

Slope (%)	Crop type	Biomass	Grain yield
	sorghum	32.95a	2250c
0-15	Teff	10.62f	1070e
	Maize	25.19b	5500a
	sorghum	19.99c	2235c
15-25	Teff	6.55g	895f
	Maize	19.27dc	3465b
	sorghum	13.7ef	2360c
≥25	Teff	4.16g	495g
	Maize	11.67ef	1890d
LSD (0.05)		3.71	139
CV (%)		19.85	13.67

Means within a column followed by same letters in superscripts are not significantly different

4. Conclusion and Recommendations

Based on this study results, most of soil nutrients significantly decreased due to soil erosion and landscape features with increasing slope steepness and unwise utilization of land. Thus higher organic matter, cation exchange capacity, soil moisture, finer soil particle, basic cation, nitrogen and phosphorus observed at foot slope and woodlot land use type. So that soil and water conservation measures like hillside trench, terracing, micro and macro basin should layout at the upper watershed side according to its suitability in the watershed. Therefore, lower slope watershed community should work cooperatively with upper and middle watershed since household in the lower watershed benefited from offsite nutrient movement to their lands. This will be happened according to this research output through allocating woodlot land use type between fragmented crops and grazing land and empowering local community through training on how they can prepare organic fertilizer from locally available materials.

References

[1] Abebayehu, A. 2010. Soil Fertility Management and Nutrient Balance in Farm Components and Socioeconomic Groups: A Case Study in Selected Sites of Gilgel Gibe Catchment. MSc Thesis, Hawassa University, Ethiopia.

[2] Alemayehu Kiflu and Sheleme Beyene, 2013. Effects of different land use systems on selected soil properties in South Ethiopia. Journal of soil science and Environmental Management. Vol.4(5), pp. 100-107.

[3] Amanuel Abate, 2011. Land Use Dynamics and Its Impact on Productivity. MSc. Thesis, Jimma University, Ethiopia. analysis. Agron. Part I, No. 9, Am. Soc. Agron. Madison, Wisconsin, USA.

[4] Ashanafi Ali, Abaynesh Esayas and Sheleme Beyene, 2010. Characterizing Soils of Delbo Wegene Watershed, Wolaita Zone, and Southwestern Ethiopia for Planning Appropriate Land Management. Journal of Soil Science and Environmental Management Vol.1 (8), pp.184-199.

[5] Azene, B., 1997. A Participatory agro-forestry approach for soil and water conservation in Ethiopia. PhD Dissertation, Wageningen Agricultural University, The Netherland.

[6] Blake, G. R., 1965. Bulk density. pp. 374-399. In: C. A. Black (ed.). Methods of soil

[7] Brady, N. C. and Weil, R. R., 2002. The nature and properties of soils. 13th ed. Prentice Hall Inc., New Jersey. 960p.

[8] Brady, N. C. and Weil, R. R., 2002. The nature and properties of soils. 13th ed. Prentice Hall Inc., New Jersey. 960p.

[9] Bremner, J. M. and Mulvaney, C. S. 1982. "Total nitrogen", In: A. L. Page, R. H. Miller and D. R. Keeny, (Eds.), Methods of Soil Analysis, American Society of Agronomy and Soil Science Society of America, Madison, pp. 1119-1123.

[10] Brubaker, S. C., Jones, A. J., Lewis, D. T. and Frank, K., 1993. Soil Properties associated with landscape positions. Soil Sci. Soc. Am. J. 57: 235-239.

[11] Butros, I. H., Awni, Y. Taimeh and Feras, M. Ziadat., 2010. Variation in soil chemical properties along toposequences in an Arid Region of the Levant.Catena 83 (2010) 34-45.

[12] Buytaert, W., J. Deckers, G. Dercon, B. D. Bierve, J. Poesen and G. Govers, 2002. impactof land use changes on the hydrological properties of Volcanic Ash Soils in South Ecuador. Soil Use and Management 18: 94-100.

[13] Buzuayehu, T., Gezahegn, A., Yigezu, A., Jabbar, M. A. and Paulos, D., 2002. Nature and causes of land degradation in the Oromiya Region: A review. Socio-economics and Policy Research Working Paper 36. ILRI (International Livestock Research Institute), Nairobi, Kenya. 82 pp. catchment of the Loess Plateau, China

[14] Central Statistical Agency (CSA), 2010. Agricultural sample survey. Statistical Bulletin 468, Octeber, 23-27, CSA, Addis Ababa, Ethiopia.

[15] Central Statistical Agency (CSA), 2010. Agricultural sample survey. Statistical Bulletin 468, Octeber, 23-27, CSA, Addis Ababa, Ethiopia.

[16] Chapman, H. D., 1965. Cation exchange capacity. In: Black et al., C. A. (eds.). Methods of

[17] Chris, J. Grose, Darren, B. Kidd, Robert M. Moreton and Susan, E. Tate., 2004. Quantifying trends in soil condition: gathering baseline data for developing targets and supporting informed land management decision making. Water and Environment, Tasamani.

[18] Day, P. R., 1965. Hydrometer method of particle size analysis. pp. 562-563. In: C. A. Black (eds.). Methods of soil analysis. Agron. Part II, No. 9. Am. Soc. Agron. degradation and restoration in the Highlands of Ethiopia: Implications for sustainable land management. Doctoral thesis. Swedish University of Agricultural Sciences, Uppsala, Sweden. 70p.

[19] Demel, T., 2001. Vegetation type and forest fire management in Ethiopia, pp 1-35. Proceedings of Round-Table Conference on Integrated Forest Fire Management in Ethiopia. September 19-21, 2001, Ministry of Agriculture, Addis Ababa.

[20] Desta, K., 2003. Farmers' perception of crop trend and soil fertility management constraints in the Kabale District, Msc Thesis, Bahir dar University, Ethiopia.

[21] Doerge, T., N. R. Kitchen and E. D. Lund, 1999. Site specific management guidelines. South Dakota State University, USA.

[22] Esteban, G., Jobba Gy and Robert, B. Jackson, 2000. The vertical distribution of soil organic carbon and its relation to climate and vegetation. Ecological Applications, 10(2), 2000, Pp. 423–436.

[23] Fabien, A. and Oliver, D. 2012. Persipective in Plant Ecology, Evolution and Systematics.

[24] Fitsum Hagos, John Pender and Nega Gebreselasie. (1999). Land Degradation in the Highlands of Tigray and Strategies for Sustainable Land Management. Addis Abeba, Ethiopia.

[25] Gebeyaw, T., 2007. Soil fertility status as influenced by different land uses in

[26] Hikmatullah, H. Subagyo and B. H. Prasetyo, 2003. Soil Properties of the Eastern Toposequence of Mount Kelimutu. Flores Island. East Nusa Tenggara and their Potential for Agricultural Use. Indonesian Journal of Agricultural Science 4(1) 2003: 1-11.

[27] Jimma Zone Agriculture and Rural Development Sector (JZARDS), 2011. Annual Report on sustainable land management strategies, Jimma, Ethiopia.

[28] Maja Sundqvist, 2011. Nitrogen and Phosphorus Dynamics Across an Elevational Gradient in aSwedish Subarctic Tundra. Doctoral Thesis, Swedish University of Agricultural Sciences, Umea.

[29] Maybar areas of south Wello Zone, north Ethiopia. M. Sc. Thesis, Haramaya University, Haramaya, Ethiopia.

[30] Muchena, F. N., 2008. "Indicators for sustainable land management in Kenya's Context". GEF Land Degradation Focal Area Indicators, ETC-East Africa. Nairobi, Kenya.

[31] Ogeh J. S. and Ukodo E., 2012. Profile Distribution of Physical and Chemical Properties in Soils of a Toposequence in Benin, Rainforest of Nigeria. Nigerian Journal of Basic and Applied Science (March, 2012), 20(1): 68-72.

[32] Pal S., Panwar P., Bhardwaj D. R., 2013. Soil quality under forest compared to other land uses in acid soil of North Western Himalaya, India Ann. For. Res. 56(1):187-198, 2013.

[33] Sheleme, B., 2011. Characterization of Soils along a Toposequence in Gununo Area, Southern Ethiopia. Journal of Science and Development 1(1) 2011, 31-41 soil analysis. Agronomy 9: 891-901. Am. Soc. Agro., Inc., Madison, Wisconsin.

[34] Sustainable land Management Programme (SLMP), 2010. Causes, consequences and solution of land degradation in Omo Nadda District: Annual report by SLMP woreda level. Nadda, Ethiopia.

[35] Taffesse, Alemayehu, S. Dorosh, Paul, A. and Sinafikeh, 2011. Crop production in Ethiopia Regional Patterns and Trends. International Food Policy Research Institute (IFPRI).

[36] Van Ranst E, Tolossa A R, Dumon M, Cornelis J-T and Deckers J (2010). On the origin of planosols – the process of ferrolysis revisited. 19th World Congress of Soil Science, Soil solutions for a Changing World, Brisbane, Australia.

[37] Van wambeke, A. R., 1962. criteria for classifying tropical soils by age. J. Soil Sci. 13.

[38] Wischmeier, W. H. and Smith, D. D., 1978. Predicting rainfall erosion losses—a guide to conservation planning. U. S. Agricultural Handbook 537. US Department of Agriculture, Washington, DC, USA. 58 pp.

[39] Wolayita zone, southern Ethiopia. M. Sc. Thesis, Hawassa University, Ethiopia.

[40] Yang Qiu, Bojie Fu, Jun Wang A., Liding Chen, 2001. Spatial variability of soil moisture content and its relation to environmental indices in a semi-arid gully.

Environmental Assessment and Monitoring Framework at the Municipality Scale

Felix Iyalomhe[1], Philipa Idogho[2]

[1]Global Change and Sustainability Research Institute, University of the Witwatersrand, Johannesburg, South Africa
[2]Department of Office Technology and Management, Auchi Polytechnic, Auchi, Edo State, Nigeria

Email address:
Iyalomhe2003@yahoo.com (Felix I.), philipaidogho@yahoo.com (Philipa I.)

Abstract: *Negative* impacts of developments on the health of humans and the environment in the Waterberg municipality of South Africa are anticipated. An understanding of such impacts demands the integration of relevant environmental monitoring and assessments (i.e. biophysical and socioeconomic) in order to examine the negative impacts before planned developments in the region. Over the years, an integrated approach to environmental monitoring has not been very successful, because of vague conceptualization and lack of clear objectives and priorities linked to the interpretation of roles and responsibilities for compliance with regulatory and management programs (e.g. South African National Environmental Management Air Quality Act No. 39 of 2004 etc.). *This* paper proposes a conceptual framework as an approach to integrate different environmental monitoring and assessments of biophysical and socioeconomic systems in the context of planned developments. It identifies relevant types of environmental monitoring and assessment and describes how these could be linked, and highlights the concept of an integrated environmental baseline not just as a basis for present and future environmental conditions, but also as a potential link between the various processes and practitioners of all monitoring and assessments; while it attempts to enhance a full understanding of changes in environmental conditions and their likely trends, drivers and impacts. *This* paper theoretically illustrate the framework using the Waterberg municipality, in order to highlight its potentials as a proactive and integrated approach to understanding changes in environmental conditions and consequent impacts on the health of humans and the environment due to anticipated changes in key environmental parameters. *Thus,* this paper argues that the framework could provide a systematic procedure to environmental monitoring and assessment, and a coherent understanding of environmental conditions before planned developments at a local scale, by informing concepts and processes that should link various monitoring initiatives and their practitioners.

Keywords: Integrated Framework, Environmental Monitoring and Assessment, Human and Environmental Health, Planned Developments, Waterberg Municipality

1. Introduction

The Waterberg municipality is described as an economic and development hub in South Africa because of its vast opportunities for industrial, agricultural and forestry developments. The present and planned coal mining fields and coal-fired power stations could have a variety of adverse as well as beneficial effects on ecological and social systems e.g. changes in the natural reserves areas (Waterberg Biosphere Reserve), ambient Air Quality, and demographic trends. Thus, the region represents an environmentally compromised area because of unsustainable trends of developments, partly due to a lack of an integrated understanding of its environmental conditions that are presently described as fragmented, incomplete and often inaccurate.

The Waterberg being a municipality in the Limpopo province possesses high reserves of mineral resources including coal, platinum and lead [27,35]. Rapid development of extensive coal mining is already taking place and is expected to increase in the next decade due to the demand for electricity. Consequently, coal mining in the region has resulted in the production of gaseous and solid air pollutants e.g. sulphur dioxide (SO_2), nitrogen dioxide (NO_2), particulate matter (PM) and lead (Pb) that are demonstrated threats to both human and environmental health. Moreover, recent understandings indicate that the area will be vulnerable to changes in ambient Air Quality and its associated impacts on humans and the environment due to unsustainable trends of developments. In view of this, the National Department of Environmental Affairs of South African under the Air Quality Act No. 39 of 2004 declared the Waterberg region a National Priority Area for air quality management. The definition of a Priority Area requires that the air quality be monitored as well as the impacts on the health of the environment and humans in order to proactively protect and enhance ambient Air Quality in South African, to prevent severe air pollution and related ecological degradation, and to ensure sustainable development [21,35]. A key aspect of the Act highlights the need for mechanisms and procedures for the establishment of national ambient air quality standards to avoid or control the effects from changes in ambient air quality on human and environmental health. To that end, the Act requires national, provincial and municipal governments to identify priority pollutants and set ambient standards with respect to these pollutants, as well as to integrate air quality planning into their integrated development plans [21].

Several environmental initiatives related to monitoring and assessments of soil, vegetation, birds, insects, and ambient air quality have been conducted in the Waterberg National Priority Areas (NPA). However, this has not led to increased public participation in the protection of air quality and improved monitoring of changes in ambient air quality, much less a complete reduction of the region's vulnerability to effects from developments on the health of humans and the environment. This is due to vague conceptualization and lack of clear objectives and priorities linked to the interpretation of roles and responsibilities for compliance with regulatory and management programs (e.g. the South African Air Quality Act No 39 of 2004). This highlights the need for an explicit integrated environmental monitoring and assessment as a benchmark to determine the trends of air pollution and possible ambient air quality standards, as well as improved knowledge of environmental conditions. Moreover, because of the Air Quality Act, numerous emission-monitoring stations have been established but there is little emphasis to date on the impacts of pollutants on human and environmental health. Thus, there is urgent need for a more integrated approach to environmental monitoring in the region in order to integrate effectively fragmented information and understanding useful for a valid knowledge of environmental conditions and/or baseline before planned developments. That will not only identify relationships and linkages, but it will also support a detailed assessment of potential biophysical and social impacts. Besides, the integration of different environmental monitoring and assessments will be premised on the definition of an interdisciplinary framework aimed at resolving the complex relationship between different knowledge domains on environmental issues. It will also promote a holistic and interactive or collaborative approach to managing interrelated environmental issues through a common objective approach [4,19,13]. However, before now, many frameworks aimed at integrated environmental assessment have apparently failed because they have not explicitly considered all relevant monitoring useful to inform environmental decision-making [23,9,30,29,14]. This results in incoherent knowledge provided by such assessments that are limited in scope, and thus undermine concerted efforts of practitioners engaged in environmental planning and management.

This paper proposes a framework as well as an approach to integrated environmental monitoring and assessment. It theoretically illustrates such to underscore its potentials for possible integration of biophysical and social impact monitoring and assessments in the context of planned developments. The paper presents four relevant types of environmental monitoring: (1) Environment monitoring to detect and document patterns of possible adverse and beneficial effects; (2) Condition monitoring to determine present conditions and attributes of ecological and human systems; (3) Trends and Drivers monitoring assessment to describe possible causal agents that could bring about changes in the status of human and ecological systems; (4) Impacts monitoring to identify potential changes in the condition of key environmental parameters under specific scenarios of trends and drivers.

This paper illustrates the framework and approach using the Waterberg NPA, and suggests that the relationship between monitoring and assessments is based on the various processes linking all assessments and engaged practitioners. It highlights the integrated environmental baseline as a means to understand the present and future environmental conditions. All assessments feed the integrated environmental baseline that links different but relevant types of weighted evidence from all monitoring and assessments. The framework presented here may help to establish a coherent understanding of environmental conditions in the Waterberg region before planned developments and thus support a proactive assessment of impacts which may occur in the future.

2. A Framework for Environmental Monitoring and Assessment

The framework (Fig.1) aims to integrate the implementation of different but related environmental monitoring and assessments by highlighting common concepts and understanding of processes. In particular, it provides a mechanism for practitioners using various environmental approaches to monitoring in order to link them conceptually. The framework theoretically analyzes environmental monitoring in the context of changes in key environmental parameters expected to occur with planned development, and is organized into two broad categories: 'Problem identification' and 'Problem characterization'. It conceptually defines relevant monitoring and their relationships (via processes) within these categories to inform a better understanding of environmental assessment and management and its complexities. The categories comprise four different environmental initiatives that are related to monitoring of trends and drivers, environmental condition and prediction. For example, changes in the biophysical system due to changes in rainfall and temperature regimes, and atmospheric composition may result in a broader range of socio-ecological consequences, such as increased deposition of acidic compounds or an increase in airborne particulates. Thus, the framework aims to incorporate relevant data from different interdisciplinary research initiatives into an integrated platform that will contribute to enhanced knowledge and a coherent understanding of potential impacts on the health of the environment and humans.

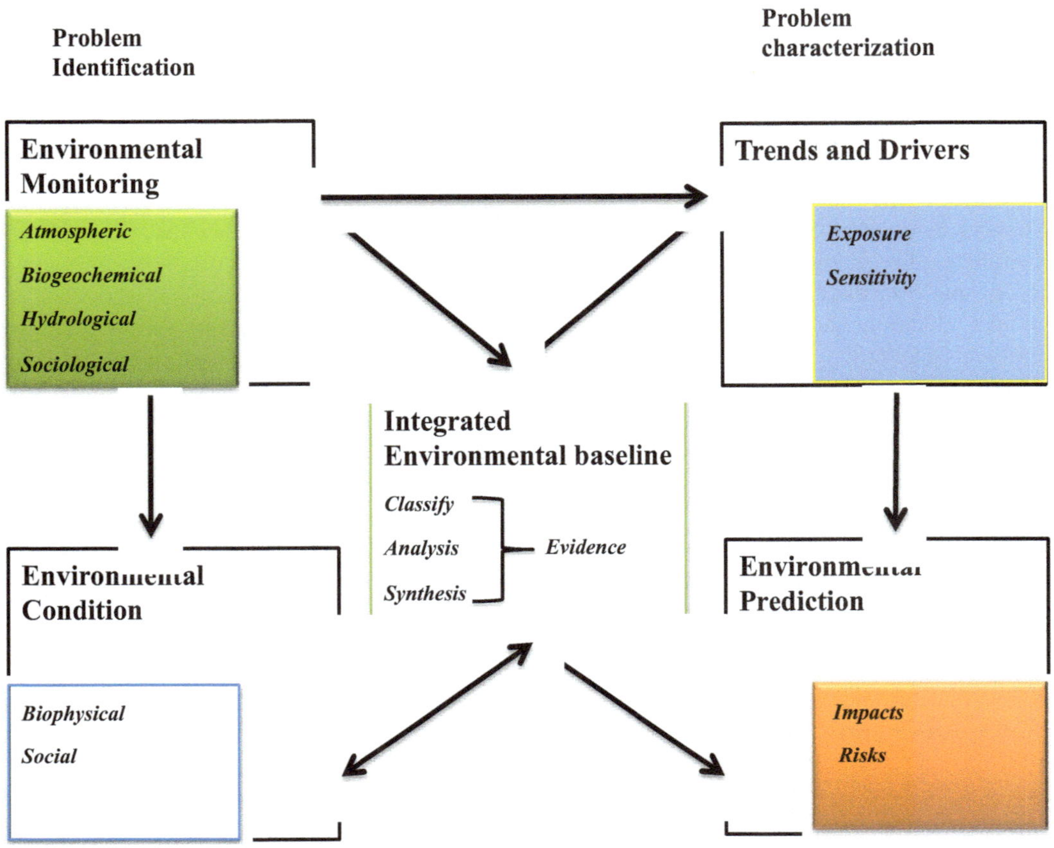

Fig. 1. A conceptual framework for integrated biophysical and socioeconomic monitoring draws on the work of these studies [31,28,21,9]. Environmental monitoring, conditions, trends and drivers, and predictions feed the integrated baseline through varying but related processes of monitoring and assessments.

3. Framework Approach for Environmental Monitoring and Assessments

This consists of categories that attempt to underscore relevant monitoring and assessments needed for potential regional environmental problems. The categories are therefore helpful in understanding how key environmental parameters have responded to historical pressures related to current developments and how these might likely respond to planned developments. Hence the framework, through these categories, establishes the changes in key environmental parameters and uses this as a basis to determine possible changes in environmental conditions and consequent effects on socio-ecological systems. Accordingly, it identifies potential target areas for monitoring of a wide range of key environmental parameters, and considers relevant outcomes in order to analyze a wide range of activities and their likely effects on natural and human systems through changes in the biophysical and social aspects of the environment. In particular, it considers relevant outcomes from the trends and drivers, and predictive monitoring assessments to carry out detailed analyses of likely effects through the modeling of relevant features.

3.1. Environmental Baseline

According to the framework in Fig. 1, integrated baseline is presupposed to support adequate evaluation of biophysical and social changes by means of aggregating relevant weighted evidence from all monitoring and assessments in order to establish a coherent understanding of the conditions of the health of the environment and humans before planned developments.

The integrated baseline is defined by weighing multiple types and pieces of weighted evidence by means of the 'weight of evidence' approach to classify, analyze and synthesize relevant evidence from the defined monitoring [3,5,32,9]. Thus, it includes qualitative and quantitative approaches to characterizing the evidence from all monitoring and associated uncertainties. This does not just reflect relevant pieces of evidence produced by individual monitoring as a weighted joint-body of evidence, but also highlights the inherent relationships between the various processes of each monitoring and their links to the joint-body of evidence, which is useful for the integrated baseline. It underscores the significance of processes of classifying, analyzing and synthesizing different bodies of evidence in each monitoring.

First, the classifying stage combines several joint-bodies of evidence from different monitoring according to criteria-guided judgment and the quality of the process used to derive the body of evidence, which is inherently flexible and transparent. Thus, it categorizes the body of evidence into significant groups that share common qualities and attributes [31]. Next, the analyzing stage evaluates the categorized, weighted body of evidence from all monitoring assessments based on specific logic and assumptions, to define different weighted joint-bodies of evidence and the relevance of each to the integrated baseline without the unnecessary repetition of evidence. Lastly, the synthesizing stage aggregates differently weighted joint-bodies of evidence in the context of associated uncertainties to define the integrated baseline in which weights express the overall quality and relevance of each body of evidence to the baseline outputs.

3.2. Environmental Monitoring

Environmental monitoring refers to the long-term, standardized measurement and observation of relevant physical, chemical, and/or biological variables designed to detect and document patterns of possible environmental change [15,2,8,37]. It involves criteria with which to identify monitoring sites and to generate a historical database, and uses this to address a particular problem (single-objective monitoring) or several problems (multi-objective monitoring). However, environmental monitoring is often limited by time and resource constraints, and in particular, a lack of clear and precise objectives that are also related to limited resources and time [1]. Despite this, it has proved to be indispensable to environmental impact assessments.

Within the framework in Fig. 1, environmental monitoring and assessments have multi-objectives because they include the observations of chemical, bio-accumulative, biological, health and ecosystem conditions etc., [33] as parallel but related activities aimed at the identification of changes within different key parameters of the environment. For example, in socioeconomic assessments, environmental monitoring can be aimed at exploring diverse social changes (e.g. perceptions and attitudes) and how these vary over time with respect to changes in the natural environment due to developments, and the expected goals and needs of people. With this, environmental monitoring provides the essential and relevant inputs to other assessments, to efficiently evaluate cumulative effects due to changes in key environmental parameters. Accordingly, environmental monitoring can feed condition, trends and drivers' assessments. However, if a particular monitoring aims to understand how and why changes in the environmental condition will occur, we need to interpret historical trends of changes (i.e. biophysical and social) in order to identify the relationship between drivers and possible adverse or beneficial effects. Environmental assessments, which aim to identify the direct and indirect effects of mining activities on socio-ecological systems, would investigate the biophysical and social changes by measuring functions of the environment and its ecosystems and attributes of human population, coupled with the relevant design and analysis of effects [15]. Thus, monitoring supports the definition of a valid baseline output that ultimately reflects the environmental condition before the occurrence of anticipated changes in key environmental parameters. For this purpose, environmental monitoring must include a wide range of insights, including the interactions and linkages of multiple monitoring and assessments within the natural and built environment, because individual variables of one system are influenced by and affect other variables in other systems [17,15].

3.3. Environmental Condition Monitoring

This examines the monitored key environmental parameters to determine the present condition and attributes of ecological and human systems. It detects variations in the condition and attributes over time, and thus provides invaluable information about trends in ecosystem function and ecological elements (structure, components and processes) with respect to spatial and temporal scales [2,22]. Accordingly, the environmental condition seems to play a crucial role in environmental impact and quality assessment, and it relies heavily on environmental monitoring. To that end, there may be comparative analyses conducted of what has happened, what is happening and what is expected, given a specific environment and its unique, potentially affected key parameters [9]. In particular, condition assessments for anticipated effects due to developments will help to establish the condition of selected key environmental parameters and to identify sensitive areas and associated threats. It will document information in a manner that allows different practitioners to assess the entire environmental conditions (biophysical and social) within a region.

Environmental conditions for changes to socio-ecological systems will involve the comparative analysis of attributes of populations, communities or ecosystems with those that would be expected based on planned developments. For both human and ecological health, monitoring data are analyzed to determine whether the frequency and severity of effects from development might be higher than presumed, given the changes in population and its attributes [9]. However, this might be undermined by a lack of adequate knowledge of the basic ecological elements and the difficulty in defining the desired standard for the environmental condition. In a typical monitoring site, for example, where human health is partly determined by effective regulatory measures, an environmental condition may be applied to examine whether these measures are observed or not. In contrast, the environmental condition for environmental health is based on the knowledge of normal ranges of variation, which are not fully understood because of the complexities inherent to key environmental parameters.

A particular environmental condition analysis can initiate predictive assessments that result in the determination of how variations in trends and drivers would alter the present condition of the environment. For example, if the present environmental condition highlights changes in ambient Air Quality due to developments in the area (extensive coal mining), trends and drivers' assessments may be necessary to characterize the various drivers e.g. priority pollutants and their likely trends over time. Environmental condition provides relevant insights into the present state of key environmental parameters, which support the assessment of potential effects as well as describing the present environmental conditions.

3.4. Trends and Drivers Monitoring

Trends and drivers describe possible causal agents that could bring about changes in key environmental parameters. In particular, this identifies possible patterns of changes, their range of causal agents and likely specific effects on the condition of the environment and humans [23, 32]. Trend and driver assessments identify and characterize multiple and often interacting drivers, which work over spatial and temporal scales in an intermittent manner, to apportion exposure to changes in trends of environmental effects. The drivers of change in ambient air quality may be found by examining several emission inventories, historical monitoring data of priority pollutants, deposition agents, etc. It is not always necessary to perform both trends and drivers' assessments because in some cases the identification of trends also serves to identify the relevant driver. However, both the trends and the drivers of change in key environmental parameters usually must be adequately determined before there can be absolute prediction of environmental effects. A possible approach to analyzing the trends and drivers of changes in the status of ecological and human systems in relation to developments calls for an inclusive analysis of a range of present developments in order to identify changes of trends within the areas. Moreover, trends and drivers should be based on an integrated approach that allows elucidation of how regional trends may be linked to specific drivers in an area. This helps to identify relevant uncertainties as well as scenarios of exposure and sensitivity needed for the characterization of potential effects on the natural and social environments. On the other hand, trend and driver assessments may depend on existing national and regional standards for environmental conditions, which might be unreliable in areas where there are no strong considerations for such standards.

3.5. Impacts Monitoring

Environmental prediction estimates potential changes in key environmental parameters under specific trend and driver scenarios. It is either aimed at the evaluation of consequences or the measurement of magnitude of consequences under scenarios of exposure and sensitivity in the context of uncertainties [24]. In some contexts, prediction evaluates historically monitored data to characterize current environmental conditions or analyses associated with causal agents to predict a range of potential effects on distinct ecological and social systems. It could provide useful insights into possible changes in the environmental baseline based on comparative analyses of environmental conditions and possible trends and drivers.

According to Fig. 1, prediction monitoring and assessments may be fed by environmental condition, trends and drivers' assessments. However, these are equally important, even though they provide different insights at different stages. For the former, the insight may serve as a basis for the evaluation of changes in the characteristics of ecological and social systems under future changes in key environmental parameters. In this case, prediction will be able to describe potential impacts on diverse socio-ecological systems. For the latter, insights may be related to exposure and sensitivity; useful for identifying those areas within the region where specific socio-ecological systems would be most impacted by changes in key environmental parameters.

Environmental prediction may particularly examine how changes in the state of ecological systems and consequent changes and impacts in social systems could be identified and characterized. However, in some cases where there are no studies of a specific ecological state, prediction could serve as a means of inferring environmental conditions, as well as providing a baseline output. In this case, the results of prediction could be sufficient to inform decision-making that is related to managing huge and obvious risks due to increased developments.

4. Applying the Framework Approach

This paper attempts to illustrate how the framework approach might relate to the management of the health of humans and the environment within the context of the South African Air Quality Act, by theoretically illustrating the various steps of the framework and how such could include information from different biophysical and social activities that are already conducted and/or being conducted within the Waterberg municipality. That includes areas that are potentially impacted by current and planned developments as well as areas that are less impacted but are located around an environmental gradient, for example, the Marapong and Onverwacht human settlements, because of their close proximity to the Eskom's Matimba and Medupi coal-fired power stations and the Grootegeluk coal field mine.

This aims to monitor and assess changes in key environmental parameters related to the quality and conditions of streams, ambient air, vegetation, soil and humans. It will also include possible trends and drivers, which can be ascribed to plan developments in order to integrate a wide range of observed data and modeled outputs that will be analyzed to describe the health of the environment and humans in the municipality.

4.1. Environmental Monitoring

The monitoring of key environmental parameters related to

biophysical and social systems in the Waterberg region may involve relevant procedures and criteria that are crucial to observe and measure the quality/status of the health of humans and the environment, especially in the hotspots (i.e. NPA) where current development activities related to coal mining and coal-fired power stations is already posing adverse as well as beneficial effects on the environment. For example, this will employ a multi-objectives monitoring approach to the condition of ambient air, streams, and the diversity of vegetation and birds, as well as social changes in the hotspots, based on long-term observations of pollutants loading (airborne particulates, sulphur dioxide and oxides of nitrogen, carbon monoxides etc.) and the examination of emissions inventories in order to identify changes in key environmental parameters and possible existing trends in biophysical and social changes. Ultimately, monitoring may help to identify priority pollutants and associated human and natural drivers as well as providing insights into the trends of these pollutants in the hotspots. Moreover, it would reveal how planned developments might affect key environmental parameters and how this could support a viable assessment of biophysical and social changes and impacts as well as promote sustainable development.

4.2. Environmental Condition Monitoring

This will reflect changes in the condition and attributes of biophysical and social systems in the hotspots according to changes in the monitored key environmental parameters. That is, it will empirically examine, for example, changes in the quality of ambient air and streams, and the perception of people about the condition of the environment, based on the monitored and observed information of pollutant loading and emission inventory in the transect sites. This will ascertain how changes in key environmental parameters due to coal-fired power stations and coal mining fields have reflected on the health of the environment and humans in the Waterberg NPA. Moreover, this will help to identify whether the frequency and severity of biophysical and social changes might be higher than assumed, considering planned developments. Condition monitoring usefully provide relevant indicators required to inform effective monitoring networks that are needed for valid knowledge of environmental conditions and the successful assessment of potential biophysical and social changes and impacts in the context of changes in key environmental parameters.

4.3. Trends and Drivers' Monitoring

This will identify and examine natural drivers (e.g. relevant ecological features, topography, climate systems) and social drivers (e.g. development projects, infrastructures, demography) within and in close proximity to transect sites in the Waterberg NPA in order to highlight, on the one hand, specific socio-ecological systems that will be more prone to changes in environmental conditions, and on the other hand, significant drivers and trends of those changes, and how these are related to changes in the health of humans and the environment. It will also identify which driver is responsible for significant trends, knowing that different trends could have similar drivers but different impacts and risks. For example, activities of coal mining fields and coal-fired power stations pose different pollutants (gaseous and solid) that interact in the atmosphere and culminate in specific impacts on the environment and humans e.g. changes in diversity and distribution of vegetation and species of birds, as well as the human respiratory system. Trends and drivers' monitoring and assessments will highlight specific areas of the Waterberg NPA where there may be multiple and interacting priority pollutants, and thus high exposure and sensitivity to changes in key environmental parameters and their consequences on daily and seasonal scales. This relates much more to hotspots where stringent and urgent measures will be needed to manage the negative effects from current and planned developments.

4.4. Impacts Monitoring

This may employ a predictive modeling approach to analyze the impacts/risks from changes in environmental conditions with respect to the identified trends and drivers, and how this will reflect on the health of humans and the environment. For example, this will measure which trend in solid and gaseous pollutants and resultant pollutant loading on soil and streams would pose the worst impacts and risks e.g. acid rain resulting from SO_2 on vegetation, soil, animals (especially in aquatic ecosystems, that is fish, insects etc.), monuments, and drinking water in the hotspots. Impacts monitoring and assessments may inform the understanding of potential impacts on the health of humans and the environment, even if there is no valid knowledge of environmental conditions before the planned industrial developments. It helps to assess and understand how impacts and risks due to changes in pollutants' loading will vary over time, and thus what level of developments will be ideal for acceptable ambient Air Quality standards in South Africa e.g. 26 000ppbv for carbon monoxide, 106ppbv for nitrogen oxide, 134 0000ppbv for sulphur dioxide [35] etc., as well as sustainable environmental decision-making in the hotspots.

4.5. Integrated Environmental Baseline

This will attempt analyze and integrate different weighted evidence produced by all monitoring and assessments, to establish a coherent understanding of environmental conditions (i.e. due to biophysical and social changes) with respect to key environmental parameters in the Waterberg NPA. This serves as a basis for the assessments of potential impacts/risks under present scenarios of trends and drivers of changes in parameters. Moreover, the baseline assessment will analyze uncertainties in the processes of all monitoring and assessments based on a systematic approach to weighing of weighted evidence in order to make sure uncertainties within criteria, value judgments and choices made by practitioners are addressed. The protection of human and environmental health will be dependent on such assessment outcome and its application. Thus, baseline assessment would provide significant indicators for effective management of vital biophysical and social systems by informing the understanding of an integrated environmental condition before planned developments.

5. Conclusion

In this paper, a framework approach was proposed to provide significant insights into the conceptual understanding of integrated environmental monitoring and assessments, in order to push for a more coherent and sustainable environmental planning and management. This is because the framework draws on a broad range of multidisciplinary expertise and approaches and their linking processes, and thus creates a mechanism for this knowledge to be shared and related in ways that allow understanding of common concepts and approaches within different environmental monitoring and assessments. This, the paper suggested, will not only make environmental monitoring and assessments achieve its full potential as a sustainable mechanism, but it will also make it more applicable and realistic.

The framework approach presents a conceptual procedure for the assessment of pertinent biophysical and socio-economic impacts due to developments, based on outputs from multi-objectives monitoring that engages practitioners of different environmental programs, and thus helps focus different expertise towards a common objective (i.e. integrated environmental baseline). The framework supports the monitoring and assessment of impacts systematically by identifying relevant trends and drivers of changes in key environmental parameters as well as likely changes in environmental conditions, which are necessary to decide whether, and how, to proceed with planned developments in particular circumstances. For example, in the Waterberg NPA, where coal mining activities and coal-fired power stations have resulted in ambient air quality being monitored as well as their impact on the environment and human health, the framework would support the assessment of impacts on the environment and humans by identifying distinct but related monitoring that can be assessed separately and then brought together again in a manner that tries not to undermine the desired integration. This would provide explicit and detailed provisions for management and a viable resource for relevant regulatory programs that are designed to abate, for example, negative impacts on the environment and humans in the Waterberg NPA.

Moreover, the framework approach seeks to inform conceptually the understanding of processes that link distinct but related environmental monitoring and assessments, including their participants, and how they could meet and interact, thereby moving the ladder of integrated environmental assessment towards the practice end rather than the principle end that is often the basis for previous multi-disciplinary assessments. Accordingly, the framework approach supports collaborative environmental monitoring and explicit assessment, even though nowadays environmental assessments are often designed for different objectives in different periods according to different approaches. In order to allow regional government as well as relevant stakeholders to have a clear understanding of changes in environmental conditions under planned developments, and thus help them to avoid some of the costly mistakes that have been made during industrial development in Europe and North America.

Finally, it should be emphasised that it is not a framework to structure biophysical and socio-economic changes that may proceed or result in impacts, but rather it is a way of identifying pertinent causal factors (i.e. trends and drivers) of impacts, in order to understanding possible changes in environmental conditions.

Acknowledgements

This study was supported by grant-B8749.R01 from the Carnegie Corporation of New York to the Global Change and Sustainability Research Institute at the University of the Witwatersrand.

References

[1] Abaza H, Bisset R, Sadler B. Environmental impact assessment and strategic environmental assessment: towards an integrated approach. UNEP/Earthprint; 2004.

[2] Beanlands GE, Duinker PN. An ecological framework for environmental impact assessment in Canada. 1983. Available from: http://epe.lac-bac.gc.ca/100/200/301/ceaa-acee/ecological_framework-e/23E.PDF

[3] Burton GA, Batley GE, Chapman PM, Forbes VE, Smith EP, Reynoldson T, et al. A weight-of evidence framework for assessing sediment (or other) contamination: improving certainty in the decision-making process. Hum Ecol Risk Assess 2002;8:1675–96.

[4] Born SM, Sonzogni WC. Integrated environmental management: strengthening the conceptualization. Environmental Management. 1995;19(2):167–81.

[5] Bryman A. Integrating quantitative and qualitative research: how is it done? Qualitative research. 2006;6(1):97–113.

[6] Canter LW. Environmental impact assessment. McGraw-Hill: New York; 1977.

[7] Chan NY, Ebi KL, Smith F, Wilson TF, Smith AE. An integrated assessment framework for climate change and infectious diseases. Environmental Health Perspectives. 1999; 107(5):329.

[8] Chapman DV. Water quality assessments: a guide to the use of biota, sediments and water in environmental monitoring. E & Fn Spon: London; 1996.

[9] Cormier SM, Suter II GW. A framework for fully integrating environmental assessment. Environmental Management. 2008;42(4):543–56.

[10] Dubé M, Munkittrick K. Integration of effects-based and stressor-based approaches into a holistic framework for cumulative effects assessment in aquatic ecosystems. Human and Ecological Risk Assessment. 2001;7(2):247–58.

[11] Edwards CA. Assessing the effects of environmental pollutants on soil organisms, communities, processes and ecosystems. European Journal of Soil Biology. 2002;38(3):225–231.

[12] Eggenberger M, Partidário MR. Development of a framework to assist the integration of environmental, social and economic issues in spatial planning. Impact assessment and project appraisal. 2000;18(3):201–7.

[13] Gough C, Castells N, Funtowicz S. Integrated Assessment: an emerging methodology for complex issues. Environmental Modeling & Assessment. 1998;3(1-2):19–29.

[14] Hisschemöller M, Tol RS, Vellinga P. The relevance of participatory approaches in integrated environmental assessment. Integrated Assessment. 2001;2(2):57–72.

[15] Humphrey CL, Faith DP, Dostine PL. Baseline requirements for assessment of mining impact using biological monitoring. Australian Journal of Ecology. 1995;20(1):150–66.

[16] Jabareen, Y. A new conceptual framework for sustainable development. Environment Development and Sustainability. 2008;10(2),197–192.

[17] Karr JR. Biological monitoring and environmental assessment: a conceptual framework. Environmental Management. 1987;11(2):249–56.

[18] Margerum RD. Integrated environmental management: moving from theory to practice. Journal of environmental planning and management. 1995;38(3):371–92.

[19] Margerum RD. Integrated environmental management: lessons from the Trinity Inlet Management Program. Land use Policy. 1999;16(3):179-70.

[20] Munns WR, Kroes R, Veith G, Suter II GW, Damstra T, Waters MD. Approaches for integrated risk assessment. Human and Ecological Risk Assessment. 2003;9(1):267–72.

[21] National Environmental Management: Air Quality Act, No. 39 of 2004. Accessed on 3 March 2014, available at http://www.environment.gov.za. Republic of South Africa (2004).

[22] Quinlan T, Scogings P. Why biophysical and social scientists can speak the same language when addressing sustainable development. Environmental Science & Policy. 2004;7(6):537–46.

[23] Rounsevell MDA, Dawson TP, Harrison PA. A conceptual framework to assess the effects of environmental change on ecosystem services. Biodiversity and Conservation. 2010;19(10):2823–42.

[24] Rothman DS, Robinson JB. Growing pains: a conceptual framework for considering integrated assessments. Environmental Monitoring and Assessment. 1997; 46(1-2): 23–43.

[25] Smith ML. Environmental impact assessments: the roles of predicting and monitoring the extent of impacts. Marine and Freshwater Research. 1991;42(5):603–14.

[26] Slootweg R, Vanclay F, van Schooten M. Function evaluation as a framework for the integration of social and environmental impact assessment. Impact Assessment and Project Appraisal. 2001;19(1):19–28.

[27] StatsSA (Statistics South Africa). Municipality Fact Sheet. Published by Statistics South Africa. Census, 2011.

[28] Stem C, Margoluis R, Salafsky N, Brown M. Monitoring and evaluation in conservation: a review of trends and approaches. Conservation Biology. 2005;19(2):295–309.

[29] Suter II GW, Vermeire T, Munns Jr WR, Sekizawa J. An integrated framework for health and ecological risk assessment. Toxicology and Applied Pharmacology. 2005;207(2):611–6.

[30] Suter II GW. Ecological risk assessment. CRC press; 2006.

[31] Suter II GW, Vermeire T, Munns WR, Sekizawa J. Framework for the integration of health and ecological risk assessment. Human and Ecological Risk Assessment. 2003;9(1):281–301.

[32] Suter II GW, Cormier SM. Why and how to combine evidence in environmental assessments: Weighing evidence and building cases. Science of the Total Environment. 2011;409(8):1406–17.

[33] Toth FL, Hizsnyik E. Integrated environmental assessment methods: Evolution and applications. Environmental Modeling & Assessment. 1998;3(3):193–207.

[34] Van der Oost R, Beyer J, Vermeulen NP. Fish bioaccumulation and biomarkers in environmental risk assessment: a review. Environmental Toxicology and Pharmacology. 2003; 13(2): 57–149.

[35] Venter AD, Vakkari V, Beukes JP, Van Zyl PG, Laakso H, Mabaso D, Tiitta P, Josipovic M, Kulmala M, Pienaar JJ, Laakso L. An air quality assessment in the industrialized western Bushveld Igneous Complex, South Africa. South African Journal of Science. 2012;108(9-10):1–10.

[36] WHO. Report on Integrated Risk Assessment. World Health Organization, Geneva, Switzerland. WHO/IPCS/ IRA/01/12; 2001.

[37] Lovett GM, Burns DA, Driscoll CT, Jenkins JC, Mitchell MJ, Rustad L, et al. Who needs environmental monitoring? Frontiers in Ecology and the Environment. 2007;5(5):253–60.

Citizen Participation Practice in Thailand Environmental Policy

Prukkanone Kittisak[1], Wang Guohua[1, 2]

[1]College of Public Administration, Huazhong University of Science and Technology, Wuhan, China
[2]The Public Opinion Researching Center, Huazhong University of Science and Technology, Wuhan, China

Email address:
kprukkanone@yahoo.com (P. Kittisak), mpawin@hust.edu.cn (Wang Guohua)

Abstract: Environmental Policy is a major mechanism for the government to drive the public administration on environmental management. In Thailand, some citizens were significantly affected from the inadequacy of Public Participation in Environmental Policy and leaded to widespread conflicts about their interests in society. Although the Constitution of the Kingdom of Thailand B. E. 2550 (2007) attempted to broadly open the opportunities on Public Participation in the environmental management but the interest conflicts were still existed. This article studied on Thailand Environmental Policy, especially examined the Citizen Participation Practice in the Environmental Policy Formulation which would contributed the further development in public administration aspect. The current national environmental policy, *the Environmental Quality Management Plan B. E. 2555-2559 (2012-2016)*, was selected as the case study and analyzed to examine Citizen Participation Practice under the policymaking context by the principle of citizen participation "Perceptions of Stakeholders and Planners: *the ladder of citizen participation*" and *the concept of environmental policymaking in government*. Related documents and researches, environmental laws and regulations, and policies were reviewed, analyzed and synthesized for appropriately conducting in the study. The results were found that Thailand had an appropriate framework on citizen participation in the Environmental Policymaking Process regarding the provisions of constitution but it was limited by the provisions of older laws and policies which were enacted before the 2007 Constitution was promulgated. All of Citizen Participation Practices were located in the category of Tokenism, where the citizens became involved but only to certain extent. This is not meaningful in term of citizen control which should be appropriately conducted in the democracy society. Thailand did not have Citizen Participation Practices in the category of Nonparticipation and the current national environmental policy open an opportunity for amendment on Environmental Legislation. These findings were benefit to the better development of Citizen Participation Practice in Thailand Environmental Policy, which would be opened the participatory opportunity on the interest conflict resolution in the future.

Keywords: Thailand Environmental Policy, Citizen Participation Practice, Policymaking Process, Public Participation

1. Introduction

Since 1932, Thailand has changed the country regime from an absolute monarchy to a constitutional monarchy by adopting a democratic regime of government with the King as Head of the State and the sovereign power belongs to the Thai Citizen [1]. Nevertheless, so far, Thai citizen still faced the problems and conflicts on their participation, especially in the environmental issues which the number of conflicts between the public and the government has obviously increased during the past few years.

This problem reflected in the statistic of the proceeding cases about public administration concerning environmental issues through Administrative Court of Thailand during 2011-2015 that citizen sued the Public Agency which found that there were 5,131 proceeded cases. In addition, regarding the Top Ten Ministries Ranking of Defendant in Administrative Court of Thailand 2011-2014, Ministry of Natural Resources and Environment (MNRE) was ranked seventh of the Defendant of Public Agency in the ministry level [2] as shown in Table 1.

Throughout Thailand's environmental management history, Thailand Political and Legal System authorized the government to manage natural resources and environment with absolute power, it was top-down management and obviously ineffective. Particularly, decisions about environmental protection and management often failed to integrate the opinions of citizen. Citizen Participation Practice was often avoided and limited. The decision-makers preferred to keep control over the decision and refused to share their authority. The citizen had few opportunities to influence the decisions. It was an important obstacle to achieve effective process and this could lead to more environmental problems and interest conflicts in Thai Social. Some citizens lost trust and developed an unfavorable attitude towards government-sponsored participation and refused to engage in the government-sponsored participatory activities [3].

Table 1. *Top Ten Ministries Ranking of the Defendant of Public Agency in Administrative Court of Thailand, 2011-2014.*

Ranking	The Defendant of Public agency (Ministry level)	Cases
1	Ministry of Interior	19,943
2	Ministry of Transport	12,490
3	Ministry of Agriculture and Cooperatives	10,676
4	Ministry of Education	8,198
5	Office of the Prime Minister	6,354
6	Ministry of Finance	5,535
7	Ministry of Natural Resources and Environment	3,183
8	Ministry of Public Health	1,477
9	Ministry of Justice	1,250
10	Ministry of Energy	1,101
	Total	70,207

The environmental and interest conflicts were acknowledged by Thai Social and there was an attempted to use the appropriate citizen participation for solving the problem through the constitution which acts as the mother law of the country. Thailand's constitution were promulgated and amended through 18 constitutions in accordance with the evolving situation of country. In particular, Constitution of the Kingdom of Thailand B. E. 2540 (1997), the 17th constitution is an initiative of Participatory Democracy in Thailand's history. The 1997 Constitution is considered as the fundamental to Thai modern political reform and recognized a wide range of basic rights, particularly in the voicing of public opinion on issues of national importance and safeguarding people's dignity [4]. This Constitution was promulgated and mandated important provisions in political reform, establishment of independent organizations, administrative decentralization, public participation in natural resources and environmental management as well as the guarantee of the community rights for the first time [5].

Although Thailand's constitution was transformed to the 18th Constitution, the Constitution of the Kingdom of Thailand B. E. 2550 (2007), still concerned on public participation in natural resources and environmental management aspect and had been modified to enhance this role, which citizen was explicitly endorsed and safeguarded the right to participate as enshrined in several sections of, the law ensuring the right to have a good environment and ensuring people participation with respect to health, laws concerning access to public documents, concerning public participation, and those concerning access to environmental justice.

The 2007 Constitution was used as a legal framework for formulating the current environmental policy of Thailand which looked seem good in principle where government formulate the environmental policy by providing the participation activities according to the constitution, but in practice, some citizen perceived that participation was merely a ritual aimed at legitimizing government decision-making and was done merely to ensure that decisions did not violate any legal provisions. As this was not meaningful participation, the country wasted substantial funds in organizing participation activities which did not have any effect on governmental policies or projects [3].

This article examined Citizen Participation Practices in Thailand's Environmental Policy Formulation Process, which the Environmental Quality Management Plan B. E. 2555-2559 (2012-2016) was considered as a case study. Therefore, the 2007 Constitution, Environmental Laws and Regulations related to Citizen Participation Practice in Thailand, including with Citizen Participation Principle and Environmental Policy Theory were reviewed, analyzed and applied for conducting in the study.

2. Literature Review

2.1. Citizen Participation and Environmental Policy Formulation

First and foremost, it is important to understand the relationship between Citizen Participation and Environmental Policy Formulation in term of the equilibrium of public management under the paradigm that humans can survive their quality of life with balancing between environmental and social capital. However, due to the social capital, the public management regarding the environmental service circumstance in actual world was fallen down to the awkward situation because the majority of natural resources and environment was being depleted both quality and quantity aspect, meanwhile the human needs were dramatically increasing. These are predicted that can severely affect both the human and environmental quality in the near future.

Generally, Attributions of the environment can be thought of as environmental capital capable of providing services which people can use [6]. People use the environment to support their demand in several aspects which are knew as an environmental service: a service that environment provides for the individuals who comprise the human race, which the valuables are depended on their characteristics that can fulfill the resources needed to the human such as; procure shelter, safety, aesthetic pleasure, spiritual sustenance and so on.

Conceptualizing the environment in terms of its ability to service the human race is an approach increasingly used by environmental policy makers. This approach is a prominent point that challenges the environmental policy makers or government takes into account the policy output that achieved the sustainable utilization concept under a thin dividing line between 'the human needs' and 'the environmental capital capable'.

In the democratic country, the environmental policy makers or government is often emphasizing democratic decision-making principles in the policy-making process to ensure that all that are affected by a given decision have the right to participate in the making of that decision. The participation is seen as an instrument for ringing public values, social objectives and preferences to decision processes that are primarily analytical [7].

Citizen participation is a process which provides private individuals an opportunity to influence public decisions and has long been a component of the democratic decision-making process. The roots of citizen participation can be traced to ancient Greece and Colonial New England. Before the 1960s, governmental processes and procedures were designed to facilitate "external" participation. Citizen participation was institutionalized in the mid-1960s with President Lyndon Johnson's Great Society programs [8].

Cogan and Sharpe [8] identify five benefits of citizen participation to the planning process: 1) Information and ideas on public issues; 2) Public Support for planning decisions; 3) Avoidance of protracted conflicts and costly delays; 4) Reservoir of good will which can carry over to future decisions; and 5) Spirit of cooperation and trust between the agency and the public.

2.2. The Principle of Citizen Participation, Perceptions of Stakeholders and Planners

This article, adopted the principle of citizen participation, Perceptions of Stakeholders and Planners, to examine Citizen Participation Practice in Thailand's environmental policy formulation process, which are described as following.

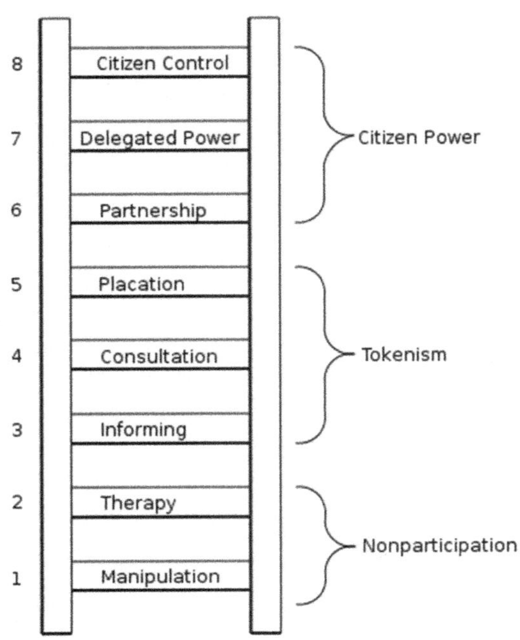

Source: Sherry R. Arnstein [9]

Figure 1. *The Ladder of Citizen Participation.*

The Perception of Stakeholders and Planners is an important consideration in the development and implementation of any public participation program. In the successful citizen involvement programs or policies, the disparity between the policy makers or government and the citizen's expectations must be stated in minimal. If expectations are different, conflict is probable [8].

To assist the policy makers in determining the perceptions of a policy's purpose and compare this with the anticipated perceptions of citizen participants, the "ladder of citizen participation" of Sherry R. Arnstein [9] is one of the best practices that can provide an overview of the different ways the public can be involved in decision-making. It describes eight levels of participation, which are divided into three main categories. These levels can be represented as a ladder, as shown in Figure 1.

2.2.1. Nonparticipation (1st - 2nd Level)

This is where the public is not directly involved and may be manipulated into thinking they are part of decision-making, where the power holders have created a phony form of participation, perhaps around a decision already made.

1st level: *Manipulation*, which is where people are "educated" and may be advised to sign proposals they believe to be in their interest.

2nd level: *Therapy*, involves the power holders "curing" the people. The power holders promise to assist the citizens and have them engage in different activities where their opinions may be "cured", and in the end accepted by the citizens.

2.2.2. Tokenism (3rd - 5th Level)

This is where the citizens become involved but only to certain extent.

3rd level: Informing, this is where the citizens are informed of what is happening. This is a one-way information process, where people receive the information in newspapers, in the media, online or by other means.

4th level: *Consultation*, in which citizens' opinions can start to affect the power holder's opinion. This is a common form of citizen participation utilized in urban planning. If consultation and information is taken into account as part of the planning process, this can be effective. However, if the consultation and information is not taken into consideration at the end of the day, this step will be of limited value and could therefore fall back into the non-participating level.

5th level: *Placation*, the level is where a citizens' opinion will start influencing the power holder's decision. At this level, citizens may be hand-picked to sit on a governing board that makes decisions on the planning process. This process is more likely to work if the board members are equally split (citizens and power holders), so the citizens cannot be outvoted in the process.

2.2.3. Citizen Power (6th - 8th Level)

This is where the citizens get to influence the decision-making directly.

6th level: *Partnership*, the power holders and citizens

create a partnership. Arnstein considers partnership relatively high on her ladder as she believes this can keep both citizens and power holders content.

7th *level*: *Delegated Power*, the citizens can start taking control, and the power holders need to start negotiating with the citizens. Compared to the example given for placation (the fifth level), the majority of the board members would be the citizens. This would mean that the power holders would need to negotiate decisions with the board members. This process is more likely to work if the board members are equally split (citizens and power holders), so the citizens cannot be outvoted in the process.

8th *level*: *Citizen Control*, it gives the citizens the power to decide. This can be achieved through referendums, but since those are often costly and difficult to arrange it would most likely slow down the process substantially. They are therefore often only utilized for larger decisions. In many cases, local authorities do not, however, give their citizens full control in such elections, but treat the results instead only as advisory for the final decision made by the city council or other such decision making bodies, citizens may be hand-picked to sit on a governing board that makes decisions on the planning process.

3. Methodology

This article is conducted by the case study research by selecting the current national environmental policy, *Environmental Quality Management Plan B. E. 2555-2559 (2012-2016)* as the case study. For the basic proposition of the case study was the application of Citizen Participation Practice under the provisions of 2007 Constitution to contribute the further development in the public administration on environmental management.

The secondary data of related documents and researches, government reports, environmental laws and regulations, and policies were reviewed, analyzed, and synthesized with the primary data from the stakeholder interviews for appropriately conducting in the study. The Policymaking Context of the Environmental Quality Management Plan 2012-2016 was simulated under *the concept of environmental policymaking in government* and *the principle of citizen participation 'Perceptions of Stakeholders and Planners: the ladder of citizen participation'* was applied to describing Citizen Participation Practice in Thailand's environmental policy formulation process.

The analytic strategy for analyzing case study evidence was *the relying on theoretical proposition*, which the theoretical orientation in this study examined Citizen Participation Practice in Thailand's Environmental Policymaking Process and described that how the formation and modification in the policy formulation process occurred, and how the relevant actors acted to contribute the provisions of 2007 Constitution.

4. Result and Discussions

4.1. Environmental Policy in Thailand - Steps, Policies and Laws Framework

The Government of Thailand had initiated the following steps to develop the environmental policy formulation in Thailand.

4.1.1. Historical Period

Before 1972, Thailand did not has the governmental organization responsible for handling environmental management in the country, therefore the common environmental policy in majority is conducted through the related acts concerning the preservation of natural resources was enacted. The examples of these acts are Forest Act 1941, Fisheries Act 1947, National Park Act 1961, Field Dyke and Ditches Act 1962, National Reserved Forest Act 1964, Petroleum Act 1971.

4.1.2. Initiating Period

In 1972, After Thailand participated in the Stockholm conference on the human environment, the government recognized the need for environmental management then established the office of National Environment Board (ONEB) to serve as the secretariat office in the central authority for coordinating environmental management of the country.

4.1.3. The Improvement and Conservation of National Environmental Quality Act 1975 (NEQ Act 1975)

This was the first and specific environmental law of Thailand which was proved inefficient as it was enacted to support the operation of the National Environment Board (NEB), an advisory body for the Prime Minister on environmental policy. Not only being empowered to carry out any implementation on its own, the National Environment Board was required to seek co-operation from other governmental agencies which was seldom successful.

4.1.4. The Enhancement and Conservation of National Environmental Quality Act 1992 (NEQ Act 1992)

After the 1991 coup d'état, Thailand was repeals and amendments of various laws concerning country development. The NEQ Act 1975 was also repealed and re-enacted as the NEQ Act 1992 with its objectives of being a basic comprehensive environmental law incorporating various aspects of environmental management.

4.1.5. The Enhancement and Conservation of National Environmental Quality Policy and Plan 1997-2016 (NEQ Plan 1997-2016)

Office of Natural Resources and Environmental Policy and Planning (ONEP) under the Ministry of Natural Resources and Environment was mandated as a governmental agency, acting the secretariat office of the National Environment Board, responsible for this policy-making process [10]. This policy was a long-term plan that legitimated to be the twenty years master plan for implementation of the national policy and plan for enhancement and conservation of environmental quality. This plan was started to conduct the policy-making process in 1996 and proclaimed in 1997 respectively.

4.1.6. The Environmental Quality Management Plan 2012-2016 (EQM Plan 2012-2016)

This is the current environmental policy of Thailand. This policy is the intermediated plan that legitimate under the NEQP 1997-2016. It was designed to be the 5 years plan for implementation of the national policy and plan for enhancement and conservation of environmental quality. Before this plan there are two plans which was conducted and proclaimed in B. E. 2542 (1999), and 2550 (2007) respectively.

For the purpose of this article, the most relevant piece of legislation that affected to the current environmental policy formulation was the 2007 Constitution. It introduced several new provisions on the issue of environmental management, promoted and supported public participation in environmental management and conservation more than any previous constitution. The five fundamental principles regarding environmental management and public participation were determined as follows:

i. The right of the public and local community to conserve and use their environment and resources sustainability
ii. The right to control and minimize the pollution that impacts their quality of life
iii. The right of the individual and community to reduce the government's sole decision-making power over natural resource management
iv. Provision of a guarantee of access to information by the public
v. Public participation

The study applied the concept of environmental policymaking in government to conceptualize the environmental policy formulation as a cycle [6], the five main components (policy environment, inputs, government, outputs, and outcomes). This concept was applied to simulate the Policymaking Context of EQM Plan 2012-2016 as shown in Figure 2.

For the purpose of this article, the context has been analyzed only the policy-formulation process by focusing on the four main components (except outcomes) those are; policy environment, inputs, government, and outputs. The rational-relationship regarding the policy-formulation process both the policy actors and their roles, and the interactions of Citizen Participation Practice under the policy context were discussed as following.

4.2. Citizen Participation Practice in Thailand Environmental Policy – An Analysis

4.2.1. Policy Environment

In this component, there were five actors that were categorized their roles into two aspects those shaped the policy environment which are consist;

i. *The political complexion of the government aspect and prevailing public political ideologies in environmental issue*

The first actor, *the 2007 Constitution* played a major role on behalf of the supreme law of Thailand which prevailed over other laws passed by parliament. *The NEQ Act 1992* was the second actor which acted as the principle of environmental law in Thailand. The third actor had the important role in this context were, *Bilateral and Multilateral Environmental Agreements (MEAs)* which bonded the international agreements created between two or multiple nations that pledged to conduct their operations in such a way that limited negative environmental impacts. This was operated in accordance with the principle of environmental management in the globalization era which the environmental issue became to the global issue.

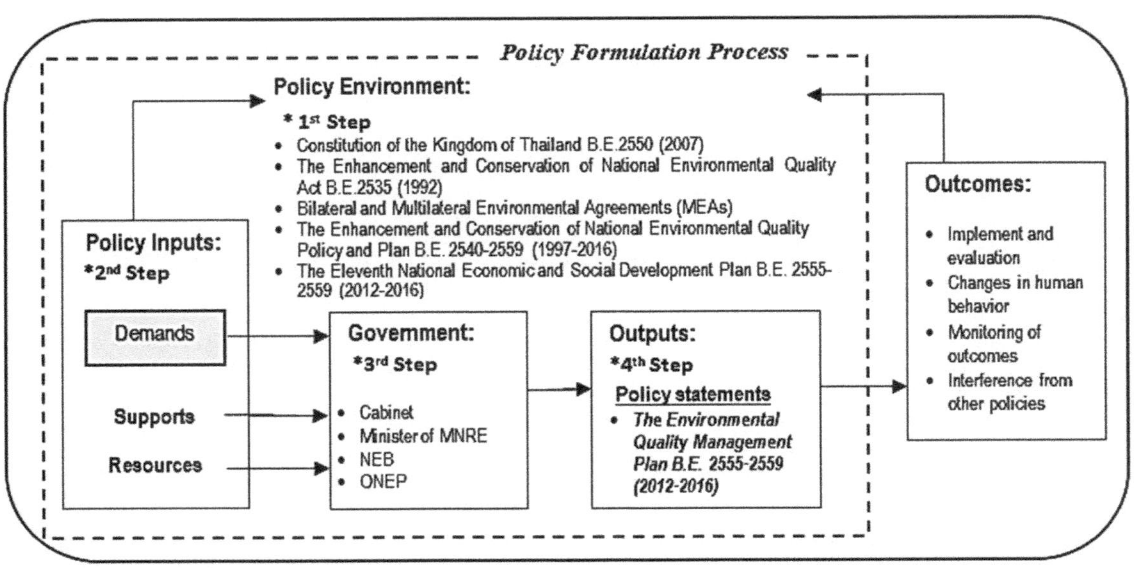

Note: * step of the policy formulation process adapted from Roberts [6]

Figure 2. The Policymaking Context of the Environmental Quality Management Plan 2012-2016.

ii. *The fundamental information about the economic and social circumstances, and the state of natural resources and environment*

The fourth and fifth actor was consisting of *The NEQ Plan 1997-2017* and *the Eleventh National Economic and Social Development Plan B. E. 2555-2559 (11th NESD Plan 2012-2016)*. They presented and interpreted the fundamental information about the economic and social circumstances [11], and the state of natural resources and environment of Thailand into public sphere through the policy framework which ONEP (the responsible agency) had to mediate these information into the environmental policy-formulation process.

4.2.2. Policy Input

Underneath this component, the three sub-components: Demands, Supports, and Resources were operated in the policy system. They configured the policy inputs on the different basis which could be described in each as follows.

i. *Demands:* 'the perceived problems affecting various interests will lead to pressure for policies to be formulated to address the problem' was a common role for the actors in this sub-component [5]. There were five actors playing this role which consist of 1) Thailand public opinion in environmental management, 2) Evaluated result of Thailand's Environmental Quality Management Plan 2007-2011, 3) The five year forecast on Political Economic Socio-cultural Technological Components, 4) The five year forecast on the environmental management scenario, and 5) The five year forecast on the environmental legislation and amendment.

ii. *Supports:* 'Public advocated and obeyed with the policy context' was the essential factor for conducting this process. In this study, there were two main actors; including 1) Citizen and 2) Government agencies, they played a major role through the comprehensive public hearings which co-operated between them and the policymaker agency as ONEP.

iii. *Resources:* This sub-component had role that emphasized on *'The important things that required implementing the policy'*. There were three actors in this role including *1) ONEP, 2) Donors, and 3) Government agencies.*

4.2.3. Government

The decision-making process within government was played by the four relevant actors which consist of *1) ONEP, 2) National Environment Board, 3) Minister of Natural Resources and Environment, and 4) The cabinet.* Their decision-making process depends on the authorization power from the NEQ Act 1992 given to government [12]. Another thing that the government must take into consideration in the decision-making process was the relevant national policy frameworks which include the NEQ Plan 1997-2017 and the 11th NESD Plan 2012-2016.

4.2.4. Outputs

The major role of this component was *'The implementations or actions that could change or affect the people's behavior to follow the policy context'*. The major variables which contributed to the successful implementation of policy was taken into account, and found that there were two actors those directly related to the successful in this policy, including *1) Citizen and 2) Government agencies.* Particularly, the governmental agencies, the law-binding agencies under Sec. 35 of the NEQ Act 1992 to take actions within their powers and functions those were necessary for effective implementation of the EQM Plan 2012-2016.

4.2.5. Citizen Participation Practice Under the Policy Context

It was considered as the context of Citizen Participation Practice in each step of the policy formulation process. The participation opportunities were analyzed and examined their context as following.

i. *1st Step*: It was focused on the framework on participation that opened the opportunities for citizen to participate in the environmental policy-formulation process. The 2007 Constitution was claimed to the scope of framework which was discussed as follows.

The 2007 Constitution had provisions affirming right and freedoms of citizen in the subject of participation on management of natural resources and environment which was covering three dimensions:

a. *The right of access data and information:* There were three relevant sections in the 2007 constitution which contributed to this study (Section 45, 56, and 57). All of Citizen Participation Practices were interpreted in term of individual level as illustrated in Figure 3.
b. *The right of participation by the citizen*: The five relevant sections were contributed to this study (Section 58, 66, 67, 287 and 290) and Citizen Participation Practices were interpreted in two forms both the individual and the community level as illustrated in Figure 4.
c. The right of access to the justice system**:** The four relevant sections represented the citizen participation which contributed to this study (Section 28, 40, 60, and 67 (Para. 3)). Citizen Participation Practices were interpreted in two forms both the individual and the community level as illustrated in Figure 5.

ii. *2nd Step:* It was focused on the process of policy input under the policy context and found that the opportunities to participate in this process was widely opened to public, the proportion of citizen that contributed in term of actors was two-third from three sub-components (except the resources).

Due to the aforementioned reason, Thailand public opinion in environmental management (from the demands) and the comprehensive public hearings (from the supports) were considered as the Citizen Participation Practice in this process.

Section 45. A person shall enjoy the liberty to express his or her opinion, make speeches, write, print, publicise, and make expression by other means....... The censorship by a competent official of news or articles before their publication in a newspaper or other mass media shall not be made ....

Section 56. A person shall have the right to know and have access to public data or information in possession of a Government agency......., unless the disclosure of such data or information shall affect the security of the State, public safety or interests of other persons which shall be protected or purport to be personal data, as provided by law.

Section 57. A person shall have the right to receive data, explanations and reasons from a Government agency...... and shall have the right to express his or her opinions to agencies concerned, for assisting further consideration of such matters.

In planning social, economic, political and cultural development, or in undertaking expropriation, town and country planning, zoning and making by laws likely to have impacts on essential interests of the public, the State shall cause to be held comprehensive public hearings prior thereto.

Figure 3. *The three relevant sections on the right of access data and information.*

Section 58. A person shall have the right to participate in the decision-making process of State officials in the performance of administrative functions which affect or may affect his or her rights and liberties.

Section 66. Persons so assembling as to be a community, a local community or a traditional community shall have the right to ..... participate in the management, maintenance, preservation and exploitation of natural resources, the environment ...

Section 67. ......Any project or activity which may seriously affect the community with respect to the quality of the environment, natural resources and health shall not be permitted, unless, prior to the operation thereof, its impacts on the quality of the environment and on public health have been studied and assessed and a public hearing process has been conducted ....

Section 287. Local residents have the right to participate in the administration of a local government organization. For this purpose, a local government organization shall also make available means for such public participation.

Section 290. A local government organization has powers and duties in connection with the promotion and maintenance of the quality of the environment as provided by law .....

Figure 4. *The five relevant sections on the right of participation by the citizen.*

Section 28. A person can invoke human dignity or exercise his or her rights and liberties in so far as it is not in violation of rights and liberties of other persons or contrary to the Constitution or good morals......

A person may have a recourse to the Court for directly enforcing the State to comply with provisions in this Chapter. Where the exercise of any particular right or liberty as recognized by this Constitution is a subject-matter of the existing law, such right and liberty shall be exercisable as provided by law.....

Section 40. A person shall have the following rights in the administration of justice:

(1) the right to have easy, expeditious, speedy and comprehensive access to justice;
(2), (3), (4), (5), (6), (7) ......

Section 60. A person shall have the right to file a lawsuit against a Government agency, a State agency, a State enterprise, a local government organization or other State authority which is a juristic person to be liable for an act or omission done by its Government official, official or employee

Section 67 (Para 3). The right of a community to bring a lawsuit against a Government agency, a State agency, a State enterprise, a local government organization or other State authority which is a juristic person for the performance of duties under this provision shall be protected.

Figure 5. *The four relevant sections on the right of access to the justice system.*

iii. *3rd Step*: The decision-making process of government was focused under the policy context and found that the NEQ Act 1992 was played a major role regarding the opportunities of citizen to participate in this process. There were some part of the NEQ Plan 1997-2017 and the 11th NESD Plan 2012-2016 that promoted the Citizen Participation Practice which could be discussed as follow.

Those relevant law and policies had provisions affirming participation of citizen in the subject of environmental policy making which was covering two dimensions:

a. The public participation in the enhancement and conservation of national environmental quality: This covered the three relevant provisions in the NEQ Act 1992, the NEQ Plan 1997-2016, and the 11th NESD Plan 2012-2016. Citizen Participation Practices were interpreted in two forms both the individual and the community level as illustrated in Figure 6.

b. The decision-making process in the environmental policy formulation: This covered the five relevant sections under the NEQ Act 1992 which all of Citizen Participation Practices were interpreted in term of the citizen's representative as illustrated in Figure 7.

The Enhancement and Conservation of National Environmental Quality Act B.E.2535 (1992)

Section 6. For the purpose of public participation in the enhancement and conservation of national environmental quality, the following rights and duties may be accorded to individual person as provided by this Act or governing law related thereto:

(1) To be informed and obtain information and data from the government service ......
(2) To be remedied or compensated by the State....
(3) To petition or lodge complaint against the offender in case of being a witness .....
(4) To co-operate and assist government officials in the performance of duty ....
(5) To strictly observe the provisions of this Act or other laws concerning...........

The Enhancement and Conservation of National Environmental Quality Policy and Plan B.E. 2540-2559 (1997-2016)

The Enhancement and Conservation of National Environmental Quality Policy (including 6 Policies)
Policy no 5. The Education and Public Relation for Environment
Strengthen the capability of communities in any level, to be strength and able to initiate the co-operation movement for the efficient in natural resource and environmental management.

The Eleventh National Economic and Social Development Plan B.E. 2555-2559 (2012-2016)

Strategy of Managing Natural Resources and the Environment to Achieve Sustainability
Strategy 5.8 Enhance natural resource and environmental management systems so they are more efficient, transparent and equitable, following these guideline:
5.8.1 Empower communities and support local rights to utilize natural resources in a balanced and sustainable manner.
5.8.2 Support popular participation and develop community abilities...... .

Figure 6. *The three relevant provisions on the public participation in the enhancement and conservation of national environmental quality.*

The Enhancement and Conservation of National Environmental Quality Act B.E.2535 (1992)

Section 7. In order to encourage public participation in the promotion and conservation of environmental quality, non-government organizations (NGOs) having the status of a juristic person .....

Section 8. The registered NGOs may propose for nomination of candidates as representatives of the private sector to be appointed by the cabinet as qualified members of the National Environment Board.

Section 12. There shall be a National Environment Board consisting of the Prime Minister as the Chairman, a Deputy Prime Minister designated by the Prime Minister as the first Vice Chairman, the Minister of Natural Resources and Environment as the second Vice Chairman, the Minister of Defense, the Minister of Finance, the Minister of Agriculture and Cooperatives, the Minister of Transport, the Minister of Information and Communications, Technology, the Minister of Interior, the Minister of Education, the Minister of Public Health, the Minister of Industry, the Secretary-General of the National Economic and Social Development Board ,the Secretary-General of the Board of Investment, the Director of the Bureau of the Budget as members ex officio and ** *members qualified in environmental matters not more than eight persons of which no less than half shall be representatives from the private sector* and the Permanent Secretary of the Ministry of Natural Resources and Environment as member and secretary.

** *The appointment of qualified members shall be made by drawing from persons who are knowledgeable and known for their expertise, contributions and experiences in the matters concerning the enhancement and conservation of environmental quality.*

Section 13. The National Environment Board shall have the power and duty as follows
(1) To submit policy and plan for enhancement and conservation of national environmental quality to the cabinet for approval.

Section 35. The Minister shall, with the approval of the National Environment Board, formulate an action plan called "Environmental Quality Management Plan" for implementation of the national policy and plan for enhancement and conservation of environmental quality determined by virtue of section 13 (1).

Figure 7. *The five relevant provisions on the decision-making process in the environmental policy formulation.*

iv. 4th Step: It focused on the policy statements of the EQM Plan 2012-2016 (case study) and found that the opportunities of citizen to participate in this process in not quite widespread. Majority of citizen participation practices were binding the government agencies. And, there were some parts of this plan that promoted the citizen participation practice which could be discussed as follows.

The policy statements were covered six strategies and Citizen Participation Practice was emphasized in the third strategy, *the natural resource and environmental management for strengthening the good governance*. The overview of this strategy was stated the practice merely the public hearing and the permission to access the public information.

Summary of Citizen Participation Practices from the analyzing on the 1st – 4th Step were integrated and compared with the ladder of citizen participation, to determine the perceptions of a policy's purpose and compare this with the anticipated perceptions of citizen participants as shown in Table 2 below.

Table 2. *Citizen Participation Practices in Thailand Environmental Policy.*

Arnstein's Ladder of citizen participation		1st Step Participation Framework	2nd Step Policy Input Process	3rd Step Decision-making Process	4th Step Implementation Process	Remark
Citizen power	*Citizen Control*	*Yes*	-	-	-	The 2007 Constitution - Sec. 28, 40, 58, 60, and 67 (Para. 3)
	Delegated Power	*Yes*	-	-	-	The 2007 Constitution - Sec. 287, 290
	Partnership	*Yes*	-	-	-	The 2007 Constitution - Sec. 66, 67
** Tokenism **	*Placation*	*Yes*	-	*Yes*	*Yes*	The 2007 Constitution - Sec. 66, 67 The NEQ Act 1992 - Sec. 7, 8, 12, 13, 35 The NEQ Plan 1997-2017 - Policy no. 5 The 11th NESD Plan - Strategy 5.8 The Policy Statements - Strategy 3
	Consultation	*Yes*	*Yes*	-	*Yes*	The 2007 Constitution - Sec. 45,56, 67 The Policy input - Public Opinion, Public Hearing The Policy Statements - Strategy 3
	Informing	*Yes*	*Yes*	-	*Yes*	The 2007 Constitution - Sec. 45, 56, 67 The Policy input - Public Opinion, Public Hearing The Policy Statements - Strategy 3
Nonparticipation	*Therapy*	-	-	-	-	-
	Manipulation	-	-	-	-	-

5. Conclusions

The study was examined Citizen Participation Practices in Thailand Environmental Policy under the case study, the Environmental Quality Management Plan 2012-2016. This could be seen that Thailand had an appropriate framework on citizen participation in the environmental policy formulation process regarding the provisions of 2007 Constitution which should have successfully implement the environmental policy in accordance with the anticipated perceptions of citizen. Unfortunately, the provisions of 2007 Constitution could not efficiently transform to the environmental policy formulation in each process.

All of Citizen Participation Practices were located in the category of *Tokenism*, where the citizens became involved but only to certain extent. Particularly, in the most important process such a decision-making process, citizens did not have the appropriate role to govern themselves which the majority of authority was given to the government through the officials and the politicians, even the NEQ Act 1992 was opened the opportunity to the selected citizens from private sector to join in the National Environment Board (NEB) but it was very few amount when compared to the proportion with the other board members (8: 16 persons), this was not meaningful in term of citizen control which should be appropriately conducted in the democracy country such Thailand.

Furthermore, due to the public administration on

environmental management in the international circumstance, the study did not found any linkage between Citizen Participation Practice and Bilateral and Multilateral Environmental Agreements (MEAs) under the policymaking context. This weak point would affected to the environmental service circumstance of citizen and inevitably leads more interest conflicts to Thai society when Thailand's government necessary to bonded with the international environmental agreements in the future.

Nevertheless, in the optimistic view, this study was shown that Thailand did not have the Citizen Participation Practices in the category of Nonparticipation and the environmental policy formulation was opened opportunity to amend and develop through the process of the environmental legislation under the implementation of EQM Plan 2012-2016 (case study). Therefore, it was necessary to properly implement the policy in this aspect to converge the purpose of citizen participation for relieving the conflicts and developing the sustainable co-operation movement in the future of Thailand's natural resource and environmental management.

References

[1] The Constitution of the Kingdom of Thailand B. E. 2550 (in Thai), August 24, 2007, Royal Gazette, Thailand. Vol. 124.

[2] ThaiPublica News (Online), 2015, Statistic of the administrative cases concerning environmental issues that citizens sue the Public Agency through the Administrative Court of Thailand between 2011-2015 (in Thai). Retrieved on October, 10, 2015 from http://thaipublica.org/2015/03/14th-anniversary-admincourt/.

[3] Thailand Environment Institute (TEI); King Prajadhipok's Institute (KPI); Sustainable Development Foundation (SDF), 2007, The Access Initiative (TAI) Assessment in Thailand: Synthesis report on "Good Environmental Governance: Public Participation Indicators for Thailand's Sustainable Development (Third Assessment) (Synthesis report under the Project Policy Strategy on Tropical Resource Base under the National Human Rights Commission of Thailand): Thailand Environment Institute (TEI).

[4] The Government Public Relations Department, 2013, Thailand's Constitutions Moving from Representative Democracy to Participatory Democracy. Retrieved on October 10, 2013 from http://thailand.prd.go.th/ebook/review/content.php ?chapterID=61.

[5] Sutthiprasid, Ganniga, 2011, Evolution of Thai Environmental Law, Administrative Court of Thailand: Trimester Newsletter. Vol. 2, April – June 2011. Retrieved November, 30, 2014 from http://www.admincourt.go.th/00_web/environment/newsletter 02.htm.

[6] Jane, Roberts, 2004, Environmental Policy, London: Routledge (an imprint of the Taylor & Francis Group).

[7] Chaisomphob, TAWEEP; Sa-nguanmanasak, JATURONG; and Swangjang, KANOKPORN, 2004, Role of Public Participation in Planning Power Plant Projects in Thailand, Thammasat International Journal for Science and Technology Vol. 9; No. 1; January-March 2004.

[8] Cogan, Sharpe and Hertberg, 1986, "Citizen Participation", Chapter 12 in The Practice of State and Regional Planning, edited by So, FrankS. II, Hand, Irving, and McDowell, Bruce D., American Planning Association, p. 283-308.

[9] Arnstein, Sherry R., 1969, A Ladder of Citizen Participation, Journal of the American Institute of Planners (JAIP), Vol. 35, No. 4, July 1969, pp. 216-224.

[10] Office of Natural Resources and Environmental Policy and Planning (ONEP) Thailand, 2013, Environmental Quality Management Plan (in Thai). Retrieved on October, 30, 2013, from http://www.onep.go.th/index.php?option=com_content& view =article&id=2761&Itemid=173.

[11] The National Economic and Social Development Board (NESDB), 2011, The Eleventh National Economic and Social Development Plan B. E. 2555 – 2559 (A. D. 2012 – 2016) (in Thai), Office of the National Economic and Social Development Board, Bangkok, Thailand.

[12] Enhancement and Conservation of National Environmental Quality Act B. E. 2535 (in Thai), 1992, Royal Gazette, Thailand. Vol. 109.

Policy Intervention to Reduce Energy Consumption and Mitigate Environmental Emission in Cement Industries of Nepal

Pradeep Singh[1], Shree Raj Shakya[2]

[1]Nepal Electricity Authority, Durbarmarg, Kathmandu, Nepal
[2]Center for Energy Studies, Institute of Engineering, Tribhuvan University, Lalitpur, Nepal

Email address:
pradeepsingh.nea@gmail.com (P. Singh), shreerajshakya@gmail.com (S. R. Shakya), shreerajshakya@ioe.edu.np (S. R. Shakya)

Abstract: Nepal is a developing country with huge potential of investment in agriculture, cement and hydropower sector. Industrial development in Nepal is at a pre-mature state and requires lot of technical and financial investment. Cement industry is one of the potential industries to grow in the future, mainly because of the reserved limestone and increasing developmental activities. This study analyzed the energy and environmental implications of implementing best available technologies in cement industries of Nepal by using Long-rand Energy Alternatives Planning system (LEAP) framework. Production capacity of cement in 2014 is estimated to be 2.46 million MT which is expected reach 25.41 million MT by 2030. The final energy demand for the base year, 2014 is 5.4 PJ. It would increase to 13.69 PJ, 16.91 PJ and 25.67 PJ in 2030, under normal (BAU), medium growth (MG) and high growth (HG) scenarios respectively. Compared to the BAU scenario, the cumulative energy demand would increase by 21.46% for MG scenario and 78.00% for HG scenario during 2014 to 2030. The CO_2 emission for the base year 2014 is estimated to be 365.40 thousand MT. It would increase to 1,540.70 thousand MT, 2,292.90 thousand MT and 4105.60 thousand MT in 2030, under BAU, MG and HG, respectively. Compared to the BAU scenario the cumulative CO_2 emission would grow as high as 78.06% under HG scenario. This indicates the need for introducing the energy efficient and low carbon technologies to address the issues related to energy supply security and environmental degradation. This study also analyzed the three policy intervention scenarios consisting of introduction of efficient technology (EFF) scenario, CO_2 emission mitigation (MIT) scenario and waste heat recovery for power generation (WHRPG) scenario. Under EFF scenario, the cumulative energy consumption would decrease by 11.67% during 2014 to 2030 as compared to the BAU scenario. Likewise, CO_2 emission would decrease by 33.64% under MIT scenario as compared to the BAU. Under WHRPG scenario, there would be cumulative electricity generation of 1,446.31 GWh worth NRs. 9.11 billion as compared to the BAU scenario during the study period. This study also indicates the need of formulating appropriate energy efficiency and climate change related policies of the country.

Keywords: Energy, Environment, Efficiency, Cement Industry, Modeling, Alternative Fuels, Waste Heat Recovery

1. Introduction

Industries, considered as backbone of the national development, are facing different energy related problems at present in Nepal. Industrial sectors in the country account for 3.3% (13369.8 TJ) of the total primary energy demand, after residential sector and transport sector in 2010 [1]. However, due to the energy crisis, the energy demand has not been adequately fulfilled. Hence, it is necessary to develop appropriate framework for addressing future energy security along with issues related to greenhouse gas (GHG) emissions and climate change.

Nepal is a developing country with huge potential of investment in the sectors like agriculture, cement and hydropower. Industrial development in Nepal is at a pre-mature state and requires lot of technical and financial investment. Cement industry is one of the potential industries to grow in the future, mainly because of the availability of untapped limestone and increasing developmental activities.

Lime deposits totaling 1.07 billion tons comprising 540 million tons of proven, 110 million tons of semi-proven and 420 million tons of feasible limestone deposits have been discovered in different areas of the country. A total of 13 cement industries with 8,450 ton per day capacity have been established and are in operation using those limestone as raw materials, while 10 cement industries are under construction [2]. Moreover, development activities and commercialization would make a greater impact on the demand of cement in the domestic market.

Also, around 26 clinker grinding industries are operating throughout the country. The grinding industries rely on the domestic and imported clinker, as raw material. For the past four years (2009 to 2012) the cumulative clinker import was around 5 million MT worth around NRs. 30 billion [3]. A total of 7 rotary plant and 9 VSK plants are in operation.

2. Literature Review

With the increasing concern toward energy security and environment issues, new technologies have been introduced and conceived in different energy intensive industries especially in the developed countries. Being an energy intensive industry there are numerous areas for increasing energy efficiency and reducing environmental emissions.

2.1. Cement Manufacturing Process

Cement manufacturing include intensive process with throughput at different stages of raw material processing. Raw materials should be mixed precisely to manufacture cement. Cement manufacturing includes first the production of clinker from the raw mix composed of mined limestone, iron ore, bauxite, clay and coal. The cement clinker requires appropriate amount of compositions of the elements calcium, silicon, aluminum and iron. All these raw materials together with the fuel ash must be combined to form the typical clinker composition as shown in Table 1.

Table 1. *Composition of dry cement manufacturing process [4].*

Elements	Composition (%)
CaO	65 ± 3
SiO_2	21 ± 2
Al_2O_3	5 ± 1.5
FeO_3	3 ± 1

In dry cement manufacturing process, which uses nearly dry raw mix containing less than 20% moisture by mass. However, in a wet process, water is added to the raw mix to form slurry and then is transported to the kiln. Raw meals are grounded, blended, pre-calcined, and burned in manufacturing cement.

In a cement manufacturing process, limestone and calcium, silicon, aluminum and iron oxides are crushed and then milled into a raw meal. This raw meal is blended in blending silos and is then heated in the pre-heating system. This will dissociate carbonate to calcium oxide and carbon dioxide. A secondary fuel is supplied to the preheating system so that temperature is sufficiently high. The meal then passed through the kiln for heating. Then a reaction takes place between calcium-oxide and other elements. This reaction will produce calcium silicates and aluminates at about 1500°C. Primary fuel is used to keep the temperature high enough in the burning zone for the chemical reactions to take place. A nodular product named clinker is produced and then allowed to leave the kiln. The clinker will be inter-ground with gypsum, limestone and/or ashes to a finer product called cement [4].

2.2. Energy Consumption in Cement Industries

Cement production is an energy intensive industry. Energy utilization in cement production accounts for 50–60% of the total production costs [5]. Thermal energy accounts for about 20–25% of the cement production cost [6]. The typical electrical energy consumption of a modern cement plant is about 110–120 kWh per ton of cement [7]. The main thermal energy is used during the clinker making process, while electrical energy issued for cement grinding and other auxiliaries [8]. Fig. 1 shows electrical and thermal energy flow in a cement manufacturing process.

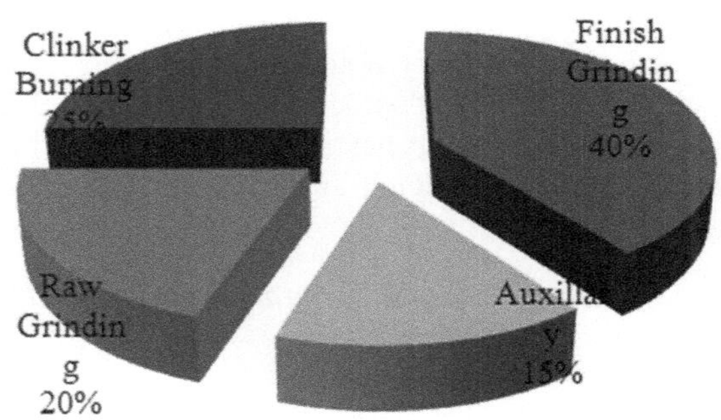

Figure 1. *Energy Distribution among Cement Manufacturing Equipment.*

2.3. Energy Analysis and Modeling in Cement Industries

Demand for cement depends on various socio-economic parameters. Cement production in China has been studied as a function of GDP growth, cement consumption per capita, non linear effect and saturation effect [9]. Similarly, study in Iran has verified the impacts of population and GDP growth on cement demand and production over the next 20 years [10]. Long-range Energy Alternative Planning system (LEAP) Framework has been extensively used to study the impact of various policies on the reduction of CO_2 emission in Chinese [9] and Iranian [11] cement industries.

Lawrence Berkeley National Laboratory (LBNL) has developed a guidebook which contains different energy efficiency improvement technologies and measures which are commercially available for the cement industries throughout the world [12]. Likewise, Madlool et. al [4] have reviewed the different energy efficiency technologies used in cement industries.

Different energy analysis techniques have been employed in industrial energy system modeling. Modeling techniques like decomposition of energy trends, econometric methods, and 'top-down' models, 'bottom-up' models and industry-specific micro-economic analyses are common in analyses of industrial energy [13]. The bottom-up CO_2 abatement cost curve (ACC) model was used for the Thai cement industry to determine the potentials and costs of CO_2 abatement, taking into account the costs and CO_2 abatement of different technologies. Different 41 CO_2 abatement technologies and measures for the cement industry were analyzed for cement industries in Thailand [14].

Ref. [15] developed an optimization model for CO_2 reduction in cement production process. The economic model on the basis of best selection strategy with the least cost was analyzed for a cement industry and found that up to 23.6% reduction in CO_2 emission per ton of cement can be achieved. Ref. [16] analyzed the co-benefit of change in local air pollutants emission while introducing CO_2 mitigating technologies and goals in the cement industries of China by using bottom up optimization model. Similarly energy efficiency improvement and CO_2 reduction potential through the use of 22 energy efficiency measures were studied for India by using forward looking bottom-up Conservation Supply Cure (CSC) model [17]. A statistical approach has been used for finding the relationship between kiln parameters with clinker quality for Spanish cement factory and found that a decrease in the kiln Sintering Temperature (standardized at 50°C) yielded produced the same level of quality in the final product, thus paving the way for reduction in petroleum coke consumption and reduction in CO_2 emissions [18].

2.4. Use of Alternative Fuels and Generation of Electricity in Cement Industries

In 2009, International Energy Agency, estimates the CO_2 emissions from cement production to be about 5% of the total global anthropogenic CO_2 emissions [19]. Different alternative fuel, used for thermal energy, emitting lesser GHGs are been recommended for the cement production. Alternative fuels being used in the cement industries can be classified as solid, liquid or gaseous state [20].

Energy intensive industries like cement release huge amount of low grade heat as a result of heat treatment and clinker burning. As a result, the exit heat, at a temperature lower than 400°C, is released as waste from the clinker cooler and pre-heater. The waste heat being released in the environment can be utilized in a cogeneration power plant established in the plant facility [21]. Different researches and studies have concluded that the waste heat can be re-used to generate electric energy, basically known as Waste Heat Recovery for Power Generation (WHRPG). WHRPG has been implemented in cement industries throughout the globe along with CDM program. The various cement capacity establishment in India [22], Thailand [23] and [24] have been benefited through the CDM projects.

3. Methodology

3.1. Data Collection

Detailed data collection questionnaire were developed and used to collect information on cement production and energy use from the surveyed plants (5 rotary plants, 4 shaft kiln and 12 clinker based plant). Similarly, the data from the baseline study of cement industry for grinding and clinker manufacturing were also used in the study [25]. The baseline study, conducted by GIZ, includes the study in 26 cement industries, and intends to increase the efficiency level of the industries by implementing different efficiency improvement measures. Hence, the compilation of the primary and survey data encompasses around 86% of cement production in Nepal.

The data forms requested for specific information on the plant lines, their age, their clinker and cement-making capacity, and annual data of clinker and cement productions. The energy consumption sheet were developed for important processes like raw quarrying, raw material grinding, additive processing, pyro-processing and clinker cooling, cement grinding and auxiliaries. Similarly, the questionnaire also requested for the information on the implementation of the recent available best practice technologies, as found in different literature. Table 2 shows the fuel consumption for the year 2010/11. Thermal energy was basically generated from coal bituminous and petroleum coke. Coal bituminous is used in pyro-processing of rotary kiln and petroleum coke in vertical shaft kiln technology. Electricity comprises the grid electricity and electricity generated by on-site diesel generators. Diesel generators produce around 26% of electricity in the base year. Likewise the clinker to cement ratio is found to be 0.88.

Table 2. *Year 2010/11 Fuel Consumption in Cement Industries [25].*

Fuel Type	Consumption
Coal (GJ)	3,836,603
Electricity (GWh)	168

3.2. Benchmarking and Energy-Saving Potential

Benchmarking is a commonly used term that generally means comparing a defined characteristic of one facility to other facilities or other 'benchmarks'. This study focuses in the energy consumption benchmarking with the international best practice. Different international energy benchmarks are considered as found in literature [4, 12, 26] and the reference from BEST-Cement[1]. A spread sheet model is generated and domestic energy consumption is compared with the international best practice using and energy intensity index (EII), calculated on the basis of facility's energy intensity and the benchmark energy intensity, as in Eq. (1) [27].

1 BEST-Cement is the cement industry energy benchmarking software package developed for China by Lawrence Berkeley National Laboratory. It can be downloaded from: http://china.lbl.gov/research/industry /benchmarking/best-cement/best-cement-china.

$$EII = 100 * \frac{\sum_{i=1}^{n} P_i * EI_i}{\sum_{i=1}^{n} P_i * EI_{i,BP}} = 100 * \frac{E_{tot}}{\sum_{i=1}^{n} P_i * EI_{i,BP}} \quad (1)$$

where:

EII=energy intensity index.

n=number of products to be aggregated;

EI_i=actual energy intensity for product i;

$EI_{i,BP}$ = best practice energy intensity for product.

P_i = production quantity for product i.

E_{tot}= total actual energy consumption for all products.

The EII is one of benchmarking method which performs analysis by comparing the production intensity of the facility to the benchmark or reference intensity. The international benchmark will have an EII of 100. In reality, plant facilities will have EII greater than 100, which shows the potential saving opportunities.

Table 3. *Fuel Energy Intensity at Base Year and Projected Years.*

Scenario	Technology	Final energy intensity by technology (GJ/tclinker)				
		2011	2014[2]	2015	2020	2030
Base Case	Rotary Kiln	4.54	4.54	4.54	4.54	4.54
	Shaft Kiln	7.89	7.89	7.89	7.89	7.89
Efficiency	Rotary Kiln	4.54	4.54	3.6	3.4	3.4
	Shaft Kiln	7.89	7.89	7.89	0	0
Best Practice	Rotary Kiln	4.54	4.54	3.6	3.4	3.4
	Shaft Kiln	7.89	7.89	7.89	0	0

3.3. Demand Forecast

The historical cement production data have been collected from the economic survey 2014/15 [2]. The data presented in the economic survey report do not seem to present correct data. Even the total production of the visited industries is more than the data presented. Hence, regression analysis is used to determine the actual demand of cement.

The end-use demand of cement is estimated using the following equation, as mentioned in different literatures [28, 29, 30].

$$ESD_{cement,t} = \left(\frac{VA_t}{VA_o}\right)^{\beta} x\ ESD_{cement,0} \quad (2)$$

where,

$ESD_{cement,\,t}$ = end use service demand in year t for cement sector,

VA_t = GDP value added in the cement sector in year t,

β = sectoral value added elasticity of demand for cement industry,

The future growth of GDP is forecasted using the regression model from the data available from various sources. The GDP growth rate is calculated at low growth rate which is considered to be 4.45% per year. Similarly, GDP growth rate for medium and high growth rate are considered to be 5.5% and 7.5% respectively. The forecasted GDP growth rate is used to generate the future growth rate of value addition of manufacturing sector. The sectoral value added elasticity for cement sector is calculated using historical data of service demand and GDP value addition in Eq. (2). The end-use cement demand, in the base year is considered to be the cement demanded by the domestic market which is fulfilled by production within the country. Unauthorized expertise opinion claims that this demand is just enough to meet around 50 percent of total domestic demand.

Twenty year historical data from 1990 to 2014 for national level cement demand [2] and manufacturing value addition (constant 2005 US$) [31] are obtained from various sources. The elasticity of demand for cement industry was determined to be 2.16, which is comparable to the one found in literature [28]. Year 2014 estimates production capacity of cement being 2.46 million metric ton which is expected to reach 10.37 million MT in 2030, with a cumulative cement demand of 96.00 million MT, from 2015 to 2030 in normal (BAU) growth rate scenario. Similarly, the cement demand would rise to 14.88 million MT in 2030 with cumulative demand of 117.14 million MT, in Medium Growth (MG) rate scenario. In case of high growth (HG) rate scenario, the cement demand rises from 2.46 million MT in 2014 to 25.41 million MT, with cumulative of 172.90 million MT. The historical and future cement demand is shown in Fig. 2.

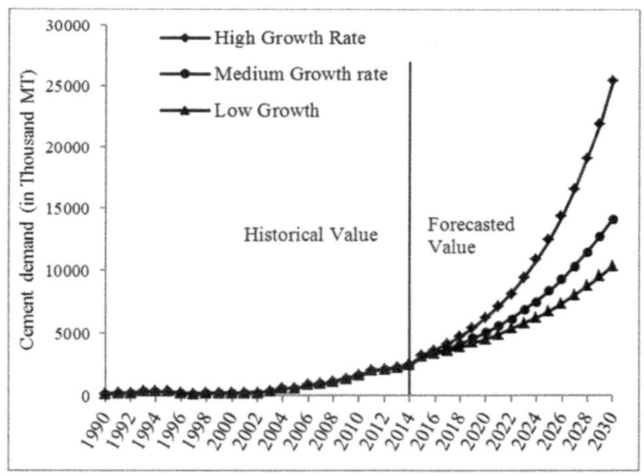

Figure 2. *Historical and Forecasted Cement Demand Projection (1990-2030).*

Since the total cummulative cement production in fifteen years accounts for 9%, 11% and 16% of the total limestone reserved, in normal, medium and high growth, respectively. Hence, the elastic growth is considered valid and is used in the modeling and scenario analysis.

3.4. Scenario Assumption

The Long-range Energy Alternatives Planning system (LEAP)[3] modeling framework is used for the scenario based modeling and analysis of potential energy savings and CO_2 emissions reduction. The scenarios have been projected up to

2 Calculated from the primary data

3 LEAP is a scenario-based energy-environment modeling tool of which scenarios are based on "comprehensive accounting of how energy is consumed converted and produced" [34]

2030. To analyze the impact of different energy efficiency and carbon reduction measures and policies, three scenarios are constructed consisting of Business as Usual (BAU) scenario, Energy Efficiency (EFF) scenario and GHG Mitigation (MIT) scenario. Scenarios have been defined on the basis of various national and international level goals and targets [19].

The BAU scenario has been considered as reference scenario without any policy intervention for energy efficiency improvement and GHG mitigation. The efficiency scenario is constructed with the objective to observe the energy consumption pattern with energy efficiency measure determined by benchmarking. On the contrary, Mitigation scenario observes the effect on GHG emission due to the inclusion of energy efficiency measures, use of alternative fuel and reduction of clinker to cement.

An average of 36 kWh of electricity can be produced per MT clinker through WHR power generation [9]. The electricity generation through WHR is calculated to forecast the electricity produced by the cement industries. Electricity production is assumed to be utilized by the industries so as to reduce the dependency on electricity supplied from the grid supply.

The average emission factor of 73.3 ton CO_2 per TJ for alternative fuels is assumed, which indicates the use of alternative fuel could reduce about 23% of overall CO_2 emissions as compared to burning bituminous coal of which assumed emission factor is 94.5 ton CO_2 per TJ [32]. The cement process emission is calculated as 0.547 ton CO_2 per clinker production [33].

3.5. General Assumption and Base Year

The following were the assumptions for the reference case (BAU) scenario

- The base year is considered to be 2014.
- The major energy intensive fuels like electricity from grid and diesel gen-sets were considered for electrical energy.
- Furnace oil and Coal bituminous or coke is considered for thermal energy.
- The calculated value of final energy demand in the base year 2014 is 5.42 PJ.

3.6. Growth Scenario

Growth scenario is created to access the energy consumption and environment emission in various growth rate of cement demand. The cement demand was determined in different growth scenarios as found out by the Eq. (2). The growth scenario is simply the representation of baseline scenario (i.e. 2014) projected up to 2030. This scenario does not incorporate efficiency improvement and emission reduction measures.

Growth scenario is categorized as Business as Usual, Medium Growth and High Growth. Each scenario resembles the energy consumption and environment emission at 4.45%, 5.5% and 7.45% growth rate of GDP (constant 2005 US$). Also, the Clinker to Cement (CC) ratio is assumed to be reduced from 0.88 in 2014 to 0.75 by 2030 [19], for all scenarios.

3.7. Efficiency Scenario

Efficiency (EFF) scenario is constructed to access the potential in reduction of energy consumption and emission by implementing the efficiency improvement measures.

This scenario presumes the efficiency improvement of cement industries by the implication of efficiency measures and reaches the best practice value of specific energy intensity. At present, different energy efficiency programs are being carried out in cement industries, with assistances from different national and international organization. Hence, it is assumed that efficiency programs will also be made viable in future; ultimately bring the efficiency to international best practice standard. Also, in future, new commissioning industries are assumed to be installed with the basic amenities of standard best practice values. It is therefore, plausible to assume that the best practice values of specific electrical energy and specific thermal energy would reach the international standard value with the succession of time. Table 4 and 5 shows the assumed gradual efficiency improvement in Specific Energy Consumption (SEC) fuel. In the efficiency scenario, the shaft kiln technology has been considered to be phased out by 2020, due to its low production volume and quality.

Since, there are three growth scenarios, corresponding to each growth scenario there will be an efficient scenario. Hence, three efficiency scenarios corresponding to three growth rates are considered.

Table 4. *Assumed efficiency improvement in SEC (fuel energy).*

Scenario	Technology	Final energy intensity by technology (GJ/MT clinker)			
		2014	2020	2025	2030
Efficiency	Rotary Kiln	4.54	4.0	3.6	3.4
	Shaft Kiln	7.9	7.9	0	0

Table 5. *Assumed efficiency improvement in SEC (electricity).*

Scenario	Technology	Final energy intensity by technology (kWh/MT cement)		
		2014	2020	2030
Efficiency	Rotary Kiln	125.5	90	90
	Shaft Kiln	214.2	0	0
	Clinker Based	48.68	35	35

3.8. Mitigation Scenario

CO_2 Mitigation (MIT) scenario analyzes the effect of substituting coal by alternative fuel in cement industries and also the penetration of waste heat recovery for power generation. This scenario considers all the efficiency measures, aforementioned in section 3.7. This scenario is developed with the opined objective of curtailing emission from cement industries, as targeted in the international level [19].

Table 6, 7 and 8 shows the penetration of alternative fuel, Waste Heat Recovery for Power Generation (WHRPG) and fuel efficiency. The penetration of WHRPG and alternative fuel is not considered in Growth and Efficiency scenario.

Table 6. *Assumed penetration of alternative fuels (in % of coal substitution).*

Scenario	2014	2015	2020	2025	2030
MIT	0	1	5	11	21

Table 7. *Assumed Penetration of Waste Heat Recovery for Power Generation (% of clinker production).*

Scenario	2014	2020	2025	2030
MIT	0	20	60	100

Table 8. *Fuel energy intensity at Base Year and Projected Years.*

Scenario(s)	Technology	Final energy intensity by technology (GJ/t clinker)			
		2014[4]	2015	2020	2030
BAU, MG and HG	Rotary Kiln	4.54	4.54	4.54	4.54
	Shaft Kiln	7.9	7.9	7.9	7.9
MIT	Rotary Kiln	4.54	4.0	3.6	3.4
	Shaft Kiln	7.9	7.9	N/A	N/A

4. Result and Analysis

4.1. Benchmarking Result

The domestic benchmarking of the rotary kiln technology has been performed and the template generated incorporates different process flow of the cement industry. Special focuses were given to the rotary kiln technology. Since, all plant do not produce cement the energy per unit clinker basis is formulated.

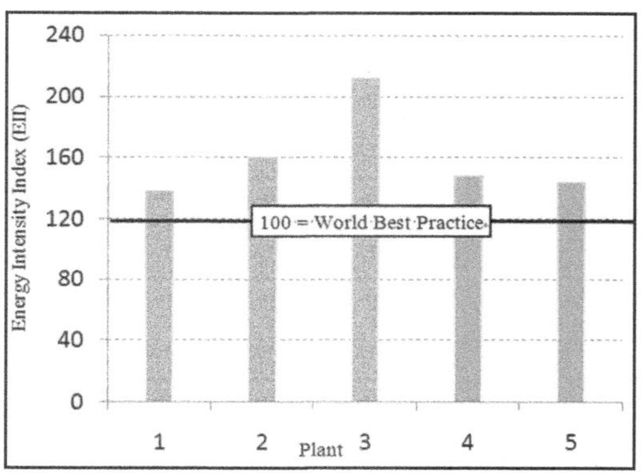

Figure 3. *Energy Intensity Index (EII) benchmarking of surveyed plant.*

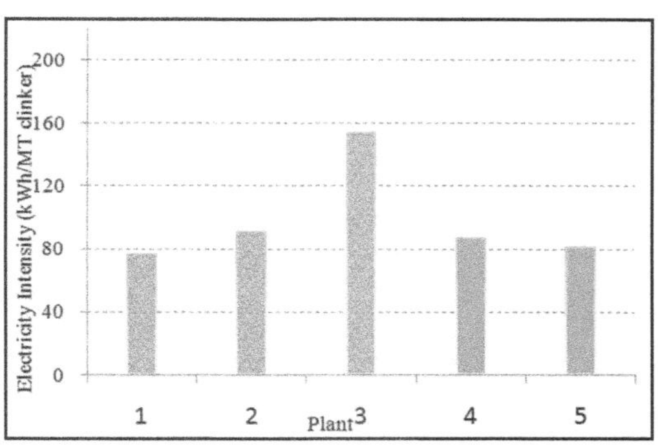

Figure 4. *Electricity Intensity in surveyed plant.*

4 Calculated from the primary data

The difference between the actual plant and its corresponding best-practice technology illustrates the technical potential for energy improvement. In order to compare the 5 plants, an energy intensity index (Eq. 1) is used to illustrate the distance between best practice and the plants improvement opportunity.

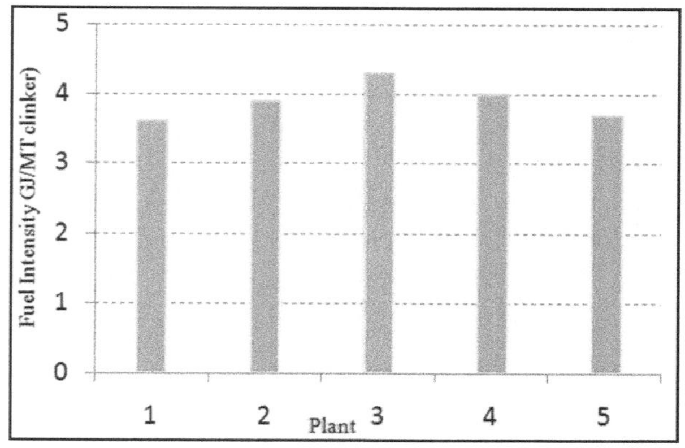

Figure 5. *Fuel Energy Intensity in surveyed plants.*

Fig. 3 shows the EII score for the 5 plants compared to the world best practice based on primary energy use. All 5 plants scored above the 100 value which hints the opportunity for efficiency improvement above the 100 value. Plant 1 scores the lowest value (138) suggesting 28% of technical potential saving. Likewise, Plant 3 scores the highest value (212) suggesting 53% of technical potential saving.

Fig. 4 and 5 shows the electricity intensity and fuel energy intensity of clinker production. The average values have been used for the scenario analysis. Plant one has the lowest electricity energy consumption of 79 kWh/MT clinker. Plant three has the highest electricity energy consumption of 154 kWh/MT clinker. Considering the cement grinding, the average electricity energy consumption for rotary kiln is 125.5 kWh/MT cement. Likewise, the average energy consumption for VSK is 214.2 kWh/MT cement.

Fig. 5 explains the fuel energy intensity in rotary kiln plants. The lowest fuel intensity is found in Plant one with 3.9 GJ/MT clinker. Similarly the highest fuel energy intensity is found in Plant 3 with 6.54 GJ/MT clinker. The average of the plant is found to be 4.54 GJ/MT clinker. Similarly for the energy consumption VSK plant is found to be 7.9 GJ/MT clinker.

In clinker based industries, the average electricity energy consumed for clinker grinding is 48.68 kWh/MT of cement.

4.2. Result of Growth Scenario

4.2.1. Energy Demand Projection

The energy demand for the base year 2014 is 5.4 PJ. The final energy demand would increase to 22.86 PJ, 31.26 PJ and 55.97 PJ in 2030, under BAU, MG and HG respectively.

The cumulative energy demand for BAU, MG and HG scenario would be 216.89 PJ, 263.43 PJ and 386.19 PJ respectively. Compared to the BAU scenario, the cumulative energy demand would be increase by 21.46% for MG scenario and 78% for HG scenario. The energy demand projection for growth scenario is shown in Fig. 6.

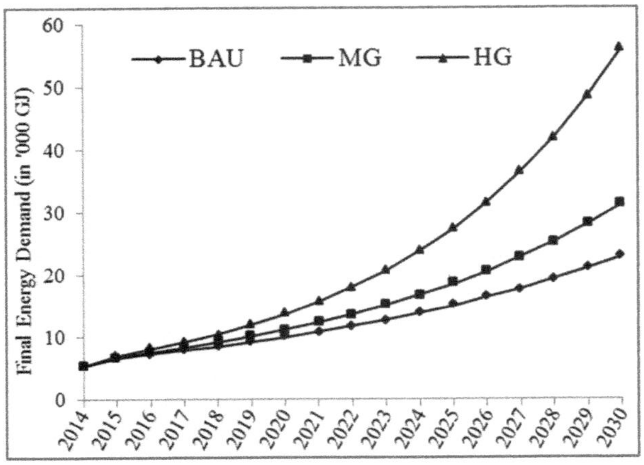

Figure 6. *Final Energy Demand Projection in BAU, MG and HG scenarios (2014-2030).*

4.2.2. Emission Projection

Likewise, the CO_2 emission for the base year 2014 is 365.40 thousand MT. The final CO_2 emission would increase to 1,540.70 thousand MT, 2,292.90 thousand MT and 4,105.60 thousand MT in 2030, under BAU, MG and HG respectively. The cumulative energy for BAU, MG and HG scenario would be 15.91 million MT, 19.32 million MT and 28.33 million MT, respectively. Compared to the BAU scenario, the cumulative emission would rise by 21.46% for MG scenario and 78.00% for HG scenario, which is in exact match with the energy consumption. The emission projection for growth scenario is shown in Fig. 7.

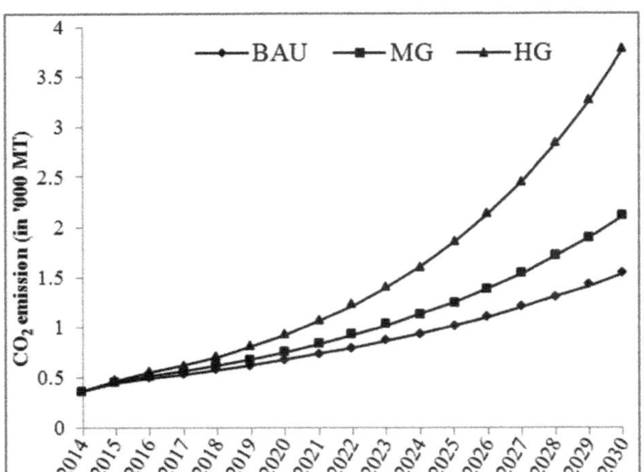

Figure 7. *CO_2 emissions in BAU, MG and HG scenarios (2014-2030).*

4.3. Result of Efficiency Scenario

The efficiency scenario is constructed as an efficiency improvement in technology employed in cement industries. The scenario can be studied as an efficiency improvement in the three growth scenario. In other words, efficiency scenario studies the effect of technological improvement in three growth scenario viz. EFF BAU, EFF MG and EFF HG scenario.

The EFF BAU, EFF MG and EFF HG scenario projects the total cumulative final energy demand to be at 191.58 PJ, 229.87 PJ and 280.32 PJ, respectively. Compared to the BAU scenario, under EFF BAU scenario, 11.67% of total cumulative energy consumption can be reduced. Likewise, in between MG and EFF MG scenarios, 12.74% of total cumulative energy can be conserved. Finally, EFF HG scenario, 27.42% of energy can be conserved compared to HG scenario.

The energy demand projections in efficiency scenario as compared with their corresponding growth scenario are shown in Fig. 8.

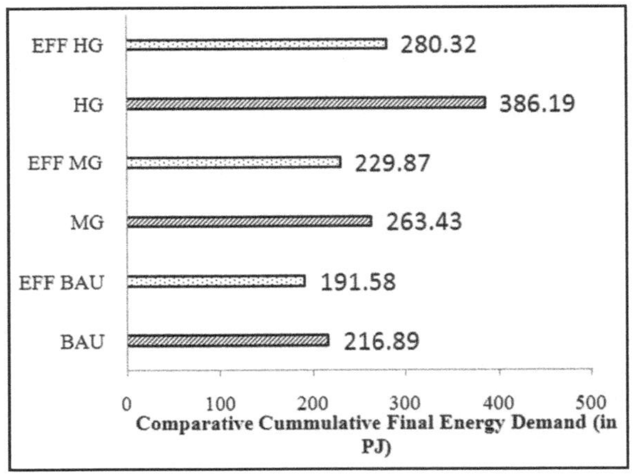

Figure 8. *Final Energy Demand in Efficiency Scenario (2014-2030).*

4.4. Result of Mitigation Scenario

The MIT BAU, MIT MG and MIT HG scenario projects the total cumulative CO_2 emissions to be at 9.70, 11.60, and 16.61 million MT, respectively. Compared to the BAU scenario, in MIT, 33.64% of total cumulative emission can be reduced. Likewise, between MG and MIT MG scenario, 34.59% of CO_2 emission can be mitigated. Finally, under MIT HG scenario, 36.20% of CO_2 emission can be reduced as compared to HG scenario.

The reduction in emission can be credited to the use of alternative fuel and phase-out of high energy intensive cement industries, Vertical Shaft Kiln Technologies.

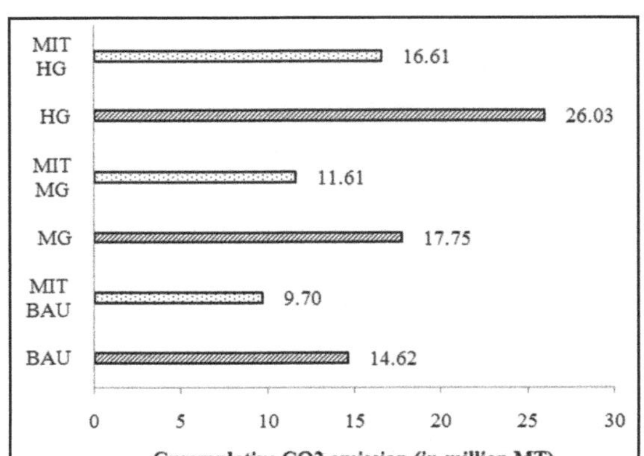

Figure 9. *Cumulative CO_2 emission comparison between respective Growth and Mitigation Scenarios (2014-2030).*

The emission projection scenarios compared with their corresponding growth scenarios is shown in Fig. 9, for BAU, MG and HG scenarios, respectively.

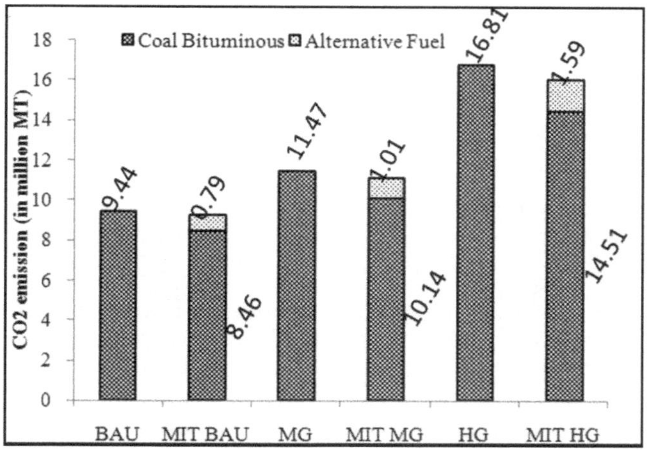

Figure 10. Cumulative CO_2 emissions- substitution of Coal Bituminous with Alternative Fuel in all scenarios (2014 -2030).

Similarly, the substitution of coal bituminous with alternative fuel has a positive impact on the reduction of CO_2 emission. Up to 4% of the total emission can be reduced, as noticed in Fig. 10.

4.5. Effect of Penetration of Waste Heat Recovery for Power Generation (WHRPG)

WHRPG is considered only in the mitigation scenarios. The production of electricity, by WHRPG in the limestone based industries is considered from 2020.

Considering scarcity of electricity in present context along with the future energy supply security, WHRPG can contribute to the power generation through waste heat studied combined with the growth scenario viz. BAU, MG and HG.

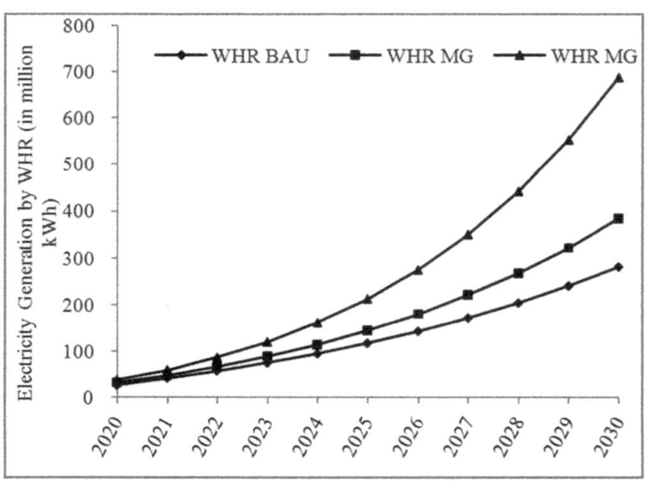

Figure 11. Electricity Generation by WHRPG in WHR BAU, WHR MG and WHR HG scenarios (2020-2030).

The subsequent growth of the production of electricity from waste heat in different growth scenarios can be accounted to the higher production of clinker. With the assumed penetration of WHRPG, for three growth scenarios, in BAU scenario cumulative of 1,446.31 GWh of electricity can be generated. Similarly, under MG and HG scenarios, 1,860.99 GWh and 2,988.35 GWh of electricity can be generated. This subsequently accounts for, around 20% of total cumulative electricity demand in cement industry.

Considering the current average electricity tariff of Nepalese Rupees (NRs.) 6.3 per kWh the total amount saving in BAU is NRs. 9.11 billion. Similarly, in MG and HG the total saving can be attained to be NRs. 11.72 billion and NRs. 18.83 billion.

4.6. Conclusion and Recommendations

In context of Nepal, the output of cement has shown a continuous growth during 1990 to 2014. The increase of the domestic demand and it fulfillment, in future perspective, expects more cement industries to commence. The growth in cement demand is studied in growth scenario as business as usual, medium growth and high growth. These growths are associated with 4.45%, 5.5% and 7.45% of GDP growth, respectively.

Log-linear regression method was used to determine the relation between value addition (VA) associated with the manufacturing industries and the end use cement demand. The elasticity of demand for cement industry was determined to be 2.16. Year 2014 estimates production capacity of cement being 2.46 million metric ton which is expected to reach 10.375 million MT in 2030, with a cumulative cement demand of 96 million MT, from 2015 to 2030 in normal (BAU) growth rate scenario. Similarly, the cement demand would rise to 14.88 million MT in 2030 with cumulative demand of 117.14 million MT, in Medium Growth (MG) rate scenario. In case of high growth rate scenario, the cement demand rises from 2.4 million MT in 2014 to 25.41 million MT, with cumulative of 172.9 million MT.

Long-range Energy Alternatives Planning system (LEAP) modeling tool was used to conduct the scenario based analysis projected for base year of 2014 up to 2030. Scenarios have been defined on the basis of various national and international level goals and objectives, which include the energy efficiency measures, waste heat recovery and use of alternative fuels.

Corresponding to the growth scenarios, energy demand were studied under the efficiency (EFF) scenario and the environmental effects were studied under mitigation (MIT) scenario. Hence, overall nine different scenarios were analyzed.

The energy demand for the base year 2014 is 5.4 PJ. The final energy demand would increase to 22.86 PJ, 31.26 PJ and 55.97 PJ in 2030, under BAU, MG and HG respectively. The cumulative energy demand for BAU, MG and HG scenario would be 216.89 PJ, 263.43 PJ and 386.19 PJ respectively. Compared to the BAU scenario, the cumulative energy demand would be increased by 21.46% under MG scenario and 78.00% under HG scenario.

Likewise, the CO_2 emission for the base year 2014 is 365.40 thousand MT. The annual CO_2 emission would increase to

1,540.70 thousand MT, 2,292.90 thousand MT and 4,105.60 thousand MT in 2030, under BAU, MG and HG respectively. The cumulative energy for BAU, MG and HG scenario would be 15.91, 19.32 and 28.33 million MT, respectively. Compared to the BAU scenario, the cumulative emission would rise by 21.46% for MG scenario and 78.00% for HG scenario, which is in exact match with the energy consumption.

The EFF BAU, EFF MG and EFF HG scenario projects the total cumulative final energy demand to be at 191.58 PJ, 229.87 PJ and 280.32 PJ, respectively. Compared to the BAU scenario, in EFF BAU scenario, 11.67% of total cumulative energy consumption can be reduced. Likewise, in MG and EFF MG scenario, 12.74% of total cumulative energy can be conserved. Finally, under EFF HG scenario, 27.42% of energy can be conserved compared to HG scenario.

Under MIT BAU, MIT MG and MIT HG scenario projects the total cumulative CO_2 emissions would be 9.70, 11.60, and 16.61 million MT, respectively. Compared to the BAU scenario, in MIT, 33.64% of total cumulative emission can be reduced. Likewise, between MG and MIT MG scenario, 34.59% of CO_2 emission can be mitigated. Finally, under MIT HG scenario, 36.20% of CO_2 emission can be reduced as compared to HG scenario.

The subsequent growth of the production of electricity from waste heat in different growth (WHRPG) scenarios can be accounted to the higher production of clinker. With the assumed penetration of WHRPG, for three growth scenarios, under BAU scenario cumulative of 1,446.31 GWh of electricity can be generated. Similarly, in MG and HG scenario, 1,860.99 GWh and 2,988.35 GWh of electricity can be generated. This subsequently accounts for, around 20% of total cumulative electricity demand in cement industry. Considering the current average electricity tariff of Nepalese Rupees (NRs.) 6.30 per kWh the total amount saving in BAU is estimated to be NRs. 9.11 billion. Similarly, in MG and HG the total saving can be attained as high as NRs. 11.72 billion and NRs. 18.83 billion, respectively.

Hence, the study opines on the following policy reformation to be implemented in cement industries of Nepal

- Categorization of energy efficiency measures at various stages of cement process, so as to meet the international best practice standards.
- Phase wise implementation of efficiency improvement and low carbon footprint programs.
- Identification and management of indigenous alternative fuel sources that can be used in cement industries to improve energy supply security.
- Formulation and implementation of the effective mechanisms for promoting the above policy intervention options in the existing industrial, energy and climate change related policies of the country.

References

[1] WECS, "Energy Sector Synopsis Report," Water and Energy Commission Secretariat (WECS), Kathmandu, Nepal, 2010.

[2] Ministry of Finance, GoN, "Economic Survey," Kathmandu, 2014.

[3] TEPC (Trade and Export Promotion Center), GoN, [Online]. Available: http://208.67.23.5/~itsoftc/projects/tepc/commoditywise.php. [Accessed 12 October 2013].

[4] N. A. Madlool, R. Saidur, M. S. Hossain and N. A. Rahim, "A critical review on energy use and savings in the cement industries," *Renewable and Sustainable Energy Reviews,* vol. 15, pp. 2042-2060, 2011.

[5] J. Wang, Y. Dai and L. Gao, "Exergy analyses and parametric optimization for different cogeneration power plants in cement industry," *Applied Energy,* vol. 86, no. 6, pp. 941-8, 2009.

[6] M. K. Singhi and R. Bhargava, "Sustainable Indian cement industry. In: Workshop on International comparison of Industrial Energy," 2010.

[7] G. G. Mejeoumov, "Improved cement quality and grinding efficiency by means of closed mill circuit modeling," 2007.

[8] L. M. Junior, "Sustainable development and the cement," 2003.

[9] K. Jing, Z. Nina, F. David, P. Lynn and Z. Nan, "Potential energy savings and CO_2 emissions reduction of China's cement industry," *Energy Policy,* vol. 45, pp. 739-751, 2012.

[10] A. S. Nastaran Ansari, "A system dynamics model for analyzing energy consumption and CO_2 emission in Iranian cement industry under various production and export scenarios," *Energy Policy,* vol. 58, pp. 75-89, 2013.

[11] F. Atabi, M. S. Ahadi and K. Bahramian, "Scenario analysis of the potential for CO_2 emission reduction in the Iranian cement industry," *World Renewable Energy Congress,* pp. 740-746, 2011.

[12] E. Worrell, C. Galitsky and P. Lynn, "Energy Efficiency Imporvement Opportunities for the Cement Industry," Berkeley, CA, 2008.

[13] L. A. Greening, G. Boyd and J. M. Roop, "Modeling of industrial energy consumption: An introduction and context," *Energy Economics,* vol. 29, no. 4, pp. 599-608, 2007.

[14] A. Hasanbeigi, C. Menke and L. Price, "The CO_2 abatement cost curve for the Thailand cement industry," *Journal of Cleaner Production,* vol. 18, no. 15, pp. 1509-1518, 2010.

[15] S. O. Ogbeide, "Developing an optimization model for CO_2 reduction in cement production process," *Journal of Engineering Science and Technology Review,* vol. 3 (1), pp. 85-88, 2010.

[16] Q. Tan, W. Zongguo and J. Chen, "Goal and technology path of CO_2 mitigation in China's cement industry: from the perspective of co-benefit," *Journal of Cleaner Production,* 2015.

[17] W. R. Morrow III, A. Hasanbeigi, J. Sathaye and T. Xu, "Assessment of energy efficiency improvement and CO_2 emission reduction potentials in India's cement and iron & steel industries," *Journal of Cleaner Production,* vol. 65, pp. 131-141, 2014.

[18] A. M. Castañón, S. García-Granda, A. Guerrero, M. P. Loreno and S. Angulo, "Energy and enviromental savings via optimisation of the production process at a Spanish cement factory," *Journal of Cleaner Production,* vol. 98, pp. 47-52, 2015.

[19] IEA/WBCSD, "Cement Technology Roadmap: Carbon emission reduction up to 2050," 2009.

[20] E. Mokrzycki and A. Uliasz-Bochenczyk, "Alternative fuels for the cement industry," *Applied Energy,* vol. 74, pp. 95-100, 2003.

[21] Zhejiang Energy Research Institute, "Waste Heat Recovery for Power Generation in Dry Cement Production," 2004.

[22] UNFCC, "Vikram Cement (VC): Energy efficiency improvemet by upgrading of preheater in cement manufacturing. CDM project design document," 2007.

[23] UNFCCC, "Siam Cement (Ta Luang) Waste Heat Power Generation Project, Thailand. CDM project design document," 2008.

[24] UNFCC, "Hebei Quzhai Cement 9000 kW Wase Heat Recovery Project - CDM PDD," 2007.

[25] GIZ, "Baseline Study of Selected Sector Industries to assess the potentials for more efficient use of energy," GIZ, Kathmandu, 2012.

[26] EPA (United States Environmental Protection Agency), "Available and emerging technologies for reducing greenhouse gas emission from the portland cement industry," 2010.

[27] A. Hasanbeigi, L. Price, H. Lu and W. Lan, "Analysis of energy-efficiency opportunities for the cement industry in Shandong Province, China: A case study of 16 cement plants," *Energy,* vol. 35, pp. 3461-3473, 2010.

[28] S. R. Shakya and R. M. Shrestha, "Transport sector electrification in a hydropwer resource rich developing country: Energy security, environmental and climate change co-benefits," *Energy for Sustainable Development,* vol. 15, pp. 147-159, 2011.

[29] R. M. Shrestha and S. Rajbhandari, "Energy and environment implications of carbon emission reduction targets: case of Kathmandu Valley, Nepal," *Energy Policy,* vol. 38, pp. 4818-27, 2010.

[30] TERI, "National Energy Map for India: Technology Vision 2030," Delhi, 2006.

[31] The World Bank, [Online]. Available: http://data.worldbank.org/indicator/NV.IND.MANF.KD. [Accessed 14 12 2014].

[32] PCC, "IPCC Guidelines for national Greenhouse Gas Inventories," 2006.

[33] CSI, "CO_2 Accounting and Reporting Standard for the Cement Industry, Version 2," 2005.

[34] SEI, 2011. [Online]. Available: http://www.energycommunity.org/documents/Leap2011UserGuideEnglish.pdf. [Accessed 2012].

Assessment of Solid Waste Management Practices and the Role of Public Participation in Jigjiga Town, Somali Regional State, Ethiopia

Yohanis Birhanu[1,*], Genemo Berisa[2]

[1]Department of Chemistry, Jigjiga University, Jigjiga, Ethiopia
[2]Department of Geography and Environmental Studies, Jigjiga University, Jigjiga, Ethiopia

Email address:
yohanisbirhanu@gmail.com (Y. Birhanu), genemo09@gmail.com (G. Berisa)

Abstract: Solid waste, which is a consequence of day-to-day activity of human kind, needs to be managed properly. Jigjiga City, like other cities in developing countries, faces problems associated with poorly managed solid waste operation. This study concerns about analysis of the city's current municipal solid waste management problems, opportunities and existing solid waste management practices and role of community participation. The survey was conducted in 03 and 05 kebeles. Different sampling methods were employed to select the study units including: stratified sampling, systematic random sampling, and purposive sampling. Though the bulk of the data collected were qualitative in nature, it was also supported by quantitative information collected through survey and secondary sources. The study discovered that there is low performance of SWM in the city mainly due to: lack of properly designed collection system and time schedule, inadequate and malfunctioning operation equipment, open burning of refuse, poor condition of the final dumpsite and less awareness creation among community which encouraged illegal dumping are the technical problem identified. Insufficient funds as well as lack of promotion on-waste reduction, recycling, absence of waste recovery, practice of energy option, waste separation and composting are among the management challenges facing the city. Social problems encountered include: lack of public awareness, unwillingness to pay, ill dumping manner (often around residence and any open spaces) and improper outlook for waste workers. Incompetence of organizations in terms of equipment required for operation and man power /staff qualifications, training and human resource developments and/or unreliable service are the institutional challenge that the city encountered in the sector. Finally, the study forwarded some important recommendations towards improving the waste management practices.

Keywords: Solid Waste Management, Disposal System, Community Participation, Jigjiga City

1. Introduction

Waste was an early problem of mankind, and a growing one that is of major concern to every nation of the world (Allende 2009). In early pre-industrial times, waste generation was not an issue as populations were smaller. Waste was disposed of in the ground where it would turn to compost to improve soil fertility. Waste management issues are coming to the forefront of the global environmental agenda at an increasing frequency, as population and consumption growth result in increasing quantities of waste. In the context of the above mentioned challenge a New Paradigm for waste management has emerged, shifting attention to resources efficiency and minimization of environmental impacts throughout the life cycle of waste management, from waste prevention to safe disposal. The primary objective of waste management is to give adequate protection to the general public and environment from harmful effects of waste.

Municipal SWM is a problem that is experienced by all counties in the world. It is an issue mostly witnessed in urban areas as a result of high surge in population growth rate and increase in per capita income thus posing a danger to environmental quality and human health (Javaheri 2006). Because of its nature, it has remained one of the major environmental problems man continues to face.

An investigation into waste management is now becoming increasingly critical in developing countries (Kyessi and Mwakalinga 2009). (Issam *et al.* 2007) report that waste management, which has generally been understudied, in developing countries is now receiving increasing attention in the literature. Municipal SWM problems have become more pronounced in recent years, as a result of inadequate collection and disposal of wastes. In most cities, wastes are not properly collected and where proper collection is ensured, only a small fraction receives proper disposals.

Urban waste management has been a challenge for municipalities and urban governments in the developing world, largely due to poor infrastructure, bureaucratic competence and limited institutional capacity of the municipalities. Municipalities throughout Ethiopia are not free of these problems as they have been facing major challenges with solid waste collection and landfill management.

According to Environmental Protection Authority (EPA) and World Bank study conducted in 2004, per capita amount of waste generated in Ethiopia ranged from 0.17 to 0.48 kg/person/day for urban areas to about 0.11 to 0.35 kg/capita/day for rural areas. The range depends on several factors such as income and season. The total generation of municipal solid waste in Ethiopia in

2003 is estimated to be 2.8 to 8.8 million tones. This can be split to approximately 0.6 to 1.8 million tons from rural areas and 2.2 to 7 million tons from urban areas (EPA/World Bank 2004).

In many of the cities in Ethiopia the municipality administration is responsible for waste collection. Though, there is a wide variation in performance in relation to waste collection in cities of Ethiopia, it has become a common business practice to have household waste to be pre-collected by individuals who are organized through formal or informal association. The pre- collected waste is then transferred into containers which are then collected by municipalities. Nevertheless, in many cities there are not enough containers to cover the population and vehicles are typically under maintenance or out of service for long periods of time. As there is very limited effort to recycle, reuse or recover the waste that is being generated; waste disposal has been the major mode of waste management practice. When considering SWM in general, it should be noted that it is rather small fraction of waste that ever reaches dump sites or landfills in Ethiopia. Some studies have shown that only 43% of waste is collected in the country are properly collected and disposed in open landfills. The remaining waste is indiscriminately disposed in drainage lines, open spaces, street sides or is informally burned.

According to Environmental Protection Authority and World Bank study conducted in 2004, per capita amount of waste generated in Ethiopia ranged from 0.17 to 0.48 kg/person/day for urban areas to about 0.11 to 0.35 kg/capita/ day for rural areas. The total generation of municipal solid waste in Ethiopia in 2003 is estimated to be 2.8 to 8.8 million tones. This can be split to approximately 0.6 to 1.8 million tons from urban areas and 2.2 to 7 million tons from rural areas (EPA/World Bank 2004). Jigjiga City, being the largest city in the Somali Regional State and one of the swiftly urbanizing centers in the country, has been grappled with an increasingly growing urban waste management problem. Taking per capita amount of waste generation in the nation and total population of the city as to CSA (2007) into account, the total generation of municipal solid waste in city in 2007 is estimated to be 21.4 to 60.4 tones.

Currently only 55% of the solid waste produced per day is collected and disposed by the Hormud SWM Company in the Sheik Ali Gure dumpsite at 6km away from the city center and the remaining 45% of the solid wastes are uncollected and dumped in unauthorized areas such as open fields, ditches, sewers, streets and many other available spaces in the city. Uncollected garbage is a serious environmental hazard for all, especially in areas where the roads are not accessible for collection by the collectors. These cause bad smells and attract various disease vectors and pests resulting in deteriorated aesthetic quality of the city.

The social waste collection service is unsatisfactory, and scenes of scattered waste are common in most part of the city. As a result, the population has the opinion that the municipal solid waste collection service is not functioning properly. As a result of this, the willingness of the population to cooperate with waste collection operation and to pay for the service is low. Solid waste collected from hospitals, residential and business areas is dumped at the dumpsite on the outskirt of the city with little incineration in the hospitals. In terms of social waste processing and recycling, little is done at all level of its management i.e. there is no source separation or sorting and this happens at disposal sites too. But some scavengers at dump sites practice an informal type of waste recovery. Other options like energy recovery and composting are not practiced as alternatives for waste recovery.

The disposal of waste has proved to be a major public health issue and a vital factor affecting the quality of the environment. This, especially in Ethiopian cities has become one of the most intractable environmental problems today. One of the main problems facing the city is open and indiscriminate dumping of refuse (authors, observation). Piles of decaying garbage are found in strategic locations in the heart of the city. Wastes in such places are obviously a source of air and water pollution, land contamination and environmental degradation.

2. Literature Review

2.1. Solid Waste Concept

Solid waste technically, any solid material disposed of as no longer useful in common usage the term has somewhat more limited meaning of solid materials thrown out by house holders those materials referred to by engineers as municipal solid waste. Solid (nonhazardous) waste is defined as any garbage, refuse, sludge from a waste treatment plant, water supply treatment plant, or air pollution control facility and other discarded material, including solid, liquid, semi liquid, or contained gaseous material resulting from industrial, commercial, mining, and agricultural operations and from community activities, but does not include solid or dissolved material in domestic sewage (Benny J. 2005). Solid wastes are wastes arise from human and animal activities, including the heterogeneous mass of garbage from the urban community as well as more homogenous accumulation comprising of countless different materials such as food wastes, packaging material such as paper, metals, plastic, glass, construction wastes, pathological wastes and hazardous wastes. Rapid population growth and urbanization in developing countries have led to the generation of enormous quantities of solid wastes and consequently environmental degradation.

2.2. Sources and Types of Municipal Solid Waste

In order to categorize what exactly municipal solid waste constitutes, there have been different attempts of categorization based on numerous classification criteria. Some of those criteria are source from which solid waste emanates, and nature of solid waste components. On the basis of the nature of items that constitute solid wastes, it can be classified into organic or inorganic, combustible or non-combustible, and putrescible or non-putrescible. With respect to source from which solid waste emanates, (Martin 2000) categorized municipal solid waste as household (residential) refuse, institutional wastes, street sweepings, commercial areas wastes, as well as construction and demolition debris. In developing countries, MSW also contains various amounts of industrial wastes from small scale industries. In these sources there are diverse types of solid wastes.

2.3. Characteristics of Municipal Solid Waste

For effective and efficient management of solid waste generated in a particular city, adequate knowledge and data about the characteristics of solid waste is essential. In order to decide or determine types of facilities required for solid waste management, best disposal options, and projecting future needs requires precise information about quantities, compositions, densities, moisture content and calorific value of solid waste produced in a city (Rushbrook 1999 in G/tsadkan 2002). Though all the above characteristics are important, for this study the researcher emphasize only on municipal SWM practices with due attention also given to the role community participation.

Solid waste generation rate: refers to the "amount of waste disposed during a given period of time and the quantification of it involves different methods: by measurement at the point of generation, through use of vehicle survey and by examination of records at the disposal facility" (UNEP 2009; cited in Zebenay 2010). The rate of solid waste generated in a given town is basically determined by demographic growth, seasonal variation, geographic location, economic development and people's attitude towards waste.

(Nashiimirimana 2004) explained the influence of economic development by comparing gross national product of developed and developing countries with their waste generation rate. The author argues that the higher the gross national product of a country result the higher the generation of waste. This means that due to difference in the level of economic performance, waste generation rate of developed countries is much greater than that of developing countries. Although developing countries have a lower rate of waste generation compared to developed countries, their quantum of waste is high owing to their higher levels of population growth and he concluded that this clearly shows impact of population size on waste generation rate. On the other hand, people's attitude towards waste can also conditioned solid waste generation rate in the form of their pattern of material use and waste handling, their interest in waste reduction and minimization, and the degree to which they refrain from indiscriminate dumping and littering (Schubeler 1996). Therefore, an accurate knowledge of quantity and rate of solid waste generation in a given area is essential for preparation and implementation of appropriate MSWM. Because it provides information on human, financial and equipment resources required for collection and transportation of waste, to enact appropriate laws on waste reduction, and establish current and future needs for solid waste disposal sites (Abel 2007).

Physical composition: refers to quantity of various material types in a particular waste stream. Just like waste generation rate, physical composition of solid waste is also extremely variable as a consequence different factors. The major once are of the following:

a) Economic level difference: higher income areas are usually producing more inorganic waste while low income areas produce relatively more organic waste.
b) Demography (difference in amount of population for example, tourist places).
c) Locations: includes abundance and type of regions natural resource, and socio-cultural factors which highly contribute for variation of waste in different areas.
d) Season: for instance during certain season's yard wastes such as grass clippings and raked leaves add greatly to solid waste (web page accessed, October 10, 2013).

Unlike various composition of solid waste in different areas, process of defining waste management practice is similar in each area i.e. by assessing waste handling processes beginning from collection, transfer to disposal present in a sample households. Conduct of waste management studies by using this method has several importances. "Some of these are: for identification of potential resource recovery activities, facilitate collection, transport and processing equipment, taking essential health, aesthetic and environmental precautions, and for monitoring changes in waste composition over time and improving waste management system" (Gidarkos, Havas and Ntzamilis 2005). Therefore, composition, generation, treatment practice study is cornerstone for successful planning and implementation of solid waste management.

2.4. Processing and Recovery

This functional element includes all techniques, equipments and facilities used both to improve the efficiency of other functional elements and to recover usable materials, conversion products, produce energy, and compost from solid wastes. In addition, it also provides several advantages. First, it can serve to reduce total volume and weight of waste material that requires collection and final disposal. Volume reduction also helps to conserve land resources since land is the ultimate sink for most waste materials. On the other side, it also reduces total transportation cost of waste to its final disposal site (Uriarte and Filemon 2008). Solid waste processing and recovery has been carried out beginning from separation and processing of wastes at the source. But, separations of mixed wastes usually occur at materials recovery facility, transfer stations, combustion facilities and disposal sites. It often includes separation of bulky items, separation of waste components by size using screens, manual separation of waste components, and separation of ferrous and non-ferrous metals. Then they enter in small and large scale industries for recovery activities. For example, organic fraction of MSW can be transformed by a variety of biological and thermal processes. The most commonly used biological transformation process is aerobic composting and, the most commonly used thermal transformation process is incineration (web page accessed, October 23, 2013).

2.5. Disposal

This is final functional element in SWM system. Disposal activities are associated with final dump of solid wastes directly to a landfill site. Today disposal of wastes by land filling or land spreading is the ultimate fate of all solid wastes whether they are residential wastes, or residual materials from materials recovery facilities. "However, in most developed countries this method is officially banned allowing only sanitary landfill for final disposal. Because sanitary landfill is not a dump it is an engineered facility used for disposing of solid wastes on land without creating nuisances or hazards to public health and environment" (Techobanglous 2002). "Though it is the most common technology around the world, conventional and environmental unfriendly methods such as open-burning, open-dumping, and non-sanitary landfill can still be used as disposal method" (UNEP 2009).

2.6. Incineration

Incineration is one option for sustainable solid waste management. It is defined as the process of burning solid waste under controlled conditions to reduce weight and volume of solid waste, and often to produce energy. This process is really waste reduction, not waste disposal, though following incineration ash must still be disposed. It is recognized as a practical method of disposing of certain hazardous waste materials (such as medical waste). Incineration can be carried out both on a small scale by individuals and on a large scale by industry. This facility does not require much area so that it is common in countries like Japan where land is scarce (web page accessed, September 5, 2010).

Generally, according to (UNEP 1996) there are considerations that we should keep in our mind when we want to choose incineration. These are: the necessary environmental controls are properly installed and maintained; the facility is properly sized and sited to fit well with other components of the MSWM and the material to be burned is combustible and has sufficient energy content.

2.7. Composting

It is a process of allowing biological decomposition of solid organic materials by bacteria, fungi, worms, insects, and other organisms in to a soil for transforming large quantities of organic materials to compost (humus like materials). "The organic materials produced by composting can be added to soil to supply plant nutrients such as nitrogen, phosphorus, potassium, iron, sulfur, and calcium, slow soil erosion, make clay soils more porous or increase water holding capacity of sandy soils" (Enger and Smith, 2008).

There are three scales at which composting has been implemented; residential level, decentralized community level, and centralized and large scale municipality level. The larger the undertaking the more capital investment is required. Most developing countries which have found success with composting revealed that composting works best when implemented at household level, with some project doing well at community level as well. At municipal level financial commitment required to maintain equipment has resulted in wide spread failures (Zerbock 2003).

In cities of developing countries, most large mixed waste compost plants have failed or operate at less than 30% of capacity. The problems most often cited for such failures include: high operation and management costs, high transportation costs, poor quality of product as a result of lack of sorting (especially plastic and glass fragments), poor understanding of composting process, and high competition from chemical fertilizers (which are often subsidized) (UNEP 1996).

2.8. Reuse and Recycle

Reuse involves cleaning and using materials over and over. In other words, it means the use of a product more than once in its original form for the same or a new purpose. It relays on items that can be used over and over instead of throw away items. This method is used to decrease the use of matter and energy resources, cuts pollution, creates local jobs, and saves money (Miller, 2007). "Reusing is more efficient and better than recycling and composting methods because cleaning and reusing materials in their present form avoids the cost of energy for remaking them in to something else" (Cunningham 2008).

In addition to reuse, recycling is also an obvious treatment of solid waste problem. It is an important way of collecting solid waste materials and turning them in to useful products. Such materials can be reprocessed in two ways: primary and secondary. "Primary recycling is a process in which original waste material is made back into the same material for example, newspapers recycled to make newsprint. In secondary recycling, waste materials are made into different products that may or may not be recyclable for instance, cardboard from waste to news papers" (Miller 2007).

Recycling is both environmental and economical issue. Many people's are motivated to recycle because of environmental concern i.e. it reduces pollution, it also save energy, space and resources, helps to protect biodiversity and reduce litter. Economically, it can save money for items like paper, metals and some plastics.

2.9. Public Participation

Public Participation may be broadly defined as the involvement of citizens in governmental decision-making processes. This ranges from being given notice of public hearings to being actively included in decisions that affect communities. It is generally a process of engaging stakeholders so that those most likely to be impacted by a particular activity can influence the outcome. *Public* refers not only to private citizens but institutions, civil society, labor unions, the Government, public officials, industrial, agricultural and trade associations, scientific and professional societies, environmental, educational and Health associations and other minority Groups (EPA 2005).

It is necessary to establish the Public Involvement Framework and identify participants or stakeholders and determine their legitimacy by social analysis (UNEP 2002). There is a wide variety of models from which to choose. The one chosen should reflect the public input required. Some models of public participation are given below.

2.9.1. The Exclusionary Model

This model indicates that the government or agency is the exclusive guardian of the people and any self-acclaimed representative of the public interest was an officious meddler. Only competitors and other institutional stakeholders were allowed to participate. This proved inappropriate for risk-oriented decision-making. There may be some usefulness in cases of rate hearings and public utility regulation. (Mc. Garity 2005) In the Caribbean, some decision-makers adopt an approach that they have been appointed or selected to make decisions for the populations who may not have the knowledge and information to make decisions for themselves.

2.9.2. The Confrontational Model

The Confrontational Model is at the other end of the spectrum to the Exclusionary Model and results from a stringent application of the Exclusionary Model. This is really not a desirable model. It is not effective in addressing matters of intense local interest but may be effective if activists can attract the sympathy of a large number of other inactive members of public. It is a way of keeping certain topics or concerns on the public agenda or getting agencies to revisit decisions already made.

When a person feels excluded or that her/his interest has not been well represented she/he can confront the agency. Sometimes there is picketing and civil disobedience e.g. in the 1980s, outraged neighbors picketed for proposed hazardous waste dumps. Also there have been shouting matches at meetings during debates on SW disposal sites or incinerators or other SWMP components being located near their homes and offices.

Confrontations are usually intended to be very public but they are not designed to be participatory and certainly are not conducive to informed dialogue about risks and mitigation. Sometimes a decision-maker agrees to speak to the activists to induce them to stop their action but they are rarely asked to participate in the actual decision-making process. This model is characterized by distrust and is restricted to local activities and usually employed by individuals or ad hoc groups. Activists, who at times participate at considerable risk to their own economic and physical well-being, seek a wider audience. Only highly emotional matters or matters of high principle are likely to justify such individual action. Purely economic interests do not often invoke the confrontational model. (Mc. Garity 2005)

2.9.3. The Adversarial Model

The Adversarial Model represents a situation where all interested groups have a right to participate by submitting facts, evidence, views and arguments. The agency assumes the role of neutral decision-maker. Based on courtroom adjudication, parties who may be represented by counsel are usually allowed testimony through experts. This is dominated by lawyers and the settlement presumes winners and losers. (Mc. Garity 2005) This model was experienced in cases where residents sought compensation from a Caribbean government in court for the health effects from an existing LF. In some cases, this is a slow and expensive process. In some cases, protesters to the location of a LF seek international attention. This can be a cheaper and more effective approach.

2.9.4. The Due Consideration Model

The Due Consideration Model is similar to the adversarial model except that the agency takes a position prior to the public hearing and invites the public to comment on their decision as well as on the issues generally. It does not adopt the procedural protection of the Adversarial Model and so is less burdensome. The agency is only required to give due consideration of outsiders and explain its chosen action. This model is better adapted to issues that are policy dominated and for which factual accuracy is not essential. Participants are less directly involved in the actual decision making process. Suspicions are easily raised that due consideration is not given to participants points of view especially when the agency adopts the option it initially proposed with little or no adjustment (Mc. Garity 2005).

2.9.5. The Mediation Model

The Mediation Model requires that representatives of groups meet together, often with the aid of a mediator or facilitator, to present facts and arguments so as to reach an agreement on the ultimate result. The agency may participate in the discussions and attempt to implement agreed solutions. Public participation may be invited at this stage while Negotiators meet until agreement is reached. This may be a useful approach in planning SWMPs as the relationship between government and residents is often confrontational. (Mc. Garity 2005)

2.9.6. The Advisory Committee Model

The Advisory Committee Model is similar to Mediation Model except that it relies heavily on scientific and technological expertise. The Agency appoints a committee of disinterested experts to advise on the technical issues and on a decision. This model seems favored by decision makers who are not scientifically trained and who do not want to "take the heat". Agency may lose control of the outcome but this may be reduced by choosing experts for the advisory committee whose views on technical issues will yield results that are in accordance with the decision-maker's policy preferences.

2.10. Measuring Success in Public Participation

Elizabeth Evans, in her paper presented at the International Association for Public Participation (IAP) in 2002, sought to address how the success of the Public Participation Process could be determined. Her paper explores the question of evaluation of consultation processes and how these may be approached. She focused on three issues:

1) What is meant by success in the context of Public Participation?
2) Whether there are meaningful consistent ways of measuring success
3) Whether it is wise to attempt such measurement (Evans 2002).

The practice of Public Participation is complex. It may be considered as a process that contributes to overall results of the business or enterprise. While it is not new, emphasis over the past five years has been on selling the idea to decision-makers that Public Participation would add value and reduce costs and risks. The problem with "success" is that those involved are likely to have different ideas of what success means. The process of Public Participation requires the invitation of stakeholders to participate and there is a need to identify: The range of stakeholders; the inputs required (financial and other resources) in information staff, training, policy etc.; processes (nuts and bolts of the programme); and outputs (products and deliverables that come out of the work done) (Evans 2002).

The range of activities that make up Public Participation is expanding rapidly and references are made to community engagement to build good will. Community participation can be a wide range of activities especially in SWM and it takes account of aspirations and views of people directly affected. It facilitates involvement of affected communities, businesses, etc. through discussions, debates, negotiations and planning. (Evans 2002) based measurement of success on four core values and associated principles previously articulated by the International Association of Public Participation (IAP) as follows:

1. *Equity* – decision-makers should provide opportunities for all those with an interest in the subject
2. *Integrity* – decision-makers must act in good faith
3. *Openness* – provision of information to the public that is accurate, honest, comprehensive, clear and accessible
4. *Accountability* – specification of the degree to which public involvement would influence decision making and accurate reports on how commitments are discharged.

2.11. Some Best Practices for Successful Public Involvement

UNEP (2002) proposed a list of best practices as follows: (1) Develop a public involvement framework as early as possible to establish the scope, timing and resource requirements necessary to support the process; (2) Identify the participants and stakeholders and establish their legitimacy and "representativeness" (using social analysis). It should be noted that not all social actors can or should be consulted on every detail of the proposed project; (3) Identify appropriate techniques of public participation/communication and provide relevant information in a form which can be easily understood (e.g. using a combination of seminars, simple written materials, visual aids and scale models can help to make the technical material accessible to the non-specialist); (4) Plan and execute events at a time and venue that will encourage the maximum attendance and free exchange of views by all interested groups. Money may be specifically allocated to help facilitate community involvement (e.g. to pay for travelling expenses or costs involved in hosting meetings and inquiries); (5) Allow stakeholders sufficient time to assimilate the information provided, consider the implications and present their views; (6) Identify mechanisms which ensure decision makers consider views and suggestions made by stakeholders - integrate findings and recommendations into the environmental assessment report, financing proposal and agreement; and (7) Ensure that responses and feedback are given on issues or concerns raised. The best practices are quite representative of the methods outlined by other commentators and form the basis on which the performance evaluation of the OECS SWMP may be based.

3. Methodology

3.1. Description of the Study Area

Jigjiga, the capital city of Somali National Regional State of East Ethiopia, is about 630km East from Addis Ababa City and 105km from Harar city of East Hararge Zone of Oromia and Harari Region. The city is astronomically located between 9° 16' 30" to 9° 24' 30" N Latitude and 42° 44' 0" to 42° 51' 0" E Longitude (Fig. 1). It has an elevation of 1609 meters a.m.s.l. The city hosts Jigjiga University.

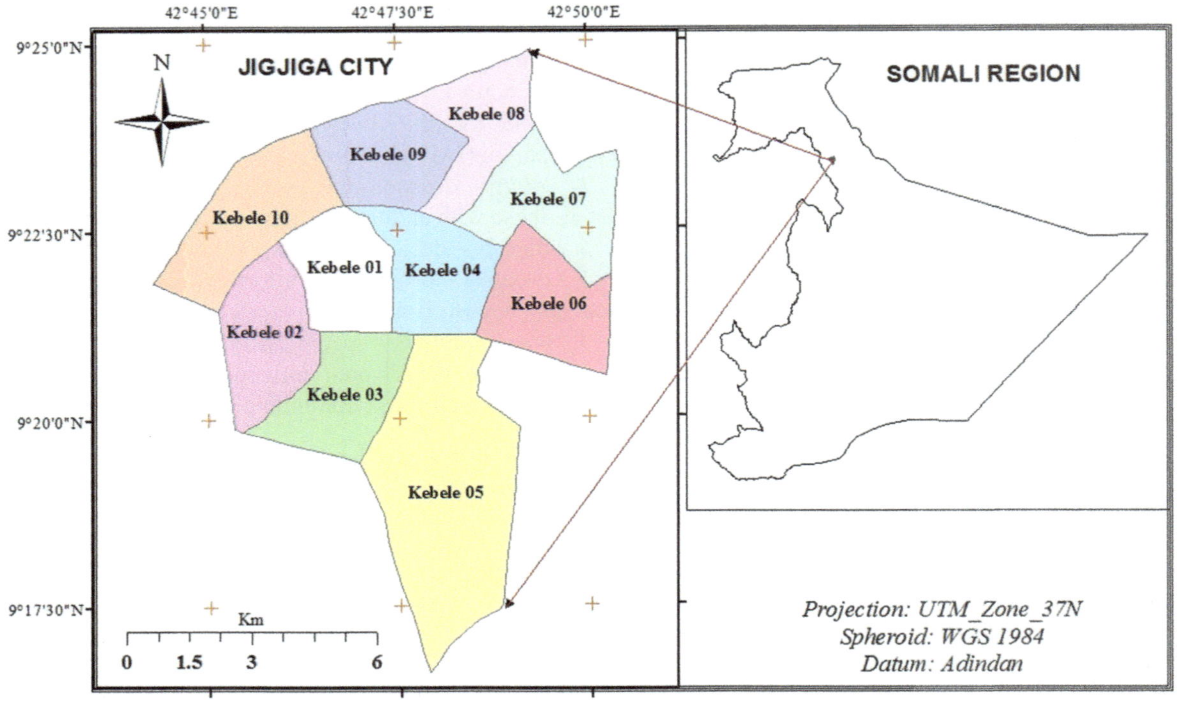

Figure 1. *Location Map of Study Area.*

Based on the recent (2007) census data from CSA of Ethiopia, the city has a population of 125,876 of which 67,128 were males and 58,745 were females. The total household count is about 23,263 with an average of size of 5.4 persons to a household. The 1994 census reported a total population of 56,821. The city has recorded a population growth of 121% in 13 years. The climate of the city is subtropical, with a mean annual temperature of 19.5 ^{0}C and annual average rainfall ranging from 150mm to 1000mm. The highest temperature is experienced between November and March and the lowest between July and September.

3.2. Methods

3.2.1. Data Source and Data Type

All required data for this study were collected both from primary and secondary sources. The primary data were collected from sampled households respondents, researchers' observation and key informants. Similarly, thorough review of all available published and unpublished documents of relevant organizations was conducted.

3.2.2. Sampling Technique

For this study, purposive sampling and stratified random sampling in two Kebeles (the smallest administrative unit), 03 kebele from the commercial center and 05 kebele from residential areas of the city were selected. Jigjiga city is purposively selected being the interests of the sponsor of the research. Stratified sampling technique was used due to the variability of the municipal solid waste generation sources. Heterogeneous units were divided into non-overlapping groups resulting in representative sampling. The numbers of samples within each stratum were determined based on respective proportional percentages of each stratum. In this study, to identify the participating households, they were stratified into two groups; residential and non- residential (commercial) categorized as 74% and 26% of total number of housing units, respectively, CSA (2007). Sample size (n) of households that participate was determined by using a sample technique (Cochran, 1977) formula, given by:

$$n = \frac{NZ^2PQ}{d^2(N-1)+Z^2PQ}$$

Where;

P=Housing unit variable (Residential houses which is 74% of N)

Q=1-P (Non residential houses which is 26% of N)

N=Total number of housing units (23,263; CSA, 2007)

Z=Standardized normal variable and valued that corresponds to 95% confidence interval equal to1.96

d=Allowable error (0.05)

Therefore, the result, n=293 was the minimum reliable sample size of housing units while five housing units were added to augment the confidence. This was done in the two selected Kebeles and the total sample size was 298 households. After this, simple random sampling was employed based on the proportional percentage of samples within each stratum to decide the number of samples that was taken from each stratum as summarized in table 1.

Table 1. Sample Selection with proportional allocation.

No.	Kebele	Number of households	Proportional percentage	Sample size
1	03 kebele	678	12.2%	36
2	05 kebele	4899	87.8%	262
Total		5077	100%	298

Source: Own Computation based on CSA, 2007 data

As showed in the table above, in 03 kebele there are about 678 households out of which 36 households samples were taken proportionally (10 households residential and 26 households commercial centers). Similarly, from 05 kebele out of about 4399 total households, 262 households samples were considered (of which 230 households were residential and 32 households were commercial centers).

3.2.3. Data Collection and Analysis

The data collection was done using checklists which were pre-tested prior to the actual field work. The bulk of the data generated for this study were qualitative. Structured questionnaire, interview guideline and field observation were used to obtain data and information for the study. The structured questionnaire was used for the selected households to have information about their sources of energy, housing conditions, onsite solid waste handling, solid waste disposal practices and other activities. Beside this, checklist was also provided for the city government's Sanitation and Beautification Department (SBD) and Hormud SWM Company to gather information on their current solid waste management. Collection of information about solid waste treatment and the role of community participation from the studied households was extended for eight consecutive days. The collection was done by two trained collectors. Equipments like digital camera and GPS were used for photographing and ground control points by the researchers during field observations. Waste Collection sites where containers are placed, market places where waste handling is expected to be difficult and open dumping sites, were also purposely used as sample units to collect qualitative information. The data were analyzed combining the information collected from secondary sources, qualitative and quantitative survey.

4. Result and Discussion

4.1. Solid Waste Composition

The result of the survey in the study area showed that municipal waste is an aggregate of all substances ready for disposal. The composition of the solid organic waste was almost homogenous in nature across the study households. As it was observed in this study, majority of the waste was of plastic origin while the animal and the industrial origin was almost none in most of the households. Of the plastic source, residues of festal (known locally as Mika) and liquid containers take the greatest portion. Overall composition includes paper, vegetable peelings, onion seed coats, broken plastic and festal, soil and dust, animal dung, grasses, used shoes, pieces of cloth, small bottles, soot, etc. The waste aggregate more frequent in the whole mass of household waste was house sweeping, which is composed of soil and dust followed by pieces of paper and vegetable peelings (plates 1).

Photo by Researchers, 15 October, 2013

Plate 1. *Open Dumpsite nearby Dr. Abdulmajid TTC.*

It can be observed from plates 1 and 2 that ash swept out from kitchens and *chatt* remnant (sticks and leaves) is more in quantity than other waste. However, it is not removed every day. When considering the solid waste production across different sources (residential and commercial, in this study), the residential origin was found to be higher (74%) while the remaining (26%) accounts for commercial and other origin.

4.2. Solid Waste Generation Rates

With regard to solid waste generation rate, even though the solid waste trend is expected to increase in general, recent and up to date data were not available. However, as per the data obtained from Hormud SWM Company, about 3150 sacks of waste each weighting from 20kg to 30kg were collected every day from about 2950 households. Computation result shows that the average waste generation rate of the city is estimated to be 26.6 kg/day/households in weight.

4.3. Solid waste Source

The result from the surveyed households reveals that MSW generation by source and type in Jigjiga City has the following distribution (Fig. 1). The composition of municipal waste varies greatly from country to country and changes significantly with time. Waste can be classified into several types. However, the following list represents a typical classification of municipal solid wastes: (1) Biodegradable waste: (like; food and kitchen waste, green waste, paper), (2) Recyclable material: (paper, glass, bottles, metals, certain plastics, batteries etc.) and (3) medical wastes.

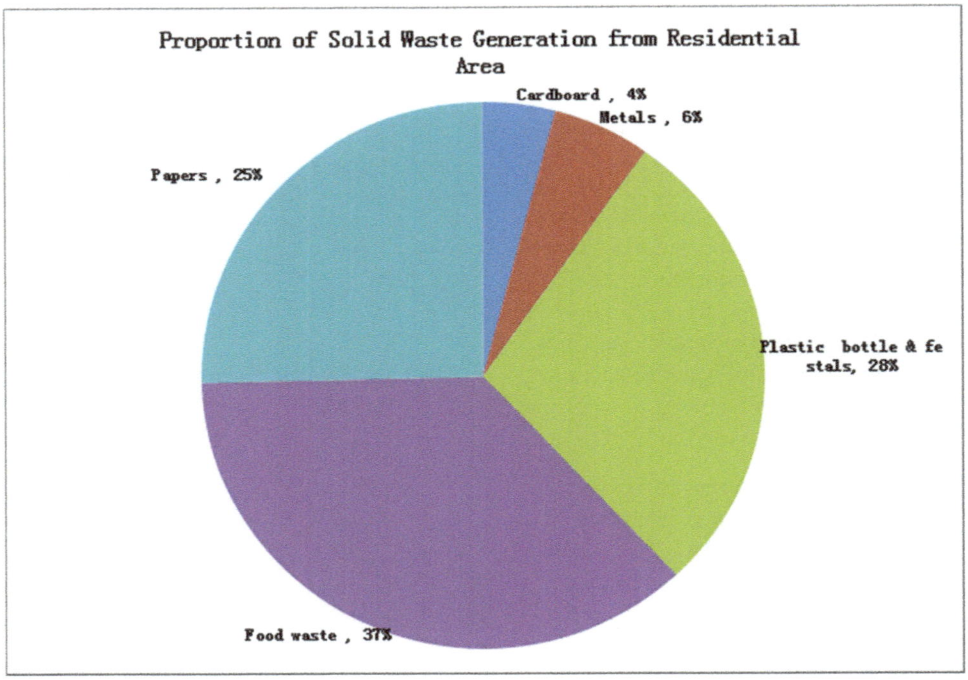

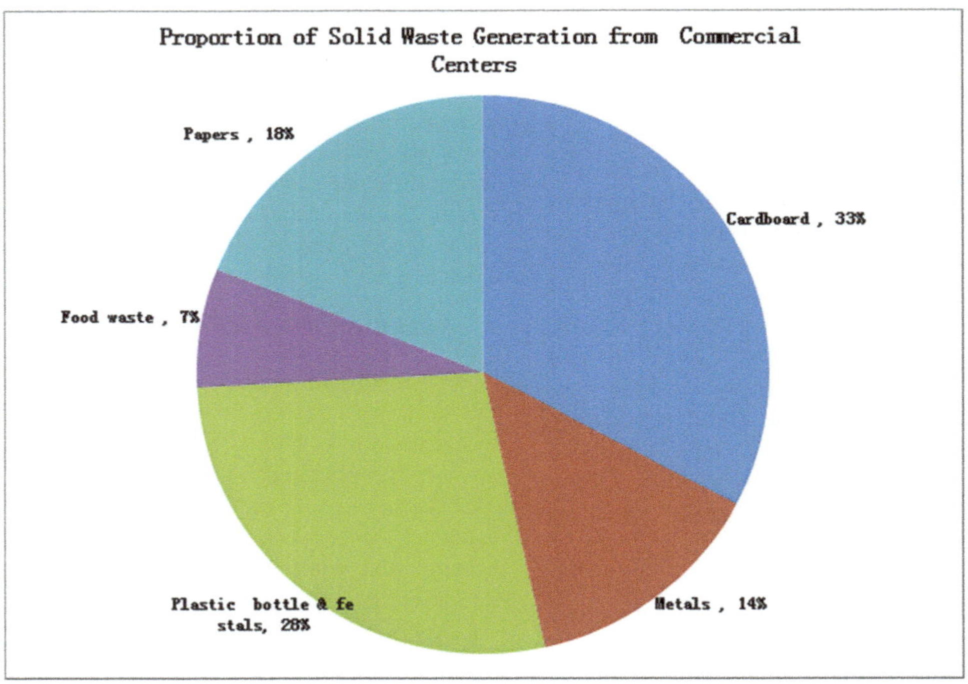

Source: A survey conducted by researchers

Figure 2. *Municipal solid waste generation by source and type of Jigjiga City.*

Fig. 2 shows percentage of municipal solid waste generations by source and type in Jigjiga City. It has been revealed that the largest constituent of residential solid waste is food-waste be 37 % and followed by plastic bottles & bags 28 %, papers 25 % and the remaining 6 % and 4 % are metals and cardboards, respectively. The situation is reverse in the case commercial area's waste composition as the highest constituent is cardboard 33 % and the least is food-waste 7 %.

However, if we see the wastes from the recycling point of view, many of the constituents are compostable and recyclable materials, and hence it indicates that recycling practices are effective SWM practice. Unfortunately, it was so far not practiced in the city, and this is why we see a lot of paper and plastic substances thrown out in river banks, ditches and open areas plate 2.

Photo by: Researchers 20 October, 2013

Plate 2. *Waste substances thrown out in unauthorized areas in Jigjiga City.*

4.4. Existing Solid Waste Management Practices

The management of Jigjiga City waste is a history of institutional trials and abandonment, with most issues remaining unsolved. The new established authorized organization on the sector, sanitation and beautification department (SBD) at the city government level was assigned the responsibility for the management in 2008. The assignment came as a result of the failure of previous authorities-the Ministry of Health and Sanitation, and the private sector to effectively manage City's solid wastes. Unfortunately, the SBD is also struggling with tight budgets, lack of qualified manpower, and experience in waste management. Uncommon to most other cities of the country, Jigjiga City waste management practice is engaged by a legalized private organization called Hormud SWM Company.

4.4.1. On Site Handling

A proper waste handling at household level has positive implication on waste management. The responsibility of household waste handling also varies between the residential and the commercial. In residential households female members of the family are responsible to handle the households cleaning and separating of the waste. The male members of the family do not participate in these activities, except at times they engage themselves when the waste is bulky and some physical help is required to transport it to dumping places. In case of the commercial areas, it is servants who collect and dispose of households wastes.

4.4.2. Sorting

Sorting is an essential component of solid waste management. It is a kind of activity which is separating different types of wastes in their respective nature. It makes waste management easy and simple. However, it should not be a onetime activity, rather should be a habit for proper and sustainable solid waste management. Separating different types of waste components helps to sort recyclable materials from non recyclable and identify decomposable (organic) materials from non decomposable. The process is also efficient in reducing the problem of landfill sites and expenses. However, as to the respondents, there is no/little separation of solid wastes in Jigjiga Town whereby over 92 % of the households, invariably confirmed the absence of solid wastes separation practices during both collection and disposal periods. The remaining few households or 8 % of the respondents replied that solid wastes were sorted into organic and inorganic wastes. Inorganic wastes like cans are usually sorted for sale. In few households, organic wastes like plant origin are sorted for the purpose of reusing in their gardens and as fuel after the waste gets dried; this is seen in residential households. Animal source (meat and bone) is used to feed domestic animals like dogs. Some households give the waste away to their neighbors who use it for different purpose like for animal feed. The purpose of separated waste is different for different surveyed households (Fig. 3).

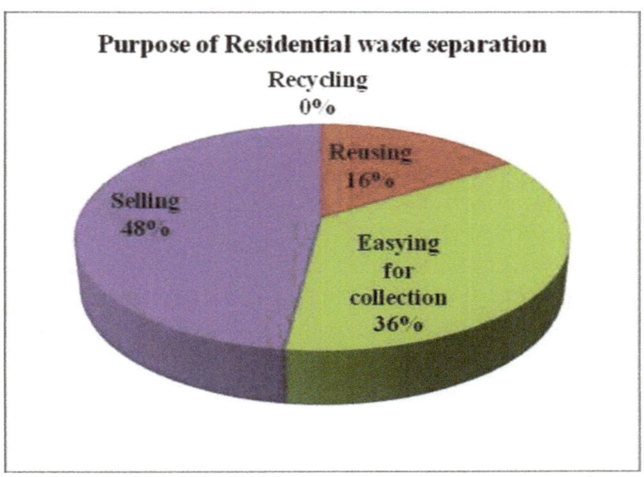

Figure 3. *Purposes of waste separation in Jigjiga Town.*

Figure 3 clearly shows that from residential households, 48 % of solid wastes are separately collected for selling, 36 % for easing collection and 16 % for reusing purposes.

Unfortunately, there is no solid waste collected separately for the purpose of recycling in Jigjiga City. This indicates that even if a great deal of recyclable solid wastes is generated in the city, it is not drawn to the attention anybody for the purpose of recycling. Absence of information from commercial households implies the solid wastes are indiscriminately collected in the area.

4.4.3. Composting

With regard to composting, the households practicing composting are very few, while a significant amount of this refuse is largely of plant origin /biodegradable/ organic waste. (Gardner 2001) and (Bezaye 2008), argued that composting is an ancient practice where more cities in the world nowadays are reclaiming the benefits of reusing solid organic waste material. It is a natural way to prepare the waste for reuse. During the study, it was observed that there are people who know at least the use of organic waste for soil fertility improvement. There are also some people who have training in composting, but are not practicing. The major limitation for not practicing compost as indicated by the respondent is lack of market and a piece of land for urban agriculture.

The key respondents were interviewed whether incineration has been practiced at secured places or not. Accordingly, as the informants, practice of incineration at secured places is confined to health institutions typically to manage medical wastes even though the practice takes place only sometimes. So far, no incineration taking place in or near the compound of households.

4.4.4. Waste Reuse

Reuse is an important factor to reduce the amount of waste to be dumped at the final disposal site. The study result indicates that there is little practice of waste reuse for different purposes like industrial waste like can for sale, plant origin waste for fuel, animal source (meat and bone) waste for domestic animals feed.

In general, as far as waste reuse is concerned, there is no formal practice in the city except some people called "*Korales*" who move through the city to buy recyclable items such as glass, plastic, tin cans, metals, shoes etc. from different houses, hotels, restaurants, repairs services and sell them to small recyclers and industries. There are also people engaged in similar types of job known locally as "*Liwach*" who go around the city and exchange used clothes and shoes with new household utensils and sell them to low-income people

4.4.5. Collection

Waste collection service is one of the chief components of municipal SWM practices. The survey showed that containers are placed in some neighborhoods. However, a number of containers are not enough with respect to waste generated and a portion of the neighborhoods are obliged to throw their refuse into an open area. Some of the households, who are not enjoying municipal services, have their own way of household waste disposal. The tools that are used for solid waste collection in both residential and commercial area are shown below in table 2.

Table 2. *Number of respondents for solid waste primary collection Tools in Jigjiga City.*

No	Tools	Number		Percent (%)	
		Residential	Commercial	Residential	Commercial
1	Metallic container	0	0	0%	0%
2	Basket	2	5	0.83%	8.6%
3	Plastic bag	0	0	0%	0%
4	Cardboard	0	0	0%	0%
5	Empty sack	238	53	99.1%	91.4%
Total		240	58	100%	100%

During the interview, almost all of the respondents from both residential and commercial households stated that the bulk of primary waste collection equipments are sacks whereas a few said that they use baskets for this purpose. As to Hormud SWM Company, in each Kebele, strategic locations are assigned for the primary collectors (mainly micro and small scale enterprises) to make the wastes ready for the motorized further collection and disposition. Most of the areas in the city are inaccessible for motorized collection; therefore, the human powered collection system is mostly used.

Generally, there are three basic types of collection equipments. These are: human powered, animal powered, and engine powered. Under the Jigjiga City situation, both human and motorized collection equipment are being used. With regard to the human aspect, transportation of wastes to the containers is possible using hands and hand pushed carts.

The current trend of storage bins reveals in the city of Jigjiga that dust bins are located nowhere on main roads. For temporary storage, households prepare different types of receptacles such as baskets, card boxes, cans, plastic bags, barrels, etc. Only 65% of the MSW generated in Jigjiga City is collected and disposed off by the Hormud SWM Company in the Shiek Aligure Open Dumpsite, with no recycled and a little composted. The rest about 35% is dumped in uncontrolled environment like on/in streets, empty spaces, and river banks.

There are different waste collection services given in the study area. Of these services, the majority use door-to-door collection services while others don't receive any services rather they dispose wastes into river systems, burn and reuse for their gardens even though they didn't mention.

As far as waste transfer in concerned, Hormud SWM Company's trucks are solely used to transfer wastes from the source to the destination. The interview with the company's manager reveals that the trucks collect wastes every day from commercial centers and every week from residential households (table 3).

Table 3. *The frequency of solid wastes collection and transfer to the junk region.*

No	Duration	Household respondents			
		Number		Percent	
		Residential	Commercial	Residential	Commercial
1	Daily	8	42	3.13%	72.70%
2	Weekly	210	16	87.50%	27.27%
3	Monthly	22	0	9.37%	0%

From Table 3 it can be realized that majority of wastes from residential households are collected using door to door service weekly (87.5 %) while the rest 9.5 % and 3 % are done so monthly and daily, respectively. Commercial households receive the door-to-door collection system more frequently as compared with residential households the former being served 73 % daily and 27 % weekly. This implies that the households are determined by their bulk of waste generation to receive a better collection of solid waste services.

In Jigjiga City, there are different actors who are directly or indirectly involved in a wide range of SWM activity including Non Governmental Organizations (NGOs), private company and Micro and Small Scale Enterprises (MSSE) are the pioneers in providing door-to- door solid waste pre-collection service in the city. NGO (the USAID) was indirectly engaged by providing equipments to collectors who deliver the service of collecting and transporting solid waste from households to temporary collection sites (Plate 3).

Photo by: Researchers 25 October, 2013

Plate 3. *Equipments donated by USAID for temporary collection service at 04 Kebele.*

The collection of waste through MSSE has provided job opportunities for many jobless youth. For instance, under the supervision of Hormud SWM Company there are 22 garbage collectors team consisting of four members each (three garbage collectors and one coordinator). The three garbage collectors are paid of 1000 birr per month while the coordinators are paid of 2000 birr per month. Apart from this, four more small scale enterprises were organized by city government on municipal SWM practices. The majority operates in the commercial households to collect solid waste using door-to-door method. The private sectors have the accumulated experience on waste collection and entrepreneurship due to their close engagement in the sector. This activity was less attractive to the government officials and easily left to be done by the informal enterprises until recently. Still the government officials have not viewed the sector as one of the employment generating fields.

Photo by: researchers on 25 October, 2013

Plate 4. *Solid wastes stayed in the containers for more than one month.*

However, as to the interview result from an SBD officer of the municipality, waste collection has gradually been attracting the attentions of the officials as one of the untapped employment generating sector. Nowadays, government has pervasively been organizing private actors (specially, MSSE) and providing initial loan to become self (indirectly government) employed. The payment for the service is first done by negotiation between waste workers who are assigned for specific location and individual households in that area. However, from experience, the government realizes that the private worker collect payment from the households while the dump the waste anywhere after they collect the waste from the households. Hence nowadays, the waste is collected by the waste workers while payment for the service is collected by the authorized private company. The payment is on the bases of the waste generating sources to collection site rating 40 to 50 birr from residential households and 100 to 500 birr per from commercial centers as per the amount of solid waste they generate. None of the data revealed the existence of government waste workers in the study area and any other place outside of the sampled kebeles. Concerning the regularity of removal (emptiness) of containers, almost all the interviewed waste workers responded that containers did not get emptied regularly. They explained that wastes in containers are usually scattered around and remain stay in the containers over a month at times (plate 4).

Other reasons identified for the scattering of waste around the container include: the lift over height to the containers, which made it difficult for women and children to empty waste in to the container; people often have to travel more than 200 meters to reach the nearest containers which means that containers are scattered and this apparently discourages households from dumping waste in the containers. The irregularity of waste removal from the containers is affecting the health of the nearby community and the surrounding environment. The households using door-to-door collection service were asked if they had seen any improvement after receiving the service. All of the households who were using the service responded that they had seen improvement in the area of solid waste situation in their surroundings.

4.4.6. Waste Disposal

Disposal at Transfer Station

Waste disposal, one of the most important management practices, needs to be carefully planned. With regard to waste disposal at transfer station, the study identified that almost all solid waste generated in households is indiscriminately disposed together i.e. there is no habit of sorting organic from inorganic waste at the household level. Huge amount of organic materials comes from the rural areas depraving nutrients from the rural soil to feed the urban population, the leftovers after consumption have no way to return to the source to build the soil, rather lost and create problems to human health and the surrounding environment in the city due to mismanagement. In this study, it was observed that disposing household waste into a river system is a common practice in the study area. The result of the study also reveals that there are households who dispose the waste in to a river/ stream, drainage system and any open place. This is especially observed in the case of *Gumruk open dumpsite* where *waste piles* seemed to have adversely impacted the nearby private properties and public institutions other than public health and environmental influences. Even though the waste disposal at this transfer station has currently been stopped, the previous pile is remaining intact since the time of this survey.

Another unauthorized open dumpsite was observed at the south-western side of *Dr. Abdulmajid Teacher Education College* on the locality termed as *CMC.* In this dumpsite, there is surface water used by the local community for washing. The worse of these dumpsites is their extreme proximity to both residential areas and public institution at a less than ten to twenty meters (plates 5). However, until the time of the observation, solid waste has been disposed in the *CMC open dumpsite* while it has currently been forbidden to dispose in the *Gumruk open dumpsite.*

Plate 5. *CMC open dumpsite and Gumruk open dumpsite.*

Regarding improper waste disposal at a transfer stations, respondents blame the rarity of inspection on the waste management condition of the households, loose management action for punishment and lack of accountability of the community members. However, they didn't deny the current relatively strong monitoring due to the establishment of government structure with respect to waste management at the city government level. The respondents further stated that the decentralization of government structure at local/kebele level for more control of waste mismanagement.

As far as the punitive measure taken against the offense improper waste disposal is concerned, in contrast to the previous periods when there was neither penalty nor inspection, the monitoring is relatively strong and monetary penalty for mismanagement of wastes is imposed these days. As to the respondent the punishment amount being is small not exceeding 40-50 birr a few households preferred paying to carrying waste to transfer station a bit away.

On the other hand, the survey showed that nobody care about the waste dumped in the river banks. Some of the waste is taken away by the river water during rainy season while the rest remains piled in the river gorge and on its banks.

Despite the punishment on individuals who dispose waste in unauthorized places, the regulations are seemed not obeyed at all by the households. This is especially pertinent to case of institution like hotels, restaurants and business institution owners. This is because business man only runs after his/her benefit since whatsoever amount of waste he/she dump anywhere, the punishment is less amount of money. This is much less than the payment they pay for pre collectors which could be one birr for informal private individuals. This, in other way round, implies punitive measure alone does not bring a significant change unless coupled with conscious community participation.

Regarding the final disposal site, currently there is one open dumpsite known as "*Shiek Aligure*" which serves as final disposal site in south- west part of the city located 6kms away from the city center. It has a surface area of about 5 hectares. The present method of disposal is crude open dumping hauling the wastes by trucks of Hormud SWM Company (plate 6).

Photo by: Researchers on 25 October, 2013

Plate 6. *Hormud truck hauling solid waste at "Shiek Aligure" Open Dumpsite.*

The major problems associated with the disposal site are: absence of spreading and leveling by bulldozer and compacting by compactor. As the result, a great deal of festal and paper origin wastes has been blowing back to the city by wind power.

Interview at "Shiek Aligure" Dumpsite

The researchers interviewed a household which is located nearby the dumpsite. Regarding the impact of waste being disposed at this site, respondents stated that *"We have been suffering from horrible increasing smell from the dumpsite that you can notice now. Smoke hardly stops; even though it stops, we are obliged to live in unsightly place and unpleasant smell. We always fear the wind-blown fire hazards to happen on our properties and lives of children. We are living here because we have no option. Even if we request everybody who may help us, nobody has drawn attention so far to us to solve the problem".*

The researchers also observed the following technical and social problems: The site is established on the open grazing land nearby urban peripheral residences. The site is characterized by lack of the following basic technical features: leachate treatment, odor or vector control, rainwater drain-off, and fencing. The area is unprotected for children, women, domestic animals and scavengers reach. Socially, it is unsafe as many human scavengers work continuously and obviously living nearby the site and interfering work operation at the site for collection of materials such as wood, scrap metals and discarded food. All of waste collected from the city is dumped in this single place without separation of even organic waste. The total area of the dumpsite and beyond is covered by smoke from the continuous burning of the dumped waste due to internal ignition by the waste itself (plate 7).

Photo by: Researchers on 15 October, 2013

Plate 7. *Continuous smoking from Shiek Aligure Dumpsite.*

Due to the indiscriminate disposal of waste in the area, the organic waste that goes to this dumpsite not only pollutes the land and water but also contributes to global warming by producing green house gases like methane if it is remain left untreated. The dumpsite being located on the intermittent stream, the catapult of polluted water flowing from waste disposal area can obviously inject hazards to livestock and people living in the lower basin of the stream.

As far as solid waste transportation is concerned, Hormud SWM Company is responsible for transporting wastes from each kebele to the final dumpsite "Skiek Aligure" by means of trucks. According to the interview held with the company, the main challenge in this regard includes renting cost of trucks and imbalanced communities' willingness and capability to pay for the waste disposal service.

4.5. Role of Community Participation

Community participation in SWM encompasses several

forms of local involvement, including: awareness and teaching proper sanitary behavior; contributing cash, goods or labor; participating in consultation, administration, and/or management functions.

Community- based waste management (CBWM) projects require institutional support and recognition in order to be successful. An integrated system including waste separation at the source, resource recovery, and composting of organic waste requires representation of waste pickers, and integration of the community to work with allied stakeholders, including representatives of waste pickers. Local leaders are often active in management of the service or maintain close contact with the municipality or community management body. Women and teens can play crucial roles, such as initiators, managers, operators, political activists, educators, and watchdogs for the community.

Public awareness and attitudes to waste can affect the whole municipal SWM system. All steps in municipal SWM starting from household waste storage, to waste segregation, recycling, collection frequency, willingness to pay for waste management services, and opposition to siting of waste treatment and disposal facilities depend on public awareness and participation. Thus, lack of public awareness and school education about the importance of proper SWM for health and well-being of people severely restricts use of community based approaches in developing countries and also crucial factor for failure of a MSWM service in developing countries (Zurbrugg 2003) in similar way the community participation in Jigjiga town is very less. The mobilization of communities was done through the local authorities and existing community leadership structures. The identified people formed themselves into either community based organization (CBOs) which are groups formed with in a local community or micro enterprises (MEs), which are defined as a service delivery or production business, usually low capital intensive consisting of an individual or up to about 20 person, either registered or operating informally in an area. These have proper leadership structures where gender equality was greatly encouraged to ensure that women take up positions of leadership.

5. Conclusion and Recommendation

5.1. Conclusion

SWM is one of the important compulsory functions of not only urban local authorities but also of rural local bodies. However, this essential service is not efficiently and properly performed by the concerned bodies of Jigjiga City resulting in sanitation, social and environmental problems. Waste management problem is complex because it involves a multitude of scientific, technical, economic and social factors. Similarly, it is observed that lack of financial resources, institutional weakness, improper selection of technology, transportation systems and disposal options, social problem associated with lethargy towards environmental cleanliness and sanitation have made this service unsatisfactory and inefficient in the study area.

5.2. Recommendations

Though there is no single readymade solution to control the waste problems, the researchers found it important to indicate issues requiring serious attention of concerned bodies. The data on generation rate and the percentage contribution of each solid waste source might have changed through time due to the change in demographic and socio-economic conditions of the residencies. Therefore, further study is needed to generate more valid data for current and future planning. Due to the complexity of the situation, cooperation among different parties/citizens should be methods of choice and industries and individuals should work as partners rather than adversary to find a long-term sustainable SWM system. The estimates indicate that over 75% of the refuse from most households is largely vegetable, biodegradable organic waste which has a potential for replacing inorganic fertilizer to save foreign exchanges earning, also strengthening and implying rural-urban-linkage.

Composting has to also be promoted by all actors who have the concern in poverty alleviation's and environmental protection. Therefore, attention should be given to reusable household wastes. Land filling site selection requires special attention; standard measures should be exercised to control contamination of surface and ground water as well as air. However, none of these practices are apparent in Jigjiga City. Being reluctant on the issue needs accountability for environmental cost/ pollution cost of water resource contamination from open dump site/, cost of closing many old dumpsites and developing a new one know and then. The existing open dumpsite is environmentally unsound and socially unaccepted.

Therefore, new landfill site is urgently needed with proper selection of the site/ environmental impact assessment and construction. Waste needs proper management. The first priority should be to segregation of wastes, preferably at the point of generation, into re-usable and non-reusable for waste reduction and changing in to financially viable material; hazardous and non-hazardous components for avoidance of hazardous containing products and waste workers safety.

Based on the research findings, the integrated SWM system which combines a range of solid waste treatment options like reusing, recycling, composting, and waste to energy transformation is recommended. Majority of the households' respondents lack awareness about the proper SWM systems. Therefore, community based waste management system should be in place to ensure sustainable and efficient alleviation from challenges of waste management.

Acknowledgements

We are grateful to Jigjiga University and SERDO for providing financial support. Our sincere thanks also goes to the Jigjiga City Administration's Sanitation and Beautification Department for unreserved cooperation to give pieces of information during data collection.

References

[1] Allende, R., 2009. Waste history in the Gambia. Thesis (MSC), University of the Gambia.

[2] Benti Getahun 2007. Jigjiga City: Migration and the Making of a Multiethnic Metropolis. Trenton, NJ: Red Sea Press.

[3] CSA 2007. Population and housing census of Ethiopia result for Somali region.

[4] Cochran, G. 1977. Sampling Techniques, 3rd edition, Wiley Series in Productivity and Applied Mathematical Statistics. New York, USA.

[5] Evans, E. 2002. "Measuring Success in Public Participation". International Association for Public Participation Conference Paper.

[6] Issam, A. 2007. Trends and problems of solid waste management in developing countries: A case study in seven Palestinian districts. Waste Management 27 (12) 1910-1919.

[7] Javaheri, H., et al., 2006. Site Selection of municipal Solid Waste Landfills Using Analytical Hierarchical Process Method in a Geographical Information Technology Environment in Giroft. Iran Journal of Environmental Health Science Engineering. 3, 177-184.

[8] Kyessi, A. and Mwakalinga, V. 2009. GIS Application in coordinating solid waste collection: the case of Sinza neighbourhood in Kinondoni Municipality, Dar Es Salaam City, Tanzania. FIG Working Week 2009.

[9] Mc Garity, T. 2005. Article entitled "*Public Participation in risk Regulation*", published on the internet arts.usask.ca/policynut/courses/mcgarity.htm.

[10] Mohamed N, Elsa Z 2003. *Waste Management Programme.* UNIDO View Document, No. 3765. Tokyo, Japan.

[11] United Nations Environmental Programme. 2002. *EIA Training Resource Manual.* Section 3, "Public Involvement".

[12] WHO 1996. *Guides for Municipal Solid Waste Management in Pacific countries. Health Cites, Health Islands Document Series. No. 6.* World Health Organization, Western Pacific Region.

[13] World Bank 2004. *Regional Guidelines on Integrated Solid Waste Management.* Prepared by an International Consortium of Expert Consultants.

[14] Yami Birke 1999. *Solid Waste Management in Ethiopia Integrated Development for water Supply and Sanitation.* Unpublished Paper.

[15] Zerayakob Belete 2002. *Analysis and Development of Solid Waste Management System of Addis Ababa City.* M.Sc.Thesis Presented to the School of Graduate Studies of Jigjiga City University.

[16] Zurbrug C. 2003. *Solid Waste Management in Developing Countries.* Retrieved from http://www.eawag.ch/organisation/abteilungen/sandec/publikationens-swm/downloads swm/basicsofSWM.pdf.

PERMISSIONS

All chapters in this book were first published in IJEPP, by Science Publishing Group; hereby published with permission under the Creative Commons Attribution License or equivalent. Every chapter published in this book has been scrutinized by our experts. Their significance has been extensively debated. The topics covered herein carry significant findings which will fuel the growth of the discipline. They may even be implemented as practical applications or may be referred to as a beginning point for another development.

The contributors of this book come from diverse backgrounds, making this book a truly international effort. This book will bring forth new frontiers with its revolutionizing research information and detailed analysis of the nascent developments around the world.

We would like to thank all the contributing authors for lending their expertise to make the book truly unique. They have played a crucial role in the development of this book. Without their invaluable contributions this book wouldn't have been possible. They have made vital efforts to compile up to date information on the varied aspects of this subject to make this book a valuable addition to the collection of many professionals and students.

This book was conceptualized with the vision of imparting up-to-date information and advanced data in this field. To ensure the same, a matchless editorial board was set up. Every individual on the board went through rigorous rounds of assessment to prove their worth. After which they invested a large part of their time researching and compiling the most relevant data for our readers.

The editorial board has been involved in producing this book since its inception. They have spent rigorous hours researching and exploring the diverse topics which have resulted in the successful publishing of this book. They have passed on their knowledge of decades through this book. To expedite this challenging task, the publisher supported the team at every step. A small team of assistant editors was also appointed to further simplify the editing procedure and attain best results for the readers.

Apart from the editorial board, the designing team has also invested a significant amount of their time in understanding the subject and creating the most relevant covers. They scrutinized every image to scout for the most suitable representation of the subject and create an appropriate cover for the book.

The publishing team has been an ardent support to the editorial, designing and production team. Their endless efforts to recruit the best for this project, has resulted in the accomplishment of this book. They are a veteran in the field of academics and their pool of knowledge is as vast as their experience in printing. Their expertise and guidance has proved useful at every step. Their uncompromising quality standards have made this book an exceptional effort. Their encouragement from time to time has been an inspiration for everyone.

The publisher and the editorial board hope that this book will prove to be a valuable piece of knowledge for researchers, students, practitioners and scholars across the globe.

LIST OF CONTRIBUTORS

Tomilola Debby Olaolu
Biochemistry Unit, Department of Biological Sciences, Landmark University, Omu-Aran, Kwara State, Nigeria

Oghenerobor Benjamin Akpor and Charity Omeche Akor
Microbiology Unit, Department of Biological Sciences, Landmark University, Omu-Aran, Kwara State, Nigeria

Onalenna Gwate
Department of Geography and Population Studies, Lupane State University, Zimbabwe

Yali E. Woyessa
Department of Civil Engineering, Central University of Technology, South Africa

David Wiberg
Water Programme, International Institute of Applied Systems Analysis, Laxenburg, Vienna, Austria

Alokwu Cyprian Obiora
Department of Religion and Philosophy, Paul University, Awka, Nigeria

Ezenwaji Emma Emeka
Department of Geography and Meteorology, Nnamdi Azikiwe University, Awka, Nigeria

Anastasia Dimitriou
Department of Education Sciences in Early Childhood, School of Education Sciences, Laboratory of Research and Communication for the Environment and Environmental Education, Democritus University of Thrace, Alexandroupolis, Greece

Dimitris Pimenidis
Department of Pre-school education and Design, School of Humanities, Msc Environmental Education, University of Aegean, Rhodes, Greece

Olawale O., Adepoju T. F. and Aikulolu S.
Chemical Engineering Department, Landmark University, Omu Aran, Kwara State, Nigeria

Oyawale F. A.
Mechanical Engineering Department, Covenant University, Otta, Nigeria

Akinmoladun A. I.
Department of Physics and Technology, Bronx Community College of the City University of New York, New York, USA

Alexander Kwaku Banson, Richard Kwasi Amankwah and Samuel Agyarko Ndur
Dept. of Minerals Engineering, University of Mines and Technology, Tarkwa, Ghana

Tao Wang
College of Geomatics, Xi'an University of Science and Technology, Xi'an, China

Mohammad Taghipour
Industrial Engineering, Science & Research Branch of Islamic Azad University, Tehran, Iran

Maryam Mahboobi, Arash Nikoeifar and Ebrahim Soofi Mowloodi
Industrial Engineering, non-profit institution of higher education, Aba, Abyek Qazvin, Iran

Manfred Fehr
Institute of Geography, Federal University, Uberlandia MG, Brazil

Mengistu Asmamaw Mengesha
Department of Biology, Debre Berhan University, Debre Berhan, Ethiopia

Mekuria Argaw Denoboba
Department of Environmental Science, Addis Ababa University, Addis Ababa, Ethiopia

Cuthbert Casey Makondo, Sydney Sichilima, Matthews Silondwa, Richard Sikazwe, Lombe Maiba and Chawezi Longwe
Environmental Engineering Department, Copperbelt University, Kitwe, Zambia

Yvonne Chiliboyi
Institute for Water Research, Rhodes University, Grahamstown, Republic of South Africa

Adel K. El-Feky
Atomic Energy Authority, Nuclear Research Center, Reactor Department, Cairo, Egypt

Santhi V.
Research Centre and P.G Department of Zoology, J. A. College, Periyakulam, Tamil Nadu, India

V. Sivakumar
P.G and Research Department of Zoology, V. O. Chidambaram College, Thoothukudi, Tamil Nadu, India

S. Jayalakshmi
Department of Marine Sciences, Annamalai University, Parangipettai, Tamil Nadu, India

R. D. Thilaga
P.G and Research Department of Zoology, St. Mary's College, Thoothukudi, Tamil Nadu, India

M. Mukilarasi
Departmemt of Computer Science, Sakthi Engineering College, Chennai, Tamil Nadu, India

Su Wen-ji
Academy of Civil and Transportation, South China University of Technology, Guangdong, China

Fu Shan-ming
School of Environmental Science and Engineering, Guangzhou University, Guangdong, Guangzhou, China

M. Kanchilakshmi and Arockiam Thaddeus
Research Centre of Zoology, Jayaraj Annapackiam College for Women (Autonomous), Periyakulam, Tamil Nadu, South India

Enoch Akwasi Kosoe and Issaka Kanton Osumanu
Department of Environment and Resource Studies, University for Development Studies, Wa, Ghana

Waleed Hussein Ali
Fayoum University, Faculty of Engineering, Department of Architecture, Fayoum, Egypt

Chien-Hua Chen
The Administrative Office of Keelung City, Taiwan, R.O.C

Siraj Beshir
Madawalabu University, Department of Plant Science, Bale Robe, Ethiopia

Mulugeta Lemeneh
International Livestock Research Institute (ILRI), Addis Ababa, Ethiopia

Endalkachew Kissi
Jimma University, Department of Natural Resource Management, Jimma, Ethiopia

Felix Iyalomhe
Global Change and Sustainability Research Institute, University of the Witwatersrand, Johannesburg, South Africa

Philipa Idogho
Department of Office Technology and Management, Auchi Polytechnic, Auchi, Edo State, Nigeria

Prukkanone Kittisak
College of Public Administration, Huazhong University of Science and Technology, Wuhan, China

Wang Guohua
The Public Opinion Researching Center, Huazhong University of Science and Technology, Wuhan, China

Pradeep Singh
Nepal Electricity Authority, Durbarmarg, Kathmandu, Nepal

Shree Raj Shakya
Center for Energy Studies, Institute of Engineering, Tribhuvan University, Lalitpur, Nepal

Yohanis Birhanu
Department of Chemistry, Jigjiga University, Jigjiga, Ethiopia

Genemo Berisa
Department of Geography and Environmental Studies, Jigjiga University, Jigjiga, Ethiopia

Index